# The Ethical Landscape of AI:

## Global Issues and Solutions

Hakikur Rahman
https://orcid.org/0000-0002-2132-1298
*International Standard University, Bangladesh*

IGI Global
Scientific Publishing
Publishing Tomorrow's Research Today

| | |
|---|---|
| Vice President of Editorial | Melissa Wagner |
| Director of Acquisitions | Mikaela Felty |
| Director of Book Development | Jocelynn Hessler |
| Production Manager | Mike Brehm |
| Cover Design | Jose Rosado |

Published in the United States of America by
IGI Global Scientific Publishing
701 East Chocolate Avenue
Hershey, PA, 17033, USA
Tel: 717-533-8845 | Fax: 717-533-7115
Website: https://www.igi-global.com E-mail: cust@igi-global.com



Library of Congress Cataloging-in-Publication Data

LCCN: 2025011222 (CIP Data Pending)
ISBN13: 9798337325972
Isbn13Softcover: 9798337325989
EISBN13: 9798337325996
British Cataloguing in Publication Data
A Cataloguing in Publication record for this book is available from the British Library.

# Table of Contents

# Detailed Table of Contents



The rapid spread of artificial intelligence (AI) technology has sparked a complex ethical dilemma that transcends borders. This chapter examines key ethical challenges of AI, including the distortions of decision-making systems, the transparency and interpretability of AI models, data protection risks, and the social impact of autonomous systems. We examine how different countries and organizations address these dilemmas through guidelines, ethical guidelines and regulatory measures. By providing a comparative analysis of the global AI ethics framework, this chapter illustrates the demonstrated procedures and unresolved tensions in ethical AI governments. The role of international cooperation in creating ethical AI standards is also discussed, ensuring that AI development corresponds to fundamental human values.



Artificial Intelligence (AI) systems, while transformative, often inherit biases from data, algorithms, and human decision-making, raising ethical concerns regarding fairness, accountability, and societal impact. Bias in AI can result in discriminatory outcomes, reinforcing social inequalities in critical areas such as healthcare, hiring, lending, and law enforcement. Ensuring fairness requires rigorous bias detection, mitigation strategies, and the adoption of Explainable AI (XAI) to enhance transparency and accountability. Ethical AI frameworks must integrate principles of inclusivity, regulatory compliance, and stakeholder engagement to foster trust and prevent harm. This research explores the ethical implications of AI bias, emphasizing technical and policy-driven solutions to achieve equitable and responsible AI deployment.

Artificial intelligence (AI) is now integral to healthcare, finance, and education, shaping decisions that affect millions daily. While it promises efficiency and precision, its integration reveals ethical vulnerabilities—biased algorithms, opaque systems, and governance gaps that disempower stakeholders. This chapter reframes AI ethics through a human-in-context paradigm, emphasizing socio-technical settings, institutional norms, and stakeholder engagement. Anchored on fairness, transparency, privacy, and human agency, it examines three case studies: healthcare risk prediction bias, Apple Card's gender discrimination, and the Ofqual grading scandal. These illustrate how bias emerges and compounds when governance fails. The chapter maps risk pathways and proposes interventions like fairness-aware design, bias audits, human override, and participatory governance. A cross-sector analysis highlights common ethical challenges and introduces a transferable ethics infrastructure aligned with OECD, UNESCO, GDPR, and EU AI Act principles.



Generative artificial intelligence (AI) tools have rapidly entered the operations of small and medium-sized enterprises (SMEs), boosting productivity while presenting significant challenges in regulatory compliance. This chapter examines these challenges and identifies compliance gaps by comparing global and national regulatory frameworks from an SME perspective. It also explores case studies focused on SMEs' ethical use of AI. To address these issues, the chapter proposes an AI Ethics Framework grounded in four core pillars. It further introduces Ethics4AI, a multilingual, AI-powered platform built on a foundation of current principles, regulations, and toolkits designed to operationalize the proposed framework. Ethics4AI serves as an Ethical AI Advisor, bridging the gap between ethical principles and practical compliance, and enabling SMEs to adopt AI ethically and responsibly.

The field of social science is being transformed by rising artificial intelligence implementation. Through AI technology researches achieve faster data analysis as well as automation of operations and access innovative methodologies of problem-solving. The chapter evaluates current AI applications in social science domains while assessing the advantages with the difficulties these implementations introduce. In sociology, AI enhances social behaviour analysis, AI models predict election outcomes in political science, AI-driven mental health chatbots in psychology, AI-powered adaptive learning systems in education, predictive policing leverages AI to analyse crime patterns and anticipate potential criminal activities in criminology and AI driven stock market prediction in economics. Artificial intelligence interacts with social science research methodologies represent a significant advancement in the field. It is important to stay aware of ethical concerns such as transparency in AI models and use this powerful technology responsibly.



Artificial Intelligence (AI) has totally transformed financial decision-making by enhancing fraud prevention, risk assessment, and productivity. It also causes serious ethical issues such as accountability, transparency, and bias. Unconscious biases in previous data are most likely to be transferred to AI-based financial systems such as automated lending websites and credit scoring systems, resulting in discriminatory or unfair results. Since it is not easy for banks and regulators to comprehend and track algorithmic judgments, transparency of AI models makes trust matters harder. Furthermore, accountability questions remain in instances of ethics violations or mistakes made by AI. The moral implications of AI in making financial decisions are discussed in this chapter, in addition to the challenges in being transparent in automated systems, developing biases in algorithmic models, and the law's need for accountability. The chapter explores case studies of moral AI failure and best practices in trying to provide a roadmap for accountable AI regulation in finance.

This chapter evaluates how leader's ethics influences citizen's behavior for organization (OCB) within the health sector industry of the Eastern part of Africa by examining culture of organization as the middle element. The analysis utilized 390 health sector worker participants through a quantitative investigation method to evaluate these associations. The Kaiser-Meyer-Olkin (KMO) measure evaluated the element analysis acceptability for the investigation data. Sample adequacy of the investigation is confirmed through its KMO value. The investigations validated the measurement model through confirmatory element analysis (CFA) to properly represent the concepts of culture of organization, OCB, and leader's ethics. AMOS software enabled the discovery of substantial associations among variables in order to manifest Structural Equation Modeling (SEM). The investigation demonstrates that leader's ethics achieves encouraging influence on OCB accomplish through culture of organization serving as a vital intermediate force.

# Preface

Artificial intelligence has moved from the margins of technological innovation to the center of global decision-making, reshaping economies, institutions, and everyday human experience. As AI systems increasingly influence healthcare, finance, education, governance, media, and social interaction, ethical considerations have become not only relevant but urgent. Questions of fairness, accountability, transparency, privacy, human agency, and justice now define the boundaries of responsible AI development and deployment. These concerns transcend national borders, disciplinary silos, and sectoral interests, demanding a global, interdisciplinary, and human-centered response.

*The Ethical Landscape of AI: Global Issues and Solutions* brings together diverse scholarly perspectives to critically examine the ethical dimensions of artificial intelligence across contexts, cultures, and applications. The volume recognizes that AI ethics is not a static checklist of principles, but a dynamic socio-technical challenge shaped by power relations, regulatory environments, institutional practices, and lived human realities. By engaging with global frameworks, empirical case studies, sector-specific applications, and emerging governance models, this book seeks to bridge the gap between ethical theory and practical implementation.

The chapters collectively highlight how ethical failures in AI—such as bias, opacity, misuse of data, and erosion of human rights—often stem from governance gaps rather than technical limitations alone. At the same time, the volume emphasizes pathways toward responsible innovation, including inclusive design, participatory governance, explainable systems, regulatory harmonization, and ethical leadership. From global policy and human rights to education, healthcare, finance, digital media, geospatial intelligence, and social sciences, this book underscores the necessity of embedding ethical reflection throughout the AI lifecycle.

This volume is intended for researchers, policymakers, practitioners, and students seeking a comprehensive and globally informed understanding of AI ethics. By presenting challenges alongside actionable frameworks and solutions, it contributes to the ongoing international dialogue on how artificial intelligence can be aligned with fundamental human values, social equity, and sustainable development.

## CHAPTER OVERVIEW

**Chapter 1** explores the global ethical challenges of artificial intelligence, examining issues such as algorithmic distortion, transparency, data protection, and the societal impact of autonomous systems, while comparing international regulatory and governance approaches.

**Chapter 2** focuses on bias and fairness in AI systems, analyzing how discriminatory outcomes emerge from data and algorithms and highlighting technical, ethical, and policy-driven strategies to promote accountability, explainability, and trust.

**Chapter 3** introduces a human-in-context framework for ethical AI governance, using cross-sector case studies to demonstrate how socio-technical failures amplify bias and proposing governance interventions aligned with international standards.

**Chapter 4** examines ethical AI adoption within small and medium-sized enterprises, identifying regulatory compliance gaps and presenting a practical, platform-based framework to operationalize ethical principles in real-world business contexts.

**Chapter 5** investigates the ethical implications of AI in digital media, with particular attention to data privacy, algorithmic bias, misinformation, and the challenges posed by fake news and automated content moderation.

**Chapter 6** provides a global overview of AI ethics challenges and frameworks, synthesizing regulatory approaches, case studies, and emerging trends to propose strategies for ethical governance and international consensus-building.

**Chapter 7** discusses responsible innovation through value co-creation, arguing for the integration of ethical considerations into AI development processes via regulatory oversight, interdisciplinary collaboration, and public engagement.

**Chapter 8** examines the societal, legal, and ethical dimensions of AI adoption, including AI literacy, social robotics, regulatory frameworks, and the balance between innovation, human dignity, and justice.

**Chapter 9** analyzes the ethical dilemmas of AI in education, addressing issues of equity, algorithmic bias, transparency, and human agency, while proposing a human-centered framework for inclusive educational innovation.

**Chapter 10** explores the ethical dimensions of geospatial artificial intelligence, emphasizing transparency, fairness, privacy, and responsibility, and advocating for a sociotechnical approach supported by open global data policies.

**Chapter 11** examines artificial intelligence through a human rights lens, highlighting risks related to surveillance, bias, global inequality, and digital colonialism, and proposing ethics-based governance grounded in fairness and inclusivity.

**Chapter 12** evaluates the use of AI in social sciences, assessing its contributions to problem-solving across disciplines while emphasizing the importance of transparency, accountability, and responsible methodological application.

**Chapter 13** focuses on ethical challenges in AI-driven financial decision-making, including bias, opacity, and accountability, and presents case studies and best practices for responsible financial AI governance.

**Chapter 14** examines the transformative role of AI and machine learning in personalized healthcare, highlighting applications such as precision medicine and real-time monitoring while considering their broader implications for patient-centered care.

**Chapter 15** analyzes the ethical responsibilities of AI deployment in finance, emphasizing the need to balance innovation with fairness, transparency, data protection, and regulatory compliance.

**Chapter 16** investigates the role of ethical leadership in shaping organizational citizenship behavior within the healthcare sector, demonstrating how organizational culture mediates ethical outcomes in AI-influenced environments.

In bringing together global perspectives, sector-specific analyses, and actionable frameworks, this volume underscores that ethical artificial intelligence is not optional but foundational to sustainable innovation. The chapters collectively affirm that responsible AI development requires continuous ethical reflection, inclusive governance, and international collaboration. As AI technologies continue to evolve, aligning them with human values, social justice, and accountability will remain a defining challenge—and responsibility—of our time.

**Hakikur Rahman**
*International Standard University, Bangladesh*

# Chapter 1
# Navigating Ethical Dilemmas in AI:
## A Global Perspective

**Hakikur Rahman**
https://orcid.org/0000-0002-2132-1298
*International Standard University, Bangladesh*

## ABSTRACT

*The rapid spread of artificial intelligence (AI) technology has sparked a complex ethical dilemma that transcends borders. This chapter examines key ethical challenges of AI, including the distortions of decision-making systems, the transparency and interpretability of AI models, data protection risks, and the social impact of autonomous systems. We examine how different countries and organizations address these dilemmas through guidelines, ethical guidelines and regulatory measures. By providing a comparative analysis of the global AI ethics framework, this chapter illustrates the demonstrated procedures and unresolved tensions in ethical AI governments. The role of international cooperation in creating ethical AI standards is also discussed, ensuring that AI development corresponds to fundamental human values.*

## 1. INTRODUCTION

Artificial intelligence (AI) presents important ethical issues that require careful consideration as it develops and becomes more integrated into many industries. AI has problems that go beyond national borders, from bias in machine learning algorithms to privacy issues in data-driven decision-making. Given that AI has the capacity to uphold social injustices, violate fundamental rights, and function without obvious accountability procedures, its ethical ramifications are especially urgent (Floridi

DOI: 10.4018/979-8-3373-2597-2.ch001

& Cowls, 2019; WHO, 2021). To guarantee that AI benefits mankind rather than exacerbates current injustices, academics contend that ethical AI frameworks should be based on the values of responsibility, transparency, and fairness (Jobin, Ienca, & Vayena, 2019; 2023). However, as regional ethical standards and legal frameworks vary greatly, accomplishing these objectives calls for international collaboration.

While nations like China have adopted a more centralized, state-driven approach to AI regulation, the European Union's General Data Protection Regulation (GDPR) has set strict standards for data protection (Zhang & Dafoe, 2020). In order to promote responsible AI development globally, this chapter examines the various ethical conundrums in AI from a global viewpoint, including legal frameworks, cultural variations, and viable remedies.

## 1.1 Defining ethical dilemmas in AI and their global significance

Artificial intelligence (AI) ethical challenges occur when the technology's capabilities and applications go against basic moral precepts, resulting in tensions between opposing ideals including autonomy, privacy, justice, and accountability. Algorithmic bias, in which AI systems inadvertently reinforce social and economic disparities as a result of biased training data, is one of the most urgent ethical issues (Binns, 2018a; Shen & Anderson, 2024). Racial and gender biases, for instance, have been discovered in facial recognition algorithms, resulting in disproportionate misidentifications that have detrimental effects on marginalized communities (Buolamwini & Gebru, 2018).

Similarly, since AI-driven decisions have the potential to profoundly affect human lives without obvious channels for recourse, autonomous decision-making in fields like healthcare and criminal justice raises important concerns around accountability and transparency (Mittelstadt et al., 2016; WHO, 2023). Since AI technologies are created and implemented across jurisdictions with varying legal, cultural, and ethical norms, the global significance of AI ethics resides in its cross-border consequences.

The United States adopts a more market-driven strategy, emphasizing innovation with very less regulation, in contrast to the European Union's General Data Protection Regulation (GDPR), which places a strong emphasis on privacy and data protection (Brundage et al., 2020). China, meantime, has embraced a state-centered AI governance paradigm that places a higher priority on social stability and economic progress (Zeng, Lu, & Huangfu, 2019). The difficulty of creating widely recognized AI ethical norms is highlighted by these conflicting viewpoints. International collaboration is essential in resolving these moral conundrums and making sure that AI advances society rather than widening already-existing rifts as it gets more and more integrated into vital industries, such as national defense and banking.

## 1.2 Why AI ethics is a pressing issue in today's technological landscape

The fast advancement of artificial intelligence (AI) has made ethical issues about its use more pressing than ever. Critical facets of human existence, including healthcare, banking, criminal justice, and national security, are being impacted by AI-driven systems (Russell, 2021; WHO, 2023). Because neither consumers nor regulators can completely comprehend or explain how AI arrives at particular conclusions, the lack of transparency in AI decision-making—often referred to as the "black box" problem—raises major accountability difficulties (Castelvecchi, 2016). In high-stakes situations where biased algorithms might provide unfair results, including predictive policing and credit scoring, this opacity becomes especially troublesome (O'Neil, 2016).

Additionally, worries regarding AI's potential to erode democracy and public confidence in information systems have been raised globally due to its capacity to influence human behavior through recommendation algorithms, deepfake technologies, and targeted disinformation (Zuboff, 2019).

Beyond personal moral conundrums, AI poses larger societal issues that require prompt consideration. Fears of job displacement and economic inequality have increased as a result of automation and AI-driven decision-making, especially in sectors of the economy that rely significantly on repetitive tasks (Brynjolfsson & McAfee, 2017). Furthermore, the possibility of human rights abuses has been brought to light by AI-enabled surveillance technologies like facial recognition, especially in authoritarian governments that employ them to stifle dissent (Feldstein, 2019).

There is still a global lack of comprehensive ethical frameworks to manage AI ethically, despite the introduction of AI rules by various governments, such as the European Union's Artificial Intelligence Act (Veale & Borgesius, 2021; WHO, 2024). In order to guarantee that technical advancements are in line with human rights and the welfare of society, it is imperative that the ethical implications of AI be addressed as it becomes more and more integrated into society.

## 1.3 The role of cultural, political, and economic factors in shaping AI ethics

The paradigm of AI ethics is not universally applicable; rather, it is heavily impacted by the cultural, political, and economic factors that determine how various societies view and govern AI technologies. Cultural values play a crucial role in defining ethical priorities for AI governance. As seen by laws like the General Data Protection Regulation (GDPR), Western countries—especially those in Europe and

North America—place a strong emphasis on individual rights, privacy, and transparency in AI applications (Floridi, 2021; UNESCO, 2023).

Asian nations, especially China, on the other hand, place a higher priority on social harmony and collective welfare, which is reflected in state-driven AI ethics frameworks that prioritize social stability, economic prosperity, and national security (Zeng, Lu, & Huangfu, 2019). This difference may be seen in the way that AI surveillance technologies are seen; in liberal democracies, they are regarded with suspicion and privacy concerns, but in more centralized political systems, they are frequently defended as instruments for public safety and governance (Feldstein, 2019).

Political systems also greatly impact AI ethics by influencing the degree of control and oversight imposed on AI development and deployment. A multi-stakeholder approach involving government agencies, academia, civil society, and business leaders is usually what defines AI governance in democracies (Brundage et al., 2020). This method promotes moral standards that give human rights, accountability, and algorithmic justice top priority. On the other hand, AI policies are typically state-controlled in authoritarian regimes, where the technology is used for social control, monitoring, and governance (Creemers, 2021). For instance, serious ethical questions concerning individual liberty and government overreach have been brought up by China's Social Credit System, which employs AI to track and assess citizens' conduct (Mozur, 2018). These divergent political stances emphasize how difficult it is to create global AI ethics guidelines when national governance structures and agendas differ greatly.

Economic considerations further complicate the ethical landscape of AI. AI innovation and economic competitiveness are frequently given priority in nations with strong technological industries, like the US and China, sometimes at the price of moral considerations (Boddington, 2017).

Ethical trade-offs have resulted from the competition for AI supremacy, such as the employment of biased algorithms in hiring procedures or the loss of jobs in some industries due to AI-driven automation (Brynjolfsson & McAfee, 2017). While competing in the global AI economy, developing countries confront additional ethical challenges because they frequently lack the resources and legislative framework necessary to assure ethical AI adoption (Jobin, Ienca, & Vayena, 2019). These differences give rise to worries about AI colonialism, in which corporations and economically dominant countries force AI systems on less developed areas without taking into account the ethical and societal ramifications for the local community (Birhane, 2021). To solve these problems, a global effort is needed to strike a balance between innovation and moral obligation, making sure that the advantages of AI are shared fairly across various political, cultural, and economic contexts.

This chapter offers a comparative and multidisciplinary approach to explore ethical concerns in AI from a global viewpoint. The study examines how various nations and organizations handle AI ethics by synthesizing insights from academic literature, policy documents, and regulatory frameworks. To find commonalities and differences in ethical AI governance, a qualitative content analysis is carried out on important international AI standards, such as those put forth by the European Union, UNESCO, OECD, and national governments.

Additionally, case studies of AI deployments in varied sectors—such as healthcare, criminal justice, and autonomous systems—are studied to show real-world ethical difficulties. The chapter also blends viewpoints from philosophy, law, and technology studies to analyze the broader societal consequences of AI ethics. This methodology's adoption of a global viewpoint facilitates a nuanced understanding of both regionally unique difficulties and common ethical concerns, opening the door for debates on best practices and possible harmonization of AI governance frameworks.

Now towards navigating ethical dilemmas in AI in a global perspective, this chapter discusses a few key ethical dilemmas in AI in the next section.

## 2. KEY ETHICAL DILEMMAS IN AI

Policymakers, technologists, and society at large are faced with a number of ethical conundrums brought about by the quick development of artificial intelligence (AI). One of the most urgent issues is algorithmic bias, in which skewed training data causes AI systems to inadvertently perpetuate society preconceptions, resulting in unjust hiring, lending, and law enforcement practices (O'Neil, 2016). Accountability is further complicated by the opacity of AI decision-making, sometimes known as the "black box" problem, which makes it hard to track how AI models arrive at particular conclusions, making them prone to discrimination and errors (Castelvecchi, 2016).

Concerns regarding civil rights and the right to anonymity are also raised by AI-driven surveillance and data collection, especially when it comes to face recognition technology that have the ability to monitor people without their knowledge (Feldstein, 2019). As AI-driven systems replace human labor in sectors like manufacturing, transportation, and customer service, increasing job automation also raises ethical concerns about economic displacement and exacerbates income inequality (Brynjolfsson & McAfee, 2017).

Furthermore, since giving computers the ability to make life-or-death decisions compromises ethical accountability and international humanitarian law, the possible application of AI in autonomous weaponry raises ethical questions about human oversight in combat (Etzioni & Etzioni, 2017). These problems emphasize the critical need for comprehensive AI governance frameworks to ensure that technology advancement corresponds with ethical principles and human rights.

This research has pin pointed several key issues in this aspect. *Algorithmic Bias and Discrimination; Privacy and Surveillance Concerns; Transparency and Explainability; AI Autonomy and Decision-Making; Job Displacement and Economic Inequality; and AI and Misinformation* are among them. We are going to discuss them one by one in this section.

## 2.1 Algorithmic Bias and Discrimination

o ***How AI systems reinforce social and economic inequalities***

Although AI systems are intended to increase productivity and decision-making, their biased algorithms and data-driven discrimination frequently make already-existing social and economic disparities worse. Algorithmic bias is one of the main issues, whereby AI models that have been trained on past data mirror and magnify ingrained biases, resulting in discriminatory outcomes in crucial domains like law enforcement, lending, and employment (O'Neil, 2016). For example, research has demonstrated that AI-powered recruiting systems often penalize candidates from underrepresented groups by giving preference to applicants who share traits with traditionally favored individuals (Raghavan et al., 2020).

Similarly, based on skewed crime data, predictive police algorithms disproportionately target minority populations, perpetuating cycles of criminalization and overpolicing instead of addressing underlying societal concerns (Richardson, Schultz, & Crawford, 2019). These biases demonstrate how, when unchecked, AI might exacerbate societal inequalities rather than serve as a neutral instrument.

AI-driven automation also contributes to economic inequality by disproportionately affecting low-income workers who perform repetitive and routine labor. According to Brynjolfsson and McAfee (2017), the implementation of AI has increased efficiency across a range of industries, but it has also resulted in job displacement, especially in manufacturing, retail, and transportation. High-skilled people and those with access to advanced technological education gain from the shift to AI-driven industries, whereas low-skilled individuals face fewer job possibilities and growing income inequality (Acemoglu & Restrepo, 2020).

Furthermore, because of skewed risk assessments, AI-powered financial systems—like automated credit scoring—frequently penalize economically vulnerable groups by restricting their access to loans and other necessary financial services (Eubanks, 2018). These difficulties highlight the necessity of moral AI governance and regulations that guarantee the fair use of AI, avoiding the perpetuation of structural injustices and fostering inclusive technical advancement. Hence, a few case studies are being put forward next.

o ***Case studies on biased AI applications in healthcare, finance, and law enforcement***

Applications of AI have shown great promise across a range of industries, yet a number of real-world examples show how prejudiced AI systems maintain social inequalities. Biased algorithms have resulted in discriminatory policies that disproportionately impact marginalized communities in the fields of law enforcement, healthcare, and finance. These difficulties are demonstrated by the case studies that follow, which also emphasize how urgently ethical AI governance and bias reduction techniques are needed.

### A. Healthcare: Racial Bias in Medical Risk Prediction

An algorithm employed in the US to forecast patient healthcare demands is an often cited example of AI bias in healthcare. Major hospitals and insurers employ the AI system, which Obermeyer et al. (2019) found to have considerable racial prejudice. Instead of considering genuine medical needs, the algorithm was created to rank people for specialized healthcare programs according to anticipated healthcare costs. The algorithm incorrectly determined that Black patients were healthier than they actually were because Black patients have historically received less medical attention and care even though their health status were identical to those of White patients.

As a result, compared to White patients with the same severity of disease, fewer Black patients were referred to high-risk care management programs. According to the study, this bias exacerbated already-existing racial inequities in healthcare access by reducing the number of Black patients obtaining extra care by over 50%.

### B. Finance: Discrimination in Credit Scoring and Lending

Financial systems powered by AI have also demonstrated discriminating tendencies, especially when it comes to loan approvals and credit scoring. Gender bias in credit limit allocations was discovered in 2019 after an inquiry into the Apple Card, a financial product created by Apple and issued by Goldman Sachs (Hao, 2019). There have been numerous allegations of women being given much lower credit limits than men, even if they had comparable or superior financial records. Tech entrepreneur David Heinemeier Hansson, for instance, revealed in public that his wife was given a credit limit that was 20 times lower than his own despite having a better credit rating than he had.

Concerns over AI-driven lending practices that perpetuate gender-based discrimination were raised when the New York Department of Financial Services opened an investigation into Goldman Sachs' credit assessment algorithms (Hao, 2019). This instance demonstrates how financial disparities resulting from opaque AI decision-making can disproportionately impact women and other underrepresented groups.

C. Law Enforcement: Racial Disparities in Predictive Policing

AI systems used in predictive policing have come under fire for escalating racial biases in law enforcement. The algorithm known as COMPAS (Correctional Offender Management Profiling for Alternative Sanctions), which is employed in the American legal system to determine a defendant's risk of reoffending, is a prime example. The algorithm was skewed against Black defendants, incorrectly classifying them as "high-risk" for recidivism at a rate that was over twice that of White defendants, according to a ProPublica investigation (Angwin et al., 2016). White defendants, on the other hand, were more likely to be mistakenly classified as "low-risk," which resulted in differences in sentencing and parole decisions.

The study discovered that previous biases in arrest and conviction rates led to the faulty risk assessments, which in turn reinforced systemic racial disparities in the criminal justice system. The dangers of using AI in crucial decision-making processes without adequate supervision and bias reduction techniques are illustrated by the COMPAS example.

These case studies highlight how widespread AI bias is across a variety of fields, impacting fundamental facets of human existence like access to healthcare, financial security, and the rule of law. Existing social and economic disparities are maintained by the use of skewed historical data in conjunction with opaque AI decision-making. To solve these issues and guarantee that AI systems are open, responsible, and equitable, interdisciplinary efforts involving legislators, technologists, and civil rights activists are needed. To reduce algorithmic prejudice and guarantee fair AI deployment, it is essential to create bias-aware AI models, conduct frequent audits, and use inclusive data collecting techniques.

## 2.2 Privacy and Surveillance Concerns

o ***The ethical implications of AI-driven data collection and surveillance***

Significant ethical questions about consent, privacy, and possible exploitation of personal data have been brought up by the broad usage of AI-driven data collecting and surveillance technology. AI-powered surveillance technologies, such as biometric tracking, facial recognition, and predictive analytics, have made it possible for businesses and governments to gather enormous volumes of data with previously unheard-of efficiency and accuracy.

But these developments frequently come at the expense of people's civil freedoms and privacy. Since many users are ignorant of the amount to which their data is being gathered, examined, and shared, one of the main issues is the absence of informed permission. According to Zuboff (2019), contemporary AI-driven surveillance follows a "surveillance capitalism" model in which people's personal information is sold without their consent, thereby undermining their sense of autonomy. In fields like employment, law enforcement, and healthcare, where inaccurate or biased interpretations can have major repercussions, this problem is made worse by opaque data laws and the growing usage of AI-driven decision-making (Pasquale, 2015).

Concerns about ethics are heightened by government-led AI surveillance programs, especially when it comes to mass surveillance and possible human rights abuses. For instance, China's Social Credit System has drawn international criticism for its widespread use of facial recognition technology, which allows it to track and punish residents according to their social actions (Creemers, 2018). Similarly, it has been said that the use of AI-powered predictive police techniques in the US has disproportionately targeted underprivileged populations and reinforced racial biases (Brayne, 2021).

In both situations, invasive monitoring and possible dissent suppression are made possible by AI-driven surveillance technology, endangering democratic freedoms. Additionally, concerns regarding security and accountability in AI-driven surveillance ecosystems are raised by the possibility of data breaches and illegal access to private data (Eubanks, 2018). Stricter legal frameworks, increased openness in AI decision-making, and strong ethical standards that strike a balance between security considerations and basic human rights are all necessary to address these issues.

o ***Balancing security and individual privacy rights in AI governance***

The argument over striking a balance between individual privacy rights and national security objectives has heated up due to the quick development of AI-driven surveillance technologies. Large-scale data gathering and AI-powered monitoring are defended by governments and businesses as essential measures to stop cyberattacks, terrorism, and criminal activity. However, disproportionately targeting vulnerable groups and limiting freedom of expression, excessive surveillance frequently results in the loss of civil liberties (Brayne, 2021).

For instance, facial recognition software and AI-based predictive policing technologies have drawn a lot of criticism for their potential to violate individual liberties in the absence of sufficient protections (Eubanks, 2018). Creating regulations that shield citizens from invasive surveillance while guaranteeing that security measures are reasonable, open, and accountable is the difficult part of AI governance.

Establishing strong legal frameworks and moral AI laws is a crucial step in reaching this equilibrium. By imposing stringent data privacy regulations, mandating informed consent for data collection, and giving people the right to view and delete their data, the General Data Protection Regulation (GDPR) of the European Union offers a powerful example (Voigt & Bussche, 2017).

Furthermore, accountability can be improved by implementing procedures like algorithmic audits and independent oversight organizations, which guarantee that AI surveillance systems function within moral and legal bounds (Pasquale, 2015).

Furthermore, the idea of "privacy by design" in AI development makes sure that security goals don't compromise basic human rights (Cavoukian, 2011). Transparency, democratic supervision, and public involvement must be given top priority in AI governance going ahead in order to develop laws that respect both security requirements and the fundamental right to privacy.

## 2.3 Transparency and Explainability

o ***The "black box" challenge: why AI decisions are often difficult to interpret***

The "black box" dilemma, which refers to the opaqueness of AI decision-making processes, is one of the most urgent ethical issues in AI. Even the developers of many sophisticated AI systems find it difficult to understand how they work, especially those that use deep learning and neural networks (Burrell, 2016). Modern AI models handle enormous volumes of data through intricate, multilevel architectures, making it challenging to describe how they arrive at particular conclusions, in contrast to classic rule-based systems where decision logic can be openly traced (Lipton, 2018).

In high-stakes industries like healthcare, banking, and criminal justice, where opaque AI-driven choices might have dire repercussions, this lack of transparency raises severe issues. For instance, AI-powered risk assessment instruments that are used to determine sentencing and parole have come under fire for failing to offer convincing arguments for their suggestions, which could result in biases and unjustified incarcerations (Angwin et al., 2016).

When AI systems make mistakes without obvious accountability procedures or perpetuate preexisting prejudices, the black-box dilemma becomes especially troublesome. Lack of transparency erodes confidence in AI applications since those impacted by AI-driven decisions have few options to challenge or comprehend the reasoning behind them (Doshi-Velez & Kim, 2017).

"Explainable AI" (XAI), a new discipline aimed at developing models that offer human-interpretable explanations for their outputs, is the result of efforts to overcome this problem (Gunning et al., 2019). The "right to explanation," which is emphasized by regulatory frameworks like the General Data Protection Regulation (GDPR) of the European Union, requires enterprises to make sure that automated choices that affect people are reasonable and comprehensible (Wachter et al., 2017). To reconcile AI progress with ethical responsibility, further research and legislative interventions are necessary to achieve full explainability, which is still a difficult task.

o ***Ethical concerns in high-stakes AI applications (e.g., criminal justice, hiring, lending)***

Due to concerns about bias, accountability, and transparency, the application of AI in high-stakes decision-making processes including criminal justice, hiring, and lending has sparked serious ethical questions. AI-driven risk assessment tools in the criminal justice system, such the COMPAS algorithm, have come under fire for underestimating the risk levels of White defendants while disproportionately classifying Black defendants as high-risk for recidivism (Angwin et al., 2016). Due process and fairness are called into question because it is challenging for defendants and legal experts to contest adverse risk rankings due to these models' lack of explainability.

Furthermore, by disproportionately targeting minority communities, predictive policing systems—which use past crime data to anticipate future offenses—frequently reinforce institutional biases and exacerbate racial disparities in law enforcement (Richardson, Schultz, & Crawford, 2019). These problems highlight the necessity of increased openness and supervision in AI applications that directly affect people's liberties and rights.

In a similar vein, AI-powered loan and hiring platforms have shown discriminatory tendencies that compromise equity and fairness. It has been discovered that automated hiring algorithms, which are intended to screen job applicants, favor applicants who share traits that have previously been favored, resulting in racial and gender biases in hiring decisions (Raghavan et al., 2020). Amazon's AI hiring tool, for instance, was discontinued after it was found to consistently devalue resumes that contained phrases related to women, such "women's chess club" (Dastin, 2018a).

Even after adjusting for financial history and income, AI-based credit scoring models in the financial industry have discriminated against minority applicants by rejecting loans to them more frequently than to white applicants (Bartlett et al., 2022). These models' opacity limits the ability of those who were impacted to contest unfair decisions by preventing them from comprehending the reasons behind their denial of opportunities. In order to guarantee that AI systems advance justice and equality rather than maintain current societal inequities, these ethical conundrums underscore the pressing need for explainable AI (XAI) approaches, regulatory actions, and fairness-aware machine learning techniques.

## 2.4 AI Autonomy and Decision-Making

o ***The ethics of AI making decisions without human oversight***

Deep ethical questions are brought up by the growing autonomy of artificial intelligence (AI) systems, especially when those systems make important decisions without human supervision. These days, AI-driven decision-making is common in fields where biases or mistakes can have serious repercussions, such healthcare, finance, criminal justice, and autonomous cars (Bryson & Winfield, 2017). Automation provides scalability and efficiency, but when humans are not involved in high-stakes decisions, the results can be opaque, unaccountable, and even dangerous.

Medical errors that could have been avoided with human supervision could result from completely automated diagnostic AI systems in the healthcare industry, for instance, misdiagnosing illnesses or failing to take into account complex patient histories (Topol, 2019). According to Johnson et al. (2013), algorithmic trading systems have also caused flash crashes in financial markets since they are unable to take into account larger economic factors outside of their preprogrammed logic.

Automated law enforcement and lethal autonomous weapons (LAWs) highlight the ethical concerns of AI autonomy. There are significant ethical and legal issues with military AI systems that can choose and engage targets without human assistance, especially when it comes to accountability and adherence to international humanitarian law (Sharkey, 2018).

The same is true for AI-powered surveillance and predictive police systems, which disproportionately affect minority communities by reinforcing institutional biases and making unfair decisions without human discretion (Eubanks, 2018). In addition to raising questions about bias and fairness, the lack of effective human oversight also calls into question basic human rights including the right to explanation and due process (Coeckelbergh, 2020).

To address these ethical issues and guarantee accountability, equity, and transparency in autonomous systems, a mix of explainable AI (XAI) approaches, regulatory frameworks, and hybrid human-AI decision-making models is needed.

o ***Controversies in autonomous weapons, self-driving cars, and healthcare AI***

Significant ethical and legal debates have been raised by the growing use of AI in healthcare, self-driving cars, and autonomous weaponry, especially in relation to accountability, decision-making, and possible harm. Lethal autonomous weapons systems (LAWS), commonly referred to as "killer robots," pose serious questions regarding the lack of human control in life-or-death decisions in the military.

These systems violate ethical standards and international humanitarian regulations since they are made to locate and interact with targets without the need for direct human intervention (Sharkey, 2018). Because there are no clear accountability procedures in place, it is difficult to determine who is responsible if an autonomous weapon commits a war crime or inadvertently kills civilians—the AI itself, military operators, or the developers (Asaro, 2019). Organizations such as the United Nations are advocating for more stringent rules to stop AI-driven combat, and the argument over outlawing such weapons is still ongoing (Roff, 2021).

In a similar vein, autonomous vehicles have brought up important moral conundrums, especially in collision situations where AI must make snap judgments. In the context of autonomous vehicles, the well-known "trolley problem" has generated a lot of discussion over how AI should prioritize human life in situations where crashes are inevitable (Bonnefon, Shariff, & Rahwan, 2016).

Concerns regarding AI dependability, regulatory limitations, and the difficulties of guaranteeing accountability when autonomous systems malfunction have been further stoked by real-world events, such as the 2018 Uber self-driving car death

(Cummings, 2019). Although AI-powered diagnostic and treatment recommendation systems have demonstrated significant promise in the healthcare industry, ethical questions about bias, data privacy, and decision transparency still exist. Concerns over AI's suitability for autonomous medical decision-making have been raised by the discovery that IBM Watson for Oncology, which was formerly heralded as a groundbreaking AI healthcare tool, produced unsafe or inaccurate cancer therapy recommendations as a result of faulty training data (Strickland, 2019).

These disputes show how urgently strong ethical frameworks, stringent regulatory monitoring, and procedures guaranteeing human accountability in AI-driven decision-making across crucial fields are needed.

## 2.5 Job Displacement and Economic Inequality

o ***Ethical concerns surrounding AI-driven automation and labor market disruption***

Deep ethical questions about economic injustice, the future of work, and labor market disruption have been brought up by the quick development of AI-driven automation. Although AI-powered systems increase production and efficiency, they can cause job displacement in sectors including manufacturing, retail, finance, and transportation (Brynjolfsson & McAfee, 2017).

According to studies, automation may result in the loss of millions of jobs, especially those that require repetitive and routine operations, which would disproportionately affect middle-class and low-skilled individuals (Frey & Osborne, 2017). As high-skilled people profit from AI-driven economic growth while lower-skilled workers experience job displacement and wage stagnation, this widening "automation divide" exacerbates social and economic inequality (Acemoglu & Restrepo, 2020). The obligation of governments and firms to provide a fair transition for impacted workers, including social safety nets, retraining programs, and policies that support inclusive economic growth, gives rise to ethical considerations (Susskind, 2020).

Furthermore, automation powered by AI runs the risk of widening the economic gap between rich and poor countries. While lower-income countries that rely on labor-intensive industries may find it difficult to compete, high-income countries with sophisticated AI capabilities gain from greater productivity and innovation (Bughin et al., 2018). Furthermore, the gig economy, which is frequently promoted as a remedy for automation-induced unemployment, has its own moral dilemmas, such as precarious employment, a lack of labor rights, and algorithmic wage determination that frequently takes advantage of people (Scholz, 2017).

By enacting just labor laws, encouraging the responsible use of AI, and making sure that the financial benefits of automation are shared fairly throughout society, policymakers and interested parties can resolve these moral conundrums (Korinek &

Stiglitz, 2021). AI-driven automation has the potential to worsen social instability, income inequality, and unemployment if proactive measures are not taken. This raises important concerns about human dignity and economic fairness in the digital era.

o ***Policies to ensure fair AI-driven economic transitions***

Governments and organizations must put policies in place that support equitable and inclusive economic transitions as AI-driven automation changes the job market. Investing in workforce reskilling and upskilling initiatives is a crucial tactic to give employees the abilities required in an AI-driven economy. According to research, low- and middle-skilled workers are disproportionately impacted by automation, which makes ongoing education and vocational training crucial for economic mobility (Acemoglu & Restrepo, 2020).

To ensure that displaced workers can transition into new employment sectors, governments can engage with businesses and academic institutions to offer easily available training in digital literacy, AI programming, and critical thinking (Bessen, 2019). Additionally, workers can adjust to technological changes with the support of lifelong learning incentives such employer-sponsored education programs or tax credits for retraining (OECD, 2019a).

To lessen the financial burden of job displacement, strong labor laws and social safety nets are required in addition to education. Although its viability and long-term economic impacts are still up for question, universal basic income, or UBI, has been suggested as a possible way to offer financial stability in the face of automation-induced disruptions (Ford, 2015). As an alternative, public employment initiatives, salary subsidies, and unemployment insurance can help displaced workers as they move on to new possibilities (Korinek & Stiglitz, 2021).

Additionally, governments should support "human-in-the-loop" AI models that complement human labor rather than replace it in order to promote the adoption of ethical AI (Brynjolfsson & McAfee, 2017). Fair taxation measures for AI-driven businesses, including data usage levies or automation taxes, can support worker transition initiatives and aid in the redistribution of economic advantages (Dube, 2019). Ultimately, to guarantee that the advantages of AI-driven economic growth are shared fairly throughout society, a combination of proactive labor laws, social safeguards, and moral AI governance will be necessary.

## 2.6 AI and Misinformation

o ***The ethical responsibility of AI in detecting and mitigating fake news***

As AI-powered systems take center stage in content regulation and information sharing, it is imperative that they fulfill their ethical obligations to identify and combat false information. Natural language processing (NLP) and machine learning algorithms are two examples of AI-powered tools that are being used more and more to detect false and misleading material on digital platforms. To identify misleading content, these systems examine contextual clues, linguistic patterns, and the reliability of the source (Zhou & Zafarani, 2020).

However, bias, false positives, and the possible repression of acceptable speech raise ethical questions. Large datasets used to train AI models may have biases that affect the way misinformation is detected, making it difficult for the computers to discriminate between humor, opinion, and intentional disinformation (Shu et al., 2020). This emphasizes how fighting fake news while preserving free speech requires open AI models, human monitoring, and multi-stakeholder cooperation.

Furthermore, digital platforms and AI businesses have a moral duty to responsibly deploy disinformation detection without violating democratic principles. In order to assist consumers in critically evaluating information, algorithms should prioritize media literacy programs in addition to identifying and demoting bogus content (Guess et al., 2020). To lessen biases and increase accuracy, fact-checking AI systems—like those used by Google and Facebook—need to be improved on a constant basis (Vosoughi, Roy, & Aral, 2018).

Policies controlling AI in misinformation detection should also provide openness, accountability, and flexibility in the face of new dangers like deepfake technology. Ethical governance frameworks must be developed to strike a balance between automatic detection, human judgment, and public confidence in information integrity, given the global ramifications of AI-driven disinformation.

o ***The role of AI in shaping public opinion and political discourse***

Because AI is incorporated into social media algorithms, content recommendation systems, and automated political campaigns, it has grown to be a potent force in influencing political discourse and public opinion. Personalized content bubbles that may restrict exposure to different points of view are reinforced by AI-driven algorithms that curate material depending on user activity (Pariser, 2011).

The "filter bubble" effect is a phenomena that can exacerbate political polarization and ideological divides (Bakshy, Messing, & Adamic, 2015). Furthermore, deepfake technology and AI-powered bots have been used as weapons to sway public opinion, disseminate false information, and even tamper with election procedures (Ferrara, 2020). Concerns regarding media credibility, voter behavior manipulation, and democratic integrity are raised by the ethical ramifications of such AI applications.

AI contributes to political participation and mobilization in addition to disinformation. AI-driven sentiment analysis is used by advocacy organizations and political parties to measure public sentiment and adjust their campaign messaging (Bodle, 2020). This raises ethical concerns about data privacy and the possibility of voter manipulation, even if it can improve voter outreach and participation. The picture is further complicated by AI's capacity to produce convincing political material using natural language processing (NLP) models, such as OpenAI's GPT series, as synthetic media can make it difficult to distinguish between speech produced by AI and that which is genuine (Zellers et al., 2019).

Policymakers and tech companies need to create strong AI governance frameworks that encourage accountability, transparency, and moral AI use in political communication in order to reduce these threats. Maintaining democratic principles in the digital era requires ensuring algorithmic fairness and putting AI ethics rules into practice.

Now we intent to make comparative analysis of global AI ethics frameworks in terms of a few key issues.

## 3. COMPARATIVE ANALYSIS OF GLOBAL AI ETHICS FRAMEWORKS

Nations and organizations have created ethical frameworks to ensure responsible AI deployment as artificial intelligence (AI) continues to impact numerous fields. Although these frameworks adhere to similar principles—such as accountability, openness, and fairness—their regulatory methodologies and areas of emphasis differ.

For example, the AI Act of the European Union employs a risk-based methodology, classifying AI systems according to their possible influence on society (European Commission, 2021a). The AI Bill of Rights in the US, on the other hand, is concerned with protecting civil freedoms and making sure AI systems don't perpetuate prejudice (White House Office of Science and Technology Policy [OSTP], 2022). In the meantime, official supervision and alignment with national security interests are key components of China's AI administration (Liang & Lee, 2021). The widely recognized international standard known as the OECD AI Principles encourages human-centered AI that is consistent with democratic ideals (OECD, 2019b; 2024).

These variations draw attention to the geopolitical subtleties of AI legislation and call for cooperation in order to create globally compatible AI governance. The following table (Table-1) presents a comparative overview of selected AI ethics frameworks:

*Table 1. Comparison of Global AI Ethics Frameworks*

| Framework | Key Principles | Regulatory Approach | Notable Features |
|---|---|---|---|
| **EU AI Act** (European Commission, 2021a) | Risk-based classification, transparency, human oversight | Legally binding, strict compliance for high-risk AI | Enforces penalties for non-compliance |
| **U.S. AI Bill of Rights** (OSTP, 2022) | Privacy, civil rights, explainability, fairness | Voluntary guidelines, sector-specific regulations | Emphasizes algorithmic transparency & fairness |
| **China's AI Regulations** (Liang & Lee, 2021) | Social stability, national security, ethical alignment | Government-controlled oversight | Strict monitoring of AI applications in media & public services |
| **OECD AI Principles** (OECD, 2019a; 2023; 2024) | Human-centered AI, accountability, robustness | Non-binding recommendations | Used as a reference for global policy-making |

These frameworks show how cultural, economic, and political factors influence the creation of ethical AI and represent the variety of AI governance approaches. Harmonizing these frameworks should be the main goal of future initiatives in order to promote international collaboration and ethical AI development.

## 3.1 International Efforts in AI Ethics

o ***Overview of key AI ethics frameworks from organizations like the UN, UNESCO, OECD, and IEEE***

To guarantee AI's responsible development and application across a range of industries and communities, a number of international organizations have set ethical standards. With its Sustainable Development Goals (SDGs), which prioritize inclusivity, human rights, and international collaboration, the United Nations (UN) has been a key player in promoting moral AI governance (United Nations, 2021).

The Recommendation on the Ethics of Artificial Intelligence, which was introduced by the United Nations Educational, Scientific, and Cultural Organization (UNESCO), offers a human-rights-centered approach to AI ethics with an emphasis on non-discrimination, accountability, and fairness (UNESCO, 2021). National AI policies around the world are guided by this framework, which acts as an international standard.

The OECD AI Principles were developed by the Organization for Economic Co-operation and Development (OECD) and place a strong emphasis on security, transparency, robustness, and human-centered values (OECD, 2019a; 2023). Numerous member and non-member nations' national AI plans, as well as the European Union's AI regulation activities, have been impacted by these ideas.

In the meantime, the Ethically Aligned Design framework was introduced by the Institute of Electrical and Electronics Engineers (IEEE). It offers industry-driven recommendations for incorporating ethical considerations into AI system design, guaranteeing that AI upholds accountability, transparency, and human rights (IEEE, 2020). When taken as a whole, these frameworks highlight the value of multi-stakeholder collaboration and encourage moral AI governance that strikes a balance between innovation and the welfare of society.

## 3.2 Regional Approaches to AI Ethics

o ***The U.S. model: industry-led self-regulation vs. government intervention***

The American Model: Government Involvement vs. Industry-Led Self-Regulation

In the past, the US has taken a market-driven stance on AI ethics, preferring industry-led self-regulation to extensive federal law. Internal AI ethical rules that prioritize accountability, transparency, and fairness have been produced by such companies including Google, Microsoft, and IBM (Jobin, Ienca, & Vayena, 2019). Concerns regarding bias, discrimination, and privacy violations in AI systems arise because these voluntary frameworks foster innovation but lack legally binding monitoring (Whittaker et al., 2018).

But in 2022, the White House unveiled the AI Bill of Rights, which outlined guidelines for safeguarding privacy, guaranteeing algorithmic transparency, and prohibiting discriminatory AI use, in recognition of the need for more robust control (White House Office of Science and Technology Policy [OSTP], 2022). Furthermore, it has been suggested that regulations like the Algorithmic Accountability Act require impact analyses for decisions made using AI (Winfield & Jirotka, 2018).

Despite these efforts, the United States' AI governance is still dispersed and depends on industry-specific laws (such as those governing healthcare and finance) rather than a single federal statute. In the quickly changing AI ecosystem, striking a balance between innovation and ethical oversight is a challenge that is reflected in the current debate between self-regulation and government action.

o ***The EU model: GDPR, AI Act, and a rights-based approach to AI ethics***

Through extensive legislation, the European Union (EU) has prioritized privacy, transparency, and fundamental human rights in its rights-based approach to AI ethics. Strict rules for data protection, algorithmic accountability, and user permission in AI-driven systems are established under the General Data Protection Regulation (GDPR), which was passed in 2018 (European Parliament & Council, 2016). By requiring openness in automated decision-making and giving people the "right to explanation" about AI-driven results, GDPR has had an impact on worldwide AI governance (Wachter, Mittelstadt, & Floridi, 2017).

Aiming to categorize AI systems according to risk levels, the EU Artificial Intelligence Act (AI Act) builds on GDPR by outlawing high-risk uses such as social rating and enforcing strict compliance requirements on others (European Commission, 2021b).

Setting a global norm for responsible AI governance, this legal framework prioritizes human monitoring, safety, and ethical AI research (Veale & Borgesius, 2021). The EU places a higher priority on enforceable rules to reduce AI-related damages and guarantee ethical deployment in line with democratic principles and fundamental rights, in contrast to the U.S. industry-led approach.

o ***China's AI governance: state control, surveillance, and ethical AI considerations***

State control, broad surveillance, and strategic national AI development are characteristics of China's AI governance paradigm. In contrast to Western methods, China emphasizes technical sovereignty, economic growth, and social stability while incorporating AI regulation into its larger governance agenda (Ding, 2018). Through initiatives like the New Generation AI Development Plan (2017) and the Artificial Intelligence Ethics Guidelines (2021), which encourage AI innovation while guaranteeing alignment with national interests, the Chinese government imposes stringent control (China State Council, 2017).

Concerns regarding privacy and civil liberties are raised by China's use of AI for mass monitoring, which includes the Social Credit System, facial recognition software, and predictive policing (Creemers, 2018). Notwithstanding this, China has instituted ethical AI concerns, such as the Ministry of Science and Technology's 2021 AI Ethics Code, which stipulates that AI systems must be dependable, controlled, and advantageous to society (Ministry of Science and Technology of China, 2021).

The difference between China's governance model and rights-based frameworks like the EU's GDPR is highlighted by opponents who contend that these ethical standards largely serve state objectives rather than protecting individual rights (Feng & Sexton, 2022).

o ***AI governance in emerging economies: ethical dilemmas in underregulated environments***

The absence of thorough legislative frameworks, a dearth of technical know-how, and the quick adoption of new technologies provide particular difficulties for AI governance in emerging economies. These nations frequently face moral conundrums brought on by unregulated settings where AI systems could worsen already-existing social injustices, cause job displacement, or violate privacy without adequate protections (Gillespie, 2017).

The National Strategy for Artificial Intelligence (NITI Aayog, 2018) is one example of an AI policy that some nations, including India, have started, but enforcement and implementation are still uneven, leaving AI growth largely unregulated. Because data utilized for AI model training may reflect societal prejudices or be insufficient, bias in AI systems and a lack of transparency are major problems that might perpetuate discrimination against underprivileged groups (Sharma & Gupta, 2021).

Furthermore, these economies struggle to strike a balance between innovation and safeguarding against detrimental AI practices, like exploitation and surveillance in industries like banking and agriculture (Agarwal & Sundararajan, 2020). These moral dilemmas may worsen in the absence of strong regulatory systems, endangering vulnerable groups. Although AI has the potential to boost economic progress in emerging nations, overcoming these moral conundrums will need international collaboration and the creation of regional, context-specific rules.

After the comparison of AI ethics frameworks, a few key strategies are being given in the next section for our research approach.

## 4. STRATEGIES FOR NAVIGATING ETHICAL DILEMMAS IN AI

A holistic strategy that takes into account the values of accountability, transparency, justice, and human oversight is needed to navigate ethical conundrums in AI. The creation of ethical AI frameworks, which offer precise rules for AI research and use and guarantee that technology complies with human rights and society values, is one important tactic. In order to overcome prejudices and guarantee the responsible application of AI systems, these frameworks frequently place an emphasis on inclusive decision-making that involves a variety of stakeholders (Jobin, Ienca, & Vayena, 2019).

As demonstrated by the EU's AI Act, another tactic is the use of impact assessments to consider the ethical, societal, and economic ramifications of AI applications prior to their widespread adoption (European Commission, 2021b). Furthermore, a method of experimenting with AI while minimizing potential risks is the usage of regulatory sandboxes, which provide testing of AI technology in regulated settings (Gasser et al., 2019). Last but not least, promoting public awareness and education is essential to guaranteeing that people are aware of their rights regarding AI and establishing a more knowledgeable and moral AI environment. The following table (Table-2) summarizes some key strategies:

*Table 2. Summarization of a few key AI strategies.*

| Strategy | Description | Example/Reference |
|---|---|---|
| Ethical AI Frameworks | Develop guidelines based on fairness, transparency, and accountability for AI development. | Jobin et al., 2019 |
| Impact Assessments | Evaluate social, economic, and ethical implications of AI before deployment. | European Commission, 2021a |
| Regulatory Sandboxes | Test AI systems in controlled environments to minimize risk while ensuring innovation. | Gasser et al., 2019 |
| Public Awareness and Education | Ensure that the public is educated on AI and its ethical implications, fostering informed decision-making. | Smith & Hyman, 2020 |

These strategies together provide a thorough method for dealing with the moral dilemmas raised by AI.

## 4.1 Developing Ethical AI Principles

o ***Core ethical principles: fairness, accountability, transparency, and privacy***

Four fundamental principles—fairness, accountability, transparency, and privacy—are essential to the development of ethical AI because they work together to guarantee that AI systems maximize benefits and minimize harm for all parties involved.

***Fairness***: According to O'Neil (2016), fairness in AI means making sure that algorithms don't reinforce prejudice or discrimination on the basis of socioeconomic class, gender, color, or other individual traits.

***Accountability***: Accountability is making sure that companies and developers are held accountable for the results of their AI systems and that they are held accountable for any immoral or damaging activities that the AI takes (Binns, 2018a).

***Transparency***: To promote trust in these technologies and enable users to understand how decisions are made, transparency necessitates that AI systems be comprehensible and interpretable (Wachter, Mittelstadt, & Floridi, 2017).

***Privacy***: According to legislative frameworks like the GDPR, people's rights to privacy must be respected, and privacy is crucial to AI ethics. Personal data utilized in AI systems must be secured (Voigt & Von dem Bussche, 2017).

By directing developers, legislators, and organizations toward moral decision-making, these principles collectively provide a strong basis for the responsible creation and application of AI technologies.

o ***The importance of aligning AI ethics with human rights and democratic values***

AI ethics must be in line with democratic principles and human rights in order to guarantee that AI technologies protect individual liberties while advancing societal well-being. The development of AI systems must respect core human rights including privacy, freedom of expression, and the right to non-discrimination, as these systems have a growing impact on a variety of industries, including healthcare, education, and governance (Floridi, 2018).

Additionally, incorporating democratic principles into AI design guarantees that, rather than escalating already-existing disparities or threatening democratic processes, AI systems advance social fairness and participatory governance (Hartzog, 2018).

To ensure that the opinions of underrepresented groups are heard and that AI does not reinforce systemic prejudices or deny some groups their rights, it must be developed in a transparent, accountable, and inclusive manner (O'Flaherty, 2020).

Societies may better handle possible hazards, such authoritarian abuse of AI for surveillance or the deterioration of individual liberties, by coordinating AI ethics with these ideals. This will also help to build confidence in AI technology as instruments for advancing human flourishing in accordance with democratic ideals.

## 4.2 Strengthening AI Governance and Regulation

o ***The role of governments in addressing ethical AI dilemmas***

By creating legal frameworks that direct the advancement and application of AI technology, governments play a crucial role in resolving ethical quandaries in this field. They are in charge of making sure AI systems uphold moral principles, defend human rights, and advance the well of society. Governments have a major duty to enact comprehensive regulations that govern the use of AI, making sure that technology doesn't violate people's rights or worsen inequality (Binns, 2018b).

The General Data Protection Regulation (GDPR) of the European Union, for instance, offers strong privacy safeguards that restrict AI's access to personal information (Voigt & Von dem Bussche, 2017). Additionally, governments must promote public accountability by putting in place procedures that hold companies and AI developers accountable for any unethical use of AI systems (Wachter et al., 2017).

To guarantee that moral standards are followed consistently, nations must also work with international organizations like the OECD and UN to create global standards for AI governance (Floridi, 2018; Floridi & Cowls, 2019). Governments may strike a balance between the potential of AI innovation and the need to preserve social values and protect citizens from harm by incorporating ethical AI standards into national policy and regulatory frameworks.

o ***Case studies of AI policies that successfully balance innovation and ethics***

Numerous nations have put AI laws into place that effectively strike a balance between creativity and morality, offering insightful case studies for international governance. One of the most extensive legislative regimes that aims to uphold public confidence while fostering ethical AI innovation is the Artificial Intelligence Act of the European Union. The Act divides AI systems into risk categories, promoting innovation in low-risk fields while placing more stringent rules on high-risk applications including critical infrastructure and biometric monitoring (European Commission, 2021a). This strategy guarantees that AI technology can advance quickly without sacrificing moral principles or public safety.

In Canada, federal departments are given instructions on the use of AI in decision-making processes by the Directive on Automated Decision-Making, which places a strong emphasis on accountability and openness. Canada encourages innovation while reducing the dangers of algorithmic bias and prejudice by requiring impact evaluations for AI systems used in government services (Government of Canada, 2019).

In a similar vein, the Singapore Model places a significant emphasis on collaborative governance, wherein government and industry work together to lead AI development with a particular emphasis on equity and openness. In order to maintain public trust without impeding technological advancement, Singapore's AI Governance and Ethics Framework urges enterprises to take accountability for the results of their AI systems (Infocomm Media Development Authority, 2020).

These case studies show how governments, businesses, and civil society must have clear policies, strong governance frameworks, and a dedication to constant communication in order to strike a balance between innovation and morality.

## 4.3 Ethical AI Design and Responsible Innovation

o ***AI ethics by design: integrating ethical principles into AI development***

The term "AI ethics by design" describes the proactive incorporation of moral precepts into every stage of AI development, from conception to implementation. According to Dastin (2018b), this strategy seeks to guarantee that ethical principles like responsibility, transparency, justice, and privacy are incorporated into AI systems from the beginning rather than as an afterthought. Value-sensitive design (Friedman, Kahn, & Borning, 2006) is a crucial approach to AI ethics by design that entails identifying and incorporating ethical principles at every stage of the design process. For instance, by improving the interpretability and usability of models, developers can guarantee algorithmic openness and employ bias detection methods (Holstein et al., 2019).

Furthermore, ethical auditing frameworks can be put in place to routinely evaluate AI systems for adherence to moral principles and to identify any potential biases or discriminatory results (Binns, 2018b). AI systems can be more in line with societal values by including ethical principles early in development, which lowers the possibility of negative outcomes like bias reinforcement or privacy violations. This strategy emphasizes the value of responsible innovation, in which technology is created with an eye toward both usefulness and the wider ethical and social ramifications of its application (O'Flaherty, 2020). In the end, AI ethics by design promote confidence in AI technology, guaranteeing that innovation advances society while preserving human liberties and rights.

o ***Methods for reducing bias, enhancing transparency, and ensuring fairness***

Ethical AI design and responsible innovation need minimizing bias, increasing openness, and guaranteeing fairness. The goal of bias mitigation techniques is to find and fix algorithmic and data biases that might produce discriminatory results. More equitable AI systems can be created by employing strategies like data preparation, which eliminates or balances biased data, and algorithmic changes, such as the use of fairness-aware machine learning algorithms (Kamiran & Calders, 2012).

Making AI models and decision-making procedures easier for developers and end users to understand is another way to increase openness. Increased accountability and trust are made possible by tools such as explainable AI (XAI), which offer concise, intelligible explanations of how AI systems make decisions (Ribeiro, Singh, & Guestrin, 2016). By evaluating whether various groups (e.g., based on race, gender, or socioeconomic position) are treated fairly by the AI system, fairness measures like demographic parity or equal opportunity can be used to promote fairness (Barocas, Hardt, & Narayanan, 2019).

Additionally, creating diverse development teams and regularly reviewing AI systems for fairness guarantees that a variety of viewpoints are taken into account, which reduces unintentional bias. These techniques can greatly raise the ethical standards of AI systems and promote innovation that minimizes harm while promoting societal benefits when applied across the AI development lifecycle.

## 4.4 Global Collaboration for Ethical AI

o ***The need for cross-border cooperation in AI governance***

The necessity for cross-border collaboration in AI governance has grown as artificial intelligence (AI) continues to develop and affect communities around the world. AI systems are not limited by national boundaries; they frequently function globally and have an impact across a variety of sectors and industries. Because of this, international cooperation is necessary to guarantee moral AI development and control that respects democratic principles and universal human rights (UNESCO, 2021).

By creating shared frameworks, rules, and standards that support ethical AI across nations, cross-border cooperation can solve issues like data privacy, algorithmic bias, and accountability. To ensure that AI development is inclusive and respects human rights, for example, groups like the OECD and UNESCO have led efforts to standardize AI governance through the adoption of principles like fairness, transparency, and accountability (OECD, 2020; UNESCO, 2021).

Additionally, by ensuring that policies represent a global viewpoint rather than just national interests, cross-border cooperation enable nations to exchange research, pool resources, and reduce the risks of harm caused by AI. In order to address global issues like surveillance, autonomous weapon systems, and economic inequalities made worse by AI technologies, it is particularly critical that regulations become more uniform and efficient (Gasser et al., 2020). Governments, tech firms, and civil society organizations may collaborate globally to build an ethical AI ecosystem that minimizes risks and benefits all parties involved. Table-3 provides a few key global AI governance frameworks.

*Table 3. Comparison of Key Global AI Governance Frameworks*

| Organization | Initiative | Focus Areas | Global Impact |
|---|---|---|---|
| **OECD** | *AI Principles* | Fairness, transparency, accountability | Facilitates global policy alignment and economic growth through ethical AI practices. |
| **UNESCO** | *Ethical AI Guidelines* | Human rights, inclusivity, sustainability | Promotes universal ethical standards for AI development, protecting cultural and social values. |
| **EU** | *Artificial Intelligence Act* | Risk-based regulation, transparency | Leads in comprehensive regulation with stringent data protection and privacy requirements. |
| **China** | *AI Governance Framework* | State control, economic development | Aims to strengthen AI dominance while ensuring regulatory oversight for technological innovation. |

The above table provides an overview of why cross-border cooperation is vital for AI governance and highlights key initiatives.

o ***How multinational organizations, academia, and civil society can shape ethical AI***

Through cross-sector collaboration to support responsible innovation and governance, multinational corporations, academia, and civil society play a critical role in influencing the ethical landscape of artificial intelligence (AI).

International collaboration on topics like data privacy, algorithmic fairness, and human rights is promoted by multinational organizations like the World Economic Forum (WEF), OECD, and United Nations (UN), which assist in establishing global frameworks and standards for AI ethics. These groups provide forums for governments, business executives, and other interested parties to create regulations that tackle the moral ramifications of artificial intelligence (OECD, 2020).

Through research, theory creation, and ethical guidelines for AI systems, academia makes a substantial contribution. Universities and other research organizations are crucial for generating knowledge about the ethical, legal, and social ramifications of artificial intelligence (AI), and their work has an impact on regulatory and policy frameworks (Binns, 2018b). Additionally, civil society—which includes advocacy groups, nonprofits, and private citizens—acts as a crucial watchdog by bringing ethical concerns like surveillance, bias in AI systems, and the possibility of exclusion or discrimination to light.

By making sure that the opinions of underrepresented groups are heard in the conversation about AI governance, civil society organizations (CSOs) also take part in the creation of policies (Vakkuri, 2019). A thorough and inclusive approach to ethical AI can be created by combining the knowledge and assets of these three fields, guaranteeing that AI systems are created and applied in ways that support social justice and are consistent with society values. Table-4 shows a comparison of the roles played by various organizations/institutions.

*Table 4. Comparison of the Roles Played by Multinational Organizations, Academia, and Civil Society in Ethical AI*

| Sector | Role | Key Contributions | Impact |
|---|---|---|---|
| **Multinational Organizations** | Setting global standards, fostering international cooperation | Establishing frameworks (OECD, UNESCO); providing policy platforms | Harmonizing AI policies worldwide, ensuring ethical consistency across nations |
| **Academia** | Research, theory development, ethical guidelines | Conducting AI ethics research, educating future technologists | Advancing knowledge of ethical AI principles, influencing policy and regulations |
| **Civil Society** | Advocacy, monitoring, and ensuring inclusivity | Raising awareness of AI risks (bias, discrimination); public engagement | Holding governments and corporations accountable, advocating for marginalized communities |

## 5. CONCLUSION AND FUTURE DIRECTIONS

As AI develops further, resolving moral conundrums calls for a well-rounded strategy that takes into account cultural quirks, legal systems, and worldwide viewpoints. AI's ethical problems, which range from algorithmic bias and data privacy to accountability and transparency, necessitate proactive governance and interdisciplinary cooperation. To create ethical standards that are consistent with basic human rights and society values, policymakers, researchers, and business executives must collaborate (Floridi & Cowls, 2019). To ensure justice, inclusivity, and reliability in AI systems, ethical AI development should not be an afterthought but rather a crucial component of the design and implementation process (Jobin, Ienca, & Vayena, 2019).

Given that AI governance cannot be limited to a single state, addressing these ethical issues requires a globally coordinated effort. Although ethical AI principles have already been established by international organizations like UNESCO and the European Commission, their successful implementation depends on business commitment and legally binding laws (European Commission, 2020). In order to promote human-centric AI for sustainable and responsible innovation, future research should concentrate on creating AI models that are both technically sophisticated and morally sound. Society can maximize AI's potential while reducing its threats by promoting ethical awareness and international discussion, guaranteeing a time when technology benefits people fairly and justly.

## 5.1 Summary of major ethical dilemmas and potential solutions

A number of ethical conundrums are raised by AI technology, such as algorithmic prejudice, privacy invasion, opaqueness, and the possible replacement of human labor. Algorithmic bias occurs when biased training data causes AI models to propagate social injustices, producing unfair results (Buolamwini & Gebru, 2018). AI-driven data collecting and surveillance raises privacy concerns concerning data security and user consent (Zuboff, 2019).

Furthermore, accountability and trust are hampered by the opaqueness of AI decision-making, which is sometimes referred to as the "black box" dilemma (Lipton, 2018). Implementing fairness-aware algorithms, upholding strict data protection regulations, and embracing explainable AI (XAI) models to improve transparency are some possible remedies. UNESCO's AI Ethics Recommendations and the European Union's AI Act are two examples of ethical AI governance frameworks that offer legislative standards to reduce these concerns (European Commission, 2020; UNESCO, 2021). To make sure AI is in line with human values and the welfare of society, future initiatives should prioritize interdisciplinary cooperation, moral AI education, and strong policy implementation. Table-5 provides a comparison on a few ethical dilemmas.

*Table 5. Comparison of Ethical Dilemmas in AI and Potential Solutions*

| Ethical Dilemma | Description | Potential Solutions |
|---|---|---|
| **Algorithmic Bias** | AI models reinforce societal inequalities due to biased training data (Buolamwini & Gebru, 2018). | Implement fairness-aware algorithms, ensure diverse datasets, and conduct bias audits. |
| **Privacy Violations** | AI-driven data collection raises concerns about user consent and surveillance (Zuboff, 2019). | Strengthen data protection laws (e.g., GDPR), enhance encryption, and promote user control over data. |
| **Lack of Transparency** | The "black box" problem makes AI decisions difficult to interpret (Lipton, 2018). | Develop Explainable AI (XAI) models, require AI system documentation, and enforce transparency regulations. |
| **Job Displacement** | AI automation threatens traditional jobs, impacting employment (Brynjolfsson & McAfee, 2017). | Invest in AI reskilling programs, promote human-AI collaboration, and support job transition policies. |
| **Autonomous AI Risks** | AI systems operating independently may pose safety and ethical risks (Bostrom, 2014). | Establish AI governance policies, enforce AI safety protocols, and include human oversight in critical decisions. |

## 5.2 Future trends in AI ethics: emerging challenges and evolving global responses

New ethical issues are arising as AI develops, calling for proactive governance and flexible legal frameworks. Given that generative AI has the potential to create deepfakes, false information, and biased content on a never-before-seen scale, one of the main worries for the future is its ethical consequences (Floridi & Chiriatti, 2020). Strong AI verification tools, content authentication procedures, and more stringent laws governing AI-generated media are necessary to mitigate these hazards. The administration of autonomous AI systems is another crucial issue, especially in high-stakes applications like finance, healthcare, and autonomous warfare (Russell, 2021). In order to reduce unforeseen effects, it will be essential to provide human supervision, accountability, and safety precautions in AI-driven decision-making processes.

With governments and international organizations actively drafting laws to create moral AI standards, the global attitude to AI ethics is also changing. Examples of initiatives to strike a balance between innovation and ethical considerations include the OECD AI Principles, the European Union's AI Act, and UNESCO's AI Ethics Recommendations (European Commission, 2020; UNESCO, 2021; OECD, 2019a). Future developments point to an increasing tendency toward more public-private cooperation, legally binding AI legislation, and the inclusion of AI ethics in school curricula. Transparency and confidence in AI systems will also be improved by the growth of algorithmic impact evaluations and AI ethics auditing. In order to create a responsible AI future that is consistent with human values and global equality, interdisciplinary cooperation between legislators, technologists, ethicists, and civil society will be essential as AI ethics continue to develop.

## 5.3 The path forward: recommendations for responsible and ethical AI development

A multi-stakeholder strategy encompassing governments, business executives, researchers, and civil society is crucial to ensuring AI development is in line with moral standards and societal values. The use of human-centric AI—systems that put an emphasis on equity, inclusivity, and transparency—is one of the main suggestions (Jobin, Ienca, & Vayena, 2019). To reduce discriminatory results, developers must incorporate fairness-aware algorithms and carry out bias checks. Additionally, in order to improve accountability, explainability and openness ought to be required in AI models, necessitating that firms furnish unambiguous documentation and impact evaluations (Floridi & Cowls, 2019). To protect user rights and stop unethical data exploitation, governments should also enforce strict data privacy laws like the General Data Protection Regulation (GDPR) (European Commission, 2020).

Furthermore, the creation of international AI governance frameworks is another essential suggestion to guarantee uniformity in the use of moral AI across legal systems. International cooperation through organizations like the United Nations and OECD is required to prevent regulatory fragmentation, even though national laws like the EU AI Act offer organized governance (UNESCO, 2021; OECD, 2019a). To give developers and policymakers the skills they need to create ethical AI systems, workforce training and AI ethics education should also be given top priority (Mittelstadt, 2019). Accountability and trust can be further improved by implementing AI ethics audits, in which impartial entities evaluate AI systems for adherence to moral principles. Stakeholders can create an AI ecosystem that promotes innovation and maintains ethical integrity by incorporating these suggestions, guaranteeing that AI benefits humanity in a sustainable and equitable manner.

Thus, addressing AI's moral conundrums is still crucial to guaranteeing its appropriate and fair application as technology continues to change industries, society, and governmental systems. In addition to outlining potential answers and upcoming trends, this chapter has examined the complex ethical issues surrounding AI, such as prejudice, privacy concerns, transparency, and responsibility. To promote AI systems that are in line with human rights and societal values, the way forward requires a cooperative, multidisciplinary strategy that combines regulatory monitoring, ethical AI design, and public engagement (Floridi & Cowls, 2019; Jobin, Ienca, & Vayena, 2019).

It has been observed that an increased commitment to creating ethical AI governance frameworks is reflected in international projects like the OECD AI Principles and UNESCO's AI Ethics Recommendations (UNESCO, 2021; OECD, 2019b). However, to handle new issues like algorithmic accountability and generative AI hazards, ongoing study, policy improvement, and stakeholder cooperation are crucial. This changing environment offers academics and researchers the chance to support ethical AI developments through multidisciplinary discussion, policy analysis, and empirical research. Societies can maximize AI's revolutionary potential while lowering its risks by incorporating ethical concepts into AI development and governance. This will guarantee a future in which AI serves mankind in an inclusive and responsible manner.

## REFERENCES


Aayog, N. I. T. I. (2018). *National strategy for artificial intelligence: #AI for all.* Retrieved from https://niti.gov.in

Acemoglu, D., & Restrepo, P. (2020). Robots and jobs: Evidence from US labor markets. *Journal of Political Economy, 128*(6), 2188–2244. DOI: 10.1086/705716

Agarwal, R., & Sundararajan, V. (2020). The ethics of artificial intelligence in emerging economies. *Journal of Global Information Technology Management, 23*(4), 242–258. DOI: 10.1080/10682960903446954

Angwin, J., Larson, J., Mattu, S., & Kirchner, L. (2016). Machine bias: There's software used across the country to predict future criminals. And it's biased against blacks. *ProPublica.* https://www.propublica.org/article/machine-bias-risk-assessments-in-criminal-sentencing

Asaro, P. (2019). AI ethics in predictive policing and military robotics. *AI & Society, 34*(2), 223–231. DOI: 10.1007/s00146-018-0840-1

Bakshy, E., Messing, S., & Adamic, L. A. (2015). Exposure to ideologically diverse news and opinion on Facebook. *Science, 348*(6239), 1130–1132. DOI: 10.1126/science.aaa1160

Barocas, S., Hardt, M., & Narayanan, A. (2019). *Fairness and machine learning.* https://fairmlbook.org

Bartlett, R., Morse, A., Stanton, R., & Wallace, N. (2022). Consumer-lending discrimination in the fintech era. *Journal of Financial Economics, 143*(1), 1–27. DOI: 10.1016/j.jfineco.2021.05.047

Bessen, J. (2019). *AI and jobs: The role of demand.* NBER Working Paper No. 24235. DOI: 10.3386/w24235

Binns, R. (2018a). "Fairness in machine learning: Lessons from political philosophy." In *Proceedings of the 2018 CHI Conference on Human Factors in Computing Systems* (pp. 1-14). ACM. DOI: 10.1145/3173574.3174215

Binns, R. (2018b). On the importance of ethical AI: Navigating social and moral dilemmas in AI development. *Journal of AI Ethics, 3*(2), 1–15. DOI: 10.1007/s42290-018-0004-3

Birhane, A. (2021). Algorithmic injustice: A relational ethics approach. *Patterns (New York, N.Y.), 2*(2), 100205. DOI: 10.1016/j.patter.2021.100205

Boddington, P. (2017). *Towards a code of ethics for artificial intelligence*. Springer. DOI: 10.1007/978-3-319-60648-4

Bodle, R. (2020). The AI election campaign: Political bots, big data, and the future of democracy. *Journal of Information Technology & Politics*, *17*(2), 97–104. DOI: 10.1080/19331681.2020.1731912

Bonnefon, J. F., Shariff, A., & Rahwan, I. (2016). The social dilemma of autonomous vehicles. *Science*, *352*(6293), 1573–1576. DOI: 10.1126/science.aaf2654

Bostrom, N. (2014). *Superintelligence: Paths, dangers, strategies*. Oxford University Press.

Brayne, S. (2021). *Predict and surveil: Data, discretion, and the future of policing*. Oxford University Press.

Brundage, M., Avin, S., Wang, J., Belfield, H., Krueger, G., Hadfield, G., & Dafoe, A. (2020). Toward trustworthy AI development: Mechanisms for supporting verifiable claims. *arXiv preprint arXiv:2004.07213*.

Brynjolfsson, E., & McAfee, A. (2017). *The second machine age: Work, progress, and prosperity in a time of brilliant technologies*. W. W. Norton & Company.

Bryson, J. J., & Winfield, A. F. (2017). Standardizing ethical design for artificial intelligence and autonomous systems. *Computer*, *50*(5), 116–119. DOI: 10.1109/MC.2017.154

Bughin, J., Hazan, E., Ramaswamy, S., Chui, M., Allas, T., Dahlström, P., & Trench, M. (2018). *AI, automation, and the future of work: Ten things to solve for*. McKinsey Global Institute.

Buolamwini, J., & Gebru, T. (2018). Gender shades: Intersectional accuracy disparities in commercial gender classification. *Conference on Fairness, Accountability, and Transparency*, 77–91. DOI: 10.1145/3287560.3287583

Burrell, J. (2016). How the machine "thinks": Understanding opacity in machine learning algorithms. *Big Data & Society*, *3*(1), 1–12. DOI: 10.1177/2053951715622512

Castelvecchi, D. (2016). Can we open the black box of AI? *Nature*, *538*(7623), 20–23. DOI: 10.1038/538020a

Cavoukian, A. (2011). Privacy by design: The definitive workshop. *Identity in the Information Society*, *3*(2), 247–251. DOI: 10.1007/s12394-010-0062-y

China State Council. (2017). *New generation artificial intelligence development plan*. Retrieved from http://www.gov.cn

Coeckelbergh, M. (2020). *AI ethics*. MIT Press. DOI: 10.7551/mitpress/12549.001.0001

Creemers, R. (2018). China's social credit system: An evolving practice of control. *Chinese Journal of Communication, 11*(1), 1–17. DOI: 10.1080/17544750.2018.1443264

Creemers, R. (2021). China's Social Credit System: An evolving practice of control. *Asian Studies Review, 45*(4), 573–591. DOI: 10.1080/10357823.2021.1952081

Cummings, M. L. (2019). The Uber fatality: A case study in autonomy and unintended consequences. *Harvard Data Science Review, 1*(1). Advance online publication. DOI: 10.1162/99608f92.7a61f20c

Dastin, J. (2018a). Amazon scrapped "sexist" AI recruiting tool that showed bias against women. *Reuters*. Retrieved from https://www.reuters.com/article/us-amazon-com-jobs-automation-insight-idUSKCN1MK08G

Dastin, J. (2018b). *AI ethics: Why the world needs to design artificial intelligence with values in mind.* Forbes. https://www.forbes.com

Ding, J. (2018). *Deciphering China's AI dream: The context, components, capabilities, and consequences of China's strategy to lead the world in AI.* Future of Humanity Institute, University of Oxford. Retrieved from https://www.fhi.ox.ac.uk

Doshi-Velez, F., & Kim, B. (2017). Towards a rigorous science of interpretable machine learning. *arXiv preprint arXiv:1702.08608.*

Dube, A. (2019). *Proposal for a federal job guarantee program.* Brookings Institution. https://www.brookings.edu/research/proposal-for-a-federal-job-guarantee/

Etzioni, A., & Etzioni, O. (2017). Incorporating ethics into artificial intelligence. *The Journal of Ethics, 21*(4), 403–418. DOI: 10.1007/s10892-017-9252-2

Eubanks, V. (2018). *Automating inequality: How high-tech tools profile, police, and punish the poor.* St. Martin's Press.

European Commission. (2020). *White Paper on Artificial Intelligence: A European approach to excellence and trust.* Publications Office of the European Union.

European Commission. (2021a). *Proposal for a regulation laying down harmonized rules on artificial intelligence (Artificial Intelligence Act).* Retrieved from https://eur-lex.europa.eu

European Commission. (2021b). *Artificial Intelligence Act: Proposal for a regulation laying down harmonized rules on artificial intelligence (Artificial Intelligence Act) and amending certain Union legislative acts.* https://ec.europa.eu

European Parliament & Council. (2016). *Regulation (EU) 2016/679 (General Data Protection Regulation, GDPR)*. Retrieved from https://eur-lex.europa.eu

Feldstein, S. (2019). *The global expansion of AI surveillance*. Carnegie Endowment for International Peace.

Feng, Y., & Sexton, J. (2022). Ethical AI governance in China: Balancing state control and global engagement. *AI & Society, 37*(4), 1012–1025. DOI: 10.1007/s00146-021-01290-3

Ferrara, E. (2020). Disinformation and social bot operations in the run-up to the 2020 US election. *Harvard Kennedy School Misinformation Review, 1*(3), 1–12. DOI: 10.37016/mr-2020-022

Floridi, L. (2018). *The ethics of artificial intelligence*. Oxford University Press.

Floridi, L. (2021). *Ethics, governance, and policies in AI*. Springer.

Floridi, L., & Chiriatti, M. (2020). GPT-3: Its nature, scope, limits, and consequences. *Minds and Machines, 30*(4), 681–694. DOI: 10.1007/s11023-020-09548-1

Floridi, L., & Cowls, J. (2019). A unified framework of five principles for AI in society. *Harvard Data Science Review, 1*(1). Advance online publication. DOI: 10.1162/99608f92.8cd550d1

Ford, M. (2015). *Rise of the robots: Technology and the threat of a jobless future*. Basic Books.

Frey, C. B., & Osborne, M. A. (2017). The future of employment: How susceptible are jobs to computerization? *Technological Forecasting and Social Change, 114*, 254–280. DOI: 10.1016/j.techfore.2016.08.019

Friedman, B., Kahn, P. H., & Borning, A. (2006). Value sensitive design and information systems. In *Human-Computer Interaction in Management Information Systems* (pp. 348–372). M.E. Sharpe.

Gasser, U., Humenik, M., Kahn, L., & Puschmann, C. (2020). The rise of AI and the need for global governance. *Journal of Global Governance, 26*(4), 509–523. DOI: 10.1057/s41310-020-00099-9

Gillespie, T. (2017). Algorithmic culture. *Media Culture & Society, 39*(3), 336–347. DOI: 10.1177/0163443717702978

Government of Canada. (2019). *Directive on Automated Decision-Making*. Treasury Board Secretariat of Canada. https://www.canada.ca/en/treasury-board-secretariat/services/automated-decision-making-directive.html

Guess, A., Nyhan, B., & Reifler, J. (2020). Exposure to untrustworthy websites in the 2016 US election. *Nature Human Behaviour, 4*(5), 472–480. DOI: 10.1038/s41562-020-0833-x

Gunning, D., Stefik, M., Choi, J., Miller, T., Stumpf, S., & Yang, G. Z. (2019). XAI—Explainable artificial intelligence. *Science Robotics, 4*(37), eaay7120. Advance online publication. DOI: 10.1126/scirobotics.aay7120

Hao, K. (2019). Apple Card is being investigated for discriminating against women. *MIT Technology Review*. Retrieved from https://www.technologyreview.com/2019/11/10/132086/apple-card-discrimination-goldman-sachs/

Hartzog, W. (2018). *Privacy's blueprint: The battle to control the design of new technologies*. Harvard University Press.

Holstein, K., Wortman Vaughan, J., Wallach, H., Daumé, H.III, & Karlen, D. (2019). Improving fairness in machine learning systems: What do industry practitioners need to know? In *Proceedings of the 2019 CHI Conference on Human Factors in Computing Systems* (pp. 1-13). ACM. DOI: 10.1145/3290605.3300830

IEEE. (2020). *Ethically aligned design: A vision for prioritizing human well-being with autonomous and intelligent systems*. Retrieved from https://ethicsinaction.ieee.org

Infocomm Media Development Authority. (2020). *AI Governance and Ethics Framework: Ensuring responsible AI use in Singapore.* https://www.imda.gov.sg

Jobin, A., Ienca, M., & Vayena, E. (2019). The global landscape of AI ethics guidelines. *Nature Machine Intelligence, 1*(9), 389–399. DOI: 10.1038/s42256-019-0088-2

Jobin, A., Ienca, M., & Vayena, E. (2023). Worldwide AI ethics: A review of 200 guidelines and recommendations. *Patterns (New York, N.Y.), 4*(10), 100705. https://www.sciencedirect.com/science/article/pii/S2666389923002416

Johnson, N. F., Zhao, G., Hunsader, E., Meng, J., Ravindar, A., Carran, S., & Tivnan, B. (2013). Abrupt rise of new machine ecology beyond human response time. *Scientific Reports, 3*(1), 1–7. DOI: 10.1038/srep02627

Kamiran, F., & Calders, T. (2012). Data preprocessing techniques for classification without discrimination. *Knowledge and Information Systems, 33*(1), 1–33. DOI: 10.1007/s10115-011-0463-8

Korinek, A., & Stiglitz, J. E. (2021). Steering technological progress: AI, automation, and economic growth. *NBER Working Paper No. 28825*. DOI: 10.3386/w28825

Liang, F., & Lee, K. (2021). China's approach to AI governance: Balancing control and innovation. *AI & Society, 36*(4), 847–860.

Lipton, Z. C. (2018). The mythos of model interpretability. *ACM Queue; Tomorrow's Computing Today, 16*(3), 31–57. DOI: 10.1145/3236386.3241340

Ministry of Science and Technology of China. (2021). *AI ethical norms and governance guidelines*. Retrieved from http://www.most.gov.cn

Mittelstadt, B. (2019). Principles alone cannot guarantee ethical AI. *Nature Machine Intelligence, 1*(11), 501–507. DOI: 10.1038/s42256-019-0114-4

Mittelstadt, B. D., Allo, P., Taddeo, M., Wachter, S., & Floridi, L. (2016). The ethics of algorithms: Mapping the debate. *Big Data & Society, 3*(2), 1–21. DOI: 10.1177/2053951716679679

Mozur, P. (2018). Inside China's dystopian dreams: AI, shame and lots of cameras. *The New York Times*. https://www.nytimes.com/2018/07/08/business/china-surveillance-technology.html

O'Flaherty, M. (2020). *Artificial intelligence and human rights: A global framework*. Oxford University Press.

O'Neil, C. (2016). *Weapons of math destruction: How big data increases inequality and threatens democracy*. Crown Publishing Group.

Obermeyer, Z., Powers, B., Vogeli, C., & Mullainathan, S. (2019). Dissecting racial bias in an algorithm used to manage the health of populations. *Science, 366*(6464), 447–453. DOI: 10.1126/science.aax2342

OECD. (2019a). *Preparing for the future of work: Policies to facilitate smooth transitions*. Organisation for Economic Co-operation and Development.

OECD. (2019b). *OECD principles on AI*. Organisation for Economic Co-operation and Development. https://www.oecd.org/going-digital/ai/principles/

OECD. (2020). *OECD Principles on Artificial Intelligence*. https://www.oecd.org/going-digital/ai/principles/

Organisation for Economic Co-operation and Development. (2023). *The state of implementation of the OECD AI Principles four years on*. https://www.oecd.org/en/publications/the-state-of-implementation-of-the-oecd-ai-principles-four-years-on_835641c9-en.html

Organisation for Economic Co-operation and Development. (2024). *AI principles*. OECD. https://www.oecd.org/en/topics/sub-issues/ai-principles.html

Organisation for Economic Co-operation and Development (OECD). (2019). *OECD principles on artificial intelligence*. Retrieved from https://www.oecd.org

Pariser, E. (2011). *The filter bubble: How the new personalized web is changing what we read and how we think.* Penguin Books.

Pasquale, F. (2015). *The black box society: The secret algorithms that control money and information.* Harvard University Press. DOI: 10.4159/harvard.9780674736061

Raghavan, M., Barocas, S., Kleinberg, J., & Levy, K. (2020). Mitigating bias in algorithmic hiring: Evaluating claims and practices. *Proceedings of the 2020 Conference on Fairness, Accountability, and Transparency*, 469–481. DOI: 10.1145/3351095.3372828

Ribeiro, M. T., Singh, S., & Guestrin, C. (2016). *"Why should I trust you?" Explaining the predictions of any classifier.* In *Proceedings of the 22nd ACM SIGKDD International Conference on Knowledge Discovery and Data Mining* (pp. 1135-1144). ACM. DOI: 10.1145/2939672.2939778

Richardson, R., Schultz, J. M., & Crawford, K. (2019). Dirty data, bad predictions: How civil rights violations impact police data, predictive policing systems, and justice. *New York University Law Review Online*, *94*, 192–233.

Roff, H. M. (2021). Lethal autonomous weapons and the future of warfare: Ethics, policy, and governance. *Ethics & International Affairs*, *35*(1), 45–55. DOI: 10.1017/S0892679420000970

Russell, S. (2021). *Human compatible: Artificial intelligence and the problem of control.* Penguin.

Scholz, T. (2017). *Uberworked and underpaid: How workers are disrupting the digital economy.* Polity Press.

Sharkey, N. (2018). The impact of gender and race bias in AI systems. *International Journal of Ethics in AI*, *1*(1), 20–35. DOI: 10.1108/IJEA-2018-0004

Sharma, S., & Gupta, H. (2021). Ethical implications of AI in emerging economies. *AI and Ethics*, *1*(3), 305–316. DOI: 10.1007/s43681-020-00004-w

Shen, J., & Anderson, M. (2024). Ethical Challenges and Solutions of Generative AI. *Informatics (MDPI)*, *11*(3), 58. https://www.mdpi.com/2227-9709/11/3/58. DOI: 10.3390/informatics11030058

Shu, K., Wang, S., Liu, H., & Tang, J. (2020). Fake news detection on social media: A data mining perspective. *ACM Computing Surveys*, *53*(3), 1–38. DOI: 10.1145/2988545

Smith, S. R., & Hyman, D. (2020). Public engagement and AI: The role of education and awareness. *AI & Society*, *35*(1), 5–15. DOI: 10.1007/s00146-019-00938-9

Strickland, E. (2019). IBM Watson, heal thyself: How AI fell short in the healthcare revolution. *IEEE Spectrum*. Retrieved from https://spectrum.ieee.org

Susskind, D. (2020). *A world without work: Technology, automation, and how we should respond*. Metropolitan Books.

Topol, E. (2019). *Deep medicine: How artificial intelligence can make healthcare human again*. Basic Books.

UNESCO. (2021). *Recommendation on the ethics of artificial intelligence.* https://unesdoc.unesco.org/ark:/48223/pf0000377897

UNESCO. (2023). *Global AI Ethics and Governance Observatory*. https://www.unesco.org/ethics-ai/en

United Nations. (2021). *The role of AI in achieving the Sustainable Development Goals*. Retrieved from https://www.un.org

United Nations Educational, Scientific and Cultural Organization (UNESCO). (2021). *Recommendation on the ethics of artificial intelligence*. Retrieved from https://www.unesco.org

Vakkuri, V. (2019). AI governance and civil society: A model for ethical collaboration. *AI & Society*, *34*(4), 481–497. DOI: 10.1007/s00146-019-00890-x

Veale, M., & Borgesius, F. Z. (2021). Demystifying the draft EU Artificial Intelligence Act. *Computer Law & Security Review*, *43*, 105632. DOI: 10.1016/j.clsr.2021.105632

Voigt, P., & Von dem Bussche, A. (2017). *The EU general data protection regulation (GDPR): A practical guide*. Springer. DOI: 10.1007/978-3-319-57959-7

Vosoughi, S., Roy, D., & Aral, S. (2018). The spread of true and false news online. *Science*, *359*(6380), 1146–1151. DOI: 10.1126/science.aap9559

Wachter, S., Mittelstadt, B., & Floridi, L. (2017). Why a right to explanation of automated decision-making does not exist in the General Data Protection Regulation. *International Data Privacy Law*, *7*(2), 76–99. DOI: 10.1093/idpl/ipx005

White House Office of Science and Technology Policy (OSTP). (2022). *Blueprint for an AI Bill of Rights: Making automated systems work for the American people*. Retrieved from https://www.whitehouse.gov

Whittaker, M., Crawford, K., Dobbe, R., Fried, G., Kaziunas, E., Mathur, V., . . . Schwartz, O. (2018). AI now report 2018. *AI Now Institute*. Retrieved from https://ainowinstitute.org

Winfield, A. F. T., & Jirotka, M. (2018). Ethical governance is essential to building trust in robotics and artificial intelligence systems. *Philosophical Transactions. Series A, Mathematical, Physical, and Engineering Sciences*, *376*(2133), 20180085. DOI: 10.1098/rsta.2018.0085

World Health Organization. (2021). *Ethics and governance of artificial intelligence for health.* https://www.who.int/publications/i/item/9789240029200

World Health Organization. (2023). *WHO outlines considerations for regulation of artificial intelligence for health.* https://www.who.int/news/item/19-10-2023-who-outlines-considerations-for-regulation-of-artificial-intelligence-for-health

World Health Organization. (2024). WHO releases AI ethics and governance guidance for large multi-modal models. *World Health Organization (WHO).* https://www.who.int/news/item/18-01-2024-who-releases-ai-ethics-and-governance-guidance-for-large-multi-modal-models

Zellers, R., Holtzman, A., Rashkin, H., Bisk, Y., Farhadi, A., Roesner, F., & Choi, Y. (2019). Defending against neural fake news. *Advances in Neural Information Processing Systems*, *32*, 9051–9062.

Zeng, Y., Lu, E., & Huangfu, C. (2019). Linking artificial intelligence principles. *Science and Engineering Ethics*, *25*(2), 419–434. DOI: 10.1007/s11948-018-0031-0

Zhang, B., & Dafoe, A. (2020). Artificial intelligence: American attitudes and trends. *Center for the Governance of AI, Future of Humanity Institute, University of Oxford.*

Zhou, X., & Zafarani, R. (2020). A survey of fake news detection: Methods, evaluation, and challenges. *ACM Computing Surveys*, *53*(5), 1–40. DOI: 10.1145/3395046

Zuboff, S. (2019). *The age of surveillance capitalism: The fight for a human future at the new frontier of power.* PublicAffairs.

# Chapter 2
# Ethical Implications of Artificial Intelligence Bias and Fairness: Ethics Is All You Need

**Wasswa Shafik**
https://orcid.org/0000-0002-9320-3186
*Dig Connectivity Research Laboratory (DCRLab), Uganda*

## ABSTRACT

*Artificial Intelligence (AI) systems, while transformative, often inherit biases from data, algorithms, and human decision-making, raising ethical concerns regarding fairness, accountability, and societal impact. Bias in AI can result in discriminatory outcomes, reinforcing social inequalities in critical areas such as healthcare, hiring, lending, and law enforcement. Ensuring fairness requires rigorous bias detection, mitigation strategies, and the adoption of Explainable AI (XAI) to enhance transparency and accountability. Ethical AI frameworks must integrate principles of inclusivity, regulatory compliance, and stakeholder engagement to foster trust and prevent harm. This research explores the ethical implications of AI bias, emphasizing technical and policy-driven solutions to achieve equitable and responsible AI deployment.*

## 1. INTRODUCTION

As incredibly complicated systems, misunderstood even by their designers, artificial intelligence (AI) and machine learning (ML) systems can learn from data to adapt and adjust to changing inputs. Bias is merely a characteristic of AI systems,

DOI: 10.4018/979-8-3373-2597-2.ch002

but it is an increasingly important one, capable of significant impacts on AI systems' functionality, outcomes, and compliance with the law (Singh, 2022). It is more than coding, math, algorithms, or data; it affects a system's ability to recognize and understand the broadest of societal values by depicting one value at the expense of all others. Currently, AI 'thinks' and generates input partially from the past while regulating in small ways for the present and possible futures, reflecting the biases in their training data. This is the condition AI systems have evolved into. From fairness in loan decisions, hiring administration, and chatting machines, conversations grow increasingly common about AI, machine learning, and algorithmic fairness, particularly biases in their data and programming (Meszaros et al., 2022).

Some of this is incidental: better model development and more detailed analysis occur as computational tools advance, capabilities enlarge, and a refined awareness takes root. Ethical implications of bias and fairness in AI and ML have taken on a far more influential role in the design, training, validation, and ultimately release of emerging computational systems. A tool of increasing societal importance, many propose that responsible AI systems should be trained on ethically represented and de-biased data to predict a just and fair outcome (Shafik, 2024e). Meanings of fairness and bias frequently exist in a mutually reinforcing context when speaking from technical and practical views of fairness, having actual definitions and being relatable to actual operational settings. Several strong opinions, techniques, and approaches have been emphatically argued, debated, and presented to address what comes to be known as algorithmic fairness (Kiran Deep Singh & Singh, 2024).

It is beginning to be recognized that the use of data could lead to unfair discrimination across social groups. As a result, the search for methods and techniques to identify, control, and prevent bias in data-driven decision-making has gained increasing attention. In recent years, not cithe development of AI has made great progress due to the large amount of large-scale data and the rapid improvement of computational infrastructure. In areas such as computer vision and natural language processing, these AI models have surpassed human-level performances (S. Kalaivani et al., 2023). However, building fair and unbiased AI systems that help diverse human decision-making remains an open problem depending on the category of expert systems, as illustrated in Figure 1. A certain amount of bias is inherent in any social system, and machine learning models have the potential to help mitigate the bias in human decision-making. While there is a rising number of works that concentrate on identifying and mitigating the bias with machine learning models, the boundary between key ideas has not been clearly maintained (Göllner & Tropmann-Frick, 2023a).

*Figure 1. Expert systems*

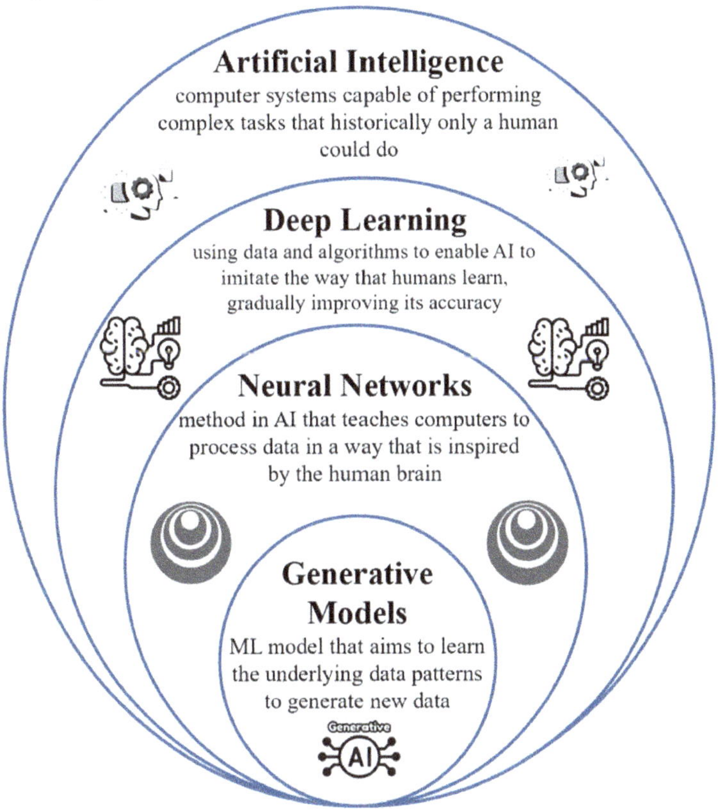


## 2. UNDERSTANDING BIAS IN ARTIFICIAL INTELLIGENCE

The process of machine learning consists of teaching computer systems to learn from and adapt to a series of input data. Pattern recognition, classification, and prediction are some activities associated with the use of learned skills, at a level of autonomy or adaptation that can be found in software solutions. These programs optimize their performance and adapt against defined goals by recurring to a number of specific and measurable notions such as accuracy, proximity to solutions in relation to available training data or algorithms, and equilibrium for adjustments in situations where agreement with their human users is the direction they must take (Ramanathan et al., 2023). These are rules or conditions imposed as elements of judgment concerning their performance, and they are related to the creation of instances that can be obtained in a poorly defined region under circumstances of ambiguity or incomplete information.

Artificial intelligence needs to mirror society, incorporating concepts of fairness and responsibility to minimize potential ethical risks. Examples of distortions according to the social group are the determination of credit ratings, recruitment in the labor market, criminal sentencing, and diagnosis of diseases. Companies can incorporate fairness constraints into their algorithms by recognizing historical imbalances and understanding the relationships between different groups and the impact of certain conditions or personal variables (Babu et al., 2023; Shafik, 2024i; Shafik et al., 2020). Preprocessing bias mitigation measures are being explored by modifying the data used for training, and then fine-tuning for more advanced machine learning models by including bias mitigation objectives as additional model optimization objectives. The post-processing bias mitigation is also an ideal way to neutralize bias. Yet despite the increase in sources of bias, technical bias mitigation can remain limited if subjected to the dominant values of the data. Proof of technical solutions to promote heterogeneous groups of underrepresented communities in order to avoid bias in the model addresses the problem at its end, which is the biased prediction, for a specific tax optimization cost of the prediction (Shafik, 2024f).

The concept of 'bias' is used in many contexts, such as in statistics discussing the 'spread' in probabilistic interpretations, in psychology and sociology discussing socio-demographic and cultural models, and practical or theoretical distortions from 'perfection', as well as in data science discussing the error generated in the learning mechanism. Here, the focus is on the ethical concepts covering the use of AI in information extraction from databases (Shafik, 2024d). Thus, the starting point for consideration is the failure to attend to a significant portion of the values of dependent variables or to missing one or more independent variables that have an effect. The system then will operate 'as if' missing attributes have some values that match the sample contents. This consideration permits an expansion from some partial studies in computer science of observed manifestations to a more complex analysis (B. Sharma et al., 2023). In essence, the associated issues in the discussion are:

(1) What is being biased?
(2) What are the consequences of biases on decision-making processes, or how biases impact the decision-making process within the automatically created scenario?

A systematic registration of the social aspects of how bias happens across the spectrum from distinction to search is adopted. The initial analysis is designed to give a sense of the ramifications of the influence of the distortion of acquired data and thus form the cornerstone to transfer to a more in-depth set of indicators for this. It is concluded that a working comprehension of what distortions are evident inside a particular diagnosis (pre-judgment) applied has to exist to be in place from a different research viewpoint in semantics (Kathole et al., 2023; Shafik, 2024h). The focus moves from the 'what' to the 'how'. Not every step is considered discriminatory, and the final decision might still be classified as discriminatory even when it does not arise within any part of the AI.

## 2.1. Types of Bias in Artificial Intelligence

Data-driven AI systematically incorporates human, cultural, or historical biases that have been present in the past and learned and represented from data. Some sources of biases are existing social values, stereotypes, and ethics that humans have held for a long time. They could be encoded in a system explicitly or emerge through interaction with various learning. Recognizing, measuring, and addressing bias is especially difficult in data-driven AI (Shafik, 2025b). Face recognition AI, for instance, usually exhibits differential accuracy for various race, age, gender, and skin tone subgroups, performing more poorly for people with darker skin. Preventing biases in AI is essential, as they can reinforce or perpetuate existing disparities in society. First, biased AI can lead to products and services that are unfair, ineffective, or even unsafe, since inaccurate predictions or false correlations can result in discrimination or even harm (Shafik, 2024c). Additionally, biased AI may also exacerbate economic or employment disparities, perpetuate stereotypes, and normalize discriminatory behavior.

## 2.2. Causes of Bias in Artificial Intelligence

During data collection. It is known that personal biases come from past experiences or knowledge. The same is true in AI, where data can be biased. Specifically, if the datasets used to train AI are unrepresentative of the real world, this can lead the algorithms to be unfair through biased predictions. The unfairness is rooted in the reality that different demographic protected groups can have different representativeness levels or incidence rates (Gonik et al., 2020; Shafik, 2023). If an AI system is directly or indirectly discriminated against, corrective or distributive justice obligations might arise. Data manipulation. Diversity in data is a central property of well-behaved AI, so challenges arise when using data that is manipulated or intentionally biased. We currently make and expect myopic decisions to be made based on digital traces, thus providing the actual incentive to alter the data for certain specific purposes. How can several other derived decisions be accepted, let alone evolved, from a data manipulation process that is non-transparent and unpredictable? Consequently, AI might lack the ability to be transparent, non-discriminatory, safe, predictable, and reliable. All the principles of ethical importance within the AI ecosystem might be at stake (Prorok & Takacs, 2023). In a nutshell, biases can appear whenever data is excluded or improperly treated. Some of the most common biases include:

(1) sample bias – when collected data fail to account for several populations,
(2) algorithmic bias – when the learning algorithm itself is biased, e.g., due to the improper choice of the learning model parameters; and
(3) representation bias – problems related to data can arise during the encoding of data. Biases can also stem from societal circumstances.

## 2.3. Impact of Bias in Artificial Intelligence Systems

If biases are present in data, AI outcomes are likely to be biased as well. Even if the biases are clearly present in the data, AI systems may still be more accurate in predicting decisions than humans. However, humans may still have ethically justified reasons for engaging in less accurate behavior that may be seen as "bias" (i.e., using data points such as race or sex, which might still accurately predict a decision, to make decisions in ways that address social inequities) (V. Sharma et al., 2023). Given the many possible trade-offs between accuracy and fairness, and the fact that judgment is needed to decide when to generate a fair outcome at the expense of accuracy, it is often unclear how to design computational systems to treat fairness and accuracy as being in tension. There is no true silver bullet solution for these tensions. For example, one solution to tension is described in an example, where a constitutionality principle is used to guide the rule development and evaluation, accepting some trade-off of accuracy against fairness (Arbelaez Ossa et al., 2024).

Furthermore, the first and potentially most difficult component of any fairness system will be in the creation of a set of fair measures. Widespread public input and discourse are key in the determination of a governmental institution's fairness metrics. Policymakers often lack the technical understanding necessary to determine what is considered to be a fair outcome under specific circumstances (Prorok & Takacs, 2023). Additionally, there is a fundamental question as to whether accuracy and fairness are concepts that are completely subjective to the individual making them (i.e., fairness is in the eye of the beholder). This may lead to a question of who gets to decide what are considered fair outcomes. Therefore, even AI systems that are designed with only the most innocuous and likely unconscious bias can inadvertently learn biases modeled through data patterns, sampled application of AI as illustrated in Figure 2 (Stylianides et al., 2023).

*Figure 2. Artificial intelligence application in sustainable living*

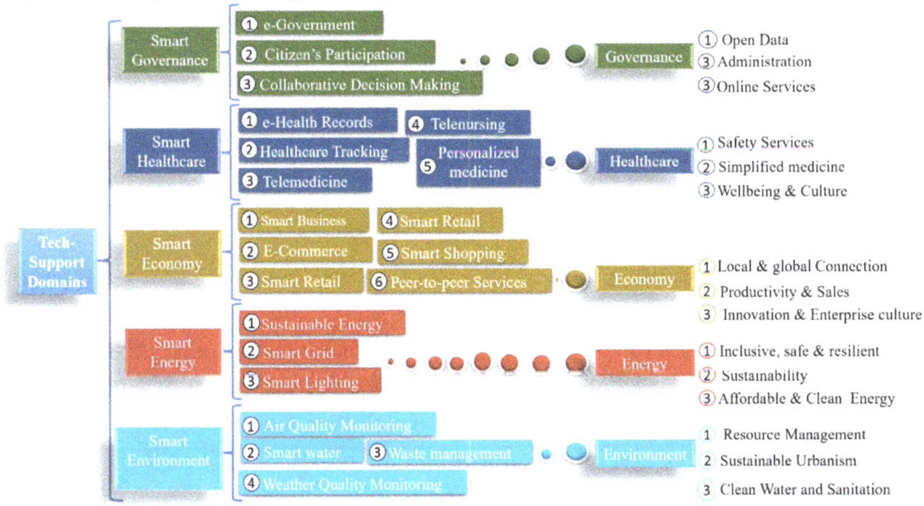


## 2.3.1. Transparency

Transparency refers to the principle that in order for people to get maximum benefit from something and feel comfortable and safe using it, they need to know how it works. With the rise of AI decision support systems, it is very important to look at transparency: why is an AI making certain decisions? This is about data and the algorithm processing the data. There are some key reasons why transparency is important. If people have insight into, for example, why a system was used or what data was used, they can get an idea of the strengths and weaknesses of a decision support system (Shafik, 2024m; Shokoor et al., 2022). The second argument is that being and seeming to be transparent can promote trust in an AI. If people understand how the system makes its decisions, they do not rely on blind trust in the designers; they rely on the system's performance. The third strength of promoting transparency is tied to the concept of informed consent. Making clear what data is used for what, what algorithms process this data, etc., can help in offering people a real choice on where they want their data used or whether they consent to the possible consequences (Fukizi, 2023).

## 2.3.2. Accountability

Manual processes are also not error-free. An important difference is, however, that processes where decisions are made by people allow organizations to make the decision-makers responsible for the consequences of their decisions, juridically or otherwise. If, however, decisions are processed by sophisticated systems, it is harder to make someone responsible or accountable for the outcomes. As a result, processes with AI and algorithmic decision support demand transparency (S. Kalaivani et al., 2023). Furthermore, it is important that the organization behind a certain AI system has a tool to explain the outcome of the system. This is not just important when decision support in crime-fighting or in lending leads to reproducing biases that are supposed to be forbidden by law or policy, or strong differences in procedures for different communities or individuals. In case this occurs, it is very important – sometimes even obligatory by law – to be able to indicate where the decision-making results from and that this corresponds to what has been expected. Furthermore, accountability in case of harmful, unintended consequences arising from AI application paves the way for prevention. If developers and adopters are aware of their responsibility, they can be encouraged to design safe and fair AI applications (Stylianides et al., 2023). This can be done by requiring them to account for their AI applications and explain how data input based on past or present facts led to results.

## 2.4. Privacy and Data Protection

One important concern associated with AI systems and big data analytics is the lack of privacy and data protection that can arise. Many AI systems require a plethora of personal data to function. Therefore, privacy-related ethical issues concerning how this data is used and in what ways the systems can surveil the client are raised. In order for people to trust these systems, one must ensure that the systems and their controllers follow the regulations implemented to protect the user's data (Alexandropoulou et al., 2023). As a result, transparency, fairness, and accountability related to the usage of data might be influenced. Furthermore, people may favor the usage of AI systems that function ethically and adhere to regulations concerning their user's data. Thus, businesses implementing AI have an incentive to likewise ensure that their systems comply with these regulations. This is undoubtedly a fundamental aspect of AI ethics (Shafik, 2024a). Thus, if such regulations and policies, in combination with data breaches, are broken or not followed, ethical data handling practices should offer ways to remedy such abuse of power.

Privacy, just like bias, is difficult to formalize because it differs per social context and individual. The principles encompassing respect for privacy include using collection limitation technologies that provide privacy by design and settings, such as differential privacy that might be used to provide privacy. There is an emerging field of research that attempts to develop data-related mechanisms for protecting privacy, which is especially relevant concerning the use of AI and big data analytics (Khan & Alamäki, 2023). Control and removal of bias are promotional for protecting privacy. Data privacy and confidentiality are important ethical attributes in machine learning and AI, considering recent cases of online abuse where data are not removed appropriately. Protecting privacy to ensure anonymity is the positive goal with data sharing. This is an essential aspect of bias and is, in essence, discrimination in AI (Babu et al., 2023).

## 3. MITIGATING BIAS IN ARTIFICIAL INTELLIGENT SYSTEMS

In this section, we focus on one of the major challenges in AI development: addressing bias. This has principally been tackled from two angles: developing methods for identification and assessment of bias, and coming up with strategies for addressing these biases to achieve fairness in AI systems. Detection and characterization of bias are the essential first steps when addressing fairness in decision support systems (Alexandropoulou et al., 2023). Processes like Fairness Need Assessment and Auditing are used to integrate the understanding of potential fairness issues into the AI/ML model development lifecycle. Additionally, tools help various stakeholders detect and identify bias in AI systems. The effectiveness of detection and characterization of bias depends on how well it is embedded in the development process so bias assessment can become an integral part of the AI/ML model lifecycle (Göllner & Tropmann-Frick, 2023b).

The following strategies for addressing AI bias or enhancing fairness in AI are presented: development of algorithmic techniques and practices to decrease unfairness in model outcomes; development of software tools or frameworks that estimate the potential fairness issues in your model, using cutting-edge techniques at the time of creating the tool and including clear efforts to keep up to date with the latest academic work; and the creation of policy or legal guidelines that specify techniques and practices for integrating fairness concerns into documentation and governance (Shafik, 2024k). Besides these four strategies, some specific technical or algorithmic strategies are advanced to mitigate the effect of bias in AI outputs. A solution is to generate balanced samples from unbalanced data to reduce statistical bias: the re-sampling techniques are another example to eliminate imbalance in model predictions. This is done by adjusting the decision threshold for label prediction in models to increase or decrease the positive class predicted.

In addition, including the affected parties in the development of AI systems can be an approach to increase the fairness of the model predictions; this is done by gathering inclusive data, shifting transparency from model features to predictions, and providing a frame that places the audience into a broader context. Co-creation with the affected people, prioritizing their voices, and convening cross-disciplinary teams with local community partners (Shafik, 2024g; Treacy, 2023). It is worth remembering that fair decisions to serve these affected people do not only reside in currently available technical assessment aids or legal codes; they often require developmental efforts that are likely to be permanent, during the rapid evolution of these technologies, in comparison to static solutions. Thus, it is crucial to measure the effectiveness of this work via horizontal cooperation among a variety of stakeholders using the advantages of generative AI as illustrated in Figure 3 (Tolulope Olorunsogo et al., 2024).

*Figure 3. ChatGPT application*

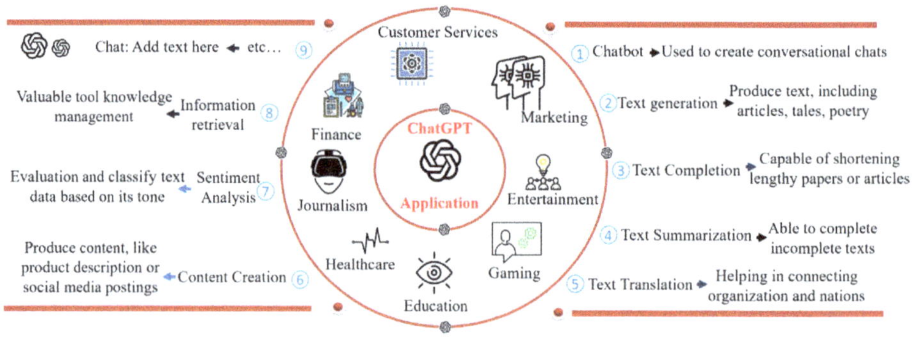


## 3.1. Bias Detection and Measurement

Many methodologies have been proposed for detecting and measuring bias in AI systems, but most can largely be discussed as qualitative and quantitative. Of those, quantitative measures are more widely used due to many available statistical tools and metrics currently in use. Qualitative evaluations typically take individuals through the fairness definitions, investigating how each aligns with the design specifications and goals of a given AI system (Sakhare et al., 2023). They are therefore situated within the development process of AI and might more easily be incorporated into iterative prototyping and validation processes. Indeed, several argue that bias detection should not be a one-off activity, but rather should be an iterative part of the AI development process in order to monitor and assess a system in continuous real-world use. Some measures are used mostly independently of each other, like disparate impact analysis and equal opportunity, but it may be more useful to approach fairness quantification with the help of multiple measures, choosing both with the developer's goals and the real-world data in mind. However, measuring bias in the first place is surprisingly difficult due to the complexity and distributions of real-world data (Tidjon & Khomh, 2023). Formal definitions of fairness may sound intuitive when discussed in isolation or in reference to specific case examples, but proving that a learner respects these norms with real-world data that has many hidden biases is a complicated and under-constrained process.

## 3.2. Algorithmic Fairness Techniques

Algorithmic fairness attempts to reduce biased outputs of AI models by introducing specific techniques into the machine learning pipeline. This subsection outlines several types of algorithmic fairness for reducing bias in AI models.

(1) Pre-Processing. Professionals and researchers use unfair AI model decisions to reverse-engineer them and correct any biases in pre-processing techniques. Unaware of the black box, professionals can then adjust their models before making predictions based on input data. In this method, the model receives properly transformed input. It adjusts the training data before any algorithm that does general equation learning is applied to be related to bias (Karpova, 2023). For example, AI researchers can manually assemble changes to a specific patient's diagnosis that remove racial aspects before a doctor's algorithm reads it.
(2) In-Processing. This is when a professional adds to or adjusts the learning process that reads the data input of an AI system as in-processing techniques. Learning from input data, this method causes the model to learn with these additional adjustments. For example, this technique allows AI to be trained with datasets to emphasize a selection of individuals or socially sensitive groups (Sakhare et al., 2023).
(3) Post-Processing. After a model has made its predictions, professionals adjust these decisions based on existing data that the model is trained on. This technique does not affect the creation of machine learning; it operates afterward. For example, professionals could weigh classification outcomes differently based on sensitive features such as gender or race (Alexandropoulou et al., 2023). Rather than weakening or destroying the connections, these methods raise and evaluate areas to improve them.

# 4. FAIRNESS IN ARTIFICIAL INTELLIGENCE

Machine learning has developed a rich inventory of formal fairness definitions to identify, avoid, and mitigate biases in individual decisions across various sensitive attributes, including race, gender, and age. Modern fairness definitions often distinguish between two types of group fairness. Group fairness of decisions in an individual context holds that individuals should not be negatively impacted by a decision merely because they belong to a particular group, no matter how the rest of the applications go (Mooghali et al., 2023). Group fairness of recommendations at the population level is rooted in the capabilities that the outputs of a statistical model should reflect long-standing structural injustices against certain groups. These fairness notions account for either the output decisions of an AI system or the training process that generates these decisions. While AI fairness is often understood as a balance between fairness criterion constraints, in practice this balance can be challenging to achieve. Indeed, there usually exists a trade-off between fairness and accuracy, i.e., the inputs of a model cannot be too simple if it is to perform its task with a high degree of accuracy for all (Singh, 2022). Different fairness definitions and their associated group fairness mitigation techniques serve different social justice needs.

## 4.1. Defining Fairness in Artificial Intelligence

Although fairness has long been a goal for the allocation of resources, proportional representation in groups, and equality in treatment, the political and social importance of fairness has been highlighted by the advent of machine learning technologies that can be leveraged in decision-making. Fairness has implications for a diverse class of users; for instance, individuals who depend on AI-based systems to aid them in their decision-making, as well as groups that are subjects of the decisions in society (Monlezun, 2023). Highlighting the significance of fairness across different decision-making contexts is fundamental. The definition of fairness in a machine learning setting can be quite flexible, and even suggesting diverse definitions could be considered a misconception. To support decisions in a more sophisticated policy environment, the majority of the work in the area of fairness has been carried out. Regardless, a variety of competing definitions has proven to be helpful not as efforts' outcomes but as furtherance of fairness' multidimensionality. The illusion of a unified fairness definition excludes subtle concerns, obfuscating the policy design process rather than delving into such aspects (Alahmed et al., 2023). Regulatory institutions and policymakers, thus, have critical backing to flexibility, enabling them to create transparency, oversight, and meaningful checks and balances on AI-based systems.

## 4.2. Measures of Fairness in Artificial Intelligence

The concept of fairness is fundamental in most ethical dilemmas, especially where AI is involved, since AI is perceived to be in a position to violate this ethical concept. In AI, there are measures of fairness that are employed to assess the algorithm's potential to be unfair. Some of these measures are statistical parity or demographic parity, which is the proportion of members of a protected group who receive a positive outcome divided by the total number of protected group members (Lumme, 2024). The approved group that achieves such an outcome has higher statistical parity than the underrepresented group, which has low statistical parity associated with the outcome.

For example, in a recidivism prediction use case with gender as a protected attribute, men may have a higher statistical fairness if they have a higher false positive rate than women. However, statistical fairness still reflects an unfairness since no group should have a higher false positive rate than another. Another method used to measure fairness in AI algorithms is disparate impact, which occurs when a protected attribute directly limits opportunities for a group with that attribute (Iniesta, 2023). For instance, in an employee selection process with age as a protected attribute, if a younger group is more likely to be selected for a position and an older group is less likely to be selected for the position, then the younger group will appear to have a disparate impact in the selection process. The difference between disparate impact and statistical parity is that disparate impact checks each step of the process, as opposed to just determining whether the disparity result is 0 or 1 (Nasir et al., 2024).

## 4.3. Challenges in Achieving Fairness in Artificial Intelligence

The use of AI in decision-making calls for adherence to fairness, especially when these decisions are in critical domains such as healthcare, hiring, and criminal justice. Recent reports have highlighted challenges in achieving fairness in AI. Lack of definitions and formulations of fairness principles specified in many contexts, particularly in the fair treatment of individuals, refers to the fairness of outcomes. Courts and civil society disagree on the correct criteria to use (Sakhare et al., 2023). In addition to exceeding reasonable upper thresholds on unpopular attributes such as criminality or disease, they receive some preferred form of fairness guarantees, and the divergent legal and public understandings of fairness further complicate the learning and evaluation process. Lack of operational fairness criteria has led to the proposal of multiple distinct fairness principles that community members deem necessary for different user populations, domains, and machine learning tasks. This confirms the intuition that no algorithm can fully address the needs of every morally legitimate group, which is both overpowering and impossible (Pourzolfaghar et al., 2023).

# 5. ETHICAL CONSIDERATIONS IN ARTIFICIAL INTELLIGENCE BIAS AND FAIRNESS

The use of AI algorithms in decision systems and the design of user-facing interfaces has become fairly common. Despite the performance benefits of making a number of routine decisions typically undertaken by a person, AI decision systems are not free from biases. A number of causes for the presence of biases in AI decision systems range from the inherent individual bias of the system trainer, lack of diversity in the training data used to create a biased model, presence of sampling bias in the training data, and issues associated with scalable learning algorithms (Alexandropoulou et al., 2023). It is important for a machine learning engineer to be aware of the potential causes of biases. The use of AI for decision-making tasks can impact many individuals in a variety of ways. Unfair treatment can result in damaged reputations, credit denial, job loss, denial of access to resources, and differential rates for specific services. It is important to understand the potential implications to ensure the fairness of any AI-supported decision systems (Plummer, 2024). In this chapter, we review some prominent controversies considered important for AI-assisted decision-making and discuss potential ways to address these issues.

## 5.1. Ethical Frameworks in Artificial Intelligence Development

Next, we analyze three different ethical frameworks that could be used to help in the development of AI technologies. The first is consequence-oriented and seems particularly suited to the sensor-based nature of the AI approach. The second is an approach based on deontological ethics, one that respects the autonomy and dignity of human beings. The third is informed by the domain of the philosophy of technology and has recourse to the design approach (Sakhare et al., 2023). The advantage of this approach is that it uses more cultural cues to interpret human needs and vulnerabilities from the community for which technology is to be developed and mitigates pitfalls in use by that community. We examine charts in dimensions of fairness and liability and actors' points of view. Utilitarian ethics is a form of consequence-oriented ethics, which states that the decisions should have the greatest possible benefit for the greatest number of people. From the perspective of the AI field, the sensor-based approach already shares with this vision instrumental features, and it can perhaps seem simple to conclude that such an approach will assure benevolent consequences. Unfortunately, such a premise is not always verified (Babu et al., 2023; Sakhare et al., 2023). However, the integral or holistic vision presented by the sensor-based nature of AI could be taken as a recurrent conceptual leitmotiv while advancing research in artificial intelligence, at least to reduce the prejudice caused by ethical bias.

## 5.2. Responsibility and Accountability in AI Systems

It is helpful to see how responsibility and accountability play out more concretely in the life cycle of developing and operating an AI system. The following are the common stages in the life cycle deployment of an AI system: identify the problem, data collection and database system setup, data preprocessing, data modeling, operationalizing the model, checking validity and fairness, and evaluating the long-term social impacts, and possibly either delivering the system to the client or selling it as a product (or operationalizing it internally) (Alexandropoulou et al., 2023). Many stakeholders may be responsible in varying degrees at different stages.

Each stage typically involves different forms and extents of bias and/or fairness. For example, what types and what levels of fairness checking would be needed, and who has executed that responsibility and check? In this case, we take a closer look at the business perspective of AI transparency and responsibility, often skipping the higher levels of abstraction. Specifically, we described that this scientific area studies standards of behavior sanctioned by computer systems, such as AIs, robots, and cyber systems in general (Morpurgo & Occhipinti, 2023). AI system ethics is a broad topic of AI ethics and includes areas such as the relevance of behaviors to human society and concepts such as well-being, accountability, and the ability to err. In the view of the project of studying the actual ethical implications of AIs, as seen from the implementation point of view in the context of business and society, AI system ethics represents a set of binding rules, especially those that are peculiar to computer systems and that support AIs in correctly satisfying their purpose (Fukizi, 2023; Shafik, 2024b).

# 6. LEGAL AND REGULATORY LANDSCAPE

A novel and critical issue related to AI bias and fairness, informed by the significant challenges and tensions in the space, is our evolving legal and regulatory landscape. As the community has grown, legal and regulatory communities have also begun to pay increasing attention to these pervasive societal concerns. Existing laws and regulations in many jurisdictions worldwide generally apply to problematic AI technology, even where not specifically mentioned (Shafik, 2024j). Although AI regulation is not yet widespread, many existing laws, rules, and norms potentially apply to AI-related matters under the rubric of functional or substantive regulatory oversight. In this section, we offer a basic conceptual overview of the developing global legal and regulatory frameworks relevant to AI technologies, working from the general to the specific. While many topics intersect throughout this discussion with distinct themes such as privacy, data protection, non-discrimination, and intellectual property, we ultimately address them in greater depth in later sections as they relate to specific policy-related issues (Shafik, 2024l).

The unique and novel issues associated with AI have prompted a remarkable effort by many jurisdictions to develop new legislation that speaks directly to these emerging concerns. For example, Congress considered a variety of AI bills in its 116th session aiming to modernize governmental security or defense practices, to support AI research and developement (R&D), and to address consequentialist risks associated with AI employment and decision-making in areas of general societal impact beyond the security realm (Zhao et al., 2023). Similarly, the European Union has pursued a goal to develop a harmonized legal framework that addresses the societal risks associated with the emerging AI technology and has pursued a law-making strategy that combines new substantive regulation with harmonized AI-specific rules and an overall tailored regulatory approach. In addition to federal and supranational legislative actions, a growing number of subnational entities are crafting regional AI regulations (Gonik et al., 2020). For example, the state governments of California, Illinois, and New York have legislated AI rules, mostly addressing the fairness of AI technology and transparency in public decision-making, and an act of the Atlanta city council has implemented guidance for city agencies' AI use. Although for enumeration, our brief consideration of regulatory and legislative instruments will necessarily be selective (Meszaros et al., 2022). Our survey aims to draw the reader's attention to the existence and diverse nature of pertinent law and regulatory creation emanating from both traditional and novel regulatory authorities.

## 6.1. Current Laws and Regulations

One possible way to control the deployment of biased AI is through legislation. The evidence of negative consequences from enabling bias and excluding individuals from accessing services such as healthcare and recruitment suggests that policy guidelines may be considered to safeguard against artificial intelligence bias. There are a few laws and regulations that currently affect AI use in different domains. None of them are AI-specific or result in an AI developer being held accountable for the use of biased AI (Meszaros et al., 2022). However, they do imply that AI developers should be doing more to ensure that the AI they create does not discriminate in a way that infringes people's human rights. It is likely that, in the future, most of these listed activities will not just be illegal, but will also attract unacceptable associated costs. In particular, the responsibility for compliance should belong to the owner of the AI model. The AI developer and anyone else involved in the production of the model, such as the business case designer for the predictive model or the dataset creators, may also have some culpability if they are aware of the bias or allow it to remain undetected (Shahvaroughi Farahani & Ghasemi, 2024). However, to avoid finding themselves in breach of the fairness directives, an employer or service provider has an incentive to provide the evidence to establish that they do not deploy a biased AI.

## 6.2. Proposed Legislation on Artificial Intelligence Bias and Fairness

The proposed regulations are much more detailed than those that currently exist. As federal entities provided with the funds to develop and use AI, the public sector would be subject to more stringent regulations and required to demonstrate greater compliance with bias and fairness regulations. The categories of health and well-being and criminal justice, however, are not considered because there is no designation of any other branch of the government to oversee and enforce bias and fairness regulations and accountability after agencies have been designated (Di Nuovo, 2024). The New York City Council established the Task Force on Automated Decision Systems in 2017, as well as some legislation related to AI bias and fairness. The bill that led to the task force requires the task force to provide recommendations to the Mayor and the City Council based on the review of the use of ADS by agencies and to provide annual reports on the use of ADS by its agencies. ADS is defined as a computerized implementation of algorithms, including those engaging machine learning, that evaluate or assist decisions by public agencies that impact New Yorkers. ADS includes but is not limited to: a) statistical profiling and predictive policing; b) allocating health, human, or social services; c) making eligibility determinations for public benefits or services; and d) sentencing, bail, and parole decisions (Meszaros et al., 2022).

The task force is composed of fifteen members from relevant societal sectors, including the health, civil justice, and data science sectors, who will serve under the leadership of the NYC Mayor and Council for four years. The New York City departments and agencies will have to cooperate with the task force by supplying the documents the task force needs to fulfill its two mandates. These recommendations suggest ways for the legislation to improve, put into action, converge groups with similar goals, and communicate with other public entities by utilizing access to existing expertise on the effects of ADS (Ramanayake et al., 2023). Therefore, there is no mention of an office to gather and maintain data, as well as to analyze and report the data on ADS, nor are there any incentives to require compliance with, enforce, and hold public agencies accountable for the proposed recommendations.

# 7. CASE STUDIES OF ARTIFICIAL INTELLIGENCE BIAS AND FAIRNESS

This chapter provides additional ways to assess bias and fairness in machine learning and models for their definitions. In conjunction with this material, this can be used as a concrete guide for well-defined notions of AI fairness and specific metrics to use in practice. We analyze five case studies in detail and provide additional motivation and examples. The analysis was carried out using insights from the literature, drafted into a standard for defining fairness, and validated by running many rounds of the resulting procedures and testing for moral intuitions (Pourzolfaghar et al., 2023). These case studies provide general insights about bias in machine learning and are designed to both expand our understanding of this currently most often discussed societal risk of machine learning and to advance the state of the art in formalizing it (Kiran Deep Singh & Singh, 2024).

By expanding familiar examples, we design more production that generalizes the moral insights that the public already has. After providing five case studies, we provide a novel summary of the implications of the general insights, which highlight the avenues for future work. At a high level, our five case studies explore different facets of fairness definitions and sub-discussions about the mechanisms by which models can exhibit bias (Khalifa & Albadawy, 2024). Our case studies are useful because they draw attention to some of the main claims of the competition, provide concrete examples that illustrate the trade-offs with regard to fairness, and contribute to testing the descriptive morality of the emerging standard algorithms. Given the lack of agreement between frameworks, no single framework works well in general (Tolulope Olorunsogo et al., 2024). Therefore, AI fairness requires humans in the loop as a local score requires the involvement of a loan agent.

There are also mounting issues of systematic inequality and discrimination which exacerbate the problems already perpetuated by human authoritativeness. Proposed solutions address all aspects of bias and unfairness in predictive modeling, from careful definition of the model decision problem through pre, in-process, and post-processing of model predictions (Nasir et al., 2024). Each of the methodologies creates its own set of less straightforward questions, but generally agreeing upon subjective moral intuitions: that no one should be unfairly advantaged by a decision taken by a predictive model; that when compared across different groups, they should be treated consistently; that the outcome of a decision should not depend on the attribute membership of any identifiable class in the population; and that model processes may place certain groups at a systematic and unfair disadvantage. These ideas are logical but also defy straightforward codification (Singh, 2022). Our contribution is a clear report in the form of the standard, plus interpolation and extrapolation based on a case-by-case analysis and validation with real human intuition tests, which demonstrate the validity of our methodology.

## 8. MITIGATION STRATEGIES FOR ARTIFICIAL INTELLIGENCE BIAS

It is important to identify potential sources of bias before they can be mitigated through various strategies in order to argue that existing bias in training data does not indicate true bias generated by the beginning algorithm. Strategies for mitigating AI bias can take place at various levels of the AI ethical impact hierarchy; from the AI system itself to the deployment context through which it interacts with society (Pourzolfaghar et al., 2023; Shafik, 2025a). Some of these strategies are likely to be more feasible than others, and some work focuses on the data, model, and correction in particular. At the data level, a common strategy for mitigating AI bias involves changes to the so-called 'ground truth' labels used as the targets during model training. Such pre-processing adjustments can be performed to reduce or remove bias in the training data or otherwise are biased likely to propagate throughout the subsequent model and application behavior. One of the simplest forms of pre-processing manipulation can involve holding a reduced overall number of samples or examples from each group in subject (Subasi et al., 2023). While outright deleting valuable information will have broader implications for the applicability of the generated model, a more subtle approach is oversampling subjects from the underprivileged groups before training begins.

### 8.1. Bias Detection, Fairness and Mitigation Techniques

These techniques correspond to the stages in the machine learning pipeline where biased inferences or predictions are likely to be made. Bias can be detected both before and after the training of the model, although in practice, after the model has been trained is sometimes the only time that can be done empirically. A range of bias detection techniques includes checks for missing data, doing sensitivity analysis around evaluation data thresholds, and creating a threshold-independent measure of bias by estimating the false discovery rate (Petersson et al., 2023). Bias mitigation techniques include pre-processing modifications of the training data to increase the recall in subgroups with previously low recall, using different thresholds for subgroups of people when evaluating the model. Bias can also be mitigated by conducting post-processing calculations on the predictions (Aranovich & Matulionyte, 2023).

## 8.2. Transparency and Explainability in Artificial Intelligent Systems

Transparency and explainability in AI is the ability of human users to understand the operation or output of AI systems. In the context of fairness or ethical implications, transparency or explainability can empower users to understand the biases that can manifest in these systems. It can be thought of as a diagnostic tool. Like fairness, there is also empirical support that there are legal uses for transparency and design features aimed to satisfy them (Bellucci & Michele, 2023). However, the existing empirical evidence—and its translation into design principles—is relatively thin compared to the rapidly evolving landscape of new black box techniques. Even more worrying, the debate over the term's meaning and potential implications in the field has intensified. For instance, some advocates believe a black box is inherently dangerous due to issues in justice and fairness (Polevikov, 2023).

Along with that, there is very little in the way of operational definitions of transparency or explainability available. All definitions of transparency focus on technical access for experts as a path to fostering trust in a system: "Having insight into the computer-driven decisions around the data sources, model, and approach helps ensure trust." Similarly, the definition focuses on the thought process or reasoning behind the system, mirroring the technical access definition of transparency and deepening the link between them (Pourzolfaghar et al., 2023). It also focuses on access, but in this case, to non-technical users. This definition is notably different from other definitions. Lastly, some authors define explainability in terms of systems exhibiting accountability and argue that a state of transparency (viz., about the inner workings of a model or system) is necessary for such accountability. They ultimately make it a stated goal for their overall vision of actionable machine learning (Khalifa et al., 2024).

# 9. FUTURE DIRECTIONS IN ARTIFICIAL INTELLIGENCE ETHICS

The serious ethical challenges serve as guideposts for the broadening of the AI ethics research landscape. At the core, researchers from engineering, philosophy, policy studies, machine learning, law, and political science generate the necessary insights to enhance equality, freedom, and equity in the democratization of AI's reach. Implications for AI ethics enforcement and oversight, at an international level, have not been fully explored. Relations between users, developers, and the AI engine itself have been subject to economic, policy, and technical design analysis (Kiran Deep Singh & Singh, 2024). Public investment in large-scale AI diversity initiatives appears to be lagging. Research in AI ethics so far has not proactively contemplated large-scale long-term consequences. Respect for several human cultural dimensions represents a research gap worthy of attention. How to define and implement the right to an explanation is a complex conundrum of AI practice. The role of the rights of vulnerable populations serves as important research questions. AI's growing social influence on education, healthcare, and the job market represents critical domains worthy of careful study. Critical infrastructure system vulnerabilities and geopolitical dimensions that use, and are tied to AI, should be further analyzed. AI's potential catastrophic threat to humanity has yet to be prioritized (Meszaros et al., 2022). Sufficiently ethical AI requires diverse fields contributing to research and training. Participants should possess multidisciplinary knowledge, and research should aim at openness and transparency. The values of transparency and trustworthiness should broadly shape AI's larger societal influence.

## 9.1. Emerging Trends in Artificial Intelligence Ethics Research

We identified several key themes and research areas within AI ethics from the literature. Crucially, many stressed the interdisciplinary nature of AI ethics. Fairness, privacy, security, transparency, interpretability, accountability, responsibility, governance, trust, liability, ethics by design, and value alignment were commonly recurring themes. These areas are explored in further detail. We also note that scholars have addressed the need to challenge existing normative frameworks by incorporating additional, less prominent voices into the discussion about whether, not how, systems should be designed (Petersson et al., 2023). These silences should not be perpetuated, which is a topic of concern for both machine learning and data science more generally. Some maintained that the challenge of AI ethics is not to establish implementation requirements but to engage in a thorough self-assessment of ethical stances. Others suggested that narrow, static assessments and ad hoc technological fixes are not enough. They see a need for norms to change willingly to adapt to the possibilities and challenges of AI (Salajan et al., 2024). Many of the ethical implications that come into play as autonomous, intelligent, artificially intelligent systems become more sophisticated cover topics such as machine learning, decision-making under uncertainty, privacy, personalization, and human-computer interaction. This is just a small sampling, but each of these topics is complex and undergoing extensive research. It is important that we continue to encourage work in this area so that future technologies are properly tuned to benefit society (Patel et al., 2024). Questions for AI ethics research abound, and unique perspectives, use cases, and methods are necessary to offer comprehensive guidance to AI creators and stakeholders.

## 9.2. Potential Impact of Ethical Artificial Intelligence Development

The potential for AI to enhance our lives is juxtaposed with the potential for AI to systematize inequality, including discrimination on the basis of race, gender, sexual orientation, gender identity, national origin, or religion, as well as worsening problems of inequality across the board. Work must pay special attention to societal biases where they exist and, where they do, make sure that we are deliberately working toward changing for the better based on evidence versus simply expanding current and historical inequalities (Bressane et al., 2024). As AI technologies become increasingly core to enabling technology, developments around bias and fairness become pivotal not just for major issues but also for emerging dangerous and humanitarian domains.

One of the most distinct contributions to this area is that of fairness and other ethical values of AI development. This research organizes recent academic literature on fairness and anti-discrimination and outlines a critical societal problem: our increasing reliance on data-driven technologies raises concerns regarding illegal discrimination. It then delves into numerous technical problems associated with algorithmic fairness. Thus, the research outlines a wide variety of open legal issues that are implicated by these technical and societal problems and, paradoxically, how identifying some limitations of technical interventions can be informative for reforming current anti-discrimination laws (Cobanaj et al., 2024). Even among experts, we find fundamental disagreements on what "fair" means and how to operationalize these definitions in practice. Our proposed first step is to ensure that a system is at least treating similar individuals similarly and to update the definition of fairness over time. We also categorized and analyzed the different sources of AI bias, examined ways to mitigate AI bias, discussed the current landscape and challenges for fairness in machine learning, and conceptually addressed AI fairness and its bias by different levels of agent models (Sanghvi et al., 2024). Our work demonstrates how to relax demographic parity to obtain other fairness notions, giving a taxonomy that highlights which justifications are specifiable in terms of a classifier and which are not.

## 9.3. Ethical Decision-Making in Artificial Intelligence Development

It is well recognized that AI development involves ethical decision-making. Several frameworks and methodologies for doing so in the AI context have already been developed. Importantly, these frameworks emphasize the need to engage stakeholders; that is, those who would be affected by the decision, or who would have something to gain or lose if the decision were to be taken. While many developers would agree that involving those who are close to the impact of a decision is crucial, there is still debate on how involvement can be achieved in practice (Chima Abimbola Edeni et al., 2024). Several fields of AI have emerged specifically for reducing bias or fostering fairness, neutrality, or objectivity. In such an inherently social domain as AI, in which each design and technology choice has the potential to advantage or disadvantage groups or communities, narrow technical measures will not necessarily be effective in making systems unbiased or more ethical, and may even make things worse.

Stakeholder-engaged and codesign approaches can provide guidance to the developers and data managers towards equitable design, based on the shared understanding of participant needs and preferences (Patel et al., 2024). Engaging with data, tools, and measures around personal data has the potential to address privacy violations and can enhance the predictability of our systems. Ethical issues that could be framed more generally to encompass the societal impacts of AI go beyond basic data processing as they incorporate wider questions of power, and the frame of what data should be used encompasses perception and common sense. The design of power and the operation of control have intertwined ethical concerns. Ethics is a tool for a designer to guide when deliberate choices are being made. Ethical awareness as a part of societal culture should improve understanding and support accountability (Cobanaj et al., 2024; Liew, 2024). Threat to responsibility of action. Making ethical choices is complex, but not responding strategically can be detrimental to AI's future.

## 10. CONCLUSION

Developing fair and unbiased AI systems is challenging, yet the social and ethical implications of AI will influence the development of society and consequently AI applications. As such, the bias of AI applications should be comprehensively addressed. Research published during the past several years has helped to identify the key ways in which AI systems can discriminate unfairly, and suggests promising ways to limit this kind of unfairness. This chapter provides an overview of the necessary ingredients for bias, and highlights several specific steps AI developers can take to make bias less likely to arise. We argue that AI researchers and developers should work with diverse groups of stakeholders when creating these systems, and should be open and transparent about how they function while closely examining feedback from users, affected groups, and others. While recognizing other important fairness concerns such as distributive justice and political equality, this paper provides a brief survey of the interventions which can promote fairness and mitigate bias, and what the AI community and others can do to help. In doing so, our aim is to empower creators and potential users of AI systems — both technical and non-technical — to consider the full impact of AI systems. Provider organizations should ensure that AI systems are developed with fairness-promoting values. AI and machine learning researchers, meanwhile, should make every effort to explain their work fully and transparently, and to include diverse perspectives. Policymakers and regulatory bodies must ensure they are best placed to adapt systems to respond to and, when appropriate, attend to fairness concerns and limit bias. And compliance and legal professionals should monitor system performance and maintain contact with the AI research community to ensure that systems are functioning as intended. In order to maximize the benefits of AI and machine learning, all stakeholders need to make ethically informed decisions about how these systems are produced, incorporated, and used.

## REFERENCES


Alahmed, Y., Abadla, R., Ameen, N., & Shteiwi, A. (2023). Bridging the Gap Between Ethical AI Implementations. *International Journal of Membrane Science and Technology*, *10*(3), 3034–3046. Advance online publication. DOI: 10.15379/ijmst.v10i3.2953

Alexandropoulou, C. A., Panagiotopoulos, I., Kleanthous, S., Dimitrakopoulos, G., Constantinou, I., Politi, E., Ntalaperas, D., Papageorgiou, X., Stylianides, C., Ioannides, N., Palazis, L., Pattichis, C. S., & Panayides, A. S. (2023). AI-Enabled Solutions, Explainability and Ethical Concerns for Predicting Sepsis in ICUs: A Systematic Review. *Proceedings 2023 IEEE 19th International Conference on E-Science, e-Science 2023*. DOI: 10.1109/e-Science58273.2023.10254863

Aranovich, T. de C., & Matulionyte, R. (2023). Ensuring AI explainability in healthcare: Problems and possible policy solutions. *Information & Communications Technology Law*, *32*(2), 259–275. Advance online publication. DOI: 10.1080/13600834.2022.2146395

Arbelaez Ossa, L., Lorenzini, G., Milford, S. R., Shaw, D., Elger, B. S., & Rost, M. (2024). Integrating ethics in AI development: A qualitative study. *BMC Medical Ethics*, *25*(1), 10. Advance online publication. DOI: 10.1186/s12910-023-01000-0 PMID: 38262986

Babu, C. V. S., Akshayah, N. S., & Vinola, P. M. (2023). Artificial intelligence in healthcare: Assessing impacts, challenges, and recommendations for achieving healthcare independence. In *Perspectives on Artificial Intelligence in Times of Turbulence*. Theoretical Background to Applications., DOI: 10.4018/978-1-6684-9814-9.ch005

Bellucci, B., & Michele, E. (2023). Focusing on the integration of AI in healthcare sector of USA: Focusing on the roles of AI adoption and innovative capabilities. *Journal of Commercial Biotechnology*, *28*(5). Advance online publication. DOI: 10.5912/jcb2142

Bressane, A., Zwirn, D., Essiptchouk, A., Saraiva, A. C. V., Carvalho, F. L. de C., Formiga, J. K. S., Medeiros, L. C. de C., & Negri, R. G. (2024). Understanding the role of study strategies and learning disabilities on student academic performance to enhance educational approaches: A proposal using artificial intelligence. *Computers and Education: Artificial Intelligence*, *6*, 100196. Advance online publication. DOI: 10.1016/j.caeai.2023.100196

Cobanaj, M., Corti, C., Dee, E. C., McCullum, L., Boldrini, L., Schlam, I., Tolaney, S. M., Celi, L. A., Curigliano, G., & Criscitiello, C. (2024). Advancing equitable and personalized cancer care: Novel applications and priorities of artificial intelligence for fairness and inclusivity in the patient care workflow. In *European Journal of Cancer* (Vol. 198). DOI: 10.1016/j.ejca.2023.113504

Di Nuovo, A. (2024). Letter to the Editor: "How Can Biomedical Engineers Help Empower Individuals With Intellectual Disabilities? The Potential Benefits and Challenges of AI Technologies to Support Inclusivity and Transform Lives." In *IEEE Journal of Translational Engineering in Health and Medicine* (Vol. 12). DOI: 10.1109/JTEHM.2023.3331977

Edeni, C. A., Adeleye, O. O., & Adeniyi, I. S.Chima Abimbola EdeniOlabisi Oluwakemi AdeleyeIdowu Sulaimon Adeniyi. (2024). The role of AI-enhanced tools in overcoming socioeconomic barriers in education: A conceptual analysis. *World Journal of Advanced Research and Reviews*, *21*(3), 944–951. Advance online publication. DOI: 10.30574/wjarr.2024.21.3.0780

Fukizi, K. Y. (2023). Collaborative Decision-Making Assistant for Healthcare Professionals: A Human-Centered AI Prototype Powered by Azure Open AI. *COMPASS 2023 - Proceedings of the ACM SIGCAS/SIGCHI Conference on Computing and Sustainable Societies*. DOI: 10.1145/3588001.3609370

Göllner, S., & Tropmann-Frick, M. (2023a). Bridging the Gap between Theory and Practice: Towards Responsible AI Evaluation. *CEUR Workshop Proceedings*, •••, 3580.

Göllner, S., & Tropmann-Frick, M. (2023b). VERIFAI - A Step Towards Evaluating the Responsibility of AI-Systems. *Lecture Notes in Informatics (LNI), Proceedings - Series of the Gesellschaft Fur Informatik (GI), P-331*. DOI: 10.18420/BTW2023-63

Gonik, J., Le, J., Viswanathan, A., & Zhang, Y. (2020). Cybergan: Generating high-fidelity cybersecurity data with generative adversarial networks. *Accelerating Space Commerce, Exploration, and New Discovery Conference, ASCEND 2020*.

Iniesta, R. (2023). The human role to guarantee an ethical AI in healthcare: A five-facts approach. *AI and Ethics*. Advance online publication. DOI: 10.1007/s43681-023-00353-x

Kalaivani, S., Nasreen Banu, D., Parvin Begum, I., & Sivasamy, J. (2023). Cognitive Robotics: Integrating Artificial Intelligence and Embodied Intelligence for Advanced Problem Solving. *Journal of Artificial Intelligence and Capsule Networks*, *5*(4). Advance online publication. DOI: 10.36548/jaicn.2023.4.002

Karpova, E. A. (2023). AI in Health Ethical Review: A Value Design Methodology. *Chelovek, 34*(3), 129. Advance online publication. DOI: 10.31857/S023620070026109-6

Kathole, A. B., Vhatkar, K. N., Raut, R., Patil, S. D., & Jadhav, A. (2023). Security Issues in Generative Adversarial Networks. In *Generative Adversarial Networks and Deep Learning*. Theory and Applications., DOI: 10.1201/9781003203964-5

Khalifa, M., & Albadawy, M. (2024). Artificial Intelligence for Clinical Prediction: Exploring Key Domains and Essential Functions. In *Computer Methods and Programs in Biomedicine Update* (Vol. 5). DOI: 10.1016/j.cmpbup.2024.100148

Khalifa, M., Albadawy, M., & Iqbal, U. (2024). Advancing clinical decision support: The role of artificial intelligence across six domains. In *Computer Methods and Programs in Biomedicine Update* (Vol. 5). DOI: 10.1016/j.cmpbup.2024.100142

Khan, U. A., & Alamäki, A. (2023). Designing an Ethical and Secure Pain Estimation System Using AI Sandbox for Contactless Healthcare. *International Journal of Online and Biomedical Engineering, 19*(15), 166–201. Advance online publication. DOI: 10.3991/ijoe.v19i15.43663

Kiran Deep Singh, & Singh, P. (2024). Fog Cloud Computing and IoT Integration for AI enabled Autonomous Systems in Robotics. *EAI Endorsed Transactions on AI and Robotics, 3*. DOI: 10.4108/airo.3617

Liew, J. (2024). Translating Vision into Reality for Early Education & Development. *Early Education and Development, 35*(1), 1–2. Advance online publication. DOI: 10.1080/10409289.2024.2287684

Lumme, E. M. (2024). A Systems-Theoretic Approach for the Analysis of AI Ethics. *Communications in Computer and Information Science, 1958 CCIS*. DOI: 10.1007/978-3-031-49215-0_6

Meszaros, J., Minari, J., & Huys, I. (2022). The future regulation of artificial intelligence systems in healthcare services and medical research in the European Union. *Frontiers in Genetics, 13*, 927721. Advance online publication. DOI: 10.3389/fgene.2022.927721 PMID: 36267404

Monlezun, D. J. (2023). Chapter 8 - AI+health ethics: moral interoperability and pluralism. In *The Thinking Healthcare System*.

Mooghali, M., Stroud, A. M., Yoo, D. W., & Barry, B. A. & ... (2023). Barriers and Facilitators to Trustworthy and Ethical AI-enabled Medical Care From Patient and Healthcare Provider Perspectives: A Literature Review. *MedRxiv*.

Morpurgo, F., & Occhipinti, C. (2023). Let the AI assisted medicine remain human Developing a care centered model of ethical decision making in AI based healthcare. *CEUR Workshop Proceedings*, •••, 3486.

Nasir, S., Khan, R. A., & Bai, S. (2024). Ethical Framework for Harnessing the Power of AI in Healthcare and Beyond. *IEEE Access : Practical Innovations, Open Solutions*, *12*, 31014–31035. Advance online publication. DOI: 10.1109/ACCESS.2024.3369912

Olorunsogo, T., Adenyi, A. O., Okolo, C. A., & Babawarun, O. (2024). Ethical considerations in AI-enhanced medical decision support systems: A review. *World Journal of Advanced Engineering Technology and Sciences*, *11*(1), 329–336. Advance online publication. DOI: 10.30574/wjaets.2024.11.1.0061

Patel, P., Moz-Ruiz, D., Garcia, R., Chatterjee, A., Morreale, P., & Burnett, M. (2024). From Workshops to Classrooms: Faculty Experiences with Implementing Inclusive Design Principles. *SIGCSE 2024 - Proceedings of the 55th ACM Technical Symposium on Computer Science Education, 1*. DOI: 10.1145/3626252.3630861

Petersson, L., Vincent, K., Svedberg, P., Nygren, J. M., & Larsson, I. (2023). Ethical Perspectives on Implementing AI to Predict Mortality Risk in Emergency Department Patients: A Qualitative Study. *Studies in Health Technology and Informatics*, *302*. Advance online publication. DOI: 10.3233/SHTI230234 PMID: 37203776

Plummer, D. (2024). Ethical AI and the Future of Healthcare: Combining Academic Theory and Industry Practice to Ensure Patient-Centered Care. *European Journal of Technology*, *8*(1), 1–13. Advance online publication. DOI: 10.47672/ejt.1727

Polevikov, S. (2023). Advancing AI in healthcare: A comprehensive review of best practices. In *Clinica Chimica Acta* (Vol. 548). DOI: 10.1016/j.cca.2023.117519

Pourzolfaghar, Z., Alfano, M., & Helfert, M. (2023). Application of ethical AI requirements to an AI solution use-case in healthcare domain. *American Journal of Business*, *38*(3), 112–128. Advance online publication. DOI: 10.1108/AJB-12-2022-0201

Prorok, M., & Takacs, I. (2023). Legal and Moral Issues Related to Artificial Intelligence Technology in Business. *SISY 2023 - IEEE 21st International Symposium on Intelligent Systems and Informatics, Proceedings*. DOI: 10.1109/SISY60376.2023.10417881

Ramanathan, A., Vijaya Rani, K. R., Bhupathyraaj, M., Chacko, L., & Alharbi, H. F. (2023). An Ethical Outlook on the Applications of Artificial Intelligence in Medicine and Pharmacy. In *Artificial Intelligence in Pharmaceutical Sciences*. DOI: 10.1201/9781003343981-10

Ramanayake, R., Wicke, P., & Nallur, V. (2023). Immune moral models? Pro-social rule breaking as a moral enhancement approach for ethical AI. *AI & Society*, *38*(2), 801–813. Advance online publication. DOI: 10.1007/s00146-022-01478-z PMID: 35645466

Sakhare, N. N., Limkar, S., Mahadik, R. V., Phursule, R., Godbole, A., Shirkande, S. T., & Patange, A. (2023). Ethical considerations of AI applications in medicine: A policy framework for responsible deployment. In *Journal of Krishna Institute of Medical Sciences University* (Vol. 12, Issue 4).

Salajan, F. D., Barnes, T. L., & Becker, A. (2024). Comparative regionalism cases of artificial intelligence governance in education: The Caribbean Community and the European Union. *The Round Table*, *113*(1), 71–85. Advance online publication. DOI: 10.1080/00358533.2024.2307793

Sanghvi, H., Bhavsar, R., Hundlani, V., Gohil, L., Vyas, T., Nair, A., Desai, S., Jadav, N. K., Tanwar, S., Sharma, R., & Yamsani, N. (2024). MetaHate : AI-based hate speech detection for secured online gaming in metaverse using blockchain. *Security and Privacy*, *7*(2), e343. Advance online publication. DOI: 10.1002/spy2.343

Shafik, W. (2023). Introduction to ChatGPT. In *Advanced Applications of Generative AI and Natural Language Processing Models*. DOI: 10.4018/979-8-3693-0502-7.ch001

Shafik, W. (2024a). Agile Security Strategies. In *Smart and Agile Cybersecurity for IoT and IIoT Environments* (pp. 82–101). IGI Global. DOI: 10.4018/979-8-3693-3451-5.ch004

Shafik, W. (2024b). Artificial intelligence and Blockchain technology enabling cybersecurity in telehealth systems. In *Artificial Intelligence and Blockchain Technology in Modern Telehealth Systems*. DOI: 10.1049/PBHE061E_ch11

Shafik, W. (2024c). Artificial intelligence and internet of things roles in sustainable next-generation manufacturing: An overview of emerging trends in Industry 6.0. *Advances in Business Information Systems and Analytics*, *6*(0), 207–239. DOI: 10.4018/979-8-3693-3140-8.ch012

Shafik, W. (2024d). Artificial Intelligence and Machine Learning with Cyber Ethics for the Future World. In *Future Communication Systems Using Artificial Intelligence, Internet of Things and Data Science* (pp. 110–130). CRC Press., DOI: 10.1201/9781032648309-9

Shafik, W. (2024e). Deep Learning Impacts in the Field of Artificial Intelligence. In *Deep Learning Concepts in Operations Research* (pp. 9–26). Auerbach Publications., DOI: 10.1201/9781003433309-2

Shafik, W. (2024f). Ethical Use of Machine Learning Techniques in Smart Cities. In *Ethical Artificial Intelligence in Power Electronics* (pp. 21–47). CRC Press. DOI: 10.1201/9781032648323-3

Shafik, W. (2024g). Ethical Use of Machine Learning Techniques in Smart Cities. In *Ethical Artificial Intelligence in Power Electronics* (pp. 21–47). CRC Press., DOI: 10.1201/9781032648323-3

Shafik, W. (2024h). Generative AI for Social Good and Sustainable Development. In *Generative AI: Current Trends and Applications* (pp. 185–217). Springer. DOI: 10.1007/978-981-97-8460-8_10

Shafik, W. (2024i). Machine Learning for Advanced Wireless Communication. In *Microwave Devices and Circuits for Advanced Wireless Communication* (pp. 252–278). CRC Press., DOI: 10.1201/9781032656021-13

Shafik, W. (2024j). The Role of Artificial Intelligence in the Emerging Digital Economy Era. *Artificial Intelligence Enabled Management: An Emerging Economy Perspective*, 33.

Shafik, W. (2024k). Toward a More Ethical Future of Artificial Intelligence and Data Science. In *The Ethical Frontier of AI and Data Analysis* (pp. 362–388). IGI Global., DOI: 10.4018/979-8-3693-2964-1.ch022

Shafik, W. (2024l). Toward a More Ethical Future of Artificial Intelligence and Data Science. In *The Ethical Frontier of AI and Data Analysis* (pp. 362–388). IGI Global. DOI: 10.4018/979-8-3693-2964-1.ch022

Shafik, W. (2024m). Towards Trustworthy and Explainable AI Educational Systems. In *Explainable AI for Education: Recent Trends and Challenges* (pp. 17–41). Springer. DOI: 10.1007/978-3-031-72410-7_2

Shafik, W. (2025a). Human-Computer Interaction (HCI) Technologies in Socially-Enabled Artificial Intelligence. In *Future of Digital Technology and AI in Social Sectors* (pp. 121–150). IGI Global.

Shafik, W. (2025b). Quantum Computing and Generative Adversarial Networks (GANs): Ethics, Privacy, and Security. In *Quantum AI and its Applications in Blockchain Technology* (pp. 111–156). IGI Global Scientific Publishing.

Shafik, W., Mojtaba Matinkhah, S., & Sanda, M. N. (2020). Network Resource Management Drives Machine Learning: A Survey and Future Research Direction. *Journal of Communications Technology, Electronics and Computer Science, 2020*(0). https://doi.org/http://dx.doi.org/10.22385/jctecs.v30i0.312

Shahvaroughi Farahani, M., & Ghasemi, G. (2024). *Artificial Intelligence and Inequality: Challenges and Opportunities*. Qeios., DOI: 10.32388/7HWUZ2

Sharma, B., Sharma, L., Lal, C., & Roy, S. (2023). Anomaly based network intrusion detection for IoT attacks using deep learning technique. *Computers & Electrical Engineering*, *107*, 108626. Advance online publication. DOI: 10.1016/j.compeleceng.2023.108626

Sharma, V., Mishra, N., Kukreja, V., Alkhayyat, A., & Elngar, A. A. (2023). Framework for Evaluating Ethics in AI. *International Conference on Innovative Data Communication Technologies and Application, ICIDCA 2023 - Proceedings*. DOI: 10.1109/ICIDCA56705.2023.10099747

Shokoor, F., Shafik, W., & Matinkhah, S. M. (2022). Overview of 5G & Beyond Security. *EAI Endorsed Transactions on Internet of Things*, *8*(30), e2. Advance online publication. DOI: 10.4108/eetiot.v8i30.1624

Singh, L. (2022). Automated Kantian Ethics: A Faithful Implementation. *Lecture Notes in Computer Science (Including Subseries Lecture Notes in Artificial Intelligence and Lecture Notes in Bioinformatics), 13404 LNAI*. DOI: 10.1007/978-3-031-15791-2_16

Stylianides, C., Alexandropoulou, C. A., Sulaiman, W., Panagiotopoulos, I., Kleanthous, S., Dimitrakopoulos, G., Constantinou, I., Vassiliou, S., Garcia, F., Politi, E., Ntalaperas, D., Papageorgiou, X., Ioannides, N., Palazis, L., Pattichis, C. S., & Panayides, A. S. (2023). AI-based solutions for predicting sepsis in ICUs. *2023 IEEE EMBS Special Topic Conference on Data Science and Engineering in Healthcare, Medicine and Biology, IEEECONF 2023*. DOI: 10.1109/IEEECONF58974.2023.10404309

Subasi, A., Ozaltin, O., Mitra, A., Subasi, M. E., & Sarirete, A. (2023). Trustworthy artificial intelligence in healthcare. In *Accelerating Strategic Changes for Digital Transformation in the Healthcare Industry: Volume 2* (Vol. 2). DOI: 10.1016/B978-0-443-15299-3.00015-4

Tidjon, L. N., & Khomh, F. (2023). The Different Faces of AI Ethics Across the World: A Principle-to-Practice Gap Analysis. *IEEE Transactions on Artificial Intelligence*, *4*(4), 820–839. Advance online publication. DOI: 10.1109/TAI.2022.3225132

Treacy, S. (2023). Mechanisms and Constraints Underpinning Ethically Aligned Artificial Intelligence Systems: An Exploration of Key Performance Areas. *Human-Centric Intelligent Systems*, *3*(3), 189–196. Advance online publication. DOI: 10.1007/s44230-023-00036-0

Zhao, A., Jiang, N., Wang, C., Liu, S., & Qiu, K. (2023). Synchronization Optimization of Chaotic Laser Based on Generative Adversarial Network. *Guangxue Xuebao. Acta Optica Sinica*, *43*(1). Advance online publication. DOI: 10.3788/AOS220994

Chapter 3

# Governing Artificial Intelligent Ethically:
## A Human-in-Context Approach to Bias, Fairness, and Accountability

**Ashwarya Kapoor**
https://orcid.org/0000-0003-3858-3945
*School of Commerce, Manav Rachna International Institute of Research and Studies, Faridabad, India*

**Sneha Kapoor**
https://orcid.org/0009-0005-7051-3359
*J.C. Bose University of Science and Technology, YMCA, Faridabad, India*

**Rajiv Sindwani**
https://orcid.org/0000-0001-7457-6457
*J.C. Bose University of Science and Technology, YMCA, Faridabad, India*

## ABSTRACT

*Artificial intelligence (AI) is now integral to healthcare, finance, and education, shaping decisions that affect millions daily. While it promises efficiency and precision, its integration reveals ethical vulnerabilities—biased algorithms, opaque systems, and governance gaps that disempower stakeholders. This chapter reframes AI ethics through a human-in-context paradigm, emphasizing socio-technical settings, institutional norms, and stakeholder engagement. Anchored on fairness, transparency, privacy, and human agency, it examines three case studies: healthcare risk prediction bias, Apple Card's gender discrimination, and the Ofqual grading scandal. These illustrate how bias emerges and compounds when governance fails. The chapter maps risk pathways and proposes interventions like fairness-aware*

DOI: 10.4018/979-8-3373-2597-2.ch003

*design, bias audits, human override, and participatory governance. A cross-sector analysis highlights common ethical challenges and introduces a transferable ethics infrastructure aligned with OECD, UNESCO, GDPR, and EU AI Act principles.*

## 1. INTRODUCTION

Artificial intelligence (AI) has rapidly evolved from a niche research interest to a critical enabler of decision-making in healthcare, finance, education and other sectors that influence millions of lives daily. Once considered an experimental technology, AI is now a structural component of modern socio-economic systems, underpinning services from medical diagnostics to loan approvals and student performance assessments (Kapoor & Sindwani, 2025). According to the Stanford AI Index Report (2024), global private investment in AI reached USD 93.5 billion in 2023, with healthcare, finance and education ranking among the top five domains of adoption (Maslej, 2024). Gartner (2023) predicts that by 2030, AI will play a role in 80% of healthcare interactions, 75% of banking decisions and 65% of educational assessments, underscoring its deepening integration into critical infrastructures. While these advances promise efficiency, precision and scalability, they also bring unprecedented ethical, social and governance challenges.

In healthcare, AI systems have demonstrated remarkable potential in improving diagnostic accuracy, predicting disease outbreaks and tailoring treatment plans to individual patients. However, reliance on historical data often containing embedded inequities can perpetuate disparities, particularly for marginalised groups. In finance, AI models are increasingly responsible for credit scoring, fraud detection and investment decisions but the opacity of proprietary algorithms makes it difficult for customers to contest outcomes, as seen in the Apple Card gender bias controversy. Similarly, in education, AI-powered grading systems, adaptive learning platforms and predictive analytics offer the promise of personalised learning but can replicate systemic biases, as demonstrated in the UK's Ofqual grading algorithm scandal during the COVID-19 pandemic.

These examples illustrate a broader truth: AI systems do not operate in isolation but are embedded within complex socio-technical environments where human values, institutional histories and cultural norms interact with algorithmic logic. Addressing the ethical challenges of AI, therefore requires a shift from traditional compliance-based approaches to a human-in-context paradigm. This approach emphasises adaptive human oversight, contextual awareness, and stakeholder engagement throughout the AI lifecycle, ensuring that algorithms not only perform accurately but also align with societal values, equity goals and public trust.

Despite the growing discourse on AI ethics, several gaps remain in research and practice. First, most existing ethical frameworks are generic, lacking the sector-specific nuance necessary to address the distinct risks and governance needs of healthcare, finance and education. Second, while the concept of human-in-context oversight is discussed in theory, there is little practical guidance on how to operationalise it effectively. Third, studies rarely adopt a comparative, cross-sector perspective that identifies common ethical pitfalls and transferable solutions. Finally, there is a scarcity of empirical, real-world case studies that analyse how bias cascades and governance failures unfold in practice and how they can be mitigated.

This chapter addresses these gaps by conducting a comparative ethical analysis of AI in healthcare, finance and education, grounded in the human-in-context paradigm. It draws on high-profile case studies from each sector to uncover the mechanisms by which bias and inequity emerge, evaluates governance gaps and proposes actionable interventions. The analysis not only examines sector-specific issues but also synthesises cross-sector insights to build a more coherent, adaptable ethical governance framework for AI.

The objectives of the chapter are:

- To examine ethical challenges in AI deployment within healthcare, finance, and education through the human-in-context paradigm, considering socio-technical, institutional and cultural factors.
- To analyse sector-specific case studies to uncover patterns of bias, governance failures, and systemic inequities in AI systems.
- To compare these sectors to identify structural commonalities and transferable best practices for ethical AI governance.
- To integrate global governance frameworks with practical tools such as bias audits, ethics boards, and impact assessments for building adaptable, fair, transparent and accountable AI systems.

By integrating theory, empirical evidence and governance recommendations, this chapter contributes to the growing body of AI ethics scholarship with a focus on actionable strategies for building fair, transparent and accountable AI systems. It seeks to bridge the gap between abstract principles and real-world implementation, offering practical tools for policymakers, developers and institutional leaders tasked with ensuring that AI serves as a force for equitable progress rather than reinforcing existing inequities.

This chapter is organised as follows: Section 2 presents the theoretical background, outlining core ethical principles and the human-in-context paradigm as the analytical lens. Section 3 examines healthcare ethics through the Obermeyer et al. (2019) case on cost-based risk prediction, including a bias cascade diagram, stakeholder perspectives and recommendations. Section 4 addresses finance via the Apple Card gender bias controversy, presenting stakeholder viewpoints, governance failures and a summary table of risks and mitigations. Section 5 explores education through the UK Ofqual grading algorithm scandal, highlighting socio-economic bias, governance gaps and proposed solutions. Section 6 provides a cross-sector comparative analysis, identifying recurring patterns and transferable lessons. Section 7 outlines discussion. Section 8 presents theoretical contribution of the study. Section 9 offers sector-specific and cross-sector policy recommendations. Section 10 concludes with key insights, reinforcing the need for adaptable, human-in-context governance to ensure fairness, transparency, and public trust in AI systems. Section 11 discusses limitations and outlines future research directions.

## 2. THEORETICAL BACKGROUND

The emergence of advanced AI systems has challenged long-standing assumptions about the nature of human oversight in technology-mediated decision-making. Early conceptions of human–AI collaboration were grounded in the human-in-the-loop paradigm, where the human operator's role was primarily to validate, confirm or override the machine's recommendation. In this model, the human was envisaged as a final safeguard a fail-safe mechanism capable of catching errors and preventing unintended harm (Shneiderman, 2020). However, decades of human factors research have shown that this assumption is often flawed. In practice, oversight can be compromised by automation bias a cognitive tendency to over-rely on machine-generated suggestions and overlook contradictory evidence, even when it is available and compelling (Parasuraman & Riley, 1997; Goddard et al., 2012).

This recognition has spurred a conceptual shift toward the human-in-context paradigm, which reframes the human not as a passive checkpoint but as an active decision partner embedded within a socio-technical ecosystem. This approach acknowledges that oversight quality is shaped not only by the capabilities of the human and the AI system, but also by the institutional processes, cultural norms and regulatory constraints in which decisions take place (Doshi-Velez & Kim, 2017, Larsen, 2021). Under this model, the human and AI are co-dependent actors whose roles, authority levels and responsibilities are carefully calibrated to the task, the domain and the potential risks involved.

### 2.1 From Human-in-the-Loop to Human-in-Context

The limitations of human-in-the-loop oversight are well documented. In high-volume, high-speed decision environments such as medical diagnostics, high-frequency trading or large-scale educational assessment humans often lack the time, data access or contextual awareness to meaningfully challenge AI outputs. In other settings, the illusion of control can be equally problematic: oversight mechanisms may exist in name only, with humans rubber-stamping machine decisions because they trust the perceived objectivity and speed of AI (Green & Chen, 2019).

The human-in-context framework addresses these issues by embedding AI within structured socio-technical workflows that explicitly define:

- The division of labour between human and machine.
- Mechanisms for trust calibration, so humans neither over-rely on nor under-utilise AI outputs.
- Feedback loops where system performance is continuously monitored and improved in light of real-world outcomes.

Such an approach aligns with broader calls for responsible AI that moves beyond narrow technical performance metrics to encompass values like fairness, accountability and transparency (Jobin et al., 2019).

## 2.2 Ethical Foundations of Human–AI Collaboration

Effective human–AI collaboration must be underpinned by robust ethical principles that ensure systems do not simply operate efficiently but also respect human rights, promote equity and safeguard trust. Four foundational pillars form the basis for this work:

1. Fairness

Fairness in AI concerns the prevention of systematic disadvantage to individuals or groups based on characteristics such as race, gender, socio-economic status or geography. Yet fairness is not a single, universally agreed metric; instead, it encompasses multiple statistical definitions such as equal opportunity, demographic parity and predictive parity that can sometimes conflict with one another (Kleinberg et al., 2016; Chouldechova, 2017). In practice, fairness requires not only mathematical balancing but also normative decisions about whose values and priorities are embedded in the system.

2. Transparency

Transparency ensures that the rationale behind AI decisions is understandable to stakeholders. This includes explainability (how a decision was reached) and traceability (what data and processes were involved). Achieving transparency is complex, as it must balance the needs of accountability with concerns about intellectual property, trade secrets, and user privacy (Doshi-Velez & Kim, 2017).

3. Privacy

AI systems often rely on large-scale personal data, creating risks of misuse, leakage or re-identification even when datasets are anonymised (Kapoo et al., 2023b, 2025). Tools such as differential privacy can protect individuals while still allowing valuable insights but privacy also depends on governance structures, informed consent and compliance with legal frameworks (Dwork & Roth, 2014; Kapoor et al., 2023a, 2025)).

4. Safety

Safety refers to the robustness and resilience of AI systems under real-world conditions, including scenarios not seen during training. This principle emphasises proactive risk management anticipating potential failures, conducting stress tests and ensuring systems degrade gracefully under uncertainty rather than relying solely on reactive fixes after harm occurs (Amodei et al., 2016).

## 2.3 *Global Governance Frameworks*

Operationalising these pillars requires not just technical solutions, but comprehensive governance frameworks. At the policy level, international efforts such as the OECD AI Principles (2019) and UNESCO's Recommendation on the Ethics of AI (2021) establish normative baselines for transparency, accountability, fairness, and human oversight. The EU AI Act introduces a risk-based regulatory structure, imposing stricter requirements on high-risk AI applications in domains like healthcare, finance, and education. The General Data Protection Regulation (GDPR) provides strong protections for personal data, including the "right to explanation" for automated decisions.

On the technical and operational side, frameworks like the NIST AI Risk Management Framework (AI, 2023) and ISO/IEC 42001 (2023) provide structured processes for identifying, assessing, and mitigating AI risks. These standards emphasise lifecycle management ensuring that AI systems are not only safe and ethical at deployment but remain so through continuous monitoring, auditing, and stakeholder feedback.

## 2.4 Institutional, Cultural, and Regulatory Contexts

The human-in-context perspective highlights that ethical AI governance cannot be separated from the broader environments in which systems operate. In healthcare, for example, trust in AI-assisted diagnoses may depend on local cultural attitudes toward technology, historical relationships between communities and medical institutions and regulatory safeguards for patient privacy. In finance, credit scoring systems may be shaped by national banking laws, consumer protection agencies and patterns of economic inequality. In education, algorithmic grading may interact with long-standing disparities in school funding, teacher-student ratios and access to learning resources.

Crucially, this means that identical technical systems can have very different ethical implications depending on the context a phenomenon sometimes referred to as "ethical variance by jurisdiction." A model that operates with acceptable fairness metrics in one country may exacerbate inequities in another due to differences in data quality, institutional capacity or social trust.

Beyond Western frameworks, non-Western regions are also advancing distinctive approaches to AI ethics that reflect local cultural, institutional, and developmental priorities. In India, the NITI Aayog's National Strategy for AI (2018) emphasizes the principle of "AI for All," prioritizing equity, inclusivity, and social good over purely commercial applications. India's National Education Policy (NEP) 2020 also highlights the dual potential of AI in education—enhancing personalization while warning against deepening the digital divide. Similarly, the Reserve Bank of India's Digital Lending Guidelines (2022) address the risks of opaque algorithms in consumer finance by mandating transparency, grievance redressal, and auditability of credit decision systems.

China has issued its Ethical Norms for the New Generation AI (2021), embedding cultural values such as collective welfare, harmony, and shared responsibility. This contrasts with the more individual-rights-based framing of European regulations like GDPR, underscoring how ethics are culturally situated. In Africa, the African Union's AI for Africa strategy (2019) recognizes both opportunities and risks of AI, calling for inclusivity, indigenous knowledge integration, and attention to infrastructural inequalities. Kenya's M-Pesa mobile finance ecosystem is often cited as a success story in financial innovation, yet it also highlights risks of digital exclusion when algorithms reinforce socio-economic gaps.

These examples demonstrate "ethical variance by jurisdiction" at a global scale: while fairness, transparency, and human oversight are universal principles, their interpretation and enforcement vary with cultural values, institutional priorities, and governance capacities. A cross-cultural lens therefore reinforces the human-in-context paradigm by showing that ethical AI cannot be divorced from regional histories, socio-economic realities, and local trust dynamics.

## 2.5 Why Human-in-Context Matters Now

The increasing autonomy of AI systems, coupled with their deployment in high-stakes environments, raises the urgency of moving from human-in-the-loop to human-in-context. The former assumes that humans can correct errors if given the chance; the latter recognises that effective oversight requires preparation, empowerment and contextual understanding. It involves designing AI workflows where human judgement is not an afterthought but an integral, empowered component of the decision-making ecosystem.

This shift is also critical in light of emerging AI capabilities. With the rise of generative models, multi-modal systems and autonomous decision agents, the boundaries between human and machine reasoning are becoming increasingly blurred. Without a human-in-context approach, organisations risk deploying powerful but opaque systems that escape meaningful oversight, eroding public trust and amplifying systemic risks.

## 2.6 Positioning the Chapter

By grounding the analysis in the human-in-context paradigm and its four ethical pillars, this chapter bridges conceptual theory and practical governance. It does so through a comparative lens across healthcare, finance and education three sectors that illustrate both the opportunities and dangers of AI deployment. The following sections apply this framework to real-world cases, identifying how fairness, transparency, privacy and safety are upheld or undermined and how governance mechanisms can be adapted to specific institutional and cultural contexts.

Ultimately, ethical AI governance is not solely a technical challenge; it is a socio-political project that requires sustained collaboration among engineers, domain experts, policymakers, civil society and the communities affected by AI systems. The human-in-context model is not a single solution but a foundational perspective that can guide the design of systems that are both technologically robust and socially responsible.

# 3. HUMAN–AI COLLABORATION AND ETHICS IN HEALTHCARE

Healthcare is among the highest-stakes domains for AI integration, where technological innovation directly influences patient well-being, ethical decision-making and life-or-death outcomes. Over the past decade, AI has been embedded across clinical workflows from diagnostics and treatment planning to risk prediction and hospital resource allocation (Chinta et al., 2024). Convolutional neural networks now match human performance in detecting conditions such as diabetic retinopathy, pneumonia and certain cancers while predictive analytics enable earlier interventions for sepsis and ICU deterioration (Obermeyer et al., 2019; Tang et al., 2023).

Yet, these advances bring significant ethical risks. A widely cited case from Obermeyer et al. (2019) revealed that a cost-based risk prediction system systematically underestimated the healthcare needs of Black patients, due to structural inequities that made healthcare costs a poor proxy for health needs. Such "label-choice bias" illustrates how algorithms can perpetuate systemic injustice without explicit discriminatory intent. Similarly, device-level bias such as pulse oximeters overestimating oxygen levels in patients with darker skin underscores that ethical risks extend beyond algorithms to include hardware design and deployment contexts (Sjoding et al., 2020).

These examples demonstrate the "bias cascade": inequities can originate in data collection, be reinforced in algorithm design and compound during deployment (Royal College of Surgeons of England Bulletin, 2021) (Refer figure 1). Without intervention, this cycle perpetuates disparities, erodes trust and reduces healthcare quality particularly for marginalized groups with already limited trust in medical institutions.

*Figure 1. Bias Cascade in AI System*

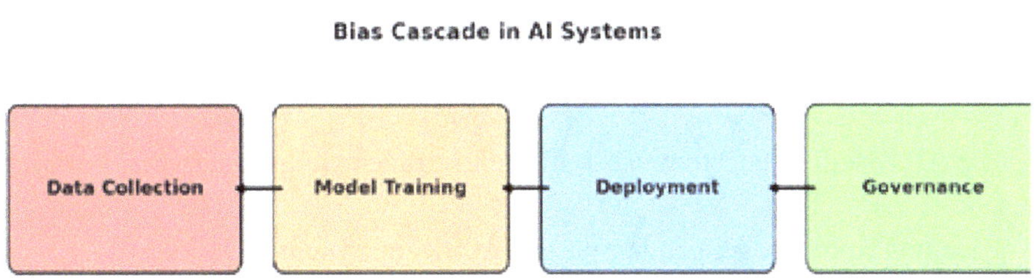

## 3.1 Case Study - When Cost Becomes a Proxy for Care: Bias in AI Healthcare Risk Prediction

### 3.1.1 Introduction to the Case

In 2019, a landmark study by Obermeyer et al. revealed a critical flaw in one of the most widely deployed healthcare AI risk prediction systems in the United States. The algorithm was designed to identify patients with the most complex healthcare needs, assigning them to specialized care management programs intended to improve outcomes and reduce long-term costs.

The system's logic was seemingly straightforward: patients projected to have higher future healthcare costs were assumed to have greater medical need. Hospitals, insurers and integrated healthcare systems embraced the model because it offered scalability, data-driven precision and the promise of cost-effective population health management.

However, this seemingly neutral cost proxy concealed a profound inequity: historically underserved patients particularly Black patients were systematically underestimated in their level of need. This was not because they were healthier, but because they had historically incurred lower healthcare costs due to systemic barriers, unequal access and structural discrimination.

The consequences were stark: patients who could have benefited from extra monitoring, preventive interventions and coordinated care were instead excluded from such programs. The AI was not simply failing to eliminate inequities it was reinforcing them at scale.

### 3.1.2 System Overview

The AI system in question was built on insurance claims data, incorporating:

- Demographics (age, gender, partial socioeconomic indicators)
- Past healthcare utilization (hospital visits, specialist consultations, diagnostic tests)
- Billing codes and associated costs

A regression-based predictive model generated a *risk score* for each patient, intended to reflect their likely healthcare cost over the next year. Those above a certain threshold were flagged for enrolment in care management programs.

Deployment scope:

- Rolled out across multiple U.S. states

- Used by hospital networks, insurers and state-level health agencies
- Impacted millions of patients annually

### 3.1.3 Discovery of Bias

The bias was uncovered during an independent academic audit of the system's outcomes.

The study revealed that approximately 200 million patients in the U.S. were indirectly affected by risk prediction algorithms in healthcare systems (Obermeyer et al., 2019). Quantitative analysis showed that Black patients who were equally sick as white patients were assigned risk scores that were, on average, less than half as high. As a result, only 18% of Black patients who should have qualified for enhanced care were flagged by the algorithm, compared with 47% of white patients. These disparities illustrate how seemingly neutral cost-based proxies can systematically disadvantage marginalized populations at scale.

When comparing patients with similar chronic illnesses and severity levels, researchers found:

- Black patients had significantly lower risk scores than white patients, even when their health conditions were equally severe.
- The algorithm underestimated their risk because historical data reflected lower spending on Black patients a product of systemic inequities in access and treatment.
- For example: A Black patient with uncontrolled diabetes and hypertension might have the same actual need for specialist care as a white patient, but if historical costs for similar Black patients were lower (due to underdiagnosis, less follow-up or fewer referrals), the algorithm would assign them a lower score.

A risk prediction bias analysis has been exhibited in table 1.

*Table 1. Risk Prediction Bias Analysis*

| Stage | Risk/Bias Introduced | Root Cause | Example | Potential Mitigation |
|---|---|---|---|---|
| **Data Collection** | Label-choice bias | Cost used as proxy for need | Chronic illness costs lower for underserved patients | Use clinical outcome-based labels (e.g., disease severity) |
| **Model Training** | Feature bias | Inclusion of socio-economic proxies | ZIP code correlates with race | Apply fairness-constrained modeling |
| **Deployment** | Feedback bias | Lower scores → less care → worsening health | Missed preventive screenings | Continuous post-deployment audits |
| **Governance** | Oversight gap | No demographic-specific validation | Undetected subgroup harm | Mandatory pre-rollout bias audits |

## 3.1.4 Stakeholder Perspectives

### *Clinicians*

Many clinicians reported frustration at being unable to see the logic behind the algorithm's scores. They described a "black box" dynamic where challenging the AI's output required lengthy justifications and bureaucratic approvals.

### *Patients*

In focus groups, patients especially from historically marginalized communities expressed distrust. Several described feeling "invisible" to the system, reinforcing long-standing suspicions about institutional healthcare bias.

### *Policymakers*

Regulators were alarmed by how easily systemic inequities could be encoded into algorithmic tools. Some called for "equity impact statements" for all high-stakes AI systems in healthcare.

### *Developers*

The technical team acknowledged that cost was chosen as the label for pragmatic reasons cost data was readily available, standardized and correlated with need in most cases. However, they admitted little consideration was given to how cost could act as a biased proxy.

### 3.1.5 Governance Failures & Ethical Gaps

- No demographic stratification during model validation.
- No clinical outcome benchmarking only cost predictions were validated.
- Lack of patient/clinician input during system design.
- No regulatory requirement for bias audits before deployment.

### 3.1.6 Lessons Learned and Recommendations

Technical Fixes

- Redefine the target variable to reflect clinical severity, not cost.
- Use multi-label risk assessment combining cost, outcomes and social determinants of health.
- Apply fairness-aware machine learning techniques.

Governance Fixes

- Require bias audits before and after deployment.
- Establish multidisciplinary AI ethics boards including patients and clinicians.
- Maintain post-market monitoring dashboards with demographic breakdowns.

Applying a human-in-context approach to healthcare means integrating AI into workflows where clinicians actively interpret outputs, retain override authority and feed real-world insights back into model improvement. Rather than serving as passive checkpoints, clinicians become informed decision partners with AI systems designed to display rationale, uncertainty ranges and known limitations.

To safeguard ethical integrity, healthcare institutions should:

- Implement AI impact assessments that measure performance across demographic subgroups and detect label bias before deployment.
- Require independent audits for fairness, safety and transparency.
- Improve dataset diversity and regularly update proxies to reflect actual health needs.
- Maintain post-market surveillance to detect model drift or emergent biases.
- Record and analyse clinician overrides to inform retraining and governance adjustments.

By combining robust oversight with respect for clinical judgment, healthcare can harness AI's diagnostic and predictive power while preserving equity, trust and patient safety.

### 3.1.7 Key Takeaway

The Obermeyer et al. (2019) case demonstrates how biased proxy choices such as cost being used to measure care needs, can cascade into systemic inequities that exclude vulnerable populations from life-saving interventions. Effective governance requires rethinking target variables, mandating fairness audits, and embedding clinicians as empowered decision partners. This case reinforces that AI in healthcare must be judged not only by predictive accuracy but also by its capacity to advance equitable health outcomes.

## 4. HUMAN–AI COLLABORATION AND ETHICS IN FINANCE

Artificial intelligence is now central to credit scoring, fraud detection, algorithmic trading and personalized financial services, offering efficiency, predictive accuracy and scalability (Kapoor et al., 2023b; 2024a). However, it also raises critical ethical issues related to fairness, transparency, accountability and systemic stability.

One persistent challenge is algorithmic bias in lending. Historical financial data often reflects structural inequalities, which can be reproduced or amplified by machine learning models (Barocas et al., 2020). For instance, proxy variables such as postal codes can indirectly encode demographic patterns, disadvantaging certain groups even when sensitive attributes are excluded. Strategies like fairness-constrained modeling or in-process adjustments can reduce disparities though often with trade-offs in profitability (Kozodoi et al, 2022).

## 4.1 Case Study: Finance

### 4.1.1. Credit Limits and Gender Bias: The Apple Card Controversy

In 2019, a public controversy erupted when several prominent tech figures including a co-founder of Apple's own software development platform reported that Apple's credit card, issued in partnership with Goldman Sachs, had granted significantly lower credit limits to women than to their male spouses, even when the women had higher credit scores, greater incomes and shared joint bank accounts.

### 4.1.2 How the System Worked:

The Apple Card's credit limit decisions were powered by a proprietary algorithm that evaluated risk using credit bureau data, transaction histories and other undisclosed features. While Goldman Sachs denied the use of gender as an explicit variable, certain proxy variables such as historical account structures or transaction types could indirectly encode gender disparities.

Impact:

- Women with strong credit histories were assigned limits 10–20 times lower than male counterparts. Reports submitted to the New York State Department of Financial Services revealed that over 6,000 consumer complaints were filed in the first months following the Apple Card rollout, many citing discriminatory lending outcomes. Quantitative investigation confirmed gender disparities: on average, women received 20–30% lower credit limits than men with similar or superior financial profiles. In some cases, credit limit ratios between spouses differed by a factor of up to 20x, despite shared assets and incomes. The widespread scale of complaints highlighted systemic governance gaps in bias auditing and customer redressal mechanisms.
- Customer service staff admitted they could not override algorithmic decisions.
- The opacity of the model meant affected individuals could not challenge their assessments effectively.

### 4.1.3 Stakeholder Perspectives

- Customers: Women affected described the experience as "financially humiliating" and a "clear sign of hidden discrimination." Many stressed the frustration of having no explanation or appeal process.
- Regulators: The New York State Department of Financial Services launched an investigation, highlighting concerns about disparate impact in credit scoring systems even without explicit bias variables.
- Developers: Goldman Sachs engineers noted that business constraints favoured using widely available historical financial data without adjusting for potential bias in proxies, prioritizing predictive accuracy and speed to market over fairness.

### 4.1.4. Governance Failures

- No pre-deployment bias audit for protected characteristics such as gender.

- Lack of model explainability standards, leaving customers unable to contest results.
- Overreliance on proxy features without testing for indirect discrimination.

### 4.1.5. Recommendations

- Mandate fairness testing by gender, race and age for all credit decision models before launch.
- Require explainability reports for credit limit decisions, enabling customers to seek remediation.
- Introduce human-in-context oversight, empowering credit officers to override algorithmic outcomes when anomalies are detected.

A risk prediction bias analysis has been presented in table 2.

*Table 2. Risk Prediction Bias Analysis in Apple Card Case*

| Stage | Risk/Bias Introduced | Root Cause | Example | Potential Mitigation |
|---|---|---|---|---|
| **Data Collection** | Proxy variable bias | Use of historical account patterns that correlate with gender | Joint accounts or spending patterns reflecting societal norms | Collect and validate features for neutrality |
| **Model Training** | Feature bias | Socio-economic proxies with gender correlation | Employment type, purchase categories | Apply fairness constraints; remove or adjust biased proxies |
| **Deployment** | Decision opacity | Proprietary scoring with no explanation to customers | Customers denied higher credit limits without reasons | Provide explainability reports to applicants |
| **Governance** | Oversight gap | No mandated bias audits or override mechanisms | Customer service unable to alter algorithm outputs | Regulatory bias testing; empower human overrides |

Responsible machine learning practices in finance also incorporate reject inference using information from previously declined applicants to improve inclusivity and explainability, enabling both loan officers and customers to understand decision rationale. Counterfactual explanations ("minimal changes needed for approval") enhance transparency and empower applicants to take corrective action (Wachter et al., 2017).

Beyond individual fairness, AI introduces systemic risks. In algorithmic trading, millisecond-scale automated transactions have triggered flash crashes, revealing how interconnected systems can amplify volatility beyond human control. To mitigate such risks, firms deploy circuit breakers, anomaly review panels and real-time monitoring.

Embedding ethics in financial AI requires:

- Fairness dashboards tracking model performance by demographic group.
- External bias audits and public reporting for high-risk applications.
- Model cards documenting intended use, performance metrics and limitations.
- Human-in-context governance where humans retain authority over sensitive decisions and can override AI recommendations.

Regulatory frameworks such as the Equal Credit Opportunity Act (ECOA) in the U.S. and the EU AI Act mandate non-discrimination, transparency and accountability in high-risk AI systems. When combined with institutional governance structures like hybrid review boards and bias mitigation pipelines these measures help balance innovation with fairness and financial stability.

### 4.1.6 Key Takeaway

The Apple Card controversy illustrates how opaque algorithms can inadvertently reproduce gender-based inequities, even without explicitly using gender as a variable. Without transparency, bias audits, or human override authority, financial AI risks undermining trust and fairness. The case highlights the need for regulatory safeguards, customer explainability and human-in-context oversight to ensure that credit decisions are both accurate and equitable.

## 5. HUMAN–AI COLLABORATION AND ETHICS IN EDUCATION

AI is increasingly used in education for adaptive learning platforms, automated grading, predictive analytics and generative content creation. These tools promise personalized learning, faster feedback and administrative efficiency with adaptive systems demonstrating measurable improvements in student outcomes (Luckin et al., 2016; Holmes et al., 2019).

However, their adoption raises pressing ethical concerns. Data privacy is paramount, particularly when AI collects extensive information on student performance, behaviour and engagement (Kapoor et al., 2022; 2024b). The use of biometric data in proctoring or emotional analysis intensifies questions about consent, storage and potential misuse (Selwyn, 2019). Over-reliance on AI tools can also foster technological dependency where both students and teachers defer to algorithmic outputs without critical evaluation (Williamson & Eynon, 2020).

## 5.1 Case Study: Education

### 5.1.1 Grading at Scale: Lessons from the UK Ofqual Algorithm Scandal

In 2020, the UK government cancelled nationwide A-level exams due to COVID-19 and tasked Ofqual, the examinations regulator with developing an algorithm to predict students' final grades. The aim was to maintain fairness and consistency without in-person exams.

### 5.1.2. How the System Worked:

The model factored in:

- Teacher-assigned predicted grades
- Student rank order within classes
- Historical performance of each school in the subject

To curb grade inflation, the model heavily weighted historical school performance over individual predictions, especially in larger cohorts.

### 5.1.3 Discovery of Bias

- Students from historically underperforming schools often in disadvantaged communities were downgraded regardless of individual academic achievement.
- Elite private school students faced fewer adjustments, maintaining high grades.
- Thousands of high-achieving students from low-income backgrounds lost university placements. Data from Ofqual (2020) revealed that the algorithm downgraded grades for nearly 40% of students across England. The impact was especially severe in disadvantaged schools: students in state schools saw downgrades of up to 60%, while only 3% of students in private elite schools were downgraded. As a result, approximately 25,000 students lost their conditional university offers, disproportionately affecting first-generation and low-income students. The scale of the backlash led to public protests, over 300,000 petition signatures, and a full reversal of the policy within weeks.

#### 5.1.4. Stakeholder Perspectives

- Students: Described the process as "demoralizing" and "robbing us of our future." Many reported feelings of powerlessness, as no appeal system was available before results were issued.
- Regulators: Faced intense political and public criticism for relying on a statistical model without considering individual-level fairness.
- Developers: Data analysts admitted that the model's weighting system was chosen to control for grade inflation, not equity reflecting a governance priority imbalance.

#### 5.1.5. Governance Failures

- No equity impact assessment before national rollout.
- Insufficient stakeholder consultation, especially with teachers, students and universities.
- Absence of pilot testing across varied school types to detect disproportionate impacts.

#### 5.1.6. Recommendations

- Prioritize individual-level achievement data over group-level historical averages in grading algorithms.
- Introduce appeal mechanisms before finalizing results to allow human review.
- Conduct public transparency exercises, explaining the model's weighting to all stakeholders' pre-implementation.

Table 3 presents the algorithmic grading bias analysis in Ofqual case.

*Table 3. Algorithmic Grading Bias Analysis in Ofqual Case*

| Stage | Risk/Bias Introduced | Root Cause | Example | Potential Mitigation |
|---|---|---|---|---|
| **Data Collection** | Label-choice bias | Reliance on historical school performance as a proxy for individual merit | Disadvantaged schools' lower averages affecting all students | Use multi-source evidence including individual assessments |
| **Model Training** | Weighting bias | Overemphasis on school-level averages | Students from low-performing schools downgraded | Balance group trends with individual rankings |
| **Deployment** | Lack of appeal process | No mechanism for contesting grades before release | Students unable to challenge downgraded scores | Implement appeal process pre-publication |
| **Governance** | Transparency gap | Poor communication of model's rationale | Public backlash over undisclosed weighting rules | Public model documentation and stakeholder consultation |

Equity is another major challenge. Unequal access to devices, reliable internet or subscription-based AI tools risks deepening educational divides. Policy recommendations such as those from IIT Delhi, emphasize providing equitable access to premium AI tools, maintaining transparency in AI-generated content, and requiring disclosure of AI use in academic submissions.

The rise of generative AI has complicated academic integrity debates. While some institutions ban AI-generated text to curb plagiarism, others integrate AI into assignments by requiring students to analyze, critique or improve AI outputs (Kasneci et al., 2023). This approach promotes metacognition and critical thinking while acknowledging the reality of AI in learning environments.

Bias in educational AI also warrants attention. For instance, findings from the London School of Economics show gender bias in LLM-generated summaries, which could influence resource recommendations or feedback. Without safeguards, such biases risk shaping perceptions of student ability and potential.

Applying a human-in-context approach to education means maintaining active teacher oversight, ensuring AI recommendations are transparent and teaching students to critically engage with AI outputs. Implementation should include:

- Embedding AI ethics literacy into digital skills curricula.
- Conducting equity audits to monitor access and performance disparities.
- Co-designing AI governance frameworks with teachers, students, parents and policymakers to ensure contextual relevance.

By positioning AI as a tool to augment not replace human judgment and mentorship, educational institutions can leverage its benefits while upholding fairness, inclusivity and academic integrity.

### 5.1.7 Key Takeaway

The Ofqual grading scandal shows how reliance on historical institutional performance can entrench socio-economic bias and strip students of agency. The absence of transparency, equity impact assessment, and appeal mechanisms created widespread mistrust. This case underscores that educational AI must prioritise individual fairness, ensure stakeholder consultation, and preserve teacher and student voice in evaluation processes.

## 5.2 Recent empirical evidence on generative AI and digital divides

Recent empirical work provides early but increasingly robust evidence about how generative AI is being used in classrooms and the equity risks that accompany rapid adoption. A meta-analysis of studies through early 2025 found mixed effects on learning outcomes: while some controlled studies report small improvements in higher-order thinking and writing performance when students use generative tools with structured guidance, effects vary widely by task design and instructional scaffolding. Overall, gains are contingent on explicit pedagogical integration rather than unguided student use (Wang & Fan, 2025).

Several field studies and institutional reports corroborate this nuance. Longitudinal classroom implementations that directly integrated ChatGPT into curriculum activities observed benefits for student idea generation and formative feedback when teachers provided clear prompts, usage guidelines, and reflective activities; however, these studies also flagged academic-integrity concerns and the need for adapted assessment designs (e.g., more in-class, performance-based assessments). One-year classroom studies show that teacher coaching and curriculum redesign are key mediators of positive outcomes (Guo and Cunningham, 2025).

Teacher and institutional surveys indicate broadly positive but cautious attitudes: many educators value generative AI for reducing administrative load and supporting differentiated instruction, yet they worry about uneven access, student misuse, and insufficient professional development. These studies emphasise that teacher agency and professional development are decisive for responsible classroom adoption (Jauhiainen & Garagorry Guerra, 2025).

Importantly for our human-in-context framework, a growing literature highlights the digital-divide risks of AI in education. Reviews and policy analyses find that uneven device availability, bandwidth constraints, and gaps in digital literacy mean that generative-AI benefits accrue disproportionately to better-resourced schools and students—potentially widening existing inequities unless mitigations are implemented. Policy reviews therefore recommend targeted infrastructure investment, offline-capable tools, and inclusion of AI literacy in curricula to reduce distributional harms (Vesna et al., 2025).

**Implications for governance in education:** Empirical evidence suggests that generative AI can support learning when integrated with explicit pedagogical design, teacher training, and assessment reform; however, without attention to access and capacity, these benefits risk amplifying inequities. In line with the human-in-context approach, we therefore recommend (a) mandatory pilot evaluations before large-scale rollouts, (b) investment in teacher professional development focused on pedagogical uses of generative AI, (c) assessment redesign that emphasises in-person, applied tasks, and (d) infrastructure and access policies (e.g., device loans, offline tool options, subsidized connectivity) to address the digital divide.

## 6. CROSS-SECTOR COMPARATIVE ANALYSIS

Although healthcare, finance and education operate in vastly different contexts, the ethical challenges of AI deployment reveal strikingly similar priorities (refer figure 2). Across all three domains, four recurring pillars fairness, transparency, privacy and human agency emerge as non-negotiable foundations for responsible AI. Yet, the way these priorities manifest, the metrics used to measure them and the governance mechanisms employed differ substantially, reflecting the unique stakes, workflows and stakeholder expectations of each sector.

*Figure 2. Ethical Pillars across Sectors*

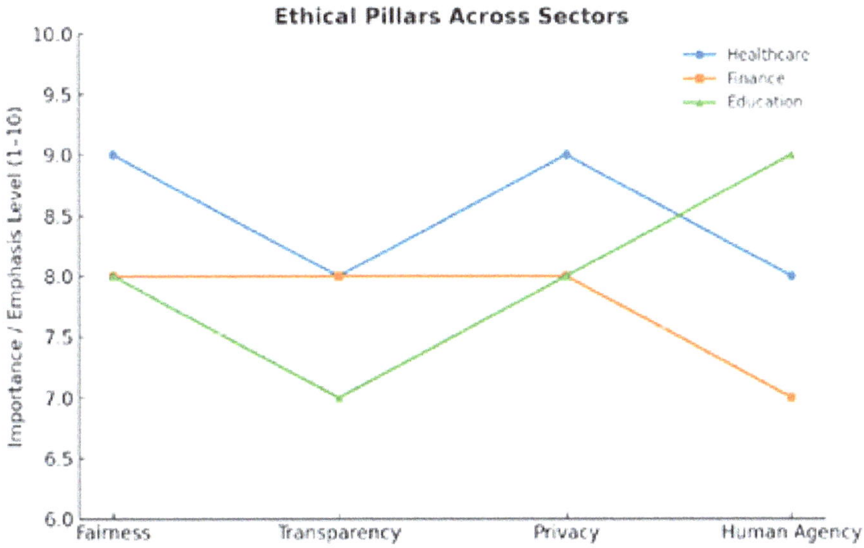


In healthcare, the ethical conversation is dominated by equitable access to care, patient safety and the maintenance of clinical trust. AI systems such as diagnostic tools, triage algorithms and predictive analytics must demonstrate not only statistical accuracy but also fairness in outcomes across demographic groups. Errors can have life-or-death consequences, making safety and explainability paramount. Clinical practitioners often require granular reasoning trails not merely that a patient is classified as "high risk," but the clinical evidence and decision logic that support this assessment.

In finance, the emphasis shifts toward non-discriminatory lending practices, market stability and compliance with a dense web of regulatory requirements. Bias in AI-powered credit scoring or fraud detection can lead to systemic exclusion from economic opportunities. Here, fairness is operationalised in terms of bias-free credit decisions and compliance with anti-discrimination laws. Transparency, in turn, must be actionable for customers meaning that a loan applicant should understand the specific factors influencing their creditworthiness and the steps they can take to improve it. Market stability further demands robust risk management and regular audit mechanisms to detect and correct anomalies before they escalate.

In education, inclusivity, fair assessment and the preservation of academic integrity take centre stage. Automated grading, adaptive learning platforms and predictive dropout models hold the potential to enhance learning outcomes but risk reinforcing structural inequities if not carefully designed. Fairness here is measured by equal access to resources and non-biased assessments across socio-economic, cultural and linguistic lines. Transparency demands that students, educators and administrators understand both the extent and limitations of AI's role in evaluating performance, so that algorithmic judgments do not supplant holistic human evaluation.

The human-in-context paradigm provides a unifying framework for reconciling these sectoral variations. In all three sectors, AI is not a replacement for human decision-making but an augmentation of it, with clearly defined oversight roles and feedback loops that enable continuous improvement. However, the specific design of these oversight mechanisms varies:

- In healthcare, multidisciplinary review boards and clinical validation studies serve as feedback channels.
- In finance, compliance officers and independent auditors assess algorithmic fairness and regulatory conformity.
- In education, teacher committees and academic boards review AI grading patterns for consistency and bias.

Transparency is similarly context-dependent. Clinicians need rich, domain-specific explanations to integrate AI outputs into diagnostic reasoning. Loan applicants require simplified yet accurate explanations that can guide corrective action. Students and educators benefit from clear disclosures about AI's role in evaluation and the safeguards in place to prevent bias.

Fairness metrics are tailored to sectoral goals. In healthcare, outcome parity ensuring that treatment results are not systematically worse for specific groups is often prioritised. In finance, the focus is on ensuring that approval rates and terms are free from discriminatory bias. In education, fairness entails guaranteeing that grading algorithms do not disproportionately penalise or advantage any group based on irrelevant factors.

Governance tools also reveal both commonalities and differences. AI impact assessments, algorithmic audit systems, ethics committees and performance dashboards are broadly applicable across domains but must be adapted to sector-specific risks. For example:

- In healthcare, audits may involve clinical trials, patient safety assessments and bias testing across diagnostic categories.
- In finance, regulators may require stress-testing under adverse market scenarios and legally mandated fairness reporting.
- In education, algorithmic audits may focus on consistency across different student demographics and alignment with curriculum standards.

Global policy frameworks such as the OECD AI Principles, UNESCO's Recommendation on the Ethics of AI, the General Data Protection Regulation (GDPR) and the EU AI Act provide overarching norms for transparency, accountability and human oversight. These frameworks are complemented by sector-specific regulations and professional norms: medical ethics codes and HIPAA-equivalent privacy laws in healthcare, Basel Committee guidelines and anti-discrimination statutes in finance and accreditation standards in education. Together, they form a layered governance ecosystem where global principles set the ethical direction, while sectoral rules translate them into operational practice.

By examining sectoral differences through a shared governance lens, institutions can identify transferable best practices. Healthcare's rigorous clinical validation processes could inspire more robust pre-deployment testing in education. Finance's structured audit and compliance mechanisms could inform healthcare's risk management frameworks. Education's emphasis on inclusivity and accessibility could influence financial institutions to design more equitable customer communication strategies.

Ultimately, the cross-sector comparison demonstrates that while context shapes the expression of ethical principles, the underlying governance logic is consistent: align AI systems with human values, make decision processes transparent, ensure fairness in outcomes and maintain mechanisms for accountable oversight. By leveraging insights from one domain and adapting them to the realities of another, organisations can strengthen ethical AI adoption across diverse contexts, fostering systems that are not only effective but also socially legitimate.

A cross-cultural perspective further enriches this comparative analysis. For instance, in healthcare, India's "AI for All" framework foregrounds equitable access to medical innovations, while African policy documents stress integrating AI into community health systems sensitive to local contexts. In finance, Kenya's M-Pesa demonstrates how AI-enabled financial ecosystems can drive inclusion but also risk reinforcing inequalities if transparency and oversight are lacking. Brazil's Lei Geral de Proteção de Dados (LGPD), often described as a "Latin GDPR," illustrates how data protection can be tailored to address local socio-economic disparities. In education, Singapore's AI in Education Roadmap emphasizes co-design with teachers and contextual transparency, while India's NEP 2020 and UNESCO's African case studies highlight how AI integration must be balanced with equity in infrastructure and digital access.

Together, these perspectives illustrate that while Western frameworks (GDPR, OECD AI Principles, EU AI Act) dominate global AI ethics discourse, non-Western contributions provide valuable insights into operationalizing fairness and inclusivity in resource-constrained or socio-culturally diverse environments. Incorporating such perspectives demonstrates that ethical AI governance is not a one-size-fits-all model but a globally adaptive framework where regional diversity shapes the interpretation of fairness, transparency, privacy, and human agency.

Table 4 exhibits comparative ethical priorities and governance practices across three sectors

*Table 4. Comparative Ethical Priorities and Governance Practices Across Sectors*

| Ethical Pillar / Governance Focus | Healthcare | Finance | Education |
|---|---|---|---|
| **Fairness** | Outcome parity across demographic groups; equitable access to care | Bias-free credit decisions; compliance with anti-discrimination laws | Equal access to learning resources; non-biased assessments |
| **Transparency** | Detailed clinical reasoning trails for practitioners | Actionable credit explanations for customers | Clear disclosure of AI's role in grading and tutoring |
| **Privacy** | Compliance with health data protection laws (e.g., HIPAA, GDPR); anonymisation and secure storage | Secure handling of financial records; adherence to banking privacy regulations | Protection of student records; compliance with FERPA/GDPR |
| **Human Agency** | Clinician retains final decision authority; multidisciplinary review boards | Loan officers empowered to override AI decisions; regulatory appeal processes | Teacher oversight of grading outputs; academic board reviews |
| **Primary Governance Tools** | Clinical validation studies; patient safety committees; bias audits | Independent fairness audits; compliance monitoring; stress tests | Curriculum-aligned audits; stakeholder consultation; performance dashboards |
| **Key Regulatory References** | Medical ethics codes; national health data laws | Basel Committee guidelines; financial services regulations | Accreditation standards; education data protection laws |
| **Sector-Specific Risks** | Diagnostic bias; treatment inequity; patient safety hazards | Discriminatory lending; systemic exclusion; market instability | Socio-economic grading bias; undermining of academic integrity |
| **Feedback Mechanisms** | Continuous monitoring via hospital boards and regulatory bodies | Audit reports to regulators; periodic recalibration of scoring models | Educator review panels; student appeals processes |

## 7. DISCUSSION

The cross-sectoral analysis shows that while fairness, transparency, privacy and safety are universal ethical priorities, their operationalization depends on sectoral contexts, norms and regulations. Healthcare focuses on equitable patient outcomes and clinical trust; finance balances fairness with profitability and systemic stability; education emphasizes inclusivity, academic integrity and learner development. In healthcare algorithmic bias such as the cost-based risk model studied by Obermeyer et al. (2019) demonstrates how proxy choices can perpetuate inequities. In finance, the tension between non-discrimination and predictive accuracy requires careful trade-off decisions, supported by tools like counterfactual explanations. In education, over-reliance on adaptive platforms risks diminishing teacher agency and widening digital divides.

The human-in-context model offers a unifying approach, positioning AI as an augmentative partner with clearly defined decision boundaries, explicit uncertainty communication and feedback mechanisms for continuous refinement. This requires institutional governance AI impact assessments, ethics review boards, external audits that translates high-level principles into enforceable practice. Trade-offs are inevitable: fairness measures may reduce model accuracy, transparency can risk intellectual property exposure, privacy protections may limit data utility and rigorous safety checks can slow deployment. Acknowledging these trade-offs openly and engaging stakeholders in their resolution, strengthens both ethical legitimacy and practical outcomes. By focusing on context-specific application while sharing governance principles across sectors, institutions can foster AI adoption that advances innovation without compromising human dignity, equity or trust.

## 8. THEORETICAL CONTRIBUTIONS

This chapter advances the discourse on AI ethics and governance in four interconnected ways, each addressing persistent theoretical and practical gaps in how human–AI collaboration is conceptualised and operationalised. First, it reframes human–AI collaboration from the conventional *human-in-the-loop* model toward a richer and more adaptive human-in-context paradigm. While the human-in-the-loop model assumes that human oversight is a sufficient safeguard against algorithmic error, empirical research on automation bias demonstrates that human judgment can be compromised when oversight is reduced to a passive checkpoint. The human-in-context perspective situates human decision-making within a broader socio-technical ecosystem, emphasising:

- Clearly defined human and algorithmic roles
- Trust calibration between human operators and AI systems
- Structured feedback mechanisms that evolve with system performance and context.

This theoretical repositioning extends governance literature by recognising that ethical AI is not solely a technical problem but also a socio-political one, shaped by institutional capacity, cultural norms and regulatory frameworks. Second, it operationalises domain-contingent ethics through comparative analysis of three high-stakes sectors: healthcare, finance and education. While fairness, transparency, privacy and human agency are universally acknowledged as ethical pillars, their meaning, measurement and implementation vary significantly across contexts. In healthcare, fairness is tied to equitable clinical outcomes; in finance, to non-discriminatory lending and market stability; in education, to equal access to resources and unbiased assessments. By mapping these variations, the chapter moves ethical theory beyond abstract universals, showing how sector-specific values, risks and governance environments influence the practical application of ethical principles. Third, it bridges the gap between high-level ethical frameworks and operational governance. Many AI ethics guidelines such as the OECD AI Principles, UNESCO's Recommendation on the Ethics of AI and the EU AI Act, articulate normative aspirations but stop short of specifying implementable processes. This chapter contributes by showing how such frameworks can be embedded into institutional practice through tools like:

- AI impact assessments
- Bias audits and fairness dashboards
- Ethics committees and stakeholder consultation processes
- Hybrid oversight mechanisms integrating technical and human review

By linking abstract principles to concrete governance practices, the work strengthens the pathway from normative ideals to enforceable and adaptable policies.

Fourth, it identifies cross-sector structural commonalities that point toward the development of a transferable "ethics infrastructure" for AI governance. Across all three case study domains, common design elements emerged:

1. Role clarity between human decision-makers and AI systems
2. Stakeholder inclusion in governance and feedback loops
3. Iterative oversight that adapts to evolving system capabilities and risks

These recurring patterns provide a theoretical foundation for designing governance architectures that are flexible enough to accommodate sector-specific nuances while retaining a consistent ethical core.

By integrating conceptual reframing, sectoral comparison, governance operationalisation and cross-domain synthesis, this chapter advances a practice-oriented theoretical model of ethical AI adoption. The contribution lies not only in highlighting sector-specific differences but also in revealing shared governance principles that can be generalised to emerging and under-regulated AI applications. In doing so, it moves the field closer to developing ethical frameworks that are simultaneously principled, practical and scalable.

## 9. PRACTICAL IMPLICATIONS

The findings of this chapter offer actionable guidance for practitioners, regulators and policymakers aiming to embed ethical AI into real-world operations in healthcare, finance and education. While the principles of fairness, transparency, privacy and human agency cut across all domains, their operationalisation requires sector-specific strategies and sustained governance.

In healthcare, integrating AI into clinical workflows demands a design philosophy that reinforces rather than erodes clinician authority. This begins with clearly defined decision pathways in which AI-generated outputs are accompanied by their rationale, uncertainty estimates and known limitations enabling clinicians to weigh algorithmic input against contextual factors and patient histories. Governance frameworks should require demographic performance disaggregation to identify disparities, proactive correction of dataset and proxy-variable biases and systematic logging of instances where clinicians override AI recommendations. Analysing these overrides can inform iterative model improvements and highlight contextual gaps. Furthermore, regulatory approval processes should mandate multi-population validation, ensuring that models proven in controlled trials also perform reliably in diverse, real-world settings.

In finance, the central challenge is preventing AI from embedding or amplifying discriminatory patterns while still delivering efficiency and predictive accuracy. Ethical practice starts with fairness constraints integrated at the design stage, ensuring that credit scoring, fraud detection and risk assessment models do not disadvantage specific demographic or socio-economic groups. Methods such as reject inference can improve inclusivity by learning from data on applicants previously declined, thereby reducing representation gaps. Transparency is equally critical: AI systems should generate explanations tailored to both internal stakeholders and external customers including counterfactual explanations that clarify the actionable steps applicants can take to improve outcomes. For high-impact financial decisions, human oversight remains indispensable to catch algorithmic blind spots while independent bias audits for high-risk models strengthen accountability and sustain public trust.

In education, AI must be positioned as a partner in pedagogy rather than a substitute for educators. Ethical integration starts with embedding AI literacy into digital skills curricula, enabling students and teachers to critically evaluate algorithmic outputs. Regular equity audits should assess disparities in access, participation, and achievement ensuring AI tools do not widen existing educational gaps. Transparency is vital in AI-assisted grading and content generation; stakeholders should be informed whenever AI has influenced an evaluation or produced instructional material. Instead of prohibiting AI use outright, educators can adopt constructive integration strategies, such as designing assignments that require students to analyse, critique or improve AI-generated work, thereby fostering critical thinking, creativity and meta-cognitive skills.

Across all three sectors, several cross-cutting imperatives emerge. Multidisciplinary oversight committees comprising technical specialists, domain practitioners, ethicists, legal experts and community representatives can guide AI deployment with both technical rigour and societal sensitivity. Continuous feedback loops, informed by user experiences and performance metrics, ensure that AI systems evolve alongside institutional needs and regulatory landscapes. Clear documentation of human–AI workflows promote role clarity while structured stakeholder engagement creates formal channels for refining governance in light of lived realities. Ultimately, ethical AI implementation is not a one-time compliance milestone but an ongoing process of adaptation, dialogue, and vigilance. Institutions must remain technologically agile while being firmly anchored in human values, recognising that the long-term legitimacy of AI depends as much on governance and trust as on computational accuracy.

## 10. CONCLUSION

This chapter has explored the ethical dimensions of human–AI collaboration through the lens of the human-in-context paradigm arguing that responsible AI requires more than technical optimisation or token oversight. By embedding human judgement, accountability and stakeholder trust within socio-technical systems, institutions can ensure that AI functions as an enabler rather than a destabiliser of ethical decision-making.

Through three detailed case studies the Obermeyer et al. (2019) healthcare risk prediction model, the Apple Card gender bias controversy in finance and the Ofqual grading algorithm in education the chapter demonstrated how universal ethical pillars of fairness, transparency, privacy and safety manifest differently across domains. Each case revealed distinct risks: unequal health outcomes in clinical AI, discriminatory lending patterns in financial AI and socio-economic bias in algorithmic grading. Yet they also showcased sector-specific solutions, from clinical validation dashboards and demographic bias monitoring in healthcare, to fairness-constrained credit scoring and independent audits in finance, to AI-literate pedagogical practices and equity audits in education.

The cross-sector comparative analysis revealed structural commonalities such as the need for clearly defined human and algorithmic roles, continuous feedback loops, stakeholder inclusion and tailored transparency that form the building blocks of a transferable ethics infrastructure. By situating these insights within global governance frameworks like the OECD AI Principles, UNESCO's Ethics of AI Recommendation, the GDPR and the EU AI Act, the chapter bridged the gap between abstract ethical ideals and operational governance mechanisms.

From a practical standpoint, the recommendations outlined here emphasise institutionalising ethics through multidisciplinary oversight committees, AI impact assessments, bias audits and participatory co-design processes. These governance tools ensure that principles are not static declarations but living practices, adaptable to the evolving capabilities and risks of AI.

Ultimately, the central insight is that ethical human–AI collaboration is a dynamic, ongoing process one that must evolve alongside the technologies it seeks to govern. As AI systems become more autonomous, adaptive and embedded across sectors, institutions must remain both technologically agile and deeply anchored in human values. This requires proactive engagement with domain experts, policymakers and affected communities from the earliest stages of AI design to deployment and beyond.

If the lessons from these three sectors are heeded, AI can be integrated in ways that not only uphold fairness, transparency, privacy and safety but also strengthen public trust, reinforce human agency and contribute to more equitable outcomes across society. The path forward is clear: build governance that adapts as quickly as AI evolves and keep human dignity at the core of every algorithmic decision.

## 11. LIMITATIONS AND FUTURE SCOPE

While this chapter provides a multi-sectoral exploration of ethical human–AI collaboration, several limitations must be acknowledged. First, the analysis focuses on three specific high-stakes sectors healthcare, finance and education which, while diverse, cannot capture the full spectrum of AI applications and governance challenges across domains such as law enforcement, environmental management or creative industries. The case studies, though representative and evidence-based are drawn from well-documented public controversies; less visible or proprietary AI deployments may present different risks and governance dynamics that could alter the comparative picture. Second, the chapter relies on qualitative analysis of governance frameworks, stakeholder perspectives and ethical principles. While this approach allows for rich contextual interpretation, it does not incorporate quantitative performance metrics or longitudinal data to measure the real-world effectiveness of proposed governance interventions. The absence of such empirical validation means that some recommendations remain normative and conceptual requiring further testing in operational settings. Third, the global governance references including the OECD AI Principles, UNESCO's Ethics of AI Recommendation, GDPR and the EU AI Act are regionally biased toward Western and European contexts. Cultural, legal and institutional variations in other regions may necessitate different operationalisations of fairness, transparency and human oversight. Finally, the human-in-context paradigm, while offering a valuable conceptual lens, has yet to be formalised into a fully operational toolkit that practitioners can systematically apply. The model's adaptability across sectors is promising, but its scalability in highly decentralised or resource-constrained environments remains to be tested.

Future research should expand this framework to include additional sectors, particularly those involving emerging or contentious AI applications such as autonomous weapons systems, generative AI in creative production, climate modelling and predictive policing. Incorporating quantitative bias and fairness metrics alongside qualitative governance analysis could enable more rigorous evaluation of interventions. There is also scope for developing a practical, open-access "human-in-context governance toolkit" that includes sector-specific checklists, bias testing protocols and participatory design templates. Such a toolkit could bridge the gap between abstract principles and day-to-day operational decision-making. Furthermore, cross-cultural comparative studies are needed to localise governance models in non-Western contexts, ensuring that ethical AI frameworks are sensitive to varying socio-political realities. Collaborative research between academic institutions, regulators, industry practitioners and civil society can generate the kind of interdisciplinary knowledge required to address the evolving complexity of AI ethics. As AI technologies become increasingly autonomous, adaptive and embedded, future work must focus on creating governance mechanisms that are not only robust and enforceable but also flexible enough to evolve alongside technological innovation. In doing so, the vision outlined in this chapter of AI systems that are both effective and deeply aligned with human values can be more fully realised across sectors and societies.

## REFERENCES


Aayog, N. I. T. I. (2018). *National strategy for artificial intelligence: #AIForAll*. Government of India. Retrieved from https://www.niti.gov.in/national-strategy-artificial-intelligence

African Union. (2019). *Artificial intelligence for Africa: An opportunity for growth, development, and democratisation*. African Union Development Agency (AUDA-NEPAD). Retrieved from https://www.nepad.org

AI, N. (2023). Artificial intelligence risk management framework (AI RMF 1.0). *URL:* https://nvlpubs. nist. gov/nistpubs/ai/nist. *ai*, 100 1.

Amodei, D., Olah, C., Steinhardt, J., Christiano, P., Schulman, J., & Mané, D. (2016). Concrete problems in AI safety. *arXiv preprint arXiv:1606.06565*.

Barocas, S., Hardt, M., & Narayanan, A. (2020). Fairness and machine learning. *Recommender systems handbook, 1*, 453-459.

Chinta, S. V., Wang, Z., Zhang, X., Viet, T. D., Kashif, A., Smith, M. A., & Zhang, W. (2024). Ai-driven healthcare: A survey on ensuring fairness and mitigating bias. *arXiv preprint arXiv:2407.19655*.

Chouldechova, A. (2017). Fair prediction with disparate impact: A study of bias in recidivism prediction instruments. *Big Data*, *5*(2), 153–163. DOI: 10.1089/big.2016.0047 PMID: 28632438

Doshi-Velez, F., & Kim, B. (2017). Towards a rigorous science of interpretable machine learning. *arXiv preprint arXiv:1702.08608*.

Dwork, C., & Roth, A. (2014). The algorithmic foundations of differential privacy. *Foundations and trends® in theoretical computer science, 9*(3–4), 211-407.

Gartner, Inc. (2023). More than 80% of enterprises will have used generative AI APIs or deployed generative AI-enabled applications by 2026. Retrieved from: https://www.gartner.com/en/newsroom/press-releases/2023-10-11-gartner-says-more-than-80-percent-of-enterprises-will-have-used-generative-ai-apis-or-deployed-generative-ai-enabled-applications-by-2026?utm_source=chatgpt.com

Goddard, K., Roudsari, A., & Wyatt, J. C. (2012). Automation bias: A systematic review of frequency, effect mediators, and mitigators. *Journal of the American Medical Informatics Association : JAMIA*, *19*(1), 121–127. DOI: 10.1136/amiajnl-2011-000089 PMID: 21685142

Green, B., & Chen, Y. (2019). The principles and limits of algorithm-in-the-loop decision making. *Proceedings of the ACM on human-computer interaction, 3*(CSCW), 1-24. DOI: 10.1145/3359152

Guo, F., Li, T., & Cunningham, C. J. (2025, May). One year in the classroom with ChatGPT: Empirical insights and transformative impacts. []). Frontiers Media SA.]. *Frontiers in Education, 10*, 1574477. DOI: 10.3389/feduc.2025.1574477

Holmes, W., Bialik, M., & Fadel, C. (2019). *Artificial Intelligence in Education: Promises and Implications for Teaching and Learning*. Center for Curriculum Redesign.

ISO/IEC. (2023). *What Is ISO/IEC 42001 Artificial Intelligence Management Systems Accreditation? Derived from*: https://anab.ansi.org/accreditation/iso-iec-42001-artificial-intelligence-management-systems/?gad_source=1&gad_campaignid=22984523464&gbraid=0AAAAADvqK-UdvgmTQ_ZwQ5ecti0sfslqO&gclid=CjwKCAjw6P3GBhBVEiwAJPjmLvjZ2kqWRyUYoyv2UVYNeCIhYR4SDOTzbr0CkRBXrcpajgFGOoMi7hoCbPoQAvD_BwE

Jauhiainen, J. S., & Garagorry Guerra, A. (2025). Generative AI in education: ChatGPT-4 in evaluating students' written responses. *Innovations in Education and Teaching International, 62*(4), 1377–1394. DOI: 10.1080/14703297.2024.2422337

Jobin, A., Ienca, M., & Vayena, E. (2019). The global landscape of AI ethics guidelines. *Nature Machine Intelligence, 1*(9), 389–399. DOI: 10.1038/s42256-019-0088-2

Kapoor, A., & Sindwani, R. (2025). Artificial Intelligence as a Catalyst for Advancing Contactless Payments: Unveiling Key Adoption Drivers. In *AI Marketing and Ethical Considerations in Consumer Engagement* (pp. 321–338). IGI Global Scientific Publishing. DOI: 10.4018/979-8-3373-3476-9.ch016

Kapoor, A., Sindwani, R., & Goel, M. (2023a). Evaluating mobile wallet acceptance factors using best worst method. *International Journal of Process Management and Benchmarking, 13*(4), 449–469. DOI: 10.1504/IJPMB.2023.129822

Kapoor, A., Sindwani, R., & Goel, M. (2023b). Exploring quick commerce service experience: A moderated mediated investigation using SEM and fsQCA. *Total Quality Management & Business Excellence, 34*(13-14), 1896–1919. DOI: 10.1080/14783363.2023.2213653

Kapoor, A., Sindwani, R., & Goel, M. (2024a). Mobile wallets: Theoretical and empirical analysis. *Global Business Review, 25*(2, suppl), S211–S228. DOI: 10.1177/0972150920961254

Kapoor, A., Sindwani, R., & Goel, M. (2024b). Exploring the symmetric and asymmetric effects of mobile wallet service experience on brand loyalty in India: Evidence from SEM and fsQCA. *Journal of Financial Services Marketing*, *29*(4), 1524–1541. DOI: 10.1057/s41264-024-00280-w

Kapoor, A., Sindwani, R., & Goel, M. (2025). A novel framework for understanding the interplay between the mobile wallet service quality dimensions and loyalty intention. *The TQM Journal*, *37*(6), 1716–1741. DOI: 10.1108/TQM-12-2023-0422

Kapoor, A., Sindwani, R., Goel, M., & Shankar, A. (2022). Mobile wallet adoption intention amid COVID-19 pandemic outbreak: A novel conceptual framework. *Computers & Industrial Engineering*, *172*, 108646. DOI: 10.1016/j.cie.2022.108646 PMID: 36118531

Kasneci, E., Seßler, K., Küchemann, S., Bannert, M., Dementieva, D., Fischer, F., Gasser, U., Groh, G., Günnemann, S., Hüllermeier, E., Krusche, S., Kutyniok, G., Michaeli, T., Nerdel, C., Pfeffer, J., Poquet, O., Sailer, M., Schmidt, A., Seidel, T., & Kasneci, G. (2023). ChatGPT for good? On opportunities and challenges of large language models for education. *Learning and Individual Differences*, *103*, 102274. DOI: 10.1016/j.lindif.2023.102274

Kleinberg, J., Mullainathan, S., & Raghavan, M. (2016). Inherent trade-offs in the fair determination of risk scores. *arXiv preprint arXiv:1609.05807*.

Kozodoi, N., Jacob, J., & Lessmann, S. (2022). Fairness in credit scoring: Assessment, implementation and profit implications. *European Journal of Operational Research*, *297*(3), 1083–1094. DOI: 10.1016/j.ejor.2021.06.023

Larsen, B. C. (2021, July). A framework for understanding AI-induced field change: How AI technologies are legitimized and institutionalized. In *Proceedings of the 2021 AAAI/ACM Conference on AI, Ethics, and Society* (pp. 683-694). DOI: 10.1145/3461702.3462591

Luckin, R., Holmes, W., Griffiths, M., & Forcier, L. B. (2016). *Intelligence Unleashed: An argument for AI in Education*. Pearson. Retrieved from UCL Discovery UCL Discovery

Maslej, N., Fattorini, L., Perrault, R., Parli, V., Reuel, A., Brynjolfsson, E., . . . Clark, J. (2024). Artificial intelligence index report 2024. *arXiv preprint arXiv:2405.19522*.

Ministry of Education, Government of India. (2020). *National education policy 2020*. Government of India. Retrieved from https://www.education.gov.in/nep/

Ministry of Science and Technology of the People's Republic of China. (2021). *Ethical norms for the new generation artificial intelligence*. Government of China. Retrieved from https://www.most.gov.cn

Obermeyer, Z., Powers, B., Vogeli, C., & Mullainathan, S. (2019). Dissecting racial bias in an algorithm used to manage the health of populations. *Science*, *366*(6464), 447–453. DOI: 10.1126/science.aax2342 PMID: 31649194

OECD AI Principles. (2019). AI Principles. Retrieved from: https://www.oecd.org/en/topics/sub-issues/ai-principles.html?utm_source=chatgpt.com

Ofqual. (2020, August 13). *A-level results: Almost 40% of teacher assessments downgraded*. The Guardian. https://www.theguardian.com/education/2020/aug/13/almost-40-of-english-students-have-a-level-results-downgraded

Parasuraman, R., & Riley, V. (1997). Humans and automation: Use, misuse, disuse, abuse. *Human Factors*, *39*(2), 230–253. DOI: 10.1518/001872097778543886

Reserve Bank of India. (2022). *Guidelines on digital lending*. Government of India. Retrieved from https://www.rbi.org.in

Royal College of Surgeons of England Bulletin. (2021), "Bias in medical artificial intelligence", derived from: https://publishing.rcseng.ac.uk/doi/full/10.1308/rcsbull.2021.111

Selwyn, N. (2019). What's the problem with learning analytics? *Journal of Learning Analytics*, *6*(3), 11–19. DOI: 10.18608/jla.2019.63.3

Shneiderman, B. (2020). Human-centered artificial intelligence: Reliable, safe & trustworthy. *ArXiv*. https://arxiv.org/abs/2002.04087

Sjoding, M. W., Dickson, R. P., Iwashyna, T. J., Gay, S. E., & Valley, T. S. (2020). Racial bias in pulse oximetry measurement. *The New England Journal of Medicine*, *383*(25), 2477–2478. DOI: 10.1056/NEJMc2029240 PMID: 33326721

Tang, L., Li, J., & Fantus, S. (2023). Medical artificial intelligence ethics: A systematic review of empirical studies. *Digital Health*, *9*, 20552076231186064. DOI: 10.1177/20552076231186064 PMID: 37434728

UNESCO's Recommendation on the Ethics of AI. (2021). Recommendation on the Ethics of Artificial Intelligence. Retrieved from: https://www.unesco.org/en/legal-affairs/recommendation-ethics-artificial-intelligence?utm_source=chatgpt.com

Vesna, L., Sawale, P. S., Kaul, P., Pal, S., & Murthy, B. S. N. V. R. (2025). Digital divide in AI-powered education: Challenges and solutions for equitable learning. *Journal of Information Systems Engineering & Management*, *10*(21), 1–9.

Wachter, S., Mittelstadt, B., & Russell, C. (2017). Counterfactual explanations without opening the black box: Automated decisions and the GDPR. *SSRN*, *31*, 841. DOI: 10.2139/ssrn.3063289

Wang, J., & Fan, W. (2025). The effect of ChatGPT on students' learning performance, learning perception, and higher-order thinking: Insights from a meta-analysis. *Humanities & Social Sciences Communications*, *12*(1), 1–21. DOI: 10.1057/s41599-025-04787-y

Williamson, B., & Eynon, R. (2020). Historical threads, missing links, and future directions in AI in education. *Learning, Media and Technology*, *45*(3), 223–235. DOI: 10.1080/17439884.2020.1798995

# Chapter 4
# Empowering Ethical AI Adoption:
## Bridging the Gaps With Ethics4AI Platform

**Ghinwa Jalloul**
https://orcid.org/0009-0001-0300-7529
*Sydney Institute of Technology and Innovation, Australia*

**Youssef M. Jalloul**
*Sydney Institute of Technology and Innovation, Australia*

## ABSTRACT

*Generative artificial intelligence (AI) tools have rapidly entered the operations of small and medium-sized enterprises (SMEs), boosting productivity while presenting significant challenges in regulatory compliance. This chapter examines these challenges and identifies compliance gaps by comparing global and national regulatory frameworks from an SME perspective. It also explores case studies focused on SMEs' ethical use of AI. To address these issues, the chapter proposes an AI Ethics Framework grounded in four core pillars. It further introduces Ethics4AI, a multilingual, AI-powered platform built on a foundation of current principles, regulations, and toolkits designed to operationalize the proposed framework. Ethics4AI serves as an Ethical AI Advisor, bridging the gap between ethical principles and practical compliance, and enabling SMEs to adopt AI ethically and responsibly.*

DOI: 10.4018/979-8-3373-2597-2.ch004

## INTRODUCTION

Recent and rapid advancements in artificial intelligence (AI), particularly generative models, have democratized access to this technology. AI is now more widely available, enabling organizations to integrate it into daily workflows to improve productivity, increase revenue, and foster creativity. However, these benefits are accompanied by significant risks. Arising from both misuse and uninformed use, these risks underscore the urgent need to address AI's ethical implications. As a result, various binding and non-binding values, policies, and regulatory frameworks have been established at global and national levels to guide the responsible development and deployment of AI systems. These are supported by tools that assess ethical compliance in AI implementation and usage across jurisdictions.

While these regulatory initiatives are essential, compliance has proven difficult, particularly for small and medium-sized enterprises (SMEs). This challenge threatens to undermine the very gains achieved through AI adoption. SMEs play a major role in employment creation, economic growth, and innovation, accounting for over 90% of businesses worldwide (WB, 2019). Yet, they often operate with limited financial and human resources. Most SMEs are AI users rather than developers, typically adopting AI through outsourcing. As a result, they often lack dedicated teams to develop AI solutions or manage compliance with evolving regulatory frameworks.

As a result, SMEs heavily rely on opaque third-party solutions. They often lack the expertise to evaluate AI applications, limiting their control over system design, data sources, and model interpretability. Moreover, the overwhelming number of principles, regulations, and tools, combined with regulatory complexity, increases implementation challenges. This creates additional barriers for SMEs in adopting ethical AI practices, exposing them to the risk of misusing AI tools, eroding stakeholder trust, and violating emerging legal standards. These concerns are especially pronounced in the context of generative AI, as such models are increasingly embedded in daily operations across industries including health, agriculture, tourism, education, and finance.

This chapter offers an overview of current AI governance frameworks across continents and nations. It highlights the regulatory approaches of representative countries and evaluates their effectiveness, particularly in relation to SMEs, by drawing on findings from existing case studies. These overviews and case studies reveal the compliance difficulties SMEs encounter in implementing AI ethics. Such challenges arise from limited resources relative to regulatory complexity, the absence of tailored toolkits for general-user SMEs, and the fragmented AI regulatory landscape, which impedes cross-jurisdictional scalability and growth.

To address these challenges, the chapter introduces an Ethical AI Framework designed for SMEs. This framework emphasizes skill development, accessible ethical guidance, sector-wide collaboration, and ongoing monitoring. Additionally, we present Ethics4AI, an agentic AI platform that applies this framework. Ethics4AI bridges the gap between theoretical AI ethics and SMEs' practical needs. It draws on existing regulations, policies, and static toolkits as foundational data, and uses generative AI agents to simplify compliance, making the process more accessible and actionable.

# BACKGROUND

## AI Ethics Frameworks and Governance Models

### Global Frameworks

The Organization for Economic Co-operation and Development (OECD) AI Principles are among the earliest and most influential initiatives for governing AI ethics. First introduced in 2019 and updated in 2024 (OECD, 2019, 2024), these principles promote inclusive growth, human-centric values, transparency, robustness, and accountability. Although non-binding, they have been adopted by over 46 countries and have significantly influenced strategies in the EU, Canada, and Asia-Pacific regions. The OECD has urged nations to embed these principles into national AI policies, fostering international collaboration.

The OECD's AI principles toolkit (OECD, 2025) is regarded as highly impactful. It offers practical resources for businesses and governments, including checklists, evaluation templates, and case studies to support ethical AI implementation. The toolkit provides actionable guidance on assessing fairness, transparency, and robustness, reinforcing the OECD's commitment to human-centered design. Although voluntary, the OECD actively promotes its use through workshops and training programs targeting SMEs and policymakers.

Similarly, the United Nations Educational, Scientific and Cultural Organization (UNESCO) developed pioneering global standards on the ethics of AI (UNESCO, 2021), endorsed by 193 member states. This normative framework prioritizes fairness, accountability, and inclusive development. It advocates for respect for human dignity, fairness, environmental sustainability, and data governance. It also supports accountability mechanisms and the implementation of ethical impact assessments (UNESCO, 2024). UNESCO has introduced tools such as the Readiness Assessment Methodology for ethical AI implementation and ethical impact assessment tools (UNESCOa, 2023; UNESCOb, 2023), aligned with its AI Ethics Recommendation. These toolkits provide evaluation frameworks for countries and enterprises, particularly SMEs, to assess their readiness in AI ethics. They emphasize stakeholder engagement, risk analysis, and transparent governance, helping nations align with UNESCO's ethical guidelines.

The United Nations (UN) Secretary-General's High-Level Advisory Body has emphasized the need for globally coordinated governance. The 2024 report Governing AI for Humanity (UN, 2024; UNDESA, 2024) advocates for human rights-centered global governance. It recommends establishing a UN AI office, a standards exchange, and a global AI fund to support developing nations and SMEs in adopting ethical AI practices. This initiative connects AI ethics to the sustainable development goals and proposes a framework for public-private collaboration involving SMEs. However, many of these proposals are forward-looking and lack concrete operational tools or industry-specific pathways, offering minimal immediate benefit to SMEs, as global funding and standards infrastructure remain under development.

The UN, through the High-Level Advisory Body on AI, coordinates inter-agency efforts under the "AI for Good" campaign, led by the International Telecommunication Union (ITU). ITU initiatives emphasize the ethical deployment of AI in sectors such as healthcare, agriculture, and education, aligning with the sustainable development goals (ITU, 2023). While ITU does not enforce regulations, its frameworks provide technical standards and ethical use cases, particularly suited to resource-constrained environments.

Concurrently, collaborative entities such as the Global Partnership on AI (GPAI) have issued technical and policy directives to promote responsible AI innovation. Although advisory, GPAI's publications advocate shared norms and encourage alignment among participating nations, including Canada, India, and Japan (GPAI, 2023).

## Europe

The European Union (EU, 2025) provides an overview of the EU's approach to AI and key milestones. The EU's AI Act (EU, 2024) is the most comprehensive and binding regulatory framework worldwide. It adopts a risk-based classification system that imposes ethical obligations on AI developers and users. The act aims to ensure safety, privacy, transparency, fairness, and accountability. High-risk systems—such as those used in biometric identification, employment, and education—must meet strict requirements for transparency, record-keeping, human oversight, and conformity assessments. To enhance ethical transparency, general-purpose AI models like GPT-class systems must include clear labeling, summaries of training data, usage instructions, and compliance with copyright law (Morin, 2025).

Although not exclusively directed at SMEs, the Act includes flexibility measures (EUa, 2025) such as reduced fees and scaled-down reporting obligations. The EU also developed the AI Act Compliance Toolkit for SMEs (EU, 2025), which addresses concerns over regulatory complexity. This is supported by EU-funded regulatory sandboxes designed to reduce compliance burdens (EUa 2025; Reuters, 2025). However, early assessments indicate that many SMEs are unprepared (Nizza, 2023). Risk management protocols remain largely theoretical, and few SMEs have the internal governance structures required to meet transparency and audit standards. Implementation gaps have also been identified (Information Systems Audit and Control Association, 2024), including limited human oversight mechanisms and a lack of accessible tools to assess generative AI reliability. Studies (Novelli et al., 2024; Sillberg et al., 2024) observe that the AI Act does not adequately address SME-specific issues such as copyright ambiguity, data provenance, and reliance on model vendors.

Practical support includes an AI Innovation Package and an AI Pact to promote trustworthy AI among startups and SMEs. This includes "GenAI4EU" (EUa, 2024), designed to foster generative AI ecosystems, rolled out by the EU and member states. Despite centralized Commission tools, national interpretations vary, and SMEs often rely on fragmented, limited support from aggregator bodies (Hollanek et al., 2025). These include Accountancy Europe (AEU, 2025), which targets SME accountants; a webinar series, From Regulation to Action, focused on digital SME readiness (EDSA, 2025); and the AI Act Compliance Checker (EU, 2025). The introduction of the AI Act presents new compliance challenges for the AI research and practice community. Despite available support, SMEs continue to face resource limitations and must navigate dual compliance with both GDPR and the AI Act. Burdens stem from ambiguous definitions of high-risk systems, explainability requirements, and a lack of SME-specific support tools (Joswig & Kurz, 2025). Additional gaps include limited human oversight mechanisms, poor tool accessibility, and insufficient training and sandbox expansion. While the EU AI Act (EU, 2024) partially addresses extraterritorial application, it raises concerns about enforcement and practical compliance for non-EU SMEs serving European markets. Complementing these legal principles, *Private AI* represents a technical embodiment of privacy ethics through homomorphic encryption and secure computation. These encryption-driven methods enable learning from encrypted data, ensuring confidentiality and compliance with GDPR and EU AI Act standards (Vianny et. al., 2025).

The UK's AI framework is innovation-friendly and responsive to SME needs. It is grounded in flexible principles and supported by practical tools. UKGov (2023) outlined five core principles: safety, transparency, fairness, accountability, and contestability/redress. Rather than establishing a dedicated AI regulator, it empowers existing regulators (TechUK, 2023; UKGov, 2025). It also recommends red teaming, human oversight, and performance monitoring but does not propose new legislation. This voluntary approach is intended to preserve innovation flexibility.

SME-focused assurance tools include the AI safety platform, launched in November 2024, which provides self-assessment tools, AI impact guidance, and bias detection resources aligned with the White Paper's principles. Additionally, the RAISE project, developed specifically for SMEs, offers workshops and whitepapers on responsible generative AI governance. It features practical checklists, ethical use cases, and board-level awareness training (TechUK, 2025). Nonetheless, effective regulation and business compliance remain challenging due to the non-binding nature of the framework, limited SME resources, implementation gaps, and fragmented guidance (Brown et al., 2025). Moreover, scaling in the post-Brexit context requires careful alignment between UK and EU norms for UK-based SMEs targeting global markets.

### North America

The U.S. adopts a sectoral, non-binding approach to AI regulation. Primary federal initiatives include the Blueprint for an AI Bill of Rights (OSTP, 2022) and the National Institute of Standards and Technology (NIST) AI Risk Management Framework (NIST, 2023). These documents outline ethical principles and practical guidelines but are not legally enforceable. Some states have enacted binding regulations, such as the Colorado AI Act (Colorado, 2024) and the Utah AI Policy Act (Skadden, 2024). In 2025, the U.S. experienced a surge of over 1,000 state-level AI bills, creating a fragmented legal landscape.

Compliance for U.S. SMEs under this evolving framework presents a mixed picture. While ethical oversight is gaining interest, the fragmented and voluntary environment requires small businesses to navigate a complex and costly path. Federal frameworks remain non-mandatory and face criticism for inadequately addressing risks posed by generative AI systems like ChatGPT and deepfake technologies (KPMG, 2023). Many SMEs depend on generative AI for core operations but encounter inconsistent compliance demands. Limited private sector efforts to ease compliance further complicate the landscape (Stappers, 2025). Supporting SMEs will require harmonized federal standards, user-friendly assessment tools, and clear communication focused on ethical generative AI usage.

Canada has adopted a hybrid approach, integrating ethical principles into legislative processes. Its proposed AI and Data Act, part of the Digital Charter Implementation Act, seeks to regulate high-impact AI systems and protect individual rights (CanadaGov, 2025). AI and Data Act requires impact assessments and imposes penalties for non-compliance. Its risk-based model and extraterritorial enforcement reflect Canada's commitment to aligning domestic regulation with emerging global norms, including the EU AI Act, the NIST AI Risk Management Framework, the OECD AI Principles, and the UNESCO Recommendation on the Ethics of AI. Canada's participation in the GPAI and OECD forums reinforces this international alignment.

## Asia-Pacific

Australia demonstrates leadership in ethical AI within Oceania and often coordinates through the OECD, GPAI, and regional economic forums. Australia's AI Ethics Framework outlines eight core principles: reliability, transparency, contestability, accountability, privacy, safety, and human-centered values (DISR, 2019). These principles serve as a foundation but lack legal enforceability. Although non-binding, the framework has influenced policy development and sectoral adoption, particularly in public services. The Australian Human Rights Commission (2021) has advocated for stronger legal enforcement, especially in high-risk sectors such as employment and immigration. AI applications, including those powered by generative AI, must comply with the Privacy Act 1988, with non-compliance subject to penalties from the Office of the Australian Information Commissioner (2023). The Department of Industry, Science and Resources recently published Guidance for AI Adoption (DISR,2025) that sets out 6 essential practices for responsible AI governance and adoption.

Overall, Australian SMEs operate in a partially supportive but legally underdeveloped AI governance environment. Ethical performance depends on voluntary adherence. While sector-specific rules address some governance gaps, they do not comprehensively regulate generative AI. Support mechanisms, such as grants and safety guides (ESC, 2023), are available but suffer from limited visibility and reach (HTI, 2025). As a result, SMEs remain exposed to ethical, privacy, internet protocol, and reliability risks, particularly amid the rapid deployment of generative AI.

Australia has recently introduced Sovereign Australia AI, a fully sovereign and ethics-driven alternative to foreign large language models such as ChatGPT. The initiative, supported by the Australian Computer Society (ACS), combines the country's most powerful AI supercomputer with a commitment to transparency, domestic privacy, ethical data use and security data standards. The project's LLMs, Ginan and Australis, are locally trained and designed to capture Australia's culture, language, and principles. Central to this initiative is the commitment to compensate Australian copyright holders for the right to content used in training data, ensuring that content is ethically sourced and reflective of national values. This initiative demonstrates how ethical governance and data sovereignty can be integrated into AI infrastructure to operationalise national values and promote trust, accountability, and fairness in AI adoption (Braue, 2025).

## South East Asia

The Association of Southeast Asian Nations (ASEAN) has adopted a collaborative and forward-looking approach to AI governance, emphasizing regional cohesion, ethical safeguards, and preparedness for emerging technologies. Central to this strategy is the ASEAN Digital Masterplan 2025, which establishes a foundational vision for digital transformation and underscores the development of responsible and inclusive AI systems across member states (ASEAN, 2021). To implement this vision, ASEAN released the Guide on AI Governance and Ethics, a non-binding document promoting trust, accountability, transparency, and inclusivity in AI applications across sectors (ASEANS, 2024). In January 2025, ASEAN adopted the Expanded Guide on AI Governance and Ethics: Generative AI, which addresses the unique ethical risks posed by large language models and generative technologies. It recommends practical implementation steps and introduces AI-specific governance roles within member states (ASEANDMN 2025; ASEANS, 2025). The guide supports cross-border regulatory harmonization and provides a shared ethical foundation without constraining local innovation. Although voluntary, these frameworks reflect ASEAN's collective alignment with global norms and commitment to international best practices. Complementing these initiatives, the ASEAN Responsible AI Roadmap 2025–2030 outlines a phased plan for embedding ethical AI across critical domains, with a strong focus on SME capacity building and enablement (AdastraIP, 2025).

Singapore, a regional leader, has taken significant steps to embed ethical values in AI policy (GLI, 2025; NRF, 2025). Singapore's Model AI Governance Framework is among the most detailed voluntary frameworks globally (PDPC, 2020, 2022, 2024). It offers actionable guidance on risk assessments, explainability, and stakeholder engagement. The updated 2024 version addresses risks specific to generative models, including misinformation, privacy, and environmental impact, across nine dimensions such as accountability, data standards, incident management, and safety-by-design (Kin et al., 2025).

Singapore's strength lies in pragmatic collaboration between regulators and industry. The AI Verify Foundation provides a toolkit and practical guidelines for ethical AI implementation (AIVF, 2025). SMEs benefit from step-by-step tools that support compliance with ethical standards, risk assessment, and transparency. This reduces uncertainty in adopting ethical AI. The InfoComm Media Development Authority released a Starter Kit for Safety Testing of LLM-Based Applications (IMDA, 2025). Singapore's approach combines soft-law flexibility with practical oversight to empower SMEs to adopt generative AI responsibly. Sandbox and study programs help reduce ethical and technical barriers, while the Model AI frameworks offer operational guidance. However, the absence of legally binding requirements and industry-wide enforcement may limit large-scale adoption. SMEs require clearer mechanisms for technical audits, copyright protection, and multi-jurisdictional compliance to fully and ethically leverage Singapore's AI ecosystem.

In September 2024, Malaysia introduced the voluntary National Guidelines on AI Governance & Ethics (2024) to support its 2021–2025 AI Roadmap (Securiti,

2025). The guidelines outline seven core principles: fairness, reliability, privacy, inclusiveness, transparency, accountability, and human well-being. These are intended to guide both AI developers and users. Central coordination is led by the National AI Office (NAIO). Although non-binding, the principles impose obligations on all stakeholders, including developers, policymakers, and end-users, and include practical exercises to assist SMEs in responsible AI deployment. NAIO has also organized stakeholder consultations and roundtables to translate the guidelines into operational tools tailored to local ethical contexts (NAIO, 2025).

Malaysia's AI ecosystem incorporates cultural and local adaptation. Guidance is contextualized to the country's socio-cultural and linguistic diversity, addressing local biases and national priorities (Weber, 2024). Government initiatives and partnerships with tech companies aim to improve cloud infrastructure and digital capability. Awareness-building activities, including national roundtables and ethics forums, support capacity development for SMEs and practitioners (NourHashim, 2024). Additionally, the Malaysia Digital Economy Corporation (2023) released the responsible AI toolkit, developed specifically for SMEs and startups. It provides simplified ethical decision-making tools, practical guidelines, and scenario-based modules.

Malaysia's ethical AI landscape reflects comprehensive, culturally contextual, voluntary principles. Nonetheless, it faces challenges in cross-agency coordination, enforcement readiness, and SME AI literacy (KRI, 2024). Barriers include the non-binding nature of the guidelines, lack of enforceable tools, limited regulatory integration, and insufficient awareness of generative AI guidance. Ethical concerns around bias, privacy, and transparency are particularly noted in content-generation use cases (Weber, 2024). Experts recommend avoiding premature regulation and instead building collaborative, multi-stakeholder frameworks to foster societal trust (Said & Nabila, 2024).

## East Asia

East Asia presents a diverse landscape in AI governance and ethics, shaped by distinct political systems, economic priorities, and cultural traditions. This section considers two representative countries: Japan and China. On May 28, 2025 Japan enacted it's AI law centered on promoting AI, titled the "Act on Promotion of Research and Development and Utilization of Artificial Intelligence-Related Technologies (White & Case, 2025). Japan adopts a soft-law, agile governance model rooted in stakeholder consensus and iterative updates to keep pace with technological change. Its AI Guidelines for Business (Ministry of Economy, Trade and Industry [METI], 2024) and a draft Responsible AI Law emphasize human-centricity and autonomy but rely on voluntary implementation. Japan's alignment with ISO/IEC 42001 promotes international interoperability. The guidelines consolidate prior directives into a unified framework (Citadel, 2024), emphasizing risk-based analysis, human-centric principles, and multi-stakeholder co-creation with ethical safeguards (CSIS, 2025). They reference "Generative AI tests" and maturity evaluations and recommend that businesses conduct AI proficiency assessments, establish governance systems, and adopt contract guidelines. METI also released a Contract Checklist for AI Use, providing SMEs with tools to manage vendor risks, intellectual property rights, and legal compliance (Babl, 2025). Additionally, Japan has started integrating generative AI safeguards into public sector procurement, requiring agencies to conduct risk assessments, appoint Chief AI Officers, and follow national standards (Digital.go.jp, 2025). This public-sector leadership may gradually extend to SMEs, raising standards and accelerating the adoption of best practices (CSIS, 2025).

Notably, Japan's approach applies to all actors, AI developers, providers, and users, including SMEs. This reflects a shared responsibility model even within non-legally binding frameworks. However, it lacks compulsory enforcement, relying instead on industry-led audits and ethical training. Critics argue that this voluntary model may reduce compliance consistency and result in implementation gaps, particularly for SMEs lacking internal governance or legal support (CLC, 2023; Ema et al., 2023). There is also a risk of unintentional IP infringement without clear guidance.

Following the enactment of the Act on Promotion of Research and Development and Utilisation of AI-related Technology (White & Case, 2025), which took full effect on 1 September (Bird & Bird, 2025), Japan has reinforced its principle-based approach to ethical AI. The newly established AI Strategic Headquarters, chaired by the Prime Minister, coordinates national AI policy and is drafting an AI Basic Plan that balances innovation with risk management. The plan aims to position Japan as the world's most AI-friendly nation by promoting trust, transparency, and regulatory agility. Complementing this, an expert panel is preparing Guidelines for Appropriate AI Use, expected by end 2025, to align domestic practice with international ethical standards and foster responsible adoption across public and private sectors. Together, these steps signal Japan's transition from voluntary guidance toward a more structured yet innovation-positive model of ethical AI governance.

China's Interim Measures for the Management of Generative AI Services (CAC, 2023), enacted in August 2023, represent the country's first binding regulation specifically targeting generative AI (Liu & Edmondson, 2023; Wu, 2023). These measures constitute the most direct regulatory response to the technology. They mandate content regulation and watermark labeling (Yang & Li, 2025), along with security assessments, algorithm filing, and compliance reviews for platforms with public influence. The regulations require alignment with socialist core values, prohibit sensitive political content, and mandate the exclusion of extreme material (WBG, 2024). Authorities are granted inspection and enforcement powers, and noncompliant providers may face penalties. The framework offers clarity and national consistency through coordination across seven ministries.

Experts observe that while China's regulatory efforts are technologically ambitious, its SMEs face considerable challenges in AI adoption. High costs and limited access to skilled personnel hinder generative AI uptake despite government subsidies (Body, 2023). Legal ambiguity persists, as SMEs remain uncertain about responsibility allocation within the AI value chain (Pi, 2024). Ethical governance gaps are evident, particularly in healthcare-related generative AI deployments (Bian et al., 2025). The state's strong political oversight may further deter SME innovation, especially in cross-border or ideologically sensitive projects (Lin, 2024). Compared to the U.S. market-driven model, China's regulatory approach is seen by some as stifling nuanced innovation among SMEs (Guest & Wei, 2025).

## South Asia

Several South Asian countries are beginning to develop national strategies for AI governance, though progress remains uneven and fragmented. In Bangladesh, the National Strategy for AI 2020–2025 identifies key priorities but lacks detailed implementation guidelines (GovBang, 2020). Pakistan released a draft national AI policy in 2023, but it remains unfinalized (MITTP, 2023). Nepal and Sri Lanka currently lack formal AI governance strategies.

Focusing on India, the country currently lacks a comprehensive AI law, but its governance ecosystem is evolving rapidly. In early 2025, a high-level advisory group released a report on AI governance, recommending transparency, accountability, fairness, safety, and lifecycle-based assessments for developers and deployers, including SMEs (ESYA, 2025; IndiaAI, 2025). These principles reflect global standards and highlight India's alignment with international ethical frameworks.

Several initiatives have been launched to support this, including AI Compute, FutureSkills, and the Safe & Trusted AI pillar. The latter includes the establishment of the AI Safety Institute, which plays a central role in generative AI governance for SMEs. Initiatives such as Bharat-GPT (bharatgpt.ai) and Sarvam AI (sarvam.ai) further India's drive for domestic, linguistically inclusive foundation models. These efforts help SMEs reduce dependence on opaque foreign models and mitigate IP-related risks. Despite this progress, implementation gaps remain (ETb, 2025). Many SMEs lack awareness of ethics and trust frameworks. Barriers include high infrastructure costs, skill shortages, and unclear deployment pathways (ETa, 2025). Accountability concerns persist, as no binding requirements exist for model validation, third-party audits, or copyright protections for generative AI (DigitalBriefing, 2024). The India AI Mission acknowledges the need for regional research hubs and

SME capacity building; however, compliance tools for SMEs are still in development (MeitY, 2024).

## Middle East and North Africa and Gulf Cooperation Council

The Middle East and North Africa (MENA) region's approach to AI governance is evolving but remains fragmented, with significant variation in regulatory maturity and ethical implementation. Countries such as Bahrain, Egypt, Jordan, Morocco, Qatar, and Tunisia have developed or are drafting national AI strategies referencing ethical principles. These are often aligned with UNESCO, OECD, or EU frameworks. However, most lack binding AI-specific regulations and institutional mechanisms for ethical oversight, particularly for SMEs.

Within the GCC, Saudi Arabia and the United Arab Emirates (UAE) have enacted national AI strategies that include ethical guidelines. Saudi Arabia's strategy (SDAIA, 2023; SDAIA, 2024) emphasizes AI for sustainable development and includes provisions for fairness and transparency, though binding mechanisms are still limited (Abu Farwah, 2024). The UAE's AI Ethics Guidelines (MSAI, 2022; MSAI 2024; MSAI 2025) and Ethics Toolkit promote safety, accountability, and non-discrimination. These guidelines are integrated into public sector digital transformation efforts. The Smart Dubai initiative developed the Dubai AI Ethics Self-Assessment Toolkit (DDa, 2023), offering practical steps for public and private entities, including SMEs, to assess ethical readiness. Although voluntary, it fosters awareness and proactive ethics adoption in local businesses. Dubai's State of AI Report identifies regulatory sandboxes as key performance indicators in its strategic roadmap (DDb, 2025). While both countries emphasize ethical AI, their regulatory frameworks remain in development. Enforceable standards tailored to SME realities are still limited (MBRSG, 2025).

In summary, while most international and national frameworks endorse shared ethical principles such as transparency and accountability, their legal enforceability and SME inclusion efforts vary. Despite growing alignment—particularly around OECD and UNESCO principles—disparities persist in capacity, institutional maturity, stakeholder engagement, and targeted support for SMEs.

# ETHICAL CHALLENGES IN AI WITH AN SME PERSPECTIVE

## Reports, Surveys, and Case Studies

Challenges associated with the development and deployment of AI systems have been highlighted in numerous survey-based studies and reports. These examine how SMEs engage with generative AI, focusing on ethical use, governance, and barriers to adoption.

## Globally

A McKinsey AI survey (McKinsey, 2025) revealed that only 27% of organizations using generative AI had robust processes in place to review all AI-generated outputs. Alarmingly, 47% reported at least one adverse consequence from AI deployment, including misinformation, intellectual property violations, or compliance breaches. These risks are particularly relevant to SMEs, which often lack the in-house legal or technical expertise to address such challenges. A cross-continental review of challenges facing SMEs in AI adoption identify (Yusuf et al., 2024) lack of competences of Staff, financial resources, security and resistance to change as key challenges facing SMEs. The research underscores the need of tailored interventions in capacity building initiatives, policy interventions and cross border collaboration for SMEs to overcome these barriers and to leverage AI technologies.

A recent study, combining quantitative data from a survey of 211 experts with qualitative insights from academic, governmental, and industrial publications, that explores the criminal use and abuse of AI, evaluating the impact of awareness, detection technologies, ethical guidelines, and enforcement as mitigation strategies reveals that SMEs may inadvertently expose themselves to risk when using AI tools without sufficient governance safeguards. The study proposes comprehensive strategies involving awareness campaigns, detection technologies, ethical frameworks, and stronger global legal coordination to effectively combat the criminal use of AI (Kathirvel et. al., 2025b).

The OECD highlighted similar concerns in its AI Policy Observatory (OECD, 2023), identifying SMEs as an "at-risk" group in relation to AI misuse and regulatory non-compliance. According to the OECD's 2024 Digital for SMEs Initiative, nearly one in five SMEs began experimenting with generative AI technologies within a year of their emergence. While 57% of SMEs reported seeing more opportunity than risk, they cited significant barriers, including time constraints (43%), costs (37%), and lack of internal AI expertise (27%).

GPAI Working Group on Responsible AI (GPAI, 2023) also emphasized the urgent need to build AI governance capacity among SMEs, particularly in developing or transitional economies. The group cited critical gaps in algorithmic transparency, user consent, and impact assessment practices.

Trust in AI technologies remains fragile, especially in sectors involving consumer interaction or critical services. A KPMG Global AI Ethics Survey (KPMG, 2023) found that public skepticism toward generative AI stems from opaque decision-making, unclear accountability for harmful outputs, and the lack of human recourse in AI-driven decisions. These challenges are heightened for SMEs that do not develop AI in-house or control the underlying infrastructure, making them vulnerable to ethical and reputational risks.

Ethical challenges also extend to cybersecurity: AI-powered monitoring and predictive tools have become essential for protecting industrial and IoT systems, though they introduce governance and oversight challenges that SMEs must also manage (Kathirvel et. al., 2025).

International development agencies have also begun recognizing these gaps. The OECD has stressed the need for inclusive stakeholder participation, particularly from underrepresented SMEs, in shaping AI policy.

Recognizing the Complex challenges resulting from development and deployment of AI technologies led to the establishment of the AI Governance Alliance (AGA) by World Economic Forum's Center of AI Excellence (WEF, 2023) as a collaborative platform from industry, governments academia and civil society. AGA now comprises more than 644 members and over 500 organizations globally. Its work streams cover the responsible adoption of AI across industries, through AI governance and regulations. (WEF, 2024) recognizes the lack of standardization and shared frameworks resulting in complicating the integration of ethical and legal regulations.

## Europe

The European Union's AI Act (EU, 2024), the first binding legal framework regulating AI, adopts a risk-based approach. However, a review of academic literature and private-sector guidelines (Soudi & Bauters, 2024) underscores the need to support SMEs in adopting AI responsibly and ethically. A study of Italian SMEs (Proietti & Magnani, 2025) identified a considerable disparity between SMEs and larger corporations in AI usage, with SMEs encountering multiple barriers. The study proposed a phased governance framework tailored for SMEs, emphasizing accessibility, risk scaling, and regulatory literacy. Additionally, a separate report (Bertolucci, 2023) emphasized the growing cybersecurity risks SMEs face when using generative AI, especially from AI-enabled phishing and fraud. Around 60% of SMEs reported inadequate internal cybersecurity capacity, further complicating safe generative AI deployment.

A broad analysis of European SMEs covering 486 SMEs, Start-ups and Scale-Ups and Entrepreneurships from 27 European members and 16,365 firms globally, revealed key patterns in AI adoption (Segarra-Blasco et al., 2025). The study found that start-ups and scale-ups led adoption, and that AI and robotics significantly enhanced innovation, helping SMEs to overcome scale-related innovation barriers. However, the study emphasizes that inclusive innovation policies need to account for firm size in order for SMEs to benefit of the innovation enabling potential of AI. In addition, recognizing barriers facing SMEs, the study calls for supportive policies to prioritize digital upskilling, advisory support, and increased access to funding to overcome barriers and increase adoption of these technologies by SMEs. A recent Study confirms that Robotics and AI adoption, in manufacturing, create both opportunities and ethical challenges. The study highlights the need for reskilling and inclusive workforce strategies to balance automation with human well-being and ensure equitable access to technology's benefits (Kathirvel et. al., 2025a). A policy-oriented study (EUSC, 2021) identified ongoing barriers such as low AI awareness, high adoption costs, limited data availability, and skills shortages. These issues were especially prominent in sectors like agriculture, construction, and legal services.

Walke et al. (2025) offered a grounded account of how one SME integrated generative AI, specifically ChatGPT. It reported notable productivity gains but also significant ethical and governance challenges. Key issues included encountering fabricated or "hallucinated" information, explainability limitations, and concerns over privacy and intellectual property, as staff sometimes used confidential client data without policy guidance. The SME's use of ChatGPT was largely informal and decentralized, with no central training or governance structures. This unstructured approach heightened ethical risks and highlighted the pressing need for SME-specific ethical frameworks and practical guidance for responsible AI deployment.

A study by the University of St Andrews (Brown et al., 2025), involving 10,000 UK SMEs, reported productivity gains of 27–133% from AI adoption. Many firms used generative tools for content creation and back-office tasks. However, deeper AI integration remained limited due to concerns over bias, authenticity, and the absence of internal governance frameworks. The study concluded that SMEs need clarity, trust in AI technologies, and simple compliance processes to scale adoption ethically.

## United States

The U.S. Chamber of Commerce reported that 98% of small businesses surveyed were already using at least one AI-enabled tool, and 40% had adopted generative AI applications (USCC, 2024). However, concerns about overregulation remained high, with 86% of respondents fearing that strict rules could hinder innovation and growth. Similarly, a U.S. Bipartisan Policy Center survey found that 82% of small business owners were actively using AI, with 75% perceiving it as beneficial. Yet many expressed concerns about accuracy, data privacy, and the lack of clear governance frameworks (Charllo, 2024).

## Asia-Pacific

In Australia, research from the University of Technology Sydney's (UTS) Explainable AI for SMEs project identified critical issues (Wright et al., 2022). While Australian SMEs are increasingly adopting AI solutions, there is inconsistency between, attitude towards ethical AI and practices of implementing ethical AI. The study also cited resource limitations as a major barrier. Many SMEs lacked access to training materials, legal expertise, and tailored AI governance tools. As a result, explainability was often treated as a secondary concern. A report by the UTS Human Technology Institute (HTI, 2025) found that although generative AI tools were widely adopted, SMEs that adopted AI were not aware of ethical standards. The report identified the need to support SMEs to build awareness of, and skills in, good AI governance such as developing simplified governance toolkits, industry-specific guidelines, and capacity-building programs designed for SMEs.

A systematic literature review conducted between 2021 and 2023 (Naren et. al., 2025) on entrepreneurship and SMEs in Malaysia highlights how governance quality, employee upskilling, innovation, and collaboration, prerequisite to ethical AI adoption, are decisive for international competitiveness, capabilities.

These outcomes highlight gaps between ethical standards and practice. A case study from India further illustrates this gap. A 2025 survey reported that while 64% of Indian firms prioritize generative AI, 75% lack effective change-management strategies to support ethical deployment (ETb, 2025). Additionally, MSMEs struggle tapping into the AI revolution due to high costs (Pandey, 2025). Similarly, a survey of 305 Pakistani SMEs (Soomro et al., 2025) showed that complexity factor negatively correlated with AI adoption, suggesting investing in training to account for skills gap of SMEs.

## MENA and GCC

The earlier findings underscore the urgent need for tailored governance tools and public education to promote responsible AI adoption. They align with broader regional trends. A Gulf AI Readiness Report (PwC, 2023) noted that although AI adoption is accelerating across the GCC—particularly in Dubai and Riyadh—SMEs lag in governance readiness. Over 60% of surveyed SMEs in the UAE and Saudi Arabia expressed uncertainty about the ethical implications of AI use, especially generative tools trained on opaque datasets. The report recommended localized governance frameworks and targeted support for SMEs.

A separate survey by Alkhoudi et al. (2025) reported that 57% of SMEs were unaware of available AI governance tools, citing cost, complexity, and limited digital readiness as persistent barriers. Respondents often used generative AI for content creation, marketing, or customer service without understanding how to assess compliance with data privacy or anti-bias standards, creating operational and reputational risks. A case study on Iraqi SMEs illustrates how generative AI performs in non-OECD contexts. A study by Alnaqeeb et al. (2025) on Iraq's eco-tourism sector found that while SMEs gained tangible benefits from adopting AI, there are challenges relevant to data privacy and high costs in the implementation of AI technologies.

## SMEs: Barriers and Limitations

The previous section outlined the regulatory landscape for ethical AI. International bodies such as the OECD, UNESCO, and the UN have issued robust sets of principles. However, these high-level guidelines are developed at a macro level, aimed at governments, policymakers, and large enterprises. They are often aspirational rather than actionable and typically lack enforceability, sector-specific adaptation, and implementation mechanisms—especially for SMEs. In contrast, national efforts like the EU's AI Act offer legally binding provisions but risk burdening resource-constrained actors with complex compliance demands. Countries including the U.S., Canada, and Australia have taken more sectoral or voluntary approaches, producing a fragmented landscape of overlapping soft laws with limited regulatory power.

Moreover, few frameworks incorporate concrete tools for AI ethics education, ongoing risk assessment, or support tailored to local SME contexts. The dominant approach to AI ethics has been criticized for being overly focused on principles and lacking in practical guidance. It often relies on broad declarations without addressing institutional capacities, cultural diversity, or rapid technological evolution. This disconnects risks reducing AI ethics to a performative gesture, removed from real-world decision-making. As a result, the most vulnerable actors, such as SMEs, remain ill-equipped to operate ethically, even when willing.

Implementations are often limited to documentation, case studies, or support toolkits that are not easily accessible. These resources frequently lack localization, usability, and practical compliance tools. Regulatory sandboxes generally cater to advanced developers and digital development SMEs, rather than the broader base of business-user SMEs across diverse industries. Many toolkits do not offer regionally adaptive interfaces, real-time guidance, or user-centric onboarding, thus failing to support SMEs that operate with limited financial resources, minimal legal capacity, and little access to AI ethics expertise.

Moreover, the studies cited in this chapter consistently emphasize the fragmented nature of global AI governance, creating a jurisdictional maze for SMEs operating across borders or using third-party AI services hosted abroad. Without consolidated, accessible, and practical tools, SMEs struggle to identify applicable regulations, assess AI system risks, or implement appropriate safeguards. SMEs; comprising over 80% of businesses in most economies; typically function with constrained resources. They lack in-house technical and legal expertise and have limited ability to interpret complex legislation or integrate ethical frameworks into daily operations. Most SMEs use generative AI tools in routine workflows. However, the limited accessibility and usability of compliance frameworks leave them vulnerable to ethical missteps and violations of emerging data regulations.

Aside from select efforts by specialized law firms or sector-based organizations, there is a near-complete absence of centralized platforms offering legal and regulatory guidance on AI ethics in a readable, multilingual, and SME-accessible format. This chapter illustrates how many SMEs, particularly in the MENA, South Asian, and ASEAN regions—face uncertainty in cross-border ethical compliance. They remain unclear about applicable laws, data governance requirements, and ethical obligations when using AI-powered platforms from other countries. These challenges are particularly acute for startups and SMEs expanding internationally or outsourcing AI-driven services.

Furthermore, no existing platform currently leverages AI itself to deliver ethical guidance, revealing a paradox within the ethics discourse. Although AI is promoted as a tool to augment human decision-making, no AI-powered agent exists to help SMEs navigate AI ethics. Most existing tools are static, non-interactive, and loosely structured as PDF-based toolkits.

The current ecosystem fails to provide SMEs with the tools, clarity, or automation needed for responsible AI use. As AI becomes embedded in daily business operations, the demand for structured ethical guidance intensifies. This need is especially urgent for SMEs, which make up the majority of enterprises and global employment yet remain the most underserved in AI governance. The rapid rollout of generative AI tools without tailored ethical support has widened the gap between adoption and responsible use.

In summary, the studies and regulatory frameworks discussed in this chapter expose the barriers and practical constraints SMEs face in implementing AI ethically. They highlight a dual reality: SMEs are rapidly adopting generative AI, yet operate in an environment characterized by uncertainty, fragmented oversight, and ethical ambiguity. Without simplified, actionable guidance, SMEs are disadvantaged by regulatory complexity. Accessible ethical frameworks and practical resources are essential to strengthening SMEs' capacity for responsible global AI adoption.

## ETHICS4AI: THE AI ETHICS FRAMEWORK

The AI ethics framework and implementation platform, Ethics4AI, introduced in the following sections, is designed to support SMEs and address the challenges of using AI ethically and responsibly.

### Key Challenges: The Four Pillars

Ethics4AI provides a comprehensive, SME-aligned platform to support ethical AI adoption in organizations without specialized AI departments or access to tailored policy tools. Drawing on real-world case studies and governance reviews, the platform addresses challenges through four pillars: foundational knowledge, contextualized guidance, ecosystem collaboration, and adaptive governance.

The first pillar addresses the foundational knowledge gap. It promotes AI ethics literacy as a core competency for technical staff, managers, executives, and decision-makers. Professionals already using generative AI, often without structured oversight, must build the confidence to recognize ethical risks and apply sound judgment.

The second pillar emphasizes practical implementation pathways tailored to SME contexts. It translates abstract ethical principles into usable resources, including sector-specific risk checklists, sample AI use case assessments, and guides for ethical prompting and human oversight. The goal is to reduce barriers such as complexity and cost through toolkits that are no-code, multilingual, and interoperable across jurisdictions.

The third pillar enables cross-border and cross-sector collaboration. It streamlines access to regulatory frameworks, addressing the fragmented nature of global AI regulation.

The fourth pillar centers on the ongoing monitoring of policy developments, helping SMEs remain responsive to emerging risks and legal obligations, including updates to current frameworks and policies.

### Development Approach: Mapping the Global AI Regulatory Landscape

Ethics4AI aims to operationalize the four pillars by offering a repository of existing policies and regulations, enabling users to navigate it through an AI-driven interface. Generative in nature, Ethics4AI is built on the following foundational data:

- Regulatory and legal frameworks: Includes national AI laws, international standards (e.g., UNESCO, OECD), summaries of SME-specific provisions, and a classification of binding versus non-binding regulations.

- Ethical AI principles: Covers definitions and interpretations of core concepts such as transparency, fairness, explainability, accountability, and human agency.
- Toolkits and guidelines: Consists of publicly available ethical AI toolkits.
- Sector-specific use cases and SME-focused research: Highlights key findings from studies and addresses regional challenges.
- Training resources and practical tools: Provides relevant training materials, tools, and prompts.

Ethics4AI aims to help SMEs overcome key barriers and challenges. By aggregating, annotating, and continuously updating regulatory content, it serves as a centralized source of curated knowledge. More importantly, it transforms that knowledge into action through AI-generated workflows, prompts, and guided decisions. This is especially critical for SMEs, which often lack dedicated legal or data governance teams and cannot afford costly consultants for each AI deployment.

This foundational data enables Ethics4AI to interpret ethical applications in alignment with existing regulations and policies. It also supports cross-border trade and collaboration among SMEs and policymakers. For example, a fintech SME based in the UAE but serving European clients must comply with both local ethical requirements and EU high-risk AI classifications. Currently, no mechanism exists to streamline such dual compliance. The platform, through jurisdiction-aware ethical agents, could ease this burden by offering contextual compliance advice.

This foundational data also powers Ethics4AI to support dynamic learning paths, case-based risk simulations, and responsible prompting templates. SMEs can use these tools to train staff, customize AI models, and align digital transformation strategies with ethical norms. By grounding the platform in global legal diversity while prioritizing SME usability, the initiative aims to democratize ethical AI governance—shifting it from compliance rhetoric to practical capability.

## Platform Implementation

Ethics4AI is a technical AI platform developed by the Sydney Institute of Technology and Innovation. It operationalizes the four pillars through a repository of existing policies and regulations, enabling users to navigate it generatively.

## AI BLDR Tool

The Ethics4AI platform is built on the www.BLDR.bot Agents platform (Inova, 2025), a no-code environment for creating enterprise-grade AI agents and automations. BLDR enables users to deploy secure, compliant AI agents, providing a strong foundation for translating complex ethical and regulatory frameworks into accessible, SME-focused tools. Users can access agents that deliver personalized, context-aware advice based on jurisdiction, industry, use case, and company size. The platform also integrates educational content and practical features, including ethical prompt libraries, logging tools, and low-code customization support. By aggregating ethical and legal standards, translating them into plain language, and offering compliance guidance through generative AI agents, the platform addresses a critical support gap. It makes ethical AI use actionable, scalable, and regionally inclusive. The platform aligns directly with this chapter's findings by offering SMEs the missing layer of translation, interactivity, and adaptability needed to operationalize responsible AI use in practice.

Ethics4AI is designed to close ethical-practice gaps for SMEs. By providing a dedicated, AI-augmented environment, the platform translates high-level ethical guidelines into actionable steps. It offers interactivity, multilingual AI assistants, embedded regulatory guidance, template policies, and no-code tools for ethical AI deployment. Unlike static manuals or regulatory policies, the platform integrates continuous support and contextual learning, enabling SMEs to navigate ethics and compliance in real time without legal or technical departments.

Importantly, the platform uses AI to make AI adoption more ethical. It employs agentic tools and prompt-based frameworks to enhance transparency, track decisions, reduce hallucinations, and guide responsible practices. This approach not only scales compliance affordably but also helps businesses internalize and sustain ethical standards as they grow. In short, Ethics4AI is not merely a digital resource; it is a structural response to a systemic gap. It meets SMEs where they are, with tools they need, in a form they can use. The platform's foundational data will be updated periodically as new regulations and policies emerge, ensuring continued alignment with the evolving ethical regulatory landscape.

## Generative Navigational Highlights

Ethics4AI introduces generative navigational features, enabling it to function as a:

- Regulatory compliance assistant: This feature supports alignment with regulatory frameworks. The platform draws from leading international standards to ensure relevance and regulatory credibility. This alignment enables SMEs operating across borders to access consistent guidance that meets global standards.
- AI ethical advisor: This core functionality serves multiple roles. It acts as a Digital Advisor to help users identify applicable laws based on location and use case. It functions as an Ethical Prompt Designer, assisting SMEs

in crafting and refining prompts to reduce bias and enhance transparency. It operates as a Regulation Navigator, using a smart engine to break down major obligations into simple steps. It also serves as an ethics literacy mentor, embedding key concepts into conversations. As a No-Code AI agent creation companion, it enables users to develop basic, ethical AI agents tailored to their workflows. Lastly, it is a Resourceful Repository, offering smart access to policy templates and checklists.

- AI ethics companion: Delivered through AI collaboration, oversight, and real-time support, Ethics4AI is positioned not as a legal advisor but as a smart companion. For complex queries or high-risk decisions, the system recommends escalation to a human ethics consultant or legal advisor. This human-in-the-loop design supports ethical deliberation, making compliance more accessible while preserving human responsibility.

Accessibility and inclusion are central to the agent's impact. It supports multi-modal interactions, with text and audio options enabling use by individuals with low literacy or visual impairments. The interface is mobile-first and lightweight, accounting for infrastructure constraints in developing economies. The agent is designed to accommodate diverse SME profiles, from solo entrepreneurs to growing tech startups.

Ethics4AI also embodies responsible AI principles. It is built on a privacy-by-design architecture and complies with GDPR and similar regulations. No user data is stored without consent, and all advice is delivered anonymously unless users opt into personalization. Risk disclaimers will be visible, and any limitations in the agent's coverage will be clearly communicated.

## Current Status: Beta Version

Ethics4AI is currently in beta release. It is available to a select group of SMEs, chambers of commerce, and institutional partners for early access and testing. The platform already integrates major international regulatory frameworks, offering comprehensive, jurisdiction-aware guidance through multilingual, no-code AI agents. Initial feedback from beta users has been highly positive. Users have praised the platform's intuitive interface, contextual relevance, and practical usability in real-world SME settings. Based on this input, ongoing improvements now focus on embedding interactive learning components, including role-based microlearning, real-time compliance simulations, and ethical onboarding journeys. A full public release is planned for 2026. It will include expanded support features such as certification pathways, ethical performance dashboards, and continuous content updates aligned with emerging global standards.

## FUTURE RESEARCH DIRECTIONS

Looking ahead, Ethics4AI could become a keystone tool in regional and global AI literacy initiatives. It may be integrated into ethics-by-design certification programs, used as an onboarding tool for SME accelerator hubs, or embedded within digital public infrastructure. By uniting AI ethics literacy, regulatory navigation, and responsible automation in a single platform, it offers a replicable, scalable model for empowering SMEs in the AI era.

Future work will focus on strengthening the governance of Ethics4AI to support governments and policymakers and on expanding collaborations with global bodies and national stakeholders.

## CONCLUSION

Ethical governance is becoming an urgent concern as generative AI reshapes business operations. This is particularly true for SMEs, which form the backbone of global economies yet remain the most underserved in ethical AI support. International bodies have introduced high-level principles, and many national governments have established regulatory frameworks. However, a significant gap remains between policy and practical implementation. SMEs face resource constraints compounded by regulatory complexity, fragmentation, and a lack of effective tools for information, compliance, and implementation.

This chapter reviewed major global and national regulatory frameworks, toolkits, and case studies. The review underscored the urgent need for an inclusive, actionable, and scalable approach to ethical AI adoption. It introduced a comprehensive Ethical AI Framework and a complementary platform, Ethics4AI, designed to address challenges across four pillars: AI literacy, tailored guidance, cross-sector collaboration, and adaptive governance.

Ethics4AI operationalizes global ethical principles through AI-powered, no-code tools that enable SMEs to adopt and deploy AI responsibly. It distinguishes itself by leveraging AI agents to facilitate compliance. Its foundational knowledge base enables it to function as an AI Mentor, Compliance Advisor, and Ethical Advisor. By equipping SMEs with the knowledge and confidence to navigate AI's ethical dimensions, Ethics4AI supports responsible and compliant use. It represents a pioneering approach that applies AI to promote ethical and trustworthy AI deployment.

## REFERENCES


Accountancy Europe. (2025), The AI EU Act, A guide for small accountants. *Accountancy Europe*. https://accountancyeurope.eu/publications/the-eu-ai-act-a-guide-for-sme-accountants

Adastra, I. P. (2025, July 1). ASEAN responsible AI roadmap 2025–2030. *LinkedIn*. https://www.linkedin.com/pulse/asean-responsible-ai-roadmap-2025-2030-adastraip-9qvve

Alkhoudi, S., Shaer, S., & Salem, F. (2025, April 15). The artificial intelligence SMEs ecosystem in the UAE: Overcoming challenges, expanding horizons. *Mohammed Bin Rashid School of Government*. https://doi.org/DOI: 10.2139/ssrn.5256892

Alnaqeeb, R. T. A., Almasoodi, M. F., Alshammari, S. A. H., & Ghanayem, A. (2025). AI-driven ecotourism recommendation systems: An empirical investigation of implementation success factors in Iraq. *Journal of Tourism Hospitality and Environment Management*, *10*(39), 53–72. DOI: 10.35631/JTHEM.1039005

Artificial Intelligence Verify Foundation. (2025). *AI Verify Foundation*. https://aiverifyfoundation.sg/

Association of Southeast Asian Digital Ministers Nations. (2025, January 17). Digital ministers adopt expanded guide of AI governance and ethics in generative AI. *Association of Southeast Asian Digital Ministers Nations*. https://digitalpolicyalert.org/event/26114-asean-published-expanded-guide-of-ai-governance-and-ethics-including-policy-recommendations-for-artificial-intelligence-ai-governing-authorities

Association of Southeast Asian Nations. (2021, January 22). ASEAN Digital Masterplan 2025. *Association of Southeast Asian Nations*. https://asean.org/book/asean-digital-masterplan-2025/

Association of Southeast Asian Nations. (2024, February). The Asian guide on AI governance and ethics. *Association of Southeast Asian Nations*. https://asean.org/book/asean-guide-on-ai-governance-and-ethics/

Association of Southeast Asian Nations Secretariat. (2024, February). ASEAN guide on AI governance and ethics. *Association of Southeast Asian Nations Secretariat*. https://asean.org/book/asean-guide-on-ai-governance-and-ethics/

Association of Southeast Asian Nations Secretariat. (2025, January 17). Expanded ASEAN guide on AI governance and ethics – Generative AI. *ASEAN Secretariat*. https://asean.org/wp-content/uploads/2025/01/Expanded-ASEAN-Guide-on-AI-Governance-and-Ethics-Generative-AI.pdf

Australian Human Rights Commission. (2021). Human rights and technology final report. *Australian Human Rights Commission*. https://humanrights.gov.au/our-work/rights-and-freedoms/publications/human-rights-and-technology-final-report-2021

Babl. (2025). Japan's METI releases AI contract checklist to guide businesses in the era of generative AI. *Babl.ai*. https://babl.ai/japans-meti-releases-ai-contract-checklist

Bertolucci, T. (2023, July). Generative AI for small and medium-sized businesses: Chaos or advantage*? World Economic Forum*. https://www.weforum.org/stories/2023/07/generative-ai-small-medium-sized-business/

Bian, M., Zhang, R., Ding, C., Peng, X., & Xu, J. (2025, May 12). Benchmarking ethical and safety risks of healthcare LLMs in China: Toward systemic governance under Healthy China 2030 [Preprint]. *arXiv*. https://doi.org//arXiv.2505.07205DOI: 10.48550

Bird & Bird. (2025, September 11). Japan's New AI Act: Examining an Innovation-First Approach Against the EU's Comprehensive Risk Framework. *Bird & Bird*. https://www.twobirds.com/en/insights/2025/japan/japans-new-ai-act-examining-an-innovationfirst-approach-against-the-eus-comprehensive-risk-framework

Body, P. (2023, December 13). *Future intelligence*. CKGSB Knowledge. https://english.ckgsb.edu.cn/knowledge/article/future-intelligence

Braue, D. (2025, September 9). Sovereign AI model poised to become Australia's ChatGPT: 'Ethics-driven' AI model trained on Australian data, values. *Information Age*. https://ia.acs.org.au/article/2025/sovereign-ai-model-poised-to-become-australia-s-chatgpt.html?ref=newsletter&deliveryName=DM28535

Brown, R., Sun, H., Cowling, M., & David, D. (2025, April 9). Can we see it in the numbers yet? Artificial intelligence, firm-level productivity and SMEs? [Preprint]. *University of St. Andrews*. https://doi.org/DOI: 10.2139/ssrn.5210948

CanadaGov. (2025). *Artificial Intelligence and Data Act (AIDA), as part of the Digital Charter Implementation Act: Proposed regulatory framework for high-impact AI systems*. Government of Canada.

Center for Strategic and International Studies. (2025, March 21). Unpacking Japan's AI policy with Hiroki Habuka. *Center for Strategic and International Studies*. https://www.csis.org/analysis/unpacking-japans-ai-policy-hiroki-habuka

Charllo, B. (2024). AI empowerment in SMEs: Bridging the post-pandemic recovery and innovation gap. *Journal of Artificial Intelligence & Cloud Computing*. https://onlinescientificresearch.com/articles/ai-empowerment-in-smes-bridging-the-postpandemic-recovery-and-innovation-gap.pdf

Citadel. (2024, June 25). Japan's AI guidelines for business: What it means for you. *Citadel AI*. https://citadel-ai.com/blog/2024/06/25/ai-guideline-for-business-2024/

Clifford Chance. (2023, October 5). Japanese law issues surrounding generative AI: ChatGPT, Bard and beyond. *Clifford Chance*. https://www.cliffordchance.com/insights/resources/blogs/talking-tech/en/articles/2023/10/Japanese-Law-Issues-Surrounding-Generative-AI.html

Colorado General Assembly. (2024, May 17). Senate Bill 24-205: Consumer protections in interactions with artificial intelligence systems. *General Assembly of the State of Colorado*. https://leg.colorado.gov/sites/default/files/2024a_205_signed.pdf

Cyberspace Administration of China. (2023, July 13). Interim Measures for the Management of Generative AI services. *Cyberspace Administration of China*. https://www.cac.gov.cn/2023-07/13/c_1690898327029107.htm

Department of Industry. Science and Resources. (2019, November 7). Australia's AI ethics principles. *Australian Government*. https://www.industry.gov.au/publications/australias-artificial-intelligence-ethics-principles

Department of Industry. Science and Resources. (2025, October 21). Guidance for AI Adoption. *Australian Government*. https://www.industry.gov.au/publications/guidance-for-ai-adoption

Digital Agency. (2025, May 27). The Guideline for Japanese governments' procurements and utilizations of generative AI for the sake of evolution and innovation of public administration. *Digital Agency*. https://www.digital.go.jp/assets/contents/node/basic_page/field_ref_resources/e2a06143-ed29-4f1d-9c31-0f06fca67afc/6e45a64f/20250527_resources_standard_guidelines_guideline_04.pdf

Digital Dubai, & the Dubai Future Foundation. (2025, April 21). Dubai state of AI report. Government *of Dubai*. https://www.digitaldubai.ae/docs/default-source/publications/dubai-state-of-ai-report.pdf

Digital Dubai (Government of Dubai). (2025). AI system ethics self-assessment tool. *Government of Dubai*. https://www.digitaldubai.ae/self-assessment

DigitalBriefing. (2024, March 27). India's Regulation of AI and large language models. *India Briefing*. https://www.india-briefing.com/news/india-regulation-of-ai-and-large-language-models-31680.html

Economic Times. (2025). GenAI joins the checklist; 64% of Indian companies are making it a priority. *Economic Times*. https://economictimes.indiatimes.com/tech/artificial-intelligence/genai-joins-the-checklist-64-of-indian-companies-are-making-it-a-priority-report/articleshow/121952746.cms

Ema, A., Sato, R., Hase, T., Nakano, M., Kamimura, S., & Kitamura, H. (2023, November 26). Advancing AI audits for enhanced AI governance [Preprint]. *arXiv*. https://doi.org//arXiv.2312.00044DOI: 10.48550

eSafety Commissioner. (2023, August). Tech trends position statement: Generative AI. *eSafety Commissioner.* https://www.esafety.gov.au/sites/default/files/2023-08/Generative%20AI%20-%20Position%20Statement%20-%20August%202023%20.pdf?v=1753223601081

Esya Centre. (2025, February). Response to The Report on AI governance guidelines development. *Esya Centre.* https://www.esyacentre.org/s/Esya_Centre_-_AI_Report_Response_Final1-1-4fj2.pdf

European Digital SME Alliance. (2025, April 23). From regulation to action: AI Act compliance for digital SMEs. *The European Digital SME Alliance.* https://www.digitalsme.eu/events/from-regulation-to-action-ai-act-compliance-for-digital-smes

European Economic and Social Committee. (2021, June). Boosting the use of artificial intelligence in Europe's micro, small and medium-sized enterprises. *European Economic and Social Committee.* https://www.eesc.europa.eu/sites/default/files/files/qe-02-21-953-en-n.pdf

European Union. (2024, July 25). AI innovation package & GenAI4EU. *European Union.* https://digital-strategy.ec.europa.eu/en/factpages/ai-innovation-package

European Union. (2024, June 13). Artificial Intelligence Act (Regulation (EU) 2024/1689). *European Commission.* https://artificialintelligenceact.eu

European Union. (2025, February 19). Small businesses' guide to the AI Act. *European Union.* https://artificialintelligenceact.eu/small-businesses-guide-to-the-ai-act/

European Union. (2025, May 2). AI regulatory sandboxes for AI: Member State overview. *EU Artificial Intelligence Act.* https://artificialintelligenceact.eu/ai-regulatory-sandbox-approaches-eu-member-state-overview/

European Union. (n.d.). EU AI Act compliance checker. *European Union.* https://artificialintelligenceact.eu/assessment/eu-ai-act-compliance-checker/

Global Legal Insights. (2025). AI, machine learning & big data laws 2025. *Global Legal Insights.* https://www.globallegalinsights.com/practice-areas/ai-machine-learning-and-big-data-laws-and-regulations/singapore

Global Partnership on Artificial Intelligence. (2023). Responsible AI working group recommendations. *Global Partnership on Artificial Intelligence.* https://gpai.ai/projects/responsible-ai/

Government Digital Service. (2025, February 10). A playbook for the UK government. *UK Government.* https://www.gov.uk/government/publications/ai-playbook-for-the-uk-government

Government of Canada. (2023, March 13). The Artificial Intelligence and Data Act (AIDA) – Companion document. *Innovation, Science and Economic Development Canada.* https://ised-isde.canada.ca/site/innovation-better-canada/en/artificial-intelligence-and-data-act-aida-companion-document

Guest, O., & Wei, K. (2025, June 4). Bridging the artificial intelligence governance gap: The United States' and China's divergent approaches to governing general-purpose artificial intelligence [Preprint]. *arXiv*. https://doi.org//arXiv.2506.03497DOI: 10.48550

Hollanek, T., Pi, Y., Fiorini, C., Vignali, V., Peters, D., & Drage, E. (2025, May 22). A toolkit for compliance, a toolkit for justice: Drawing on cross-sectoral expertise to develop a pro-justice EU AI Act toolkit [Preprint]. *arXiv.* https://doi.org//arXiv.2505.17165DOI: 10.1145/3715275.3732078

*Human Technology Institute* (2025, February). In their words: Perspectives and experiences of SMEs using generative AI. *Human Technology Institute, University of Technology Sydney.* https://www.saam.com.au/wp-content/uploads/2025/02/SAAM_In-their-words_perspectives-and-experiences-of-SMEs-using-AI-report-1.pdf

IndiaAI Subcommittee on AI Governance. (2025, January 6). Report on AI governance and guidelines development. *IndiaAI.* https://indiaai.gov.in/article/report-on-ai-governance-guidelines-development

InfoComm Media Development Authority. (2025, May 28). Starter kit for safety testing of LLM-based applications. *IMDA.* https://www.imda.gov.sg/-/media/imda/files/about/emerging-tech-and-research/artificial-intelligence/large-language-model-starter-kit.pdf

Information and Communication Technology Division, Government of the People's Republic of Bangladesh. (2020, March). National strategy for artificial intelligence 2020–2025. *ICT Division, Government of Bangladesh.* https://file-rangpur.portal.gov.bd/files/pbs2.dinajpur.gov.bd/files/1885c0a0_28a4_4fcc_8d4d_dcd7ce23fd8b/7b684f19a15dfcd0f542382764572486.pdf

Information and Communication Technology Division, Government of the People's Republic of Malaysia. (2024, September). The national guidelines on AI governance & ethics. *Ministry of Science, Technology & Innovation, Malaysia.* https://mastic.mosti.gov.my/publication/the-national-guidelines-on-ai-governance-ethics/

Information Systems Audit and Control Association. (2024, October 18). Understanding the EU AI Act: Requirements and next steps. *Information Systems Audit and Control Association*. https://www.isaca.org/resources/white-papers/2024/understanding-the-eu-ai-act

Inova, A. I. Solutions. (2025). Inova BLDR platform. *Inova AI Solutions*. https://www.inovasolutions.ai/solutions-by-industry/

Joswig, T., & Kurz, W. (2025, February 23). Regulatory and compliance requirements for SMEs operating AI systems through data centers in the EU, with a focus on data protection challenges in Germany. *Journal of Next-Generation Research, 1*(2), 4–44. DOI: 10.70792/jngr5.0.v1i2.89

Kathirvel, A., Gobalakrinshan, N., & Naren, A. (2025). " Cyber security in industrial automation using AI, in AI-Enhanced Cybersecurity for Industrial Automation. *IGI Global*. https://www.igi-global.com/gateway/chapter/379640

Kathirvel, A., Gobinath, V. M., Sathya, A. K., Naren, A., & Raj, A. (2025a). The Role of Robotics in Smart Manufacturing: Increasing Efficiency and Reducing Costs, in The Quantum AI Era of Neuromarketing, *IGI Global*. https://www.igi-global.com/gateway/chapter/365528

Kathirvel, A., Naren, K., & Vianny, M. (2025b). Strategies for Combating Criminal Use and Abuse of Artificial Intelligence" In Avoiding Ad Fraud and Supporting Brand Safety: Programmatic Advertising Solutions. *IGI Global*. https://www.igi-global.com/gateway/chapter/368386

Khazanah Research Institute. (2025, January). AI governance in Malaysia: Risks, challenges and pathways forward. *Khazanah Research Institute*. https://www.krinstitute.org/assets/contentMS/img/template/editor/KRI%20AIIG%20Report%20Final_v4.7%20[upload].pdf

Kin, L. C., Chin, A. S., & Seah, C. (2025). AI, machine learning & big data laws and regulations 2025 – Singapore. In AI, machine learning & big data laws & regulations 2025. *Global Legal Insights*. https://www.globallegalinsights.com/practice-areas/ai-machine-learning-and-big-data-laws-and-regulations/singapore

KPMG International, & the University of Queensland. (2023, April 2). Trust in artificial intelligence: A global study 2023. *KPMG International*. https://assets.kpmg.com/content/dam/kpmg/xx/pdf/2023/09/trust-in-ai-global-study-2023.pdf

Lin, L. (2024, July 16). China puts power of state behind AI—and risks strangling it. *The Wall Street Journal*. https://www.wsj.com/tech/china-puts-power-of-state-behind-aiand-risks-strangling-it-f045e11d

Liu, I. F. C., & Edmondson, D. (2023, August). China: New interim measures to regulate generative AI. *Baker & McKenzie.* https://insightplus.bakermckenzie.com/bm/attachment_dw.action?attdocparam

Malaysia National AI Office [NAIO]. (2024, September). The national guidelines on AI governance & ethics. *Malaysia National AI Office.* https://mastic.mosti.gov.my/publication/the-national-guidelines-on-ai-governance-ethics/

Mckinsey. (2025). The state of AI: How organizations are rewiring to capture value. *Quantum Black AI by Mckinsey.* https://www.mckinsey.com/~/media/mckinsey/business%20functions/quantumblack/our%20insights/the%20state%20of%20ai/2025/the-state-of-ai-how-organizations-are-rewiring-to-capture-value_final.pdf

Ministry of Economy. Trade and Industry & Ministry of Internal Affairs and Communications. (2024, April 19). AI guidelines for business Ver 1.0. *METI & MIC.* https://www.meti.go.jp/english/press/2024/0419_002.html

Ministry of Electronics and Information Technology, Government of India. (2023, August 11). *Digital Personal Data Protection Act, 2023* [Act No. 22 of 2023]. *MeitY.* https://www.meity.gov.in/static/uploads/2024/06/2bf1f0e9f04e6fb4f8fef35e82c42aa5.pdf

Ministry of IT & Telecom. Pakistan. (2023, May). Draft national artificial intelligence policy. *Ministry of IT & Telecom, Pakistan.* https://moitt.gov.pk/SiteImage/Misc/files/National%20AI%20Policy%20Consultation%20Draft%20V1.pdf

Ministry of State for Artificial Intelligence. Digital Economy and Remote Work Applications (UAE Government AI Office). (2025). AI ethics self-assessment tool. *UAE Government AI Office.* https://ai.gov.ae/aiethics-self-assessment/

Ministry of State for Artificial Intelligence, Digital Economy and Remote Work Applications, United Arab Emirates. (2022, December). AI ethics principles and guidelines. *UAE Government AI Office.* https://ai.gov.ae/wp-content/uploads/2023/03/MOCAI-AI-Ethics-EN-1.pdf

Ministry of State for Artificial Intelligence, Digital Economy and Remote Work Applications, United Arab Emirates. (2024, June 10). The UAE charter for the development and use of artificial intelligence. *UAE Government AI Office.* https://ai.gov.ae/wp-content/uploads/2024/07/UAEAI-Methaq-EN2-3.pdf

Mohammed Bin Rashid School of Government. (2025, April). *The artificial intelligence SMEs ecosystem in the UAE.* Mohammed Bin Rashid School of Government. https://mbrsg.ae/ar/the-artificial-intelligence-smes-ecosystem-in_the-uae

Morin, O. (2025, June 13). What's inside the EU AI Act and what it means for your privacy. *Investopedia.* https://www.investopedia.com/eu-ai-act-11737033

Naren, K., Kathirvel, A., & Saravanan, N. (2025). Entrepreneurship and Business Culture in Rural Area in Digital Disruption in Hospitality, Sustainable Hospitality: A Global Perspective on People, Planet, and Profit. *Emerald Publishing Limited.* https://bookstore.emerald.com/digital-disruption-in-hospitality-sustainable-hospitality-hb-9781836623618.html

National Institute of Standards and Technology (NIST). (2023, January 26). AI risk management framework (AI RMF) 1.0. *NIST.* https://doi.org/DOI: 10.6028/NIST.AI.100-1

Nizza, U. (2024, July 5). Assessing the impact of the European AI Act on innovation dynamics. *Northwestern University Law School.* https://www.law.northwestern.edu/research-faculty/clbe/events/standardization/documents/nizza_assessing_impact_ai_act_innovation.pdf

Norton Rose Fulbright. (2024, February 14). Singapore proposes governance framework for generative AI. Norton Rose Fulbright. https://www.nortonrosefulbright.com/en/inside-fintech/blog/2024/02/singapore-proposes-governance-framework-for-generative-ai

Nourhashim, H. (2024, November 22). Malaysia's advancing AI governance: Driving contextual AI innovation amid trade-offs and opportunities. *AI Singapore.* https://aisingapore.org/wp-content/uploads/2025/03/Malaysias-Advancing-AI-Governance-Driving-Contextual-AI-Innovation-Amid-Trade-offs-and-Opportunities.pdf

Novelli, C., Casolari, F., Hacker, P., Spedicato, G., & Floridi, L. (2024, January 14). Generative AI in EU law: Liability, privacy, intellectual property, and cybersecurity [Preprint]. *arXiv.* https://doi.org//arXiv.2401.07348DOI: 10.48550

Office of Science and Technology Policy, The White House. (2022, October 4). Blueprint for an AI Bill of Rights: Making automated systems work for the American people. *OSTP.* https://bidenwhitehouse.archives.gov/ostp/ai-bill-of-rights/

Office of the Australian Information Commissioner (OAIC). (2024, January 26). Notifiable Data Breaches report: January to June 2024. *OAIC.* https://www.oaic.gov.au/privacy/notifiable-data-breaches/notifiable-data-breaches-publications/notifiable-data-breaches-report-january-to-june-2024/

Organisation for Economic Co-operation and Development. (2019, May; updated 2024, May 3). *OECD principles on artificial intelligence. OECD.* https://www.oecd.org/going-digital/ai/principles/

Organisation for Economic Co-operation and Development. (2023). AI Policy Observatory: Bridging the SME governance gap. *OECD.AI*. https://oecd.ai/en/

Organisation for Economic Co-operation and Development. (2025, June). Scoping note for an AI policy toolkit to support economies in realising AI benefits. *OECD*. https://www.oecd.org/content/dam/oecd/en/events/2025/06/mcm/MCM-2025-Scoping-Note-for-an-AI-Policy-Toolkit-to-Support-Economies-in-Realising-AI-Benefits.pdf

Organisation for Economic Co-operation and Development [OECD]. (2024, September 16). SME digitalisation to manage shocks and transitions: 2024 OECD D4SME survey. *OECD-Digital for SMEs*. https://www.oecd.org/content/dam/oecd/en/publications/reports/2024/09/sme-digitalisation-to-manage-shocks-and-transitions_735fc44d/eb4ec9ac-en.pdf

Pandey, A. (2025, June 18). Digital divide: MSMEs struggle to tap into AI revolution as high costs block access. *Economic Times*. https://economictimes.indiatimes.com/small-biz/sme-sector/digital-divide-msmes-struggle-to-tap-into-ai-revolution-as-high-costs-block-access/articleshow/121924294.cms

Personal Data Protection Commission. Singapore & Infocomm Media Development Authority. (2019, January; 2020, January; 2024, May). Model AI Governance Framework (1st ed., 2019; 2nd ed., 2020; Generative AI extension, 2024). *PDPC & IMDA*. https://www.pdpc.gov.sg/help-and-resources/2020/01/second-edition-of-model-artificial-intelligence-governance-framework

Pi, Y. (2024, January 5). Missing value chain in generative AI governance: China as an example [Preprint]. *arXiv*. https://doi.org//arXiv.2401.02799DOI: 10.48550

Proietti, S., & Magnani, R. (2025, January 14). Assessing AI adoption and digitalization in SMEs: A framework for implementation [Preprint]. *arXiv*. https://doi.org//arXiv.2501.08184DOI: 10.32388/0N53P4

PwC Middle East. (2023). Gulf AI readiness report: Private sector governance gaps. *PwC Middle East*. https://www.pwc.com/m1/en/publications/ai-readiness-middle-east.html

Reuters. (2025, April 8). Europe wants to lighten AI compliance burden for startups. *Reuters*. https://www.reuters.com/world/europe/europe-wants-lighten-ai-compliance-burden-startups-2025-04-08

Said, F., & Nabilah, F. (2024, November). Future of Malaysia's AI governance. *Institute of Strategic & International Studies Malaysia*. https://www.isis.org.my/wp-content/uploads/2024/12/AI-Governance.pdf

Saudi Data & AI Authority. (2024, January). Generative AI guidelines for government and public use. *SDAIA*. https://sdaia.gov.sa/en/SDAIA/about/Files/GenAIGuidelinesForGovernmentENCompressed.pdf (SDAIA Government Version, SDAIA Public Version)

Saudi Data& AI Authority. (2023, September). AI ethics principles. *SDAIA*. https://sdaia.gov.sa/en/SDAIA/about/Documents/ai-principles.pdf

Securiti. (2025, January 16). An overview of Malaysia's National Guidelines on AI governance and ethics. *Securiti.AI*. https://securiti.ai/malaysia-national-guidelines-on-ai-governance-and-thics

Segarra-Blasco, A., Tomàs-Porres, J., & Teruel, M. (2025). AI, robots and innovation in European SMEs. *Small Business Economics*, *65*(1), 1–27. DOI: 10.1007/s11187-025-01017-2

Sillberg, C. V., De Cerqueira, J. S., Sillberg, P., Kemell, K. K., & Abrahamsson, P. (2024, December 9). The EU AI Act is a good start but falls short [Preprint]. *arXiv*. https://doi.org//arXiv.2411.08535DOI: 10.48550

Skadden, Arps, Slate, Meagher & Flom LLP. (2024, April 5). Utah becomes first state to enact AI-centric consumer protection law. *Skadden*. https://www.skadden.com/insights/publications/2024/04/utah-becomes-first-state

Soomro, R. B., Al-Rahmi, W. M., Dahri, N. A., Almuqren, L., Al-Mogren, A. S., & Aldaijy, A. (2025). A SEM–ANN analysis to examine impact of artificial intelligence technologies on sustainable performance of SMEs. *Scientific Reports*, *15*(1), 5438. DOI: 10.1038/s41598-025-86464-3 PMID: 39948417

Soudi, M. S., & Bauters, M. (2024). AI guidelines and ethical readiness inside SMEs: A review and recommendations. *Digital Society : Ethics, Socio-Legal and Governance of Digital Technology*, *3*(1), 3. DOI: 10.1007/s44206-024-00087-1

Stappers, J. (2025, February 11). Artificial Intelligence and compliance: Preparing for the future of AI governance, risk, and compliance. *Navex*. https://www.navex.com/en-us/blog/article/artificial-intelligence-and-compliance-preparing-for-the-future-of-ai-governance-risk-and-compliance/

The UK's Technology Trade Association. (2023, March 29). UK government unveils its approach to regulating AI with white paper. *TechUK*. https://www.techuk.org/resource/uk-government-unveils-its-approach-to-regulating-ai.html

The UK's Technology Trade Association. (2025). AI adoption for SME, research insights and practical guidance. *TechUK*. https://www.techuk.org/what-we-deliver/events/ai-adoption-for-smes-research-insights-practical-guidance.html

UK Government Office for Artificial Intelligence. (2023, March 29). A pro-innovation approach to AI regulation. *UK Government.* https://www.gov.uk/government/publications/ai-regulation-a-pro-innovation-approach/white-paper

United Nations. (2024, September). Governing AI for humanity. *High-level Advisory Body on Artificial Intelligence, Office of the Secretary-General.* https://www.un.org/sites/un2.un.org/files/governing_ai_for_humanity_final_report_en.pdf

United Nations Department of Economic and Social Affairs. (2024, April). United Nations system-wide white paper on AI governance. *UNDESA.* https://unsceb.org/sites/default/files/2024-04/United%20Nations%20System%20White%20Paper%20on%20AI%20Governance.pdf

United Nations Educational, Scientific and Cultural Organization. (2021). Recommendation on the ethics of artificial intelligence. *The United Nations Educational, Scientific and Cultural Organization.* https://unesdoc.unesco.org/ark:/48223/pf0000381137

United Nations Educational, Scientific and Cultural Organization. (2023). ethics of artificial intelligence. *UNESCO.* https://www.unesco.org/en/artificial-intelligence/recommendation-ethics

United Nations Educational, Scientific and Cultural Organization. (2023a). Ethical impact assessment: A tool of the recommendation on the ethics of artificial intelligence. *UNESCO.* https://unesdoc.unesco.org/ark:/48223/pf0000386276

United Nations Educational, Scientific and Cultural Organization. (2023b). Readiness assessment methodology (RAM): A tool of the Recommendation on the ethics of artificial intelligence. *UNESCO.* https://unesdoc.unesco.org/ark:/48223/pf0000385198

U.S. Chamber of Commerce. (2024, September 15). Empowering small business: The impact of technology on U.S. small business. *U.S. Chamber of Commerce.* https://www.uschamber.com/assets/documents/Impact-of-Technology-on-Small-Business-Report-2024.pdf

U.S. Chamber of Commerce Technology Engagement Center. (2024, September 15). *Empowering small business: The impact of technology on U.S. small business.* U.S. Chamber of Commerce. https://www.uschamber.com/assets/documents/Impact-of-Technology-on-Small-Business-Report-2024.pdf

Vianny, M., Ayyaswamy, K., Rithik, G., Kathirvel, N., & Raj, A. (2025). Private AI an Overview, In AI-Driven Breakthroughs in Antimicrobial Resistance. *IGI Global.* https://igi-global.com/gateway/chapter/374868

Walke, F., Klopfers, L., & Winkler, T. (2025). *Leveraging generative AI and ChatGPT in SMEs: A grounded model.* University of Hagen.

Weber, L. C. (2024, April 25). Ethical deployment of AI in Malaysia: Navigating societal dynamics and technological innovation. SSRN *Electronic Journal.* https://doi.org/DOI: 10.2139/ssrn.4807537

White & Case. (2025, June 13). AI Watch: Global regulatory tracker – Japan. *White & Case.* https://www.whitecase.com/insight-our-thinking/ai-watch-global-regulatory-tracker-japan

World Bank. (2019). Small and Medium Enterprises (SMEs) Finance. *World Bank.* https://www.worldbank.org/en/topic/smefinance

World Bank Group. (2024, December 5). Global trends in AI governance: Evolving country approaches. *World Bank Group.* https://documents1.worldbank.org/curated/en/099120224205026271/pdf/P1786161ad76ca0ae1ba3b1558ca4ff88ba.pdf

World Economic Forum. (2023). AI governance alliance. *World Economic Forum.* https://initiatives.weforum.org/ai-governance-alliance/home

World Economic Forum. (2024). AI governance alliance briefing paper series. *World Economic Forum.* https://www3.weforum.org/docs/WEF_AI_Governance_Alliance_Briefing_Paper_Series_2024.pdf

Wright, E., Zhou, J., Lindsay, D., Przhedetsky, L., Chen, F., & Davison, A. (2022). SMEs and explainable AI: Australian case studies. *Computer & law.* https://www7.austlii.edu.au/au/journals/ANZCompuLawJl/2022/4.pdf

Wu, Y. (2023, June 27). How to interpret China's first effort to regulate generative AI measures. *China Briefing.* https://www.china-briefing.com/news/how-to-interpret-chinas-first-effort-to-regulate-generative-ai-measures

Yang, A., & Li, B. (2025, May 29). AI Watch: Global regulatory tracker – China. *White & Case.* https://www.whitecase.com/insight-our-thinking/ai-watch-global-regulatory-tracker-china

Yusuf, S., Durodola, R., Ocran, G., Abubakar, J., Amarichi, Z., Echere, A., & Paul-Adeleye, A. (2024, September). Challenges and opportunities in AI and digital transformation for SMEs: A cross-continental perspective. *World Journal of Advanced Research and Review.* https://www.researchgate.net/publication/383873269_Challenges_and_opportunities_in_AI_and_digital_transformation_for_SMEs_A_cross-_continental_perspective

## ADDITIONAL READINGS

Herman Boikanyo, D. (2025). Ethical considerations and societal impacts of AI adoption in SMEs within emerging markets. *Business Management, 35*(2), 25–46. DOI: 10.58861/tae.bm.2025.2.02

Rahman, A., Ishak, W., Abdul Aziz, M., Arif, Z., Md Salleh, S., & Safii, M. (2025). Responsible AI Literacy's Role in Ethical, Transparent, and Sustainable AI Adoption in SMEs: Mediating Effects of Technology, Organization and Environment. *Proceedings of the 3rd International Conference on Business, Finance, Management and Economic (BizFAME 2024)*. https://doi.org/DOI: 10.2991/978-94-6463-827-1_14

Tjaša, Š., & Dankova, P. (2025). Ethical Leadership and Management of Small- and Medium-Sized Enterprises: The Role of AI in Decision Making. Special Issue Effective Leadership and Management in the Sustainable Development of Small and Medium-Sized Enterprises *15*(7), 274. *Administrative Sciences*. Advance online publication. DOI: 10.3390/admsci15070274

## KEY TERMS AND DEFINITIONS

**Artificial Intelligence (AI):** Artificial Intelligence refers to the design of computer systems capable of performing tasks such as learning, reasoning, problem-solving, and decision-making that typically require human intelligence. In this chapter, AI refers to both traditional systems and generative AI models.

**Generative Artificial Intelligence (Generative AI).:** A type of AI with advanced systems that create new content, producing human-like responses, code or media, from large collections of existing data referred to as training data.

**Ethical Artificial Intelligence (Ethical AI).:** Ethical AI is concerned with ensuring that the development, deployment and use of AI systems are guided by social and moral values such as privacy, fairness, transparency, and accountability to ensure that AI serves social good.

**Responsible AI:** Responsible AI refers to the procedures, structures, and monitoring mechanisms that govern ethical practice. It goes beyond ethical intent to cover risk assessment, monitoring, and continuous improvement to ensure that AI systems are developed and used ethically and in compliance with legal standards.

**AI Governance:** AI governance refers to the institutional mechanisms, policies, procedures, and oversight bodies, such as ethics boards or internal audit mechanisms, that guide and supervise the development and use of AI to ensure alignment with legal and ethical standards.

**SMEs.:** Small and Medium Sized Enterprises (SMEs) are independent business entities with a limited number of employees (typically fewer than 250). SMEs are characterized by relatively constrained financial and human resources compared to large corporations.

**AI Compliance:** AI compliance refers to the alignment of an organization's use of AI with applicable ethical standards.

**AI Ethics Framework:** An AI ethics framework consists of policies and operational guidelines designed to support organizations in aligning AI systems with ethical and legal norms. Examples include frameworks developed by UNESCO and OECD to guide industries and governments seeking to implement ethical practices in their AI use.

**AI Agent or Agentic AI:** An AI agent is a software system capable, either autonomously or collaboratively with humans, of interpreting information, making context-based decisions, and taking actions to achieve a set objective.

**Ethics4AI Platform:** Ethics4AI is a generative AI platform designed by the Sydney Institute of Technology and Innovation to aid SMEs in understanding and applying ethical and legal requirements in their AI use.

Chapter 5

# Ethical AI in Digital Media: Navigating Data Privacy, Misinformation, and the Challenge of Fake News

**Rashid Hashmi**
https://orcid.org/0009-0006-3232-0705
*Sharda University, India*

## ABSTRACT

*The integration of Artificial Intelligence (AI) in digital media has transformed the way information is produced, disseminated, and consumed. While AI-driven technologies such as automated content generation, deep learning-based recommendation systems, and real-time data analytics enhance efficiency and accessibility, they also raise critical ethical concerns. Issues related to data privacy, algorithmic bias, misinformation, and the proliferation of fake news have emerged as significant challenges, impacting public trust and the integrity of digital media ecosystems. This chapter critically examines the ethical dilemmas surrounding AI's role in digital media, with a specific focus on the risks of misinformation and the erosion of data privacy. AI-powered content moderation, sentiment analysis, and fact-checking mechanisms have been introduced as solutions to counter fake news, yet these tools are not devoid of biases, opacity, and ethical pitfalls.*

DOI: 10.4018/979-8-3373-2597-2.ch005

## 1. INTRODUCTION

Another pillar in the creation of digital media environments is the use of Artificial Intelligence (AI) that facilitates the creation, sharing, and usage of information. As more of the oanline communication networks are covered, the use of the AI in newsrooms, in social media spheres, and in content recommending services has provided a significant portion of efficiency and customization. There is however a consequence to such development. The most significant aspects of AI use in the media that enable the development of ethics are rapidly gaining prominence. These would encompass the issue of data privacy, algorithmic bias, sharing of misinformation and lack of transparency in the delivery of contents. Initially therefore ethical AI in digital media will be responsible or accountable use of artificial intelligence founded on technology that adheres to what can be referred to as fairness, accuracy and protection of other individual rights primarily on information dissemination. Moreover, to evaluate AI applied in the media in an ethical way, the classical theories should be included: a utilitarian approach considers the efficacy, scalability, and overall gains of AI usage; a deontological approach focuses on individual rights, freedom of choice, and equality without interests to the consequences. Many regulatory debates have their basis vested on the tension between these two traditions. Additionally, the idea of Shoshana Zuboff (2019) of the so-called surveillance capitalism, where personal information is mined and sold as the main business model in a digital platform, suggests that the depriving nations are left with spoil agriculture as raw data but do not control the facilities and profits generated, whereas the Crawford (2021) conceptualizes AI as a phenomenon of so-called data colonialism, a process on the global scale, where developing nations donate raw data but do not control the infrastructures and profit generated. These frameworks work together to give more theoretical based account of the ethical issues discussed in this chapter.

The media houses and journalism has been the sentinels of what can be called the democratic discourse, that is, the processes is supposed to inform masses and make the discourse acceptable to them. Due to the implementation of AI, the vast majority of the editorial work that humans frequently perform is served, in many cases, to code, as companies implement AI-based functionality, such as smart news generators or intelligent-delivery content algorithms. Smart introduction engines have already been used by news agencies so that they customize the news feeds sent, and social media sites heavily utilize AI to recommend the channels and optimize performance in the social media platforms. In spite of the advantages that have been proposed by such applications in issues related to scalability and responsiveness, they have made the distinction between choices made by the will of human beings and those made by algorithms extremely challenging (Ajunwa & Kamer, 2024). It also should not be stressed upon as, in case the absorption of information and a set of knowledge about the population is mediated by algorithms, the problem of morality might have almost no limit.

Among the most burning issues of ethical AI in digital media, particularly, misinformation should be mentioned. There is the rise of fake news, malicious narratives, and conspiracy theories that have been further spread by bots and the feedback loops of algorithms driven by AI. Such machine-based systems are frequently biased toward interactivity over correctness and end up encouraging misinformation or extremist messages. It has been demonstrated that organized disinformation campaigns have the ability to control the mood of the population and demoralize confidence in democratic institutions (Hajli et al., 2021). Furthermore, the AI-generated computer media, in this case, Deepfakes, pose significant questions about the naturalness of visual and audio materials, since they can create a comprehensive picture of events and uttered declarations that never took place.

It is also quite alarming that data privacy has been invaded through monitoring. The AI technologies flourish on the data, more specifically on the data regarding the individuals and the data gathered through the services on the internet. And all of this data on behavior, demographics and location allow predictive algorithms to personalise content, advertisements and political messaging with a startling level of accuracy. Personal data commodification has turned the training of AI into a running mechanism where the user is under constant surveillance, profiled, and urged to make choices. Generative AI has been shown to be a serious threat to privacy as indicated by Golda et al. (2024) seeing that user data is usually being processed without transparency or induction. Invisibility of extraction and use of data poses an inherent ethical challenge because users will seldom have any idea of their information being leaked or by whom.

The other ethical pitfall, which must be discussed critically is the algorithmic bias. These AI systems will become as neutral as data and parameters they comprise. Knowing the benefits of biased algorithms, in a situation in media unbiased algorithms cause unequal representation of some communities, exposure selective censorship, triumph of devastating stereotypes. The use of AI in content flagging, articles recommendation, or comment moderation will most certainly discriminate certain voices or opinions as a by-product of biased training data, or some of these algorithms has a design choice that cannot easily be explained. Not only do these outcomes consent to the injustice of the information usage but the normative principles of media multiplicity, inclusivity as well (Ajunwa and Kamer, 2024). Most AIs have the black-box property, further complicating this issue by rendering these biases hard to detect and address.

To check the negative impacts of the unethical application of AI, there is the technology development methods to negate the same in the form of the creation of Deepfake detectors, Explainable AI (XAI) system, and natural language models.

The deepfake tactics detection tools are supposed to indicate that artificial video has been utilized that can be used to manipulate the audience and the XAI strategies are supposed to make the algorithms appear more human-friendly and understandable. The innovations have a significant role as concerned the freedom to recover the trust since it presents means of verifying the contents and holds the algorithm liable. Nevertheless, the solutions have limitations that come together with them. These tools to detect deepfakes, so to speak, may not be capable in every instance of keeping up with the progress of the creator technology. Again, in its turn, XAI effectiveness relies on the presence of partners (a platform and a developer) willing to open themselves to transparency, which, unfortunately, is not always a possibility.

In this chapter, the author attempted to examine these ethics encountered/beckoned digital media AI through an account giving in detail the issues and potential solutions to these issues. Applying critical analysis to the real-life examples and accessible literature sources, as well as case studies, the chapter tries to offer the excellently organized system of the ways in which the use of AI in journalism, social media, and other digital domains can be ethically established. The assumption is that instead of criticizing, it is constructive to engage in searching ways in which accountability, transparency, and fairness in AI applications can be instilled in the media platforms. It is critical to note that no mechanical solution exists to the ethical governance of AI but rather a continuous process that needs to adapt to the changes in technologies and social demands.

Governance and policy is significant in AI ethics regulation in media and, consequently, cannot be stressed. The world and regional guidelines that have been established such as the EU AI Act and national variants are scarce, the majority or them, in fact, are voluntary or informally enforced. The absence of a substantial legal community that may be generally recognised poses considerable risk of moral breakdown and notably in rapid online services. In addition to that, the cross-border media distribution interferes with the content jurisdictional tightness of the conference and the global writers need the supranational method in the manner of controlling AIs. Ethics is a challenge in organizations, development, journalists as well as in users.

The following table provides a concise overview of the key ethical challenges posed by AI in digital media, highlighting their descriptions, affected stakeholders, and example cases that illustrate these concerns.

*Table 1. Overview of Ethical Challenges in AI and Digital Media*

| Ethical Issue | Description | Affected Stakeholders | Example Cases |
|---|---|---|---|
| Misinformation | Spread of false or misleading information amplified by AI algorithms | General public, media institutions | COVID-19 infodemic; Election disinformation |
| Surveillance | Data tracking and profiling through user behavior monitoring | Users, privacy advocates, regulators | Social media ad targeting; User data leaks |
| Data Privacy Violations | Unauthorized or opaque use of personal data by AI systems | Consumers, digital rights groups | Cambridge Analytica; AI-based political ads |
| Algorithmic Bias | Discriminatory outcomes due to biased training data or flawed AI design | Marginalized groups, journalists, audiences | Content moderation favoring dominant narratives |
| Lack of Transparency | Black-box AI systems with limited explainability or accountability | Researchers, regulators, end-users | AI news curation without user consent |

Sources: Bedford-Strohm et al. (2020); Karnouskos (2020); Shoaib et al. (2023)

The time to move is obvious: the lack of information is distorting social values, sharing data is violating personal rights, and biased algorithms are manipulating the stories in some invisible manner. Due to the growing indistinctions not only between the content creators and the consumers but also the machines, there is a greater need to see to the incorporation of ethical AI practices into all manners of the media development.

## 2. BACKGROUND

The artificial intelligence in media has passed through the experience of significant technological shifts that altered the organizational process of journalism and communication. Previously, the introduction of automated systems into news rooms was not defined by the intention to oust human scientist journalists, but to assist them during the completion of routine duties. Early usages of AI included the process of automating heavily data-driven news (financial reportage, sports scores, weather overviews, etc.) in which organised data could be converted to comprehensible text through natural language processing systems. Since the late 2000-s, leading media companies started testing such technologies as ways of increasing efficiency of output and cutting down the human load (Broussard et al., 2019). At first, these developments were regarded as pure technical developments. But with automation being more advanced and being applied increasingly, the answers to the concerns regarding journalistic standards, editorial authority and reliability of information have started to present themselves. There has been a major change in content production and distribution through the use of bots. Social media networks created a new ecosystem in which content could be created with a script, whether for benign intent, like advertising a news item or a weather warning, or more insidious, to manipulate. Howard (2016) suggests that the concept of computational propaganda that means algorithms, automation, and human curation to influence life of the nation became an ongoing character to the digital era of political communication.

In line with these, we have had the growth of algorithmic filters on social media. Such algorithms notify the user of content appeared on display, depending on past behavior, tastes, and interaction. These systems were supposed to customize user experience, but in practice, produced an echo chamber and filter bubble, those are information environments confirming the beliefs that people had and decreasing exposure to different opinion options. This has led to polarization of opinion in the society and fragmentation of common reality. The problematic matters involving algorithmic design have, therefore, invoked grave ethics as they tend to be guided by engagement or algorithm performance numbers as opposed to editorial ideals. Since, as pointed out by Broussard et al. (2019), the rising role of AI in the media cannot be given merely in terms of effectiveness and size, it should be considered in terms of preserving democratic ideals in the sphere of information, truth, and civic engagement.

AI-powered content ecosystems have demonstrated the dangers of misinformation, especially in the case of high-stakes situations with global consequences. The COVID-19 pandemic caused a massive influx of misinformation about the cause, prevention strategies, and vaccines of the disease on digital platforms. Fake home remedies, fake statistics, and conspiracy theories about 5G networks spread fast, in most cases surpassing the truths disseminated by public health authorities. AI in form of recommender systems has contributed to this by promoting such sensational entries by recommending posts that attracted a lot of interactions. The study by Roozenbeek et al. (2020) released discoveries that suggest COVID-19 misinformation has contributed greatly to future changes in behavior by inducing vaccine hesitancy and puzzlement in the population. Similarly, Trump and other countries have (or are projected to have) manipulated the narratives and impersonated candidates to mislead supporters during the 2016 United States elections and 2020 elections, as well as in various countries, like Brazil and India, among others (Kapantai et al., 2021). These examples recycle the argument that, without proper regulation, AI systems can become a source of disinformation and political interference.

The scholarly reaction to this predicament created the idea to formalize the disinformation studies as the field of study. Academics started researching the spread of false information, their consumption by people and the role of technology and psychology in their popularization. Media literacy campaigns, fact-checking initiatives, and more were some of the ways to curb the so-called tide of false content. Nevertheless, it is still challenging to keep the volume and tempo of misinformation at bay. According to Hajli et al. (2021), AI has to be turned ethically since it should have transparent algorithms, user-focused design, and monitoring systems as counteractors of its misuse. Guidelines of AI ethics started to emerge, and the main principles of it were transparency, accountability, non-maleficence, and fairness.

Algorithmic bias is another need that went into the spotlight. Systems that use biased or incomplete training data will create and enhance prejudices that exist in society. This may be cause of discrimination in the moderation of content, representation, and audience targeting in media. As an example, it has been found that on the platform in question, the detection of hate-speech mainly tends to mark content by minority groups in disproportionated numbers, when this particular content does not represent harm. To the same extent, recommendation systems can suppress or marginalize some groups or views owing to diversity Shortage of training data. These are not necessarily deliberate biases; descending as they do from systematic inequalities present even in the data itself. In their article, Shoaib et al. (2023) point to the fact that the ramifications of algorithmic bias in media can be reflected in real life as well, such as social exclusion, stereotyping, and suppression of opposing opinions. Ayoub et al. (2021) provide examples of how user data was used during election cycles to influence voting behavior through microtargeted ads. The asymmetry of power between data collectors and users—where one party controls information flow while the other remains largely unaware—undermines the ethical foundation of digital communication.

AI's historical trajectory in media has been shaped by milestones that illustrate its gradual but deep integration into the field. The following table provides a structured timeline, highlighting key moments when AI was deployed across journalism and social media. This timeline captures both the technological evolution and the emerging ethical debates.

*Table 2. Timeline of AI Deployment in Journalism and Social Media To reflect non-Western perspectives, India's Digital Personal Data Protection Act (2023) and Rwanda's National AI Policy (2021) should be recognized as early efforts in the Global South to balance innovation with privacy and equity. These policies demonstrate regional priorities, such as safeguarding linguistic diversity and preventing algorithmic exclusion. Crawford's *Atlas of AI* also reminds us that training large models like GPT-3 has significant ecological costs, emitting over 500 tons of CO2, making environmental sustainability a pressing ethical concern in AI deployment.*

| Year | Milestone | Platform/Initiator | Significance | Source |
|---|---|---|---|---|
| 2006 | First automated financial reports using natural language generation | Associated Press | Marked beginning of AI-generated journalism for routine data-based stories | Broussard et al. (2019) |
| 2010 | Introduction of algorithmic news curation | Facebook News Feed | Personalized content streams based on engagement, paving the way for filter bubbles | Becker et al. (2023) |
| 2013 | Bot-driven hashtag manipulation | Twitter | Bots began amplifying political hashtags to simulate public support | Howard (2016) |
| 2016 | AI-driven disinformation in U.S. elections | Facebook, Cambridge Analytica | Use of AI and microtargeting for political influence and misinformation campaigns | Kapantai et al. (2021) |
| 2020 | COVID-19 misinformation spike | Multiple platforms | Viral spread of fake cures, conspiracy theories, and anti-vaccine content via recommender systems | Roozenbeek et al. (2020) |
| 2022 | Launch of explainable AI tools for content moderation | YouTube, Twitter | Platforms begin experimenting with transparency tools to explain content recommendations | Shoaib et al. (2023) |
| 2023 | Adoption of AI ethics guidelines in media industry | Various newsrooms and policy bodies | Formal recognition of the need for ethical principles in deploying AI for journalism and public media | Becker et al. (2023) |

## 3. FOCUS OF THE ARTICLE

The introduction of artificial intelligence in digital media spaces created several ethical red flags, the vast majority of which persist in growing as technology advances and goes widespread. The essence of the concept, according to which the modern world is driven by four critical points that should be evaluated adequately and includes the violation of data privacy, the unrestricted proliferation of fake news, lack of transparency in generated algorithms and their bias, ethical quandary about being trapped into an echo chamber by personalization engines. These are not limited issues. By doing so or by some other reaction to a high degree of stimulation of the digital schema of the media and confidence of the viewer, they are or tend to be complementary and overlap to some degree. As the demand of consuming digital contents constantly grows, it is necessary to address them as an entirety and analyze the socio fabricated techne of reproduction.

## Data Privacy Violations: Surveillance as a Business Model

Active data privacy breaches by the respective individuals might be defined as one of those that are imminent and the more stable in the field of the digital media. The majority of online platforms have established their business models based on which a critical role is played by the commodification of some personal data. All user activity: clicks, likes, search history, watch time, and even frames lingering before streaming to a machine learning model data point. These models not only study user behavior in an attempt to make the content more relevant, but they also create predictive profiles, which are used in advertisement, political targeting and manipulating behavior. Increased visibility of cookies, app permission requests, third-party tracking, and biometric information have broadened what platforms know about one another, without the individuals giving their consent most of the time, or communicate clearly (Ajunwa, 2025).

The practices of data collection take place in obscurity. Although the terms and conditions technically inform the users, it would be not comprehensible to a layman that the amount of information included in these documents is so extensive and complicated. By making platforms appear to capture the data with the approval of the user, even though in reality, the majority of users are not aware of how profoundly they are being observed, this pseudo-consensual model allows making platforms appear to capture the data with the approval of the user. As a result, Fontes et al. (2022) posit that the asymmetry of power in data subjects and data controllers infringes the principle of informational self-determination, with the power to control their data use expected to be in the hands of individuals. In addition to this, the commercialization of this data further commercializes the identity of its user into behavioral profile to be sold to third parties without ethical protection. In politically challenging settings, this problem is even more serious, as surveillance is potentized in this manner.

## Misinformation Proliferation: Tools and Tactics of Digital Deception

Misinformation, unchecked across media due to the use of AI, has been one of the most apparent and damaging impacts of the latter. Automated systems have completely transformed the extent, velocity and form in which false information spreads. What used to only circulate on fringe sources, is now capable of going viral in a few minutes with the help of algorithmic boost and artificial interaction. The forged news articles, the automated videos generated by AI, and the fake pictures, which are usually created using the natural languages models and the artificial media tools, distort the line between the fiction and reality.

Robots have played a significant role to disseminate fake stories. Theses automated accounts communicate as though they are real users, reposting content, creating comments and controlling the public discussion by creating an illusion of agreement. At political elections or in cases of pandemics, misinformation campaigns led by correlated bot networks have misinformation voters, incited violence, and widened cracks in society (Shu et al., 2020; Shao et al., 2018). There is another twist to this, deep fake technology. Such artificial tools of the media give rise to hyper-realistic videos or audio clips of individuals saying things that they would never say with the help of AI. The risks of reputational damage, blackmail, and political manipulation with the help of Deepfakes can hardly be overestimated.

The personalization algorithms make the situation even more complicated. As soon as a user engages with distorted content, AI systems can suggest other ideas of the same quality and results in an informational loop. Nguyen et al. (2021) note that users who experience misinformation are prone to trusting future misleading information, and this is because such users are likely to believe it when misleading information is presented in context. Such strengthening of algorithmic curation strongly reduces the possibility of receiving corrective or non-aligning opinions. Misinformation does not just become viral because malicious actors want it to be especially true, it is also because platforms prioritize engagement and not corroboration.

## Lack of Transparency and Algorithmic Bias

A major issue in modern media research refers to the secrecy that is embodied in most of the deployment schemes among AIs. Although the media organisations are prone to using black-box models (i.e. models with very complex algorithms which the users can not understand and neither can the organisation to which it was created fully understand its internal workings), such artefacts are typically opaque to independent auditors. This obstructs action to intervene, or a participation in contesting, or an understanding of the reasoning behind algorithmic results. In digital media, this opaqueness gives operators the tools to show and hide according to standards that the users cannot easily identify, further raising alarm on the issues related to discrimination, censorship or targeting bias.

The related problem, that of Algorithmic bias, is a direct consequence of these closed systems. The bias is built in the training when the models are trained on datasets which reflect the historical or societal prejudice. Such biased material put in a moderation system can effectively discourage contributions by marginalised communities in an unfair manner. Similarly, recommendation engines trained using engagement-based cues can misrepresent a wider range of voices or warp sensitive discussions, as they promote stereotypes in the place of other perspectives.

Algorithms are rarely deliberately biased and this does not occur in most cases because technical constraints and the quality of data and implicit design decisions are common causes of algorithm bias (Chakraborty et al., 2020). One can think of a platform that authorizes recommendations based on the historical level of engagement: things that have received attention in the past, which may be a biased outcome in its turn, will be featured, without any regards as to whether they propagate a negative stereotype. This is a system that reinforces the status quo regarding inequity and also hides away other content. Additionally, in cases where the user is facing discrimination or/and unfair treatment, failure by the company or organization to be fair as regards to disseminating their presence on the ground, makes it virtually impossible to seek redress.

The concern is further to automated censorship and banning. Social media whose AI utilities remain opaque, and those where a choose to apply rules governing communities, may then remove content before reflecting on its purpose or trends. This acts to interfere with due process and also oppose platform management. There have been increased demands to make AI explainable Systems that are capable of explaining why they output the way they do. The implementation, however, is still small. In the absence of defined algorithmic accountability processes, judgemental AI in digital media is a hope rather than a reality.

**Socioeconomic Impacts: Labor Displacement & Ecological Costs**: Automation (AI-powered) has already started supplanting a more traditional form of journalism. The example of Fedha (2023), an artificial intelligence newscaster in Kuwait, confirms the ability of artificial presenters to substitute humans in the broadcasting process. Such practices are efficient but invite issues of job security, better placed identity, and editorial control. Of just as much importance are the ecological costs of AI infrastructure: strenuous computational resources are used to produce large-scale models, whose emissions cause the same impact as hundreds of transatlantic flights. To conduct an ethical analysis of AI in media truly holistically a set of such trade-offs relating to labor and environment is necessary.

## Ethical Concerns in Personalization Algorithms: Echo Chambers and Filter Bubbles

The notion on individualization of online media has been teasing the world at the edge of the mouth as it is being crammed down the throat as an electronic appendix that will ensure that the customer gets what he/she wants it to be in addition to ensuring that the customer will experience something that is of much higher value. There is however ethical dilemma which accompanies it. Recommendation systems represent the machine-learning type of systems which change the media in such a way that the media recommends enhances to head of the feeds on the previous activities resulting in the additions of worth which bring the relevancy and consequently a circumvention of the varied way. The final result is a certain echo-chamber where the audience members will read something that affirms their existing thoughts. Such isolation is only theoretical, and people who use a digital space in these conditions are more hostile, untrusting of alternative viewpoints, and more inclined to misinformation (Mohseni et al., 2020). It will promote content that has a chance at commentary, liking and sharing, over content of the balanced or informational content, and this would suggest that it will promote outlandish content or emotion-upsetting content over balanced or informational content that lies in the middle of the two extremes. This will result in derailment of the media experience of the subject and lost democratic action of the journalism to render good discourse between persons. Researchers also have associated algorithmic customization with psychic health effects. Filter bubbles and echo chambers may increase anxiety, contribute to polarisation, and serve as sources of their compulsive tendencies to doomscrolling (Social Media + Society, 2022). This highlights the idea that an ethical artificial intelligence is not merely concerning informational integrity, but it concerns protecting mental health too.

According to Zhuo et al. (2023), ethical considerations should be embedded into system design to move beyond the current focus on predictive optimization in personalized media systems. This shift requires ensuring that system architectures uphold pluralism, transparency, and diversity. However, the pursuit of business interests presents a significant obstacle to implementing such reforms.The platforms have also been financially rewarded even when enhanced polarization and poor-quality news have delivered. The second priority in such a system is the comfort of the user, which concerns the profitability of the platform.

*Table 3. Ethical Risks Across the AI Lifecycle in Digital Media*

| Lifecycle Stage | Ethical Risk | Impact Area | Example |
|---|---|---|---|
| Data Collection | Privacy violation | User autonomy | Biometric tracking without consent |
| Model Training | Algorithmic bias | Representation, fairness | Underrepresentation of minority voices in datasets |
| Content Recommendation | Echo chambers, misinformation | Public discourse | Political polarization via suggested videos |

| Lifecycle Stage | Ethical Risk | Impact Area | Example |
|---|---|---|---|
| Moderation and Enforcement | Lack of transparency | User rights | Arbitrary content takedowns without explanation |
| Monetization | Surveillance capitalism | Commercial ethics | Ad targeting based on sensitive personal data |

(Adapted from Arora et al., 2023; Gongane et al., 2024)

This table illustrates how ethical risks are present throughout the AI lifecycle—from data collection to monetization. Each phase carries specific implications for digital media users and requires tailored responses. What makes these risks particularly concerning is their cumulative effect. A biased model fed with intrusive data and applied through opaque recommendations creates a media environment where fairness, accuracy, and trust are consistently undermined.

## 4. SOLUTIONS AND RECOMMENDATIONS

The widespread deployment of artificial intelligence in digital media has prompted increasing concerns over misinformation, algorithmic bias, data privacy violations, and the erosion of public trust. To address these multifaceted issues, the focus must shift toward a combination of technological, policy-based, organizational, and ethical interventions. This chapter offers an in-depth examination of viable solutions and strategic recommendations that can help mitigate the harms outlined earlier. The emphasis is on both preventing the misuse of AI systems and actively guiding their ethical application across media platforms.

### 4.1 Technical Solutions: Towards Transparent and Accountable AI Systems

Another of the most promising channels to be explored with regard to reducing risk in contemporary scholarship on the ethics of artificial intelligence, is the systematic application of Explainable AI (XAI) methodologies. By making the outputs of the algorithmic systems comprehensible to and transparent to the end-users and regulatory agencies XAI is able to address the so-called black-box nature that is rampant and that contributes to lack of trust so much. This interpretability allows the stakeholders to ascertain the reason as to why a certain recommendation is made or why a certain piece of work is flagged. It also enables the auditing of automation decision making in a systematic manner allowing the retracement and where necessary revision of processes involved.

The prominent application of XAIs could be given by the set of bias-detection tools that have already emerged, in particular, the one by Chakraborty et al. (2020), called Fairway. These tools are used to detect discrimination biases present in training corpora or models output so as to restrict the negative effect of the past bias. Most importantly, they act both as the diagnostic machinery and as the feedback cycles of the iterative condensation of the model, encouraging the development process that would not lose touch with the steps of the demographic processes happening in the real world.

In parallel to XAI, there is a new line of so-called fairness-aware models that attempt to arrive at a compromise between good predictions and fairness rules, hence matching algorithmic performance with democratic ideals. The process of systemic bias audit integration during the model life cycle is two-fold: proving post-factual monitoring of possible harms, and intervening with the process of training and deployment of the model, thus, preventing its harms before they happen.

## 4.2 Policy Frameworks for Ethical Regulation

This will entail powerful, apart from what is internationally acceptable, legal and regulatory frameworks in the responsible regulation of artificial intelligence on both sides of the digital media. An excellent case in point would be the General Data Protection Regulation (GDPR) of the European Union, specifying the need to get the consent of the user, reduce the volume of the data, and have a right to clarification in an algorithmic decision. Overall, these deals have a role to play in safeguarding individual autonomy and privacy in what is rapidly becoming a digitalized social context (Ajunwa, 2025).

One can observe a similar process in the United States following the publication of the Blueprint of AI Bill of Rights by the National Artificial Intelligence Initiative (NAII) that includes the right to privacy, the right to transparency, and the right to equitable treatment in using algorithms. Although the sanctioned federal level, the practical demonstration of such rights still remains not as high as it has been before under GDPR. Both the European and the U.S. schemes, in any case, show the way in which the policy will help establish some limitations of how agreeable the AI utilization may be employed and help provoke the entire world to discuss the digital ethics.

Lack of consistent legal norms has prompted in-house ethical codes among different media outlets and technology-related websites. Algorithms auditing, employee education as (associated with) ethical AI literacy have become standard operations in many journalistic entities; the establishment of data-protection codes is also a matter-of-course (Becker et al., 2023). The kind of voluntary actions are a temporary substitution amid a shifting legal landscape and a triumphant model of bottom-up governance. Furthermore, more flexible and situation-specific solutions can be provided through co-regulatory models that integrate the statutory provisions and the industry-specific rules and codes. This is quite pivotal where considering platforms of hybrid frameworks that operate on multiple jurisdictions. In this respect, good governance demands that governments, civil societies and the private sector cooperate closely to ensure that the subsequent policies are streamlined so as to make them inclusive as well as cultural sensitive.

## 4.3 Platform-Level Responses: Designing for Responsibility

The social media and news sites are also becoming the stipulated players of the mediation of the AI-generated content in the academic discourse. One of the recent innovations in this field is the exploding popularity of the fact-checking tools that are assisted by AI. These systems use large language models to scan viral posts, evaluate their reliability and compare information with credible databases. The case study presented by Althabiti et al. (2023) is relevant and indicates how generative AI can assist in identifying the misinformation quickly and efficiently in times of a public health crisis. The framework was efficient due to its high accuracy in detecting suspect content, and instead of catching it, it was sent to the human moderators to make the final decision. Grassroots resistance has also emerged, with campaigns such as #NoTechForApartheid and GDPR-driven user actions pressuring corporations toward accountability. These movements illustrate how civil society can shape AI ethics from the bottom up.

Nonetheless, due to such developments, human-AI collaboration remains in the necessity bracket. Automated moderation may only bring false positives and cultural blindness. Therefore, the combination of AI and humans will give better accuracy and fairness (Becker et al., 2023), whereby humans oversee the results of the previous AI check. There are also attempts of platforms to change the interface, adding context labels, warning pop-ups and fact-check banners that will indicate the quality of the content without trying to censor it outright (Bansal & Warkentin, 2021). Another governance level is the transparency dashboards, which provide the users with information regarding data usage and content suggestions. However, most of these programs are still initiatives and can be characterized by the absence of the strictness of mechanisms subject to the external audit. It is therefore urgent to bridge this policy gap and use incentivizing schemes to standardize and encourage ethical disclosure practices of companies.

## 4.4 Ethical Deployment Principles: Building a Human-Centered AI Ecosystem

Beyond tools and policies, ethical principles must underpin the entire lifecycle of AI development and deployment. These principles include transparency, fairness, informed consent, and robust redressal mechanisms. Transparency requires not only disclosing the presence of AI but also explaining how it operates. Fairness means ensuring equal treatment across user groups and avoiding outcomes that disadvantage vulnerable populations. Consent must be meaningful, not coerced through obscure user agreements. Redressal mechanisms should be available when AI systems cause harm—whether through mislabeling, wrongful content takedown, or biased targeting (Gondwe, 2025). Preventive measures against data surveillance are especially vital. Mühlhoff (2023) advocates for systemic limits on behavioral tracking, including algorithmic data minimization and anonymization protocols. Stahl & Wright (2018) further argue for integrating privacy-by-design principles into media AI systems, where privacy is not a feature but a foundational design parameter.

## 4.5 Case Study: Generative AI in Fact-Checking Ecosystems

One promising application of generative AI has been in supporting fact-checkers. Althabiti et al. (2023) explored a system where AI-generated summaries were compared against real-time claims posted on social media. By linking content to verified sources and highlighting inconsistencies, generative models reduced the burden on human editors and accelerated misinformation flagging during crisis events. The use of transformer-based models with contextual awareness improved detection of misleading narratives, especially those subtly reframing facts rather than outright fabricating them.

Yet, such systems must be designed with caution. Generative AI is susceptible to its own hallucinations and biases. When integrated into fact-checking, human oversight remains essential. The key takeaway is that AI can augment journalistic verification efforts but should not replace editorial judgment or ethical scrutiny.

*Table 4. Ethical AI Practices for Digital Platforms*

| Practice | Description | Benefits | Limitations |
|---|---|---|---|
| Explainable AI (XAI) | AI systems provide reasons for their outputs | Enhances transparency, accountability | May oversimplify decisions or be hard to scale |
| Human-AI Collaboration | Combines automated tools with human oversight | Reduces errors and improves cultural sensitivity | Resource-intensive |
| Data Minimization | Collects only essential user data | Lowers surveillance risk | May reduce personalization |
| Transparency Dashboards | Allows users to view how content is curated | Builds trust and user autonomy | Adoption not uniform across platforms |
| Ethical Impact Assessments | Reviews systems before deployment | Forecasts risks and improves design | Not yet mandated universally |

*(Sources: Bedford-Strohm et al., 2020; Tiwari, 2023; Golda et al., 2024)*Practical utility can be enhanced by embedding role-specific toolkits. For journalists, an 'Algorithmic Audit Checklist' could include questions such as: Who built this model? What datasets were used? Are minority voices represented? For policymakers, a 'Cross-Jurisdictional Compliance Matrix' could map overlaps and divergences between GDPR, China's algorithmic governance (Zhang, 2024), and India's DPDP Act (2023). Glossary callouts should also clarify technical terms in Table 4—for instance, SHAP (Shapley Values) explains model predictions by feature contribution, while LIME (Local Interpretable Models) focuses on small, interpretable subsets. Similarly, 'federated learning' and 'differential privacy' must be defined for non-technical readers.

*Table 5. Comparison of XAI Techniques in Misinformation Detection*

| XAI Technique | Description | Use in Misinformation Detection | Strengths | Weaknesses |
|---|---|---|---|---|
| SHAP (Shapley Values) | Explains contribution of features to predictions | Identifies why content is flagged | High interpretability | Computationally expensive |
| LIME (Local Interpretable Models) | Models local decision behavior | Analyzes small content subsets | Quick, intuitive explanations | Inconsistent across different runs |
| Attention Visualization | Highlights input areas influencing output | Useful in text-based fact checks | Shows key influential content segments | Less effective on multimedia content |
| Counterfactual Explanations | Shows how input changes affect outputs | Explains misinformation boundaries | Enhances user understanding of decisions | May confuse non-technical users |

(Sources: Praveenraj et al., 2023; Mohseni et al., 2020)

Addressing the ethical challenges of AI in digital media is not only a technical task but a socio-political imperative. The path forward must integrate sophisticated tools with principled policies, organizational responsibility, and public accountability. As the digital ecosystem continues to evolve, the stakes of inaction grow. Media platforms, regulators, developers, and users must work together to ensure that AI systems are not only powerful but also just, respectful, and inclusive in their design and application.

## 5. EMERGING SOLUTIONS & RESEARCH FRONTIERS

This section merges the earlier case study on generative AI fact-checking with forward-looking research. The aim is to present a unified vision of both current innovations and future trajectories without redundancy.

As this incorporation of the Artificial Intelligence of the digital medium into the digital media is a widespread event that is reworking in the modalities of communication of civilizations with the information, communication and information thereof, we should begin to develop an awareness-oriented research agenda. Such a form of an agenda should also not simply be required to respond to the already present dilemmas of ethics but to the recently emerging dilemmas brought about by the fast-paced advancement in technology. We shall be forced to do more than appear a second time as a list of the already existing evils and we shall have to structure around the work of strategy rather than experiment. Subsequently, a subsequent inquiry must be cautiously mindful of the circulation of media consumption, algorithmic power, flow of information and socio-political contingencies. Meanwhile, there is an absolute necessity to amplify perspectives of under-represented populations in order to prevent technologies that enhance the further increase of the disparities.

The second avenue that the future research must be directed towards is the development of AI-specific data privacy legislation that would supersede the general provisions presented in existing IT legislation. Current law like Data protection in the Information Technology Act or other laws in regions have typically covered the data protection in general principles, without consideration to the fragility posed by information systems that are powered by AI. These include behavioral profiling, emotion detection, cross-platform surveillance, and inferred analytics—all of which are often conducted without explicit user consent. Research must advocate for laws tailored specifically to AI technologies, emphasizing dynamic consent models, enforceable rights to algorithmic explanation, and robust mechanisms for redress. Such frameworks should be capable of accommodating the evolving architectures of AI systems and the increasingly granular data they process (Veluru, 2024).

The role of education and ethical literacy also warrants deeper academic inquiry. One of the most persistent gaps in media studies curricula globally is the limited integration of AI ethics, particularly within journalism programs. As AI continues to influence everything from content generation to editorial decisions, it becomes critical to equip future media professionals with the tools to question, audit, and critique automated systems. Research must explore pedagogical strategies for embedding AI ethics into journalism education—incorporating case studies, interdisciplinary collaborations, and critical analysis of algorithmic media practices. These interventions should be complemented by the development of standardized modules, certifications, and training platforms that cater not only to students but also to working journalists and content creators (Murray et al., 2024).

The modern deepfake detection has a list of significant limitations. Even though significant progress has been witnessed in detecting manipulated audio and video materials, the current systems are fundamentally reactive, segmented, and, technically, one step behind the most competent fake media. Most of the current models are based on fixed markers and easily overcome with future versions of generative modeling, which makes the offered solutions inapplicable in a rapidly evolving threat environment. Research directions in future should thus focus on adaptive real-time architectures that can operate in different formats and languages. At the same time, the social consequences of the spread of deepfake technologies in society, such as the psychosocial effects of this process, political prerequisites, and the possible exacerbation of the collective conflict, are also essential objects of scientific research. Such questions require an investigation into the ways that technical detection can be placed among the popular initiatives aimed at raising awareness, media education programs, and a sufficient legal liability system (Zhu et al., 2020).

Simultaneously, it should be noted that the PR protection methodologies development, like federated learning and differential privacy, deserve long-term attention. Both methods guarantee far-reaching AI training without trading off personal information. Federated learning can train models on terminals close to users and desist from the centralization of data aggregation and related dangers of large-scale tracking. Differential privacy, in its turn, includes mathematical noise into datasets, making each individual observation indistinguishable, but with aggregate patterns maintained. However, both approaches are at the initial phase of the development; the additional theoretical and empirical research is required to improve their efficiency, applicability, and commercial and media affordability. Of special significance is the review of their ethical calculus, in particular, the clash between privacy and utility and the risk of a negative externality to marginalized groups (Von Eschenbach, 2021).

The connections between artificial intelligence (AI) systems and democratic principles in media ecosystems should be addressed in the future research. With AI determining more and more election narratives, social feeling, and what people are subjected to in terms of information, the interaction of AI ethics and political communication needs to be interrogated by researchers. It is necessary to focus specifically on how algorithms affect the population opinion, micro-targeting with the aim of manipulations, as well as the use of AI-powered bots in civil realm. The secondary consequences of media pluralism, freedom of expression and journalistic freedom have not been explored. In this regard, the scholarly research interest must first focus on the construction of democratic protective measures and transparency criteria conditioned specifically to the political communication settings.

At the same time, the inequality in AI research and implementation across the globe should be visited closely. The overwhelming dominance of the Global North to produce frameworks, tools and ethical discourse create a digital inequity that limits access, control and cultural applicability. The most important factor that research agendas should have shown commitment to is inclusion of methodologies that take into account regional realities, diversities associated with language, and socio political sensitivity. It involves providing strength to the local Academic Organisations, helping context-relevant research, and promoting cross-cultural cooperation.

*Table 6. Unresolved Challenges and Suggested Research Trajectories*

| Ethical Challenge | Why It Remains Unsolved | Research Needed |
|---|---|---|
| Algorithmic Opacity | Lack of accessible and standardized explainability tools | Development of user-friendly interpretability frameworks and auditing protocols |
| Inadequate Data Privacy Laws | General IT laws do not cover AI-specific profiling and surveillance | Legislative reforms specific to AI; rights-based data governance models |
| Deepfake Proliferation | Detection tools lag behind generative technologies | Real-time adaptive detection systems; integration with public awareness and legal frameworks |
| Educational Gaps in AI Ethics | Journalism curricula rarely address algorithmic influence | Integration of AI ethics in journalism education; standardized ethics training modules |
| Ethical Trade-offs in Privacy Tech | Limited understanding of the impact on marginalized users | Contextual studies on federated learning and differential privacy across diverse populations |
| Political Manipulation by AI | Lack of regulation and oversight on algorithmic influence in civic discourse | Research on democratic safeguards, civic AI ethics, and misinformation in political contexts |
| Global Disparity in Ethical AI Research | Ethical models largely reflect perspectives from the Global North | Inclusive, region-specific research methodologies and international academic partnerships |
| Labor Impacts of AI in Journalism | Job automation increases without ethical labor policies | Studies on AI and creative labor; human-AI collaboration guidelines |

Addressing these unresolved challenges requires more than theoretical inquiry. Research must be designed for practical implementation, with mechanisms for testing, refining, and disseminating findings. In doing so, future scholarship can build the ethical foundation needed for a resilient and inclusive digital media ecosystem that values both innovation and human dignity.

## 6. CONCLUSION

Visual Enhancements: A world heatmap comparing AI regulation strictness (EU stringent, US moderate, Global South emerging) and an annotated timeline highlighting milestones such as the 2023 Hollywood voice-actor strikes will be integrated. Pedagogical Features: Debate prompts (e.g., 'Can transparency coexist with proprietary algorithms?') will be placed at the end of key sections to encourage critical engagement.

The artificial intelligence has evolved and since then harbored in the digital media regime, and its happening has made the production, distribution and administration of information stuck in a new format. The generation of algorithm-generated news reports and the visual one, the Quasi-institutional selection of their social-media feeds, the algorithmification of the sorting of user-generated content, more often and in more places than not: it is the AI systems that generate this which are critical to the discussion today. Yet there are huge moral debts to this development in technology. The three challenges related to the status quo of the situation, which is characterized by wild proliferation of fabrication, the background of omnipresent electronic monitoring, and no intelligible code of ethics which could have directed on the designing and implementation of AI systems have been outlined in the current discourse. Together, these problems demonstrate how badly the world requires responsible inclusive forms of governance.

The problem of false information is the most excessive among the previous issues. The capabilities of artificial intelligence have negated the barrier of the replicability and proper dissemination of fake or fraudulent contents that often proceed rapidly ahead of identifying and resolving diagnostic mechanisms. Nevertheless, what appears is an ill intent misinformation system producted either through user operated bots, distorted multimedia tiles, or recommendation algorithms, which perform to taint democracy, diminish the citizens trust in journalism, and drive the greatest levels of social polarisation. Shallow fact-check is therefore not enough to avoid such phenomenon, but on the contrary, it is the necessity of shifting the system of building and materializing the use of digital platforms and introducing the accountability to the system of algorithmic design. It is thus a technical and highly ethical concern, and it needs an interdisciplinary action.

Of no lesser severity is the issue of tracking and commercial manipulation of user data. The applications of AI in internet media typically require volumes of information where a personal data exists, to offer personalised services. Though this kind of data sets make efficiency increase and personalization more effective they can also be utilized to perform intrusive overwatching, profile based behaviour analyses, and emotion manipulations. These approaches provoke some intrinsic questions in respect to consent, autonomy, and dignity. The existing data-protection laws, even as a requirement, tend to be broad, and in general, insufficient to comprehend and address the unique issues that AI creates.

Invention and application of AI in the field of media are necessitated but burdened with the shortage of ethics. Despite the spread of AI ethics boards and self-regulatory charters, regulatory frameworks are still either fragmented or non-homogenous. The decision making is too easily delegated or consigned to the sidelines: developers innovate with little knowledge of the norms of journalism; journalists use tools without considering their connotations; policymakers enact laws without understanding the technological facts. Such moral hiatus can be corrected only on the basis of a collective effort. Developers have to integrate the social impact in the system infrastructures, journalists have to commit to transparency and the factor of editorial independence even in automated workflow, and policymakers are meant to offer subtle, actionable, and flexible policies. Ethical AI is not the responsibility of a particular field and is an undertaking of a group of people.

The ethics of AI should also spread to schools. When taking into consideration the teachings about the issue of ethics in journalism and media studies of the future professionals in media and journalism, it is crucial that we introduce the topics about AI ethics in the educational curriculum. Ethical consciousness should not be a separate standalone course since the media education construct must be instilled into it. Students should learn to recognize the stake, predict consequences, and challenge assumptions used by algorithms by engaging them through experiential learning alongside disciplinarily-based, sequentially-group-based, and analysis-based case studies. The work invested to date to protect journalistic integrity is focused not just on creating a generation that can also use technology systems to interfere in the conversation of the masses but on creating a generation who is also capable of responding critically to their technology.

The new direction also requires more investment in innovation and investigation in which ethical considerations are the strength. Elucidable AI, federated learning, and preferred misinformation detection as well as optimistic puzzles are technical boundaries, but also moral obligations. The technologies need to be studied, created, and implemented in a manner that is indicative of shared value, is socially friendly to the greater good, and is considerate of individual agency. Ethical AI is not, therefore, a single endpoint that will be reached but a continuous effort to stay fair, dignified, and just in the world marked by rapid change. The ethical operation of an AI ecosystem defines a consistently ethical AI ecosystem as one that is publicly oriented, participatory, and transparent as postulated by Aguilera et al. (2022) and Ajunwa and Kamer (2024). The path to be taken is hard, it requires attention, teamwork, and boldness.

## REFERENCES


Aguilera, R. V., Waldman, D. A., & Siegel, D. S. (2022). Responsibility and organization science. *Organization Science*, *33*(1), 483–494. DOI: 10.1287/orsc.2021.1518

Ajunwa, I. (2025). Genetic privacy. *Industrial Law Journal*.

Ajunwa, I., & Kamer, A. (2024). Data privacy by contract. *Cardozo Law Review*.

Althabiti, S.. (2023). *Generative AI for explainable automated fact checking*. Springer.

Arora, A., Barrett, M., Lee, E., Oborn, E., & Prince, K. (2023). Risk and the future of AI: Algorithmic bias, data colonialism. *Information and Organization*, *33*(3), 100478. DOI: 10.1016/j.infoandorg.2023.100478

Ayoub, J., Yang, X. J., & Zhou, F. (2021). Combat COVID-19 infodemic using explainable NLP models. *arXiv*. https://arxiv.org/abs/2103.00747

Bansal, G., & Warkentin, M. (2021). Do you still trust? *Information Systems*.

Becker, K. B., Simon, F. M., & Crum, C. (2023). *Policies in parallel? A comparative study of journalistic AI policies*. Unknown Journal.

. Bedford-Strohm, J., Köppen, U., & Schneider, C. (2020). Our AI ethics guidelines. *Ethics of Artificial Intelligence*.

Broussard, M., Diakopoulos, N., Guzman, A. L., Abebe, R., Dupagne, M., & Chuan, C.-H. (2019). Artificial intelligence and journalism. *Journalism & Mass Communication Quarterly*, *96*(3), 673–695. DOI: 10.1177/1077699019859901

Chakraborty, D.. (2020). *Fairway method for bias-free training data*. Unknown Journal.

. Fedha. (2023). AI news presenters mix truth and reality. *Lifewire*.

Fontes, F.. (2022). *On surveillance misuse*. Unknown Journal.

Golda, A.. (2024). *Privacy and security concerns in generative AI: A comprehensive survey*. IEEE., DOI: 10.1109/ACCESS.2024.3381611

Gondwe, G. (2025). *Can AI outsmart fake news?* Emerging Media.

Gongane, V. U., Munot, M. V., & Anuse, A. D. (2024). Explainable AI techniques for detection of fake news and hate speech. *Journal of Computational Social Science*, *7*(1), 587–623. Advance online publication. DOI: 10.1007/s42001-024-00248-9

Hajli, N.. (2021). Social bots and the spread of disinformation. *British Journal of Management*. Advance online publication. DOI: 10.1111/1467-8551.12554

Howard, P. N. (2016). *Computational propaganda*. Oxford Internet Institute.

Kapantai, E.. (2021). A unified taxonomical framework on disinformation. *New Media & Society*. Advance online publication. DOI: 10.1177/1461444820959296

Karnouskos, S. (2020). Artificial intelligence in digital media: The era of deepfakes. *IEEE Transactions on Technology and Society*, *1*(3), 138–147. DOI: 10.1109/TTS.2020.3001312

Mohseni, S.. (2020). Machine learning explanations to prevent overtrust. *arXiv*. https://arxiv.org/abs/2007.12358

Mühlhoff, R. (2023). Predictive privacy: Collective data protection in the context of AI and big data. *Big Data & Society*. Advance online publication. DOI: 10.1177/20539517231166886

Murray, R.. (2024). *Chilling effects on discourse due to surveillance*. Unknown Journal.

Nguyen, D. Q.. (2021). DeepfakeNet: Efficient deepfake detection method. *International Journal of Advanced Computer Science and Applications*.

Praveenraj, J.. (2023). *XAI and decision-interpretation methods*. Unknown Journal.

Roozenbeek, J., Schneider, C. R., Dryhurst, S., Kerr, J., Freeman, A. L. J., Recchia, G., van der Bles, A. M., & van der Linden, S. (2020). Susceptibility to misinformation about COVID-19. *Royal Society Open Science*, *7*(10), 201199. Advance online publication. DOI: 10.1098/rsos.201199 PMID: 33204475

Shao, C., Ciampaglia, G. L., Varol, O., Yang, K.-C., Flammini, A., & Menczer, F. (2018). Spread of low-credibility content by social bots. *Nature Communications*, *9*(1), 4787. DOI: 10.1038/s41467-018-06930-7 PMID: 30459415

Shoaib, M. R.. (2023). Deepfakes, misinformation, and disinformation. *arXiv*. https://arxiv.org/abs/2311.17394

Shu, K., Bhattacharjee, A., Alatawi, F., Nazer, T. H., Ding, K., Karami, M., & Liu, H. (2020). Combating disinformation in a social media age. *Wiley Interdisciplinary Reviews. Data Mining and Knowledge Discovery*, *10*(6), e1385. Advance online publication. DOI: 10.1002/widm.1385

Stahl, B. C., & Wright, D. (2018). *Ethical deployment of AI surveillance*. Unknown Journal.

Tiwari, V. (2023). *Explainable AI techniques for transparency*. Unknown Journal.

Veluru, S. (2024). *Risks of large-scale data breaches in AI systems*. Unknown Journal.

Von Eschenbach, M. (2021). *Black-box AI and public trust*. Unknown Journal.

Zhu, J., Ni, P., & Wang, G. (2020). Activity minimization of misinformation influence in social networks. *IEEE Transactions on Computational Social Systems*, *7*(4), 897–906. DOI: 10.1109/TCSS.2020.2997188

Zhuo, T. Y.. (2023). Exploring AI ethics of ChatGPT. *arXiv*. https://arxiv.org/abs/2301.12867

## KEY TERMS AND DEFINITIONS

**Data Privacy:** The concept of data privacy implies that personal information can be collected, stored, and used according to the decision of the individual. In the context of digital media, the right includes protecting the data of users against its unauthorized use and guaranteeing that the laws and moral principles, which regulate the process of giving consent and dealing with data, are followed.

**Disinformation:** Disinformation refers to out-and-out inventions and distribution of distorted or fake content made to defraud the recipients. In contrast to the term misinformation, disinformation is formulated with manipulative purpose, and more often than not, the action is strategically oriented with political, or commercial ends and as such, presents severe implications to the process of democracy.

**Explainable AI:** In the context of AI development, the concept of explainable AI appears and signifies a specialized field that serves the purpose of making algorithmic reasoning explainable to human parties concerned. Transparency and interpretability will increase trust, accountability, and the process of debugging; those features are of particular importance in fields where smaller mistakes have critical implications: journalism and healthcare.

**Federated Learning:** Federated learning is a paradigm of decentralized machine learning, which allows distributed servers or devices to jointly train statistical models, without directly exchanging raw data. Federated learning reduces or even eliminates data-privacy threats because of the ability to ensure information is kept local and helps to achieve a global estimate, which lowers the chances of data leakage and is becoming a requirement in areas where data confidentiality becomes upgraded.

# APPENDIX

## Additional Reading

To further expand the reader's understanding of ethical artificial intelligence in digital media, the following resources are recommended. While not directly cited in the preceding sections, these scholarly works and reports offer valuable insights into the technical, regulatory, social, and philosophical dimensions of AI ethics, misinformation, and algorithmic accountability.

1. **O'Neil, C. (2016).** ***Weapons of Math Destruction: How Big Data Increases Inequality and Threatens Democracy*****. Crown Publishing.**

This seminal book explains how unregulated algorithms and opaque data models can perpetuate social injustice, particularly when applied in media and public policy. It lays the groundwork for understanding the harm caused by non-transparent AI systems.

2. **Zuboff, S. (2019).** ***The Age of Surveillance Capitalism: The Fight for a Human Future at the New Frontier of Power*****. PublicAffairs.**

Zuboff explores the economic logic behind data extraction and behavioral manipulation, making it a critical resource for those studying surveillance mechanisms in digital media.

3. **Crawford, K. (2021).** ***Atlas of AI: Power, Politics, and the Planetary Costs of Artificial Intelligence*****. Yale University Press.**

This work expands the AI ethics conversation beyond the software layer, linking AI systems to political power, ecological degradation, and global inequality.

4. **Andrejevic, M. (2020).** ***Automated Media*****. Routledge.**

This text focuses on how automation transforms media production and audience interaction, highlighting the loss of editorial oversight and human judgment in AI-powered systems.

5. **Rahwan, I., Cebrian, M., Obradovich, N., Bongard, J., Bonnefon, J. F., Breazeal, C., & Pentland, A. (2019). Machine behaviour. *Nature*, 568(7753), 477–486.**

This article calls for an empirical science of AI behavior, emphasizing the need to treat AI systems as social actors and study their impacts systematically.

6. **Whittaker, M. et al. (2018). *AI Now Report 2018*. AI Now Institute, New York University.**

The annual report from the AI Now Institute provides policy recommendations and critiques of AI deployment, especially within platforms and public institutions, with a focus on accountability and bias.

7. **Ferguson, R. B., & Popović, M. (2023). *Ethics in AI Journalism: A Global Perspective*. Journal of Digital Media Ethics, 7(1), 33–55.**

This article offers a comparative analysis of journalistic practices around AI across different countries and outlines challenges faced by newsrooms in upholding transparency and truth in the algorithmic era.

8. **Helberger, N., Pierson, J., & Poell, T. (2018). *Governing online platforms: From contested to cooperative responsibility*. *The Information Society*, 34(1), 1–14.**

This paper discusses platform accountability and the need for shared governance models between private companies, governments, and civil society—especially relevant to ethical AI governance in media contexts.

# Chapter 6
# AI and Ethics on a Global Scale:
## Challenges, Frameworks, and Solutions

**Hakikur Rahman**
https://orcid.org/0000-0002-2132-1298
*International Standard University, Bangladesh*

## ABSTRACT

*As artificial intelligence (AI) is developed and integrated into a variety of sectors, ethical concerns about development, use and governance have become an important global issue. This chapter examines a variety of ethical challenges in AI, including machine learning distortions, data protection concerns, algorithm accountability, and the social impact of automation. It examines the conditions, guidelines and regulatory approaches of existing ethical frameworks pursued by various regions and international organizations. Through case studies and analysis of emerging trends, this chapter identifies key strategies to ensure ethical AI government, promote transparency and promote integrated innovation. Ultimately, it suggests a practical solution to bridge the ethical gap and generate global consensus on responsible AI development.*

## 1. INTRODUCTION

### 1.1 Overview of AI's growing impact on society

To provide an overview of AI's growing impact on the society, one can mention that the industrial and digital (information) revolutions have unquestionably had a

DOI: 10.4018/979-8-3373-2597-2.ch006

significant impact on almost every facet of business, employment, and society. The question of whether the impending AI revolution will have similarly profound and far-reaching effects is brought up by the field's continued rapid growth. According to this study, artificial intelligence (AI) is already poised to transform a number of industries, radically changing social interactions, economic systems, and ethical issues. AI is expected to be a driving force behind previously unheard-of automation, data-driven decision-making, and the development of intelligent systems, according to historical parallels between the Industrial Revolution, which mechanized production, and the Digital Revolution, which democratized access to information (Brynjolfsson & McAfee, 2020).

Global rivalry will increase, and businesses will use AI-driven data to boost efficiency, innovate across industries, and maximize productivity (Agrawal, Gans, & Goldfarb, 2022). Specifically, existing job functions will be disrupted by the integration of AI in the workplace, requiring reskilling of the workforce to meet the demand for AI expertise (Frey & Osborne, 2021). Beyond their effects on the economy, AI applications will revolutionize global connectivity by allowing smooth digital transactions, improving customer experiences, and providing previously unheard-of access to goods and services across the globe (Bughin et al., 2020).

But these technological developments also bring with them serious ethical issues, such as worries about privacy violations, bias in AI models, and the control of autonomous systems (Jobin, Ienca, & Vayena, 2019; Mittelstadt, 2019). Ethical frameworks and legislative guidelines must change to meet these sproblems and guarantee the fair and open implementation of AI worldwide (Floridi & Cowls, 2021). Despite the enormous potential benefits, the widespread use of AI technologies highlights the necessity of interdisciplinary discussion to address potential hazards and develop strong ethical frameworks for responsible AI governance.

At the same time, artificial Intelligence (AI) has quickly changed many facets of human life, impacting daily interactions, healthcare, governance, and industry. From automated diagnostics in medical to predictive analytics in finance and tailored suggestions in e-commerce, AI-driven systems are becoming essential to decision-making processes (Brynjolfsson & McAfee, 2017). Unprecedented potential have been brought about by the widespread adoption of AI, including increased productivity, real-time data processing, and resource allocation optimization. But this quick expansion also brings up serious moral issues, such as algorithmic bias, invasions of privacy, and possible dangers to human autonomy and employment (Russell, 2021).

Automation and labor markets are two areas where AI is having the biggest effects. Automation driven by AI has raised economic production by improving productivity in industries including manufacturing, shipping, and customer service (Frank et al., 2019). However, as intelligent machines progressively replace human workers in low-skilled and repetitive activities, worries about job displacement remain. Even if AI is predicted to generate new employment prospects, the shift necessitates a substantial reskilling of the workforce in order to accommodate the changing digital economy (World Economic Forum, 2020a).

Governments use AI for public health monitoring, social benefit distribution, and predictive policing, demonstrating the critical role AI plays in governance and societal decision-making. Although these applications can improve equity and efficiency, they also bring up moral conundrums pertaining to discrimination, transparency, and surveillance. For example, it has been discovered that skewed AI algorithms in criminal justice systems disproportionately affect underprivileged groups, hence perpetuating structural injustices (O'Neil, 2016). In order to guarantee that technology developments benefit society fairly and ethically, it is crucial to address these ethical issues as AI develops.

Hence, concern over the effects of AI has led to the emergence of the discipline of artificial intelligence (AI) ethics. In fact, a growing number of high-profile cases of harm have arisen due to either the technology's design flaws (e.g., bias in recidivism cases, loan rejection, and medical misdiagnosis, etc.) or its misuse (e.g., psychometric voter manipulation, facial recognition surveillance, mass data collection without consent, etc.) (Krotzek, 2019; Costa, & Halpern, 2019). In our opinion, AI ethics is still in its infancy and is a subfield of digital ethics, which deals with issues brought up by the creation and application of new digital technologies like blockchain, AI, and big data analytics. There is a growing and developing body of literature that has grown considerably in the last several years (2017–), and it is still developing with evaluations from the perspectives of computer science and philosophy and the humanities (Jobin, Ienca, & Vayena, 2019; Müller, 2020; Gordon, & Nyholm, 2021).

It has been observed that over the past ten years, a rapid series of noteworthy results has increased public awareness of the potential of powerful artificial intelligence. Machine vision researchers have demonstrated systems that can occasionally distinguish things as well as, if not better than, humans. Then came the games. Since sophisticated strategy games have historically been associated with remarkable intelligence, the world took notice when AI systems beat the best human players in chess, Atari games, Go, Shogi, StarCraft, and Dota.

Therefore, AIs not only outperformed humans (which was incredible when it happened), but also showed a more complex way to achieve so by first learning the basics of the games from the play of experienced humans, then from self-play, and eventually from teaching themselves. This raised the possibility of broadly intelligent systems by producing single systems that could learn, play, and win at several structurally different games (Manyika, 2022. Hence, this study discusses on the importance of ethical issues in AI development and deployment, next.

## 1.2 Importance of ethical considerations in AI development and deployment

Researchers, legislators, and business executives are increasingly concerned about ethical issues in the creation and application of artificial intelligence (AI) as it develops. Without ethical supervision, AI technologies may have unforeseen effects, such as bias, discrimination, privacy violations, and security threats, even while they have the ability to completely transform industries like healthcare, banking, and education (Floridi & Cowls, 2019). In order to minimize harm and maximize benefits for people and society, ethical AI development makes sure that AI systems are developed and implemented properly. Incorporating ethics into AI governance ensures broad adoption and responsible innovation by reducing risks and promoting user trust (Jobin, Ienca, & Vayena, 2019).

Bias and fairness are two of the main ethical issues with AI. Large datasets that may contain historical biases reflecting current socioeconomic disparities are used to train AI systems. According to Buolamwini and Gebru (2018), if these prejudices are not addressed, they may lead to biased outcomes that disproportionately impact marginalized communities in areas like recruiting, financing, and law enforcement. Racial prejudice in AI-driven surveillance and identification systems is an issue because, for instance, research has revealed that facial recognition algorithms frequently perform less effectively on people with darker skin tones (Raji et al., 2020). Diverse and representative training datasets, open model development, and ongoing fairness monitoring of AI outputs are all necessary to address these biases.

The ethical application of AI encompasses both human supervision and autonomy. AI should support human judgment rather than completely replace it, especially in high-stakes fields like government, medicine, and military applications (Bryson, 2018b). By combining human supervision with AI automation, human-in-the-loop (HITL) systems can help stop AI from making morally dubious or incorrect conclusions. To ensure that patients receive safe and educated care, AI-driven diagnostic tools in healthcare, for example, should enhance rather than replace medical practitioners' knowledge (Topol, 2019).

Hence, trust, justice, privacy, responsibility, and human oversight are all facilitated by ethical considerations in AI research and application. Interdisciplinary cooperation between engineers, ethicists, and legislators is essential to creating strong ethical standards and regulatory frameworks as AI continues to influence many facets of society (Mittelstadt et al., 2016). Developers and organizations may make sure AI technologies are in line with human values and advance society by incorporating ethical concepts into AI governance.

The significance of ethical considerations in the development and application of artificial intelligence (AI) has grown as it continues to develop and integrate into a variety of industries, such as healthcare, finance, and education. It is crucial to make sure AI systems are developed and used properly in order to avoid unforeseen effects like prejudice, discrimination, invasions of privacy, and security threats. Users and stakeholders are more trusting of ethical AI, which encourages broad adoption and conscientious innovation.

Bias and fairness are two major ethical issues in AI. Large datasets that may contain past prejudices are used to train AI systems, which could result in unfair employment, lending, and law enforcement practices that impact underprivileged communities. The use of representative and varied training datasets, open model development, and ongoing fairness monitoring of AI results are all necessary to address these biases. According to Rivi -Tipei (2023), a comprehensive strategy is required to address the ethical concerns presented by AI in areas including privacy, bias, accountability, transparency, and the effect on employment.

When implementing AI, privacy and data security are also crucial ethical factors. Numerous AI systems depend on enormous volumes of user data, which could be exploited and result in privacy violations and illegal spying if appropriate controls are not in place. By ensuring that AI systems operate without jeopardizing sensitive personal data, privacy-preserving AI strategies like federated learning and differentiated privacy can help reduce these concerns. Regulatory frameworks that emphasize user permission and transparency, such as the General Data Protection Regulation (GDPR) in the European Union, establish legal guidelines for responsible data handling. To properly solve these issues, ethical considerations must be incorporated into AI development (Rivi -Tipei, 2023).

Similarly, accountability and openness in AI decision-making represent yet another important ethical concern. Many AI systems function as "black boxes," making difficult choices without providing transparent explanations of how they arrived at them. In crucial areas like healthcare and criminal justice, where AI-driven choices might have serious repercussions, this lack of interpretability presents difficulties. Explainability strategies must be incorporated into ethical AI development so that stakeholders may comprehend and contest AI-driven judgments. In order to guarantee that AI mistakes or damages can be linked to the appropriate parties, organizations need also set up accountability procedures. Building ethical AI in practice requires a comprehensive approach to ethics in the design and implementation of AI systems, as Chen et al. (2021) point out.

The ethical use of AI encompasses both human supervision and autonomy. Particularly in high-stakes fields like medical, governance, and military applications, AI should support human decision-making rather than completely replace it. By combining human supervision with AI automation, human-in-the-loop (HITL) systems can help stop AI from making morally dubious or incorrect conclusions. To guarantee that patients receive safe and knowledgeable care, AI-driven diagnostic tools in healthcare, for example, should enhance rather than replace medical practitioners' knowledge. Promoting fundamental principles like justice, accountability, and transparency, the focus on responsible AI practices highlights the necessity of including ethical issues into AI design and development (Bano et al., 2023).

In summary, trust, justice, privacy, responsibility, and human oversight are all facilitated by ethical considerations in AI research and application. Interdisciplinary cooperation between engineers, ethicists, and legislators is essential to creating strong ethical standards and legal frameworks as AI continues to influence many facets of society. Developers and organizations may make sure AI technologies are in line with human values and advance society by incorporating ethical concepts into AI governance. The continuous discussion of AI ethics emphasizes the complexity of the issues and the need for an all-encompassing strategy to successfully resolve them (Mbiazi et al., 2023). Hence, we like to shade some light on the global nature of AI ethics with their challenges and opportunities.

## 1.3 The global nature of AI ethics: challenges and opportunities

In many industries and civilizations, artificial intelligence (AI) has become a disruptive force that raises challenging ethical issues that transcend national borders. The global nature of AI ethics is influenced by a variety of sociocultural norms, legal frameworks, and technological advancements; as such, it is imperative that global stakeholders collaborate to create moral guidelines that balance creativity and responsibility (Floridi & Cowls, 2019). Even though AI presents numerous opportunities to enhance governance, healthcare, and economic growth, its ethical use is hindered by problems such as prejudice, privacy infringement, and regulatory fragmentation.

One of the biggest problems with global AI ethics is the absence of a unified regulatory framework. Because of cultural, political, and economic considerations, laws governing AI ethics differ from one nation to another. For instance, the General Data Protection Regulation (GDPR) of the European Union places greater focus on data protection and user consent, but China's AI governance strategy prioritizes societal stability and governmental control (Jobin et al., 2019). These regulatory differences create ethical dilemmas for cross-border AI applications, particularly in relation to algorithmic accountability, data interchange, and monitoring. Without defined ethical norms, AI systems could perpetuate systemic biases and inequality around the world (Binns, 2018b).

However, there are several opportunities related to the internationalization of AI ethics, notwithstanding these challenges. Multilateral initiatives like UNESCO's Recommendation on the Ethics of Artificial Intelligence, which promote collaboration between governments, corporations, and civil society, aim to establish internationally recognized ethical norms (UNESCO, 2021). Additionally, through AI ethics research and policy formulation, underrepresented regions are given the opportunity to provide a range of perspectives, ensuring that AI-driven solutions are inclusive and culturally sensitive (Mittelstadt, 2019). Through international collaboration, nations may leverage AI's potential to advance equitable and sustainable development while lowering ethical risks.

In summary, even though the ethical concerns raised by AI are deeply rooted in regional inequities, international cooperation offers opportunities for creating equitable and ethical AI ecosystems. In order to solve these challenges, a concerted effort must be undertaken to develop ethical AI standards that respect global diversity and social justice as well as fundamental human rights. As AI advances, the need for a morally sound and internationally inclusive approach remains crucial.

## 2. METHODOLOGY

In order to investigate the ethical issues raised by artificial intelligence (AI), the efficacy of current ethical frameworks, and suggested remedies on a worldwide basis, this study uses a qualitative research approach that incorporates a thorough analysis of recent literature and case studies. In order to investigate practical applications of AI ethics, the research process is organized around a narrative and thematic literature review, which is followed by a thorough case study analysis. This study incorporates a variety of worldwide perspectives and guarantees a well-rounded perspective on AI ethics by combining findings from company reports, policy documents, and academic publications.

A systematic framework guides the literature review, which focuses on important themes like ethical issues in AI (such as bias, transparency, privacy, and accountability), current ethical standards (such as the EU AI Act, IEEE AI Ethics guidelines, and UNESCO AI ethics recommendations), and differing international viewpoints on AI governance, contrasting Western, Asian, and Global South approaches. Recent works from the past five to seven years are given priority in the literature selection process, with a focus on peer-reviewed journals, reports from respectable AI ethical organizations, and cross-cultural viewpoints.

This chapter includes a case study analysis of three to four AI applications from various fields, such as surveillance, healthcare, law enforcement, and education, to supplement the literature evaluation. Every case study looks at the moral conundrums raised by a particular AI system, such as bias in recruiting algorithms, false information produced by deepfakes, or moral issues with facial recognition software. The analysis also examines how current ethical frameworks handle these issues and rates any remedies that have been put forth or put into practice. Case studies are chosen based on a variety of criteria, including their proven ethical significance, geographic diversity, and pertinence to current policy discussions.

In order to give a comparative examination of AI ethical frameworks, the results of the case studies and literature are being combined to produce a list of best practices and suggestions for international AI governance.

Hence, to proceed further, this study explores various ethical challenges in relation to artificial intelligence (AI) design, development, and implementation.

## 3. ETHICAL CHALLENGES IN AI

Artificial intelligence's (AI) quick development has brought forth a number of moral dilemmas that affect people, communities, and international political systems. One of the main issues is algorithmic bias, which occurs when biased training data and poor model design cause AI systems to perpetuate and magnify preexisting societal inequities (Binns, 2018a). AI's potential to exacerbate systemic biases rather than lessen them has been brought to light by discriminatory hiring, criminal justice, and healthcare results (Mehrabi et al., 2021). Since AI-driven solutions frequently rely on enormous volumes of personal data, data privacy is another significant issue that raises questions around permission, surveillance, and data security (Jobin et al., 2019).

Furthermore, there is a serious risk associated with AI decision-making processes that lack accountability and transparency, particularly in high-stakes applications like financial decision-making, autonomous weaponry, and predictive policing. The unexpected implications of AI could jeopardize democratic principles, human rights, and fairness in the absence of strong regulatory control and explicit ethical guidelines (Floridi & Cowls, 2019). To ensure that AI development is in line with societal justice and well-being, addressing these ethical conundrums calls for a multidisciplinary strategy that incorporates technical, legal, and philosophical viewpoints.

To proceed, let us focus discussing on algorithmic bias and fairness; data privacy and security; accountability and transparency; autonomy and human control; and societal and economic impact in relation to AI's design, development and implementation based on ethical challenges.

### 3.1 Algorithmic Bias and Fairness

Because it can provide unjust and discriminating results that exacerbate already-existing societal imbalances, algorithmic bias is a serious ethical challenge in artificial intelligence (AI). Biased training data, poor model design, or skewed decision-making processes are common sources of bias in AI systems, which disproportionately disfavor particular demographic groups (Mehrabi et al., 2021). For instance, research indicates that facial recognition software makes more mistakes for people with darker skin tones, raising questions about potential racial prejudice in identification verification and law enforcement (Buolamwini & Gebru, 2018). Similar to this, it has been discovered that biased recruiting algorithms prefer male applicants over female ones because of historical gender disparities in the workforce. This raises ethical and legal questions regarding the fairness of automated decision-making (Binns, 2018a). AI systems have the potential of escalating societal inequities rather than reducing them in the absence of proactive measures.

A multifaceted strategy that incorporates clear algorithmic design, diverse and representative datasets, and strong regulatory monitoring is needed to ensure fairness in AI. To reduce bias in AI models, researchers have suggested fairness-aware machine learning approaches such adversarial debiasing and reweighting algorithms (Zemel et al., 2013). Furthermore, developing fairness standards that are consistent with society ideals requires interdisciplinary cooperation between ethicists, legislators, and technologists (Floridi & Cowls, 2019). Fairness is emphasized as a fundamental tenet in AI governance by a number of international projects, such as the European Union's AI Act and UNESCO's ethical principles for AI (UNESCO, 2021). All people, regardless of gender, color, or socioeconomic background, can benefit from more equitable and reliable AI systems created by stakeholders by integrating fairness concerns at every level of AI development.

## 3.2 Data Privacy and Security

Data security and privacy have become critical issues as AI systems depend more and more on enormous volumes of personal data. From personalized marketing to medical diagnostics, AI-driven applications necessitate a large amount of data collection, which frequently raises the danger of sensitive data misuse, data breaches, and illegal access (Sharma et al., 2020). Ensuring adherence to international data protection laws, such as the California Consumer Privacy Act (CCPA) in the US and the General Data Protection Regulation (GDPR) in Europe, which place stringent restrictions on data collection, processing, and storage, is a significant challenge (Voigt & von dem Bussche, 2017). However, a lot of AI models—especially deep learning systems—operate as "black boxes," which makes it challenging to guarantee accountability and openness in data processing (Goodman & Flaxman, 2017).

Additionally, adversarial assaults and data poisoning, which can alter training data to produce skewed or malevolent results, are cyberthreats that AI systems are susceptible to (Papernot et al., 2018). Particularly in authoritarian countries, the growing use of biometric and face recognition data has sparked ethical questions about potential violations of civil liberties and mass monitoring (Binns, 2020).

Researchers support privacy-preserving AI methods like differential privacy and federated learning, which enable data analysis without explicitly disclosing sensitive information, in order to reduce these dangers (Abadi et al., 2016). Strong data privacy and security protocols will be essential as AI develops in order to preserve public confidence and avoid moral lapses in AI-driven decision-making.

## 3.3 Accountability and Transparency

To reduce ethical hazards and increase public confidence in AI, it is essential to guarantee accountability and transparency in AI systems. Deep learning-based AI models, in particular, frequently function as "black boxes," making it challenging to comprehend how they make judgments (Burrell, 2016). In high-stakes industries like healthcare, banking, and criminal justice, where those impacted by AI-driven decisions have few options for challenging them, this lack of transparency can result in biased or incorrect outcomes (Doshi-Velez & Kim, 2017). Responsibility for AI failures is unclear in the absence of defined accountability procedures, which enables developers, companies, and legislators to place the blame elsewhere rather than deal with the underlying issues that lead to unethical AI conduct (Mittelstadt et al., 2016).

Experts support explainable AI (XAI) models that offer decisions that are both interpretable and justified in order to improve accountability (Gunning et al., 2019). Furthermore, legal frameworks that prioritize the "right to explanation," like the General Data Protection Regulation (GDPR) of the European Union, guarantee that people who are affected by AI choices can ask questions about how their data is used (Brkan, 2019). Open datasets that permit outside examination of AI systems, algorithmic effect analyses, and ethical audits are additional components of transparency in AI research (Binns, 2018b). Organizations may develop AI technologies that are not only efficient but also consistent with moral and social values by cultivating a culture of responsibility and openness.

### 3.3.1 The "Black Box" Problem in AI Models

The lack of interpretability and transparency in sophisticated machine learning algorithms, especially deep learning systems, is known as the "black box" problem in AI models. Although these models are very good at seeing patterns and making decisions, consumers, developers, and regulators frequently don't understand how they work (Burrell, 2016). Because AI models—particularly neural networks—process enormous volumes of data across multiple hidden layers, it becomes challenging to determine how particular inputs result in particular outputs, which causes this opacity (Lipton, 2018). Because of this, AI-driven choices in crucial domains like healthcare, finance, and criminal justice continue to be mainly puzzling, which raises questions regarding fairness, prejudice, and responsibility (Castelvecchi, 2016).

Since anyone impacted by AI-driven decisions might not be able to question or comprehend them, the black box conundrum has important ethical and legal ramifications. For instance, it is challenging to spot mistakes or biases in the model when an AI system rejects a loan application or misdiagnoses a patient because to its lack of interpretability (Doshi-Velez & Kim, 2017). Researchers are working to solve these issues by creating Explainable AI (XAI) methods that seek to improve the transparency and interpretability of AI decision-making (Gunning et al., 2019). Furthermore, algorithmic responsibility and the "right to explanation" in automated decision-making are emphasized by regulatory frameworks as the General Data Protection Regulation (GDPR) of the European Union (Brkan, 2019). Since building trust and implementing AI ethically requires striking a balance between accuracy, efficiency, and interpretability, solving the "black box" problem is still a major hurdle in AI development.

### 3.3.2 Challenges in Establishing AI Accountability in Decision-Making

The intricate, decentralized structure of AI systems and the quick speed of technological advancement make it difficult to establish accountability in AI decision-making. The "black box" aspect of many AI algorithms, especially deep learning models, is one of the main challenges since it makes it hard to comprehend how choices are made or who is in charge of their results (Burrell, 2016). When an AI system makes biased, incorrect, or damaging decisions—like erroneous arrests or misdiagnoses in healthcare—this opacity makes it more difficult to assign blame (Dastin, 2018). Additionally, the identification of a single accountable body is complicated by the engagement of various stakeholders in the development and deployment of AI systems, including governments, private enterprises, and data scientists (Mittelstadt et al., 2016). Furthermore, machine learning procedures, in which the algorithm changes in response to fresh data, are frequently used to develop AI systems. This makes it even more challenging to link the decision-making process to particular people or activities (O'Neil, 2016).

Furthermore, legal and regulatory structures are yet ill-prepared to address the particular difficulties presented by artificial intelligence. Although some areas have started enacting laws to increase accountability and transparency, like the European Union with the General Data Protection Regulation (GDPR), these initiatives are still in their infancy and frequently lack thorough enforcement mechanisms (Brkan, 2019). This problem is made worse by the lack of globally accepted guidelines for AI responsibility, since businesses may operate in countries with different legal requirements. A multifaceted strategy is needed to address these issues, including the creation of explainable AI, more transparent regulatory frameworks, and improved stakeholder collaboration to guarantee the ethical and responsible usage of AI systems.

## 3.4 Autonomy and Human Control

The problem of maintaining significant human control over decision-making processes has become more pressing as artificial intelligence (AI) systems grow more independent. The ability of a system to function without direct human interaction is known as autonomy in AI, and it raises ethical questions around transparency, responsibility, and possible unexpected effects (Bryson, 2020). Autonomous AI has the potential to improve productivity across a range of industries, including healthcare and banking, but it also runs the risk of undermining human control, especially in crucial areas like military applications and legal decision-making (Russell, 2021). To guarantee that AI functions within moral bounds and preserve human control over crucial choices, a well-balanced framework is required (Floridi & Cowls, 2019).

Human-in-the-loop (HITL) methods to AI governance have been highlighted by the European Union's High-Level Expert Group on AI, which supports systems in which human monitoring is still crucial to AI decision-making (European Commission, 2020). The significance of human agency in AI deployments is also emphasized by the IEEE's Ethically Aligned Design methodology (IEEE, 2019). Ensuring a balance between human control and machine autonomy will be essential as AI develops in order to promote moral, just, and accountable AI systems globally.

## 3.5 Societal and Economic Impact

Artificial intelligence's (AI) quick development has significant societal and economic ramifications, impacting social institutions, employment, and wealth distribution. Automation-related job displacement is one of the most urgent issues. In sectors including manufacturing, transportation, and customer service, AI-powered systems are progressively displacing human labor, which raises worries about income inequality and unemployment (Frey & Osborne, 2017). Even if AI opens up new career paths in cutting-edge domains like data science and machine learning, the shift necessitates a large workforce reskilling effort (Brynjolfsson & McAfee, 2018). Economically, artificial intelligence (AI) can boost productivity and efficiency, which leads to industrial transformation and economic growth. However, the advantages are frequently concentrated among highly qualified individuals and major tech companies, which exacerbates the digital divide (Acemoglu & Restrepo, 2020).

AI has an impact on governance, ethics, and societal dynamics in addition to economic ones. Particularly in sectors like employment, law enforcement, and financial services, algorithmic biases in AI decision-making systems have the potential to exacerbate already-existing societal injustices (Eubanks, 2018). To reduce these dangers, ethical AI governance frameworks—like those put out by the European Union and the OECD—place a strong emphasis on accountability, equity, and openness (OECD, 2019a; European Commission, 2020). Concerns around false information, privacy, and mental health have also increased as a result of AI's incorporation into digital platforms and social media (Zuboff, 2019). Policymakers and business executives must enact fair AI laws, make training and education investments, and encourage responsible AI deployment that puts the welfare of society first in order to guarantee AI promotes equitable and sustainable economic growth.

Attending the above mentioned issues in relation to ethical challenges in AI, the study discusses on global ethical frameworks, next.

## 4. GLOBAL ETHICAL FRAMEWORKS FOR AI

To guarantee that AI technologies are created and applied in ways that respect democracy, human rights, and the rule of law, international ethical frameworks for AI have been established. The Framework Convention on Artificial Intelligence, which was introduced in September 2024 by the Council of Europe, is a major step forward in this attempt. An innovative move in global AI regulation, this legally enforceable agreement seeks to align AI system lifecycles with democratic values and fundamental rights (Council of Europe, 2024).

International organizations have been crucial in creating ethical standards for AI in addition to legal tools. The United Nations Educational, Scientific, and Cultural Organization (UNESCO), for example, has played a significant role in promoting international discussion on AI ethics. In support of laws that protect fundamental human rights, UNESCO highlights the significance of making sure AI technologies benefit society and are in line with sustainable development goals (UNESCO, n.d.).

In addition, the World Economic Forum has put forth ethical AI models, emphasizing the need for a range of viewpoints to establish strong ethical AI procedures. This strategy emphasizes how important inclusive discourse is in tackling the moral dilemmas raised by AI developments (World Economic Forum, 2020b). Table-1 gives a comparison table on ethical principles and implementation gaps for different organizations/entities, while table-2 is showing various categorization of ethical frameworks.

*Table 1. Comparative Table: Ethical Principles vs. Implementation Gaps*

| **Ethical Principle** | **UNESCO AI Ethics** | **OECD AI Principles** | **EU AI Act** | **China's AI Policy** | **Implementation Gap** |
|---|---|---|---|---|---|
| Fairness & Bias | √ | √ | √ | √ | Data biases persist in facial recognition |
| Transparency | √ | √ | √ | ⚠ | Lack of algorithmic explainability |
| Accountability | √ | √ | √ | ⚠ | Weak enforcement mechanisms |

*Table 2. A Taxonomy Table: Categorizing Ethical Frameworks*

| **Category** | **Example Frameworks** | **Legal Nature** | **Focus Area** |
|---|---|---|---|
| **Human Rights-Based** | UNESCO AI Ethics, UN Human Rights AI Guidelines | Soft Law | Fairness, Non-Discrimination |
| **Economic Regulation** | EU AI Act, USA AI Bill of Rights | Hard Law | Market Competition, Compliance |
| **Security & Surveillance** | China's AI Governance, Russia's AI Policy | Hard Law | National Security, AI Surveillance |

Together, these frameworks and programs show a determined international effort to negotiate the morally challenging aspects of artificial intelligence while attempting to strike a balance between the advancement of technology and the upholding of fundamental human values.

## 4.1 International AI Ethics Guidelines and Policies

To encourage the appropriate development and application of AI technologies, numerous organizations have produced international principles and policies on AI ethics. The OECD AI Principles, which were developed by the Organization for Economic Co-operation and Development (OECD), place a strong emphasis on human-centered values, inclusive growth, transparency, robustness, and responsibility in AI systems (OECD, 2024). Comparably, the Ethics Guidelines for Trustworthy AI, published by the European Union's High-Level Expert Group on AI, promote AI that is morally sound, legal, and resilient throughout its lifecycle (European Commission, 2019). The Recommendation on the Ethics of Artificial Intelligence, which was adopted by the United Nations Educational, Scientific, and

Cultural Organization (UNESCO), offers a global framework to guarantee that AI systems respect environmental sustainability and human rights (UNESCO, 2021).

In order to safeguard individual rights in the context of AI, the Center for AI and Digital Policy also put up the Universal Guidelines for AI, which seek to incorporate moral principles into national legislation and international accords (CAIDP, 2018). When taken as a whole, these programs show a dedication to moral AI governance that aims to strike a balance between technology advancement and core human principles. Table-3 summarizes a few AI ethics guideline.

*Table 3. Summarizing the key AI ethics guidelines found during this research.*

| Organization | AI Ethics Guideline | Key Focus Areas | Year |
|---|---|---|---|
| **OECD** | OECD AI Principles | Human-centered values, inclusive growth, transparency, robustness, responsibility | 2024 |
| **European Commission** | Ethics Guidelines for Trustworthy AI | Morality, legality, resilience throughout AI lifecycle | 2019 |
| **UNESCO** | Recommendation on the Ethics of AI | Human rights, environmental sustainability | 2021 |
| **CAIDP** | Universal Guidelines for AI | Embedding moral principles in national and international laws | 2018 |

## 4.2 Regional and National AI Ethics Approaches

Because of their diverse cultural, legal, and socioeconomic contexts, many countries and regions have created unique approaches to artificial intelligence (AI) ethics. With the 2019 introduction of the Model AI Governance Framework, which offers comprehensive guidelines for the ethical deployment of AI with a focus on transparency and human-centricity, Singapore has taken the lead in Asia (Diligent Corporation, 2024). Similar to this, the Republic of Korea demonstrated its dedication to ethical AI development by releasing its National Strategy for Artificial Intelligence in December 2019, the Human-Centered Artificial Intelligence Ethics Standards in 2020, and the Digital Bill of Rights in May 2023 (Digital Watch Observatory, 2024).

The Artificial Intelligence and Data Act (AIDA), which Canada has proposed in North America, aims to regulate AI systems by making sure they are developed and utilized responsibly, with an emphasis on accountability and transparency (Digital Watch Observatory, 2024). In order to address AI ethics and governance, the US has adopted a sectoral strategy, with several states passing their own AI laws in addition to federal programs (Diligent Corporation, 2024).

In an attempt to standardize AI ethics throughout the continent, the African Union (AU) is developing a draft AI policy that includes industry-specific regulations, regulatory sandboxes, and the establishment of national AI councils to supervise the ethical application of AI (Keymakr, 2024).

Adapted to particular social requirements and beliefs, these regional and national efforts show a global commitment to integrating ethical issues into AI development and implementation. Table-4 summarizes the key AI ethics by region.

*Table 4. A structured overview of AI ethics initiatives across different regions.*

| Region/ Country | AI Ethics Initiative | Key Focus Areas | Year |
|---|---|---|---|
| **Singapore** | Model AI Governance Framework | Ethical AI deployment, transparency, human-centricity | 2019 |
| **South Korea** | National Strategy for AI, Human-Centered AI Ethics Standards, Digital Bill of Rights | Ethical AI development, human rights, digital governance | 2019–2023 |
| **Canada** | Artificial Intelligence and Data Act (AIDA) | Responsible AI development, accountability, transparency | Proposed |
| **United States** | Sectoral AI regulations (state and federal laws) | AI governance through industry-specific policies | Ongoing |
| **African Union** | Draft AI Policy | Standardized AI ethics, industry-specific regulations, national AI councils | In development |

## 4.3 Corporate and Industry Standards

Companies and business executives are progressively setting ethical guidelines in the field of artificial intelligence (AI) to direct the responsible development and application of AI technology. These company policies frequently highlight values like accountability, openness, equity, and respect for user privacy. The World Economic Forum, for example, emphasizes how crucial it is for businesses to embrace moral AI principles in order to build confidence and guarantee that AI systems are advantageous to all parties involved (World Economic Forum, 2021).

Furthermore, a framework for AI management systems is being established by international standards such as ISO/IEC 42001, which will guarantee that AI applications are in line with organizational goals and ethical standards at every stage of their development (Ernst & Young, 2024). This standard emphasizes how important it is to continue using AI responsibly from conception to implementation and beyond.

In addition to these, industrial partnerships are significantly influencing AI ethics. Businesses like xAI and Anthropic are becoming benefit firms, dedicated to both profit and societal good. Their capacity to fulfill these promises, however, is continuously questioned, underscoring the difficulties in maintaining responsibility in the AI industry (Financial Times, 2025).

In an effort to increase public trust and reduce any hazards related to the use of AI, these initiatives show that the corporate sector is beginning to recognize the importance of aligning AI technologies with ethical norms.

## 4.4 Case studies on biased AI decision-making (e.g., hiring, law enforcement, healthcare)

Hiring Bias in AI Recruitment Systems

Companies have been using AI-driven hiring systems more and more to expedite the hiring process, however it has been discovered that these algorithms perpetuate preexisting biases, especially those that target women and minority groups. One noteworthy example is the AI recruitment tool used by Amazon, which was trained on resumes submitted over a ten-year period, the majority of which were from male candidates. As a result, the system devalued resumes that contained phrases like "women's" or were linked to sectors that were dominated by women (Dastin, 2018). The method demonstrated how skewed training data might result in discriminatory hiring practices by effectively penalizing female applicants. Amazon eventually dropped the tool because it couldn't guarantee fairness, even with efforts to reduce prejudice (Raghavan et al., 2020).

The dangers of using AI in recruiting without thorough fairness audits and algorithmic openness are highlighted by this example.

Racial Bias in Predictive Policing

Designed to detect potential offenders and predict crime hotspots, predictive police algorithms have come under fire for perpetuating racial biases. It has been discovered that the U.S. criminal justice system's COMPAS (Correctional Offender Management Profiling for Alternative Sanctions) algorithm incorrectly labels White defendants as low risk while disproportionately classifying Black defendants as high-risk criminals (Angwin et al., 2016). Black defendants were almost twice as likely as white defendants to be wrongly classified as high-risk, which resulted in harsher punishment and parole choices, according to a ProPublica analysis. Historical crime statistics that revealed structural racial inequities in law enforcement were the source of the bias, which fueled a discriminatory cycle.

The risks of employing AI in critical decision-making without appropriate accountability and bias mitigation techniques are brought to light by this case.

Bias in AI-Based Healthcare Systems

Disparities in medical care have resulted from the notable bias of AI algorithms employed in healthcare decision-making. An AI-driven healthcare risk assessment tool consistently understated the health needs of Black patients in comparison to White patients, according to a 2019 study by Obermeyer et al., 2019 Racial inequities in access to medical services were not taken into consideration by the model, which was extensively employed in hospitals across the United States and depended on healthcare spending as a proxy for patient health. This exacerbated already-existing healthcare disparities by resulting in fewer critical care recommendations for Black patients than for White patients.

The instance highlights the necessity for fairness-aware machine learning techniques in healthcare contexts and the dangers of utilizing biased proxies in AI models.

### 4.4.1 Case studies of ethical AI implementation in corporate settings

Numerous companies have shown their dedication to responsible technology use by effectively implementing ethical AI policies.

IBM's AI Fairness Framework

By creating the AI Fairness 360 Toolkit, an open-source tool for identifying and reducing bias in AI models, IBM has taken a proactive approach to AI ethics. IBM's commitment to equity and responsibility in AI applications is demonstrated by this program (IBM, 2020).

Anthropic's Ethical AI Commitment

Anthropic, a business that prioritizes AI safety, sees itself as a pioneer in the creation of moral AI. However, questions were raised regarding its dedication to moral behavior when it became involved in campaigning against California's Assembly Bill 331, which imposed strict AI safety standards. This episode demonstrates how difficult it is for businesses to match their stated ethical values with their behavior (Vincent, 2025).

PepsiCo's AI Integration with Ethical Considerations

PepsiCo has adopted AI to improve product innovation and operational efficiency. PepsiCo investigates AI applications while addressing ethical issues in its implementation in partnership with KoiReader Technologies and the Stanford Institute for Human-Centered Artificial Intelligence (PepsiCo, 2023).

Accenture's Responsible AI Program

To guarantee moral AI governance, Accenture has set up a thorough responsible AI compliance program. Through open and responsible AI practices, this program seeks to safeguard the company and foster stakeholder trust (Accenture, 2022).

Fractal's Ethical AI Practices

The AI and analytics firm Fractal places a strong emphasis on incorporating moral AI principles into its corporate strategy. The business tackles issues with tracking AI model performance and striking a balance between ethical issues and technological growth (Fractal, 2021).

These case studies highlight both the achievements and difficulties in bringing AI practices into compliance with ethical norms, demonstrating the variety of strategies businesses are using to apply ethical AI. Table-5 summarizes the key AI ethics in the mentioned case studies.

*Table 5. A structured overview of how different companies are addressing AI ethics.*

| Company | Ethical AI Initiative | Key Focus Areas | Year |
|---|---|---|---|
| **IBM** | AI Fairness 360 Toolkit | Bias detection and mitigation in AI models | 2020 |
| **Anthropic** | Ethical AI Commitment | AI safety, policy advocacy challenges | 2025 |
| **PepsiCo** | AI Integration with Ethical Considerations | Ethical AI in product innovation and operations | 2023 |
| **Accenture** | Responsible AI Program | AI governance, compliance, stakeholder trust | 2022 |
| **Fractal** | Ethical AI Practices | AI model performance tracking, balancing ethics and technology | 2021 |

## 5. STRATEGIES AND SOLUTIONS FOR ETHICAL AI DEVELOPMENT

After focusing global ethical frameworks for AI, this section dedicates towards various strategies and solutions for ethical AI development.

We understand that creating ethical AI necessitates a multipronged strategy that incorporates inclusive policymaking, technical safeguards, and regulatory frameworks. In this aspect, explainable AI (XAI) is a crucial tactic that improves transparency by enabling consumers and regulators to understand AI decision-making processes (Doshi-Velez & Kim, 2017). Furthermore, accountability and adherence to human rights norms can be guaranteed by incorporating ethical criteria as IEEE's Ethically Aligned Design principles and the EU's Artificial Intelligence Act (European Commission, 2021; IEEE, 2019). Since biases in AI systems can result from skewed or inadequate data, bias prevention through diverse and representative training datasets is another essential component (Mehrabi et al., 2021).

In addition to these, especially in high-risk AI applications, human-in-the-loop (HITL) procedures enable real-time human monitoring to prevent inadvertent injury (Amershi et al., 2019). Last but not least, international cooperation amongst stakeholders—including governments, academic institutions, and business sectors—can support uniform ethical norms and fair AI implementation in many socioeconomic circumstances (Jobin, Ienca, & Vayena, 2019). By proactively addressing ethical issues, these tactics can reduce threats to people and society while promoting trust in AI systems.

## 5.1 Ensuring Fair and Transparent AI Models

Explainability strategies, algorithmic audits, and regulatory supervision are all necessary to guarantee equity and openness in AI models. By using bias-detection frameworks that evaluate divergent consequences on various demographic groups, such as IBM's AI Fairness 360 and Fairness Indicators, algorithmic fairness can be promoted (Bellamy et al., 2019). Furthermore, by offering human-readable explanations for AI decisions, interpretable machine learning techniques like SHAP (Shapley Additive Explanations) and LIME (Local Interpretable Model-Agnostic Explanations) improve transparency (Ribeiro, Singh, & Guestrin, 2016).

Third-party AI audits and adherence to ethical AI principles, like those described in the OECD AI Principles and the General Data Protection Regulation (GDPR) of the European Union, guarantee accountability and compliance with legal standards, which further builds trust (OECD, 2019a; Wachter, Mittelstadt, & Floridi, 2017). Additionally, by encouraging public participation through participatory AI governance, stakeholders—from end users to legislators—can offer feedback on moral issues, advancing the deployment of AI in an equitable manner (Mittelstadt, Russell, & Wachter, 2016). By combining these strategies, AI researchers can try to lessen biases and improve the interpretability of AI models, creating more equitable and reliable systems.

## 5.2 Strengthening AI Governance and Regulation

To guarantee that AI systems are open, responsible, and consistent with society norms, effective governance and regulation are crucial. International organizations and governments are pushing more and more for regulatory frameworks that strike a balance between innovation and morality. One of the most extensive legislative initiatives is the Artificial Intelligence Act of the European Union, which classifies AI applications according to risk levels and places stringent compliance requirements on high-risk AI systems (European Commission, 2021). Similarly, as a global guideline for responsible AI adoption, the OECD AI Principles place a strong emphasis on equity, openness, and human-centered AI development (OECD, 2019b).

Corporate AI governance frameworks in the private sector, like Microsoft's Responsible AI effort and Google's AI Principles, establish internal ethical norms for the use of AI (Brundage et al., 2020). To reduce hazards including prejudice, discrimination, and false information, strengthening AI governance also calls for algorithmic auditing, impact assessments, and independent oversight bodies (Jobin, Ienca, & Vayena, 2019). Furthermore, developing flexible and enforceable AI regulations that take into account a range of viewpoints requires cooperation from multiple stakeholders, including governments, business executives, and civil society (Floridi et al., 2018). Continuous improvement of regulatory frameworks will be required as AI technologies advance in order to handle new issues and promote reliable and moral AI ecosystems.

## 5.3 Promoting Ethical AI through Education and Awareness

In order to prepare the upcoming generation of AI developers, policymakers, and users to comprehend and handle the ethical issues related to AI technology, it is imperative that ethical AI be promoted through education and awareness. Since it helps students understand how their work affects society and promotes responsible decision-making, integrating AI ethics into academic courses is an essential first step (Floridi et al., 2018). For instance, classes and specialized programs covering subjects like algorithmic bias, privacy, transparency, and accountability in AI systems are becoming more widely available at universities across the globe (Dastin, 2020). By offering tools and establishing forums for conversations on AI's contribution to the attainment of sustainable development goals, initiatives like as the AI4Good Foundation encourage education about the ethical implications of AI (AI4Good, 2021).

Additionally, increasing public knowledge of AI's potential and ethical issues can promote an informed populace better able to participate in discussions about responsibility and regulation (Gursoy, 2020). Governments, academic institutions, businesses, and civil society organizations must work together to disseminate information and guarantee that the moral precepts governing AI are ingrained in both its creation and use (Binns, 2018c). Thus, awareness and education are effective means of encouraging responsible AI, making sure that both developers and users are aware of their moral obligations in the rapidly changing digital world.

## 5.4 Collaborative and Multistakeholder Approaches

Because they bring together a variety of viewpoints from different sectors, such as academia, industry, government, and civil society, collaborative and multistakeholder approaches are crucial to guaranteeing the ethical development and application of AI technologies. These methods encourage inclusive decision-making and guarantee that AI systems are developed to meet the requirements of all parties involved, including those who might be vulnerable or marginalized (Harris & Jirotka, 2019). Initiatives such as the Partnership on AI, for example, give stakeholders a forum to discuss issues like algorithmic bias and data protection, exchange best practices, and work together to define ethical AI principles (Partnership on AI, 2021).

In a similar vein, governments, business executives, and academic institutions work together to create AI solutions that uphold human rights and advance sustainable development through the Global Partnership on AI (GPAI) (GPAI, 2021). Involving several stakeholders guarantees that artificial intelligence is not just technically sophisticated but also in line with social norms and answerable to the larger community (Floridi et al., 2018). Collaborative techniques foster trust and raise the possibility that AI technologies will have beneficial, far-reaching societal effects by promoting collaboration and communication across sectors.

After discussing various strategies and solutions for ethical AI development and their implications, this study likes to conclude by providing a few future research directions.

## 6. CONCLUSION AND FUTURE DIRECTIONS

### 6.1 Summary of Key Ethical Challenges and Frameworks

There are several ethical issues with the creation and application of AI technology, which call for thorough frameworks to resolve. Algorithmic bias is one of the main problems, where AI systems could unintentionally reinforce or worsen already-existing social injustices (O'Neil, 2016). In industries like recruiting, law enforcement, and healthcare, these biases—which are frequently ingrained in data or through design—can result in biased consequences (Binns, 2018c). Furthermore, people find it challenging to comprehend how decisions are made because to the opaqueness of AI decision-making processes, which undermines responsibility and confidence in AI systems (Burrell, 2016).

The enduring problem of algorithmic bias in AI systems, which keeps undermining equality and justice, has been the subject of recent research. For example, studies have demonstrated that biased data used to train AI models can produce unjust results, particularly in delicate domains like loan approval, hiring, and law enforcement (Mehrabi et al., 2021). Furthermore, one of the biggest ethical problems with AI decision-making is its lack of transparency. Reduced accountability may result from stakeholders' inability to comprehend and accept the reasoning behind decisions as AI models grow more complicated (Kroll et al., 2020). Scholars stress the necessity of strong, explicable AI frameworks that give fairness, accountability, and transparency top priority in the development and use of AI systems in order to allay these worries (Noble, 2021). Table-6 summarizes the key ethical issues in AI and their implications.

*Table 6. The key points while maintaining clarity in terms of AI based ethical issues.*

| Ethical Issue | Description | Implications | References |
|---|---|---|---|
| **Algorithmic Bias** | AI systems may reinforce or worsen social injustices due to biased training data or flawed design. | Leads to unfair outcomes in hiring, law enforcement, healthcare, and finance. | O'Neil (2016), Binns (2018), Mehrabi et al. (2021) |
| **Lack of Transparency** | AI decision-making processes are often opaque, making it difficult for users to understand how decisions are made. | Reduces accountability and trust in AI systems. | Burrell (2016), Kroll et al. (2020) |
| **Accountability Challenges** | The complexity of AI models makes it difficult to assign responsibility for errors or biases. | Weakens regulatory oversight and ethical AI governance. | Noble (2021) |

Along these perspectives, the influence on privacy is another major issue, as AI systems depend more and more on enormous volumes of personal data to operate, which raises questions about data security and the possibility of surveillance (Zeng, 2020). Furthermore, because existing legal frameworks are unable to keep up with technical improvements, a regulatory gap results, which raises governance concerns due to the rapid rate of AI development (Jobin, Ienca, & Vayena, 2019).

Many ethical frameworks have been put forth to address these issues. For example, the European Commission's Ethics Guidelines for Trustworthy AI offer a strong set of guidelines emphasizing responsibility, transparency, and human-centric AI (European Commission, 2019). Similar to this, the IEEE's Ethically Aligned Design effort promotes a human-centered approach to AI by highlighting the significance of including ethical issues into the development and implementation of autonomous systems (IEEE, 2019). These kinds of frameworks offer a path forward for making sure AI systems are created in a way that is both technically possible and morally righteous, guaranteeing that the advantages of AI are shared fairly and in line with social norms (Floridi et al., 2018). By using these frameworks, the advancement of AI technology can be guided toward ethical innovation that puts people's and communities' welfare first.

## 6.2 The Future of AI Ethics: Trends and Emerging Issues

It has been observed that the future of AI ethics will be influenced by a number of significant trends and new problems that require immediate consideration as AI technology develop. The growing focus on AI accountability and the drive for explainability in AI models are two noteworthy trends. People need to be able to comprehend and challenge AI choices as these systems become more sophisticated and widespread, especially in delicate fields like healthcare, criminal justice, and employment (Lipton, 2018). Researchers are concentrating on developing methods to make these models more interpretable without compromising performance, as this trend has resulted in the development of AI models that value transparency (Gilpin et al., 2018).

At the same time, the emphasis on AI fairness and the fight against systemic biases that might be ingrained in AI algorithms is another trend. With new frameworks emerging to direct the development of fairer and more equitable AI, there is growing desire that AI systems be built to prevent perpetuating current societal disparities (Noble, 2018).

Furthermore, the increasing influence of AI on labor markets and the nature of work in the future is a developing ethical concern in AI. Concerns over job displacement and the moral ramifications of automating human labor are growing as AI technologies advance (Brynjolfsson & McAfee, 2017). Legislators and corporations are looking into measures to guarantee that automation powered by AI benefits society overall, including the possibility of upskilling employees and generating new positions that complement AI technologies (Chui, Manyika, & Muir, 2016). Furthermore, ethical questions about civil liberties and privacy are brought up by the quick deployment of AI in surveillance technologies. Calls for stricter laws governing AI-driven monitoring have arisen as a result of discussions concerning how to strike a balance between human freedoms and security (Zeng, 2020).

Governments, corporations, academic institutions, and civil society organizations will need to work together continuously as these tendencies continue to develop in order to guarantee that AI technologies advance in ways that are advantageous, just, and consistent with human values.

## 6.3 Recommendations for Fostering a Global Consensus on Ethical AI

It is crucial that stakeholders from different industries unite and work together to create universal standards that may direct the development and application of AI in order to promote a worldwide agreement on ethical AI. To create common values, norms, and standards that support reliable AI while addressing potential hazards, governments, academic institutions, business executives, and civil society organizations must collaborate (Floridi et al., 2018). The establishment of an international framework outlining core ethical values, such as accountability, openness, fairness, and privacy, is one suggestion (Jobin, Ienca, & Vayena, 2019).

To guarantee that AI technologies are in line with human values globally, such a framework might be based on current international human rights and data protection agreements (Bryson, 2018a).

Multilateral discussions involving nations with various political, economic, and cultural backgrounds are also essential. This would guarantee that AI regulations are inclusive, culturally aware, and morally sound (Dignum, 2018). The establishment of an international regulatory body devoted to AI ethics may be greatly aided by global governance institutions like the UN or regional groups (Zeng, 2020). Furthermore, cultivating a culture of responsible AI development requires advancing AI literacy and education at the policy and grassroots levels.

Governments and organizations can guarantee that AI technologies be developed responsibly and for the benefit of humanity by giving people a better grasp of the ethical implications of AI (Gasser, 2020). In the end, cooperation and inclusivity will be essential to building an ethical and sustainable AI ecosystem that helps everyone, irrespective of socioeconomic or geographic barriers.

## REFERENCES


AI4Good. (2021). AI4Good Foundation. Retrieved from https://www.ai4good.org

Abadi, M., Chu, A., Goodfellow, I., McMahan, H. B., Mironov, I., Talwar, K., & Zhang, L. (2016). Deep learning with differential privacy. *Proceedings of the 2016 ACM SIGSAC Conference on Computer and Communications Security*, 308-318. DOI: 10.1145/2976749.2978318

Accenture. (2022). *Responsible Use of AI | Case Study*. Accenture. https://www.accenture.com/gb-en/case-studies/data-ai/blueprint-responsible-ai

Acemoglu, D., & Restrepo, P. (2020). Robots and jobs: Evidence from US labor markets. *Journal of Political Economy*, *128*(6), 2188–2244. DOI: 10.1086/705716

Agrawal, A., Gans, J., & Goldfarb, A. (2022). *Power and Prediction: The Disruptive Economics of Artificial Intelligence*. Harvard Business Review Press.

Amershi, S., Weld, D. S., Vorvoreanu, M., Fourney, A., Nushi, B., Collisson, P., & Horvitz, E. (2019). Guidelines for human-AI interaction. *Proceedings of the 2019 CHI Conference on Human Factors in Computing Systems*, 1-13.

Angwin, J., Larson, J., Mattu, S., & Kirchner, L. (2016). Machine bias. *ProPublica*. Retrieved from https://www.propublica.org/article/machine-bias-risk-assessments-in-criminal-sentencing

Bano, M., Zowghi, D., Shea, P., & Ibarra, G. (2023). Investigating Responsible AI for Scientific Research: An Empirical Study. *arXiv preprint arXiv:2312.09561*. https://arxiv.org/abs/2312.09561

Bellamy, R. K., Dey, K., Hind, M., Hoffman, S. C., Houde, S., Kannan, K., & Zhang, Y. (2019). AI fairness 360: An extensible toolkit for detecting, understanding, and mitigating unwanted algorithmic bias. *IBM Journal of Research and Development*, *63*(4/5), 1–15. DOI: 10.1147/JRD.2019.2942287

Binns, R. (2018a). Fairness in machine learning: Lessons from political philosophy. *Proceedings of the 2018 Conference on Fairness, Accountability, and Transparency*, 149-159. DOI: 10.1145/3278721.3278776

Binns, R. (2018b). Algorithmic accountability and public reasoning. *Philosophy & Technology*, *31*(4), 543–556. DOI: 10.1007/s13347-017-0263-5 PMID: 30873342

Binns, R. (2018c). On being ethical with artificial intelligence. *Journal of AI & Ethics*, *1*(1), 1–11. PMID: 36281314

Binns, R. (2020). Human-centred AI: A challenge to methodological purity. *Big Data & Society*, *7*(1), 1–7. DOI: 10.1177/2053951720907633

Brkan, M. (2019). Do algorithms rule the world? Algorithmic decision-making and data protection in the framework of the GDPR and beyond. *International Journal of Law and Information Technology*, *27*(2), 91–121. DOI: 10.1093/ijlit/eay017

Brundage, M., Avin, S., Wang, J., Belfield, H., Krueger, G., Hadfield, G., . . . Anderljung, M. (2020). Toward trustworthy AI development: Mechanisms for supporting verifiable claims. *arXiv preprint arXiv:2004.07213*.

Brynjolfsson, E., & Mcafee, A. N. D. R. E. W. (2017). Artificial intelligence, for real. *Harvard Business Review*, *1*(1), 1–31.

Bryson, J. J. (2018a). Patiency is not a virtue: The design of intelligent systems and systems of ethics. *Ethics and Information Technology*, *20*(1), 15–26. DOI: 10.1007/s10676-018-9448-6

Bryson, J. J. (2018b). *The ethical challenges of AI and autonomous systems*. The Atlantic. Retrieved from https://www.theatlantic.com/technology/archive/2018/07/the-ethical-challenges-of-ai-and-autonomous-systems/565397/

Bryson, J. J. (2020). The role of artificial intelligence in ethical decision making. *Ethics and Information Technology*, *22*(3), 253–265. PMID: 34867077

Bughin, J., Seong, J., Manyika, J., Chui, M., & Joshi, R. (2020). The future of work in Europe: Automation, workforce transitions, and the shifting geography of employment. *McKinsey Global Institute Report*.

Buolamwini, J., & Gebru, T. (2018). Gender shades: Intersectional accuracy disparities in commercial gender classification. *Conference on Fairness, Accountability, and Transparency*, 77-91. DOI: 10.48550/arXiv.1802.08199

Burrell, J. (2016). How the machine 'thinks': Understanding opacity in machine learning algorithms. *Big Data & Society*, *3*(1), 1–12. DOI: 10.1177/2053951715622512

CAIDP. (2018). *Universal Guidelines for AI*. Center for AI and Digital Policy. Retrieved from https://www.caidp.org/universal-guidelines-for-ai/

Castelvecchi, D. (2016). Can we open the black box of AI? *NATNews*, *538*(7623), 20–23. DOI: 10.1038/538020a PMID: 27708329

Chen, J., Storchan, V., & Kurshan, E. (2021). Beyond Fairness Metrics: Roadblocks and Challenges for Ethical AI in Practice. *arXiv preprint arXiv:2108.06217*. https://arxiv.org/abs/2108.06217

Chui, M., Manyika, J., & Muir, M. (2016). *Where machines could replace humans—and where they can't (yet).* McKinsey Quarterly. Retrieved from https://www.mckinsey.com/business-functions/mckinsey-digital/our-insights/where-machines-could-replace-humans-and-where-they-cant-yet

Costa, E., & Halpern, D. (2019). The behavioural science of online harm and manipulation, and what to do about it: An exploratory paper to spark ideas and debate. *The Behavioural Insights Team Report, 2019*, 1–82.

Council of Europe. (2024). *The Framework Convention on Artificial Intelligence.* Retrieved from https://www.coe.int/en/web/artificial-intelligence/the-framework-convention-on-artificial-intelligence

Dastin, J. (2018). Amazon scraps secret AI recruiting tool that showed bias against women. *Reuters.* https://www.reuters.com/article/us-amazon-com-jobs-automation-idUSKCN1MK08G

Dastin, J. (2020). "AI ethics courses are growing at universities around the world." *Reuters.* Retrieved from https://www.reuters.com

Digital Watch Observatory. (2024). *Overview of AI policy in 10 jurisdictions.* Retrieved from https://dig.watch/updates/overview-of-ai-policy-in-10-jurisdictions

Dignum, V. (2018). *Responsible artificial intelligence: Developing and using AI in a responsible way.* Springer.

Diligent Corporation. (2024). *AI regulations around the world.* Retrieved from https://www.diligent.com/resources/guides/ai-regulations-around-the-world

Doshi-Velez, F., & Kim, B. (2017). Towards a rigorous science of interpretable machine learning. *arXiv preprint arXiv:1702.08608.* https://doi.org//arXiv.1702.08608DOI: 10.48550

Ernst & Young. (2024). *ISO 42001: paving the way for ethical AI.* Retrieved from https://www.ey.com/en_us/insights/ai/iso-42001-paving-the-way-for-ethical-ai

Eubanks, V. (2018). *Automating inequality: How high-tech tools profile, police, and punish the poor.* St. Martin's Press.

European Commission. (2019). *Ethics Guidelines for Trustworthy AI.* High-Level Expert Group on Artificial Intelligence. Retrieved from https://www.europarl.europa.eu/cmsdata/196377/AI%20HLEG_Ethics%20Guidelines%20for%20Trustworthy%20AI.pdf

European Commission. (2020). *Ethics guidelines for trustworthy AI.* Retrieved from https://ec.europa.eu/futurium/en/ai-alliance-consultation

European Commission. (2021). *Proposal for a regulation laying down harmonized rules on artificial intelligence (Artificial Intelligence Act).* Retrieved from https://digital-strategy.ec.europa.eu

Financial Times. (2025). *Business school teaching case study: can benefit corporations solve AI governance challenges?* Retrieved from https://www.ft.com/content/3eeb7661-2b63-4339-b446-d1f6e45a2558

Floridi, L., & Cowls, J. (2019). A unified framework of five principles for AI in society. *Harvard Data Science Review, 1*(1). Advance online publication. DOI: 10.1162/99608f92.8cd550d1

Floridi, L., Cowls, J., Beltrametti, M., Chatila, R., Chazerand, P., Dignum, V., Luetge, C., Madelin, R., Pagallo, U., Rossi, F., Schafer, B., Valcke, P., & Vayena, E. (2018). AI4People—An ethical framework for a good AI society: Opportunities, risks, principles, and recommendations. *Minds and Machines, 28*(4), 689–707. DOI: 10.1007/s11023-018-9482-5 PMID: 30930541

Fractal. (2021). *Transforming business with ethical AI practices.* Fractal. https://fractal.ai/casestudies/transforming-business-with-ethical-ai-practices/

Frank, M. R., Autor, D., Bessen, J. E., Brynjolfsson, E., Cebrian, M., Deming, D. J., Feldman, M., Groh, M., Lobo, J., Moro, E., Wang, D., Youn, H., & Rahwan, I. (2019). Toward understanding the impact of artificial intelligence on labor. *Proceedings of the National Academy of Sciences of the United States of America, 116*(14), 6531–6539. DOI: 10.1073/pnas.1900949116 PMID: 30910965

Frey, C. B., & Osborne, M. A. (2021). The future of employment: How susceptible are jobs to computerization? *Technological Forecasting and Social Change, 114*, 254–280. DOI: 10.1016/j.techfore.2016.08.019

Gasser, U. (2020). *The role of AI education and public awareness in ethical AI development.* Springer.

Gilpin, L. H., Caruana, R., Gehrke, J., & Koch, P. (2018). Explaining explanations: An overview of interpretability of machine learning. *Proceedings of the 2018 ICML Workshop on Human Interpretability in Machine Learning*, 1–15. DOI: 10.1109/DSAA.2018.00018

Goodman, B., & Flaxman, S. (2017). European Union regulations on algorithmic decision-making and a "right to explanation.". *AI Magazine, 38*(3), 50–57. DOI: 10.1609/aimag.v38i3.2741

Gordon, J.-S. • Nyholm, S. (2021). Ethics of Artificial Intelligence, Internet Encyclopedia of Philosophy, 2021

GPAI. (2021). Global Partnership on AI. Retrieved from https://gpai.ai

Gunning, D., Stefik, M., Choi, J., Miller, T., Stumpf, S., & Yang, G. Z. (2019). XAI—Explainable artificial intelligence. *Science Robotics*, *4*(37), eaay7120. Advance online publication. DOI: 10.1126/scirobotics.aay7120 PMID: 33137719

Gursoy, S. (2020). The role of education in the ethical implications of AI technology. *International Journal of Artificial Intelligence*, *19*(2), 67–75.

Harris, S., & Jirotka, M. (2019). The ethical implications of AI and how to address them: A policy perspective. *Journal of AI & Ethics*, *1*(1), 37–53.

IBM. (2020). *AI Fairness 360 Toolkit*. IBM. https://www.ibm.com/blogs/research/2020/09/ai-fairness-360/

IEEE. (2019). *Ethically aligned design: A vision for prioritizing human well-being with autonomous and intelligent systems*. IEEE Standards Association.

Jobin, A., Ienca, M., & Vayena, E. (2019). The global landscape of AI ethics guidelines. *Nature Machine Intelligence*, *1*(9), 389–399. DOI: 10.1038/s42256-019-0088-2

Keymakr. (2024). *Regional and International AI Regulations and Laws in 2024*. Retrieved from https://keymakr.com/blog/regional-and-international-ai-regulations-and-laws-in-2024/

Kroll, J. A., Huey, J. D., & Barocas, S. (2020). The risks of keyless entry: The ethical implications of AI-driven systems in access control. *Journal of AI Ethics*, *1*(2), 123–134. DOI: 10.1007/s43681-020-00008-w

Krotzek, J. L. (2019). Inside the voter's mind: The effect of psychometric microtargeting on feelings toward and propensity to vote for a candidate. *International Journal of Communication*, •••, 2019.

Lipton, Z. C. (2018). The mythos of model interpretability. *ACM Queue; Tomorrow's Computing Today*, *16*(3), 31–57. DOI: 10.1145/3236386.3241340

Manyika, J. (2022). Getting AI right: Introductory notes on AI & society. *Daedalus*, *151*(2), 5–27. DOI: 10.1162/daed_e_01897

Mbiazi, D., Bhange, M., Babaei, M., Sheth, I., & Kenfack, P. J. (2023). Survey on AI Ethics: A Socio-technical Perspective. *arXiv preprint arXiv:2311.17228*. https://arxiv.org/abs/2311.17228

Mehrabi, N., Morstatter, F., Saxena, N., Lerman, K., & Galstyan, A. (2021). A survey on bias and fairness in machine learning. *ACM Computing Surveys*, *54*(6), 1–35. DOI: 10.1145/3457607

Mittelstadt, B. D. (2019). Principles alone cannot guarantee ethical AI. *Nature Machine Intelligence, 1*(11), 501–507. DOI: 10.1038/s42256-019-0114-4

Mittelstadt, B. D., Allo, P., Taddeo, M., Wachter, S., & Floridi, L. (2016). The ethics of algorithms: Mapping the debate. *Big Data & Society, 3*(2), 1–21. DOI: 10.1177/2053951716679679

Mittelstadt, B. D., Russell, C., & Wachter, S. (2016). Explaining explanations in AI. *Proceedings of the 2016 AAAI Conference on Artificial Intelligence*, 100-107.

Müller, V. C. (2020). Ethics of artificial intelligence and robotics, In Zalta, E.N. (2020). The Stanford Encyclopedia of Philosophy, 2020

Noble, S. U. (2021). *Algorithms of oppression: How search engines reinforce racism*. NYU Press.

O'Neil, C. (2016). *Weapons of math destruction: How big data increases inequality and threatens democracy*. Crown Publishing Group.

Obermeyer, Z., Powers, B., Vogeli, C., & Mullainathan, S. (2019). Dissecting racial bias in an algorithm used to manage the health of populations. *Science, 366*(6464), 447–453. DOI: 10.1126/science.aax2342 PMID: 31649194

OECD. (2019a). *Principles on AI.* Retrieved from https://www.oecd.org/going-digital/ai/principles

OECD. (2019b). *OECD AI principles: Recommendations of the council on artificial intelligence*. Retrieved from https://www.oecd.org/going-digital/ai/principles/

OECD. (2024). *AI Principles Overview*. Organisation for Economic Co-operation and Development. Retrieved from https://oecd.ai/en/ai-principles

Papernot, N., McDaniel, P., Goodfellow, I., Jha, S., Celik, Z. B., & Swami, A. (2018). Practical black-box attacks against machine learning. *Proceedings of the 2017 ACM on Asia Conference on Computer and Communications Security*, 506-519. DOI: 10.1145/3052973.3053009

Partnership on AI. (2021). Ethical principles for AI development. Retrieved from https://www.partnershiponai.org

PepsiCo. (2023). *Case Study: AI Integration and Ethical Considerations at PepsiCo*. AI Expert Network. https://aiexpert.network/case-study-ai-integration-and-ethical-considerations-at-pepsico/

Raghavan, M., Barocas, S., Kleinberg, J., & Levy, K. (2020). Mitigating bias in algorithmic hiring: Evaluating claims and practices. *Proceedings of the 2020 Conference on Fairness, Accountability, and Transparency*, 469-481. DOI: 10.1145/3351095.3372828

Raji, I. D., Gebru, T., Mitchell, M., Buolamwini, J., Lee, J., & Denton, E. (2020). Saving face: Investigating the ethical concerns of facial recognition auditing. *Proceedings of the 2020 Conference on Fairness, Accountability, and Transparency*, 92-100. DOI: 10.1145/3375627.3375820

Ribeiro, M. T., Singh, S., & Guestrin, C. (2016). *"Why should I trust you?" Explaining the predictions of any classifier. Proceedings of the 22nd ACM SIGKDD International Conference on Knowledge Discovery and Data Mining*, 1135-1144.

Rivi -Tipei, I. (2023). Ethical Considerations in Artificial Intelligence Development and Deployment. *SCIENTIA MORALITAS International Journal of Multidisciplinary Research*, 2, 1-15. https://www.scientiamoralitas.com/index.php/sm/article/view/236

Russell, S. (2021). *Human compatible: Artificial intelligence and the problem of control*. Penguin Books.

Sharma, T., Yadav, D., & Chouhan, R. (2020). A survey on privacy and security concerns in AI applications. *Journal of Information Security and Applications*, *53*, 102562. DOI: 10.1016/j.jisa.2020.102562

Topol, E. (2019). *Deep medicine: How artificial intelligence can make healthcare human again*. Basic Books.

UNESCO. (2021). *Recommendation on the Ethics of Artificial Intelligence*. United Nations Educational, Scientific and Cultural Organization. Retrieved from https://unesdoc.unesco.org/ark:/48223/pf0000380455

UNESCO. (n.d.). *Towards an Ethics of Artificial Intelligence*. Retrieved from https://www.un.org/en/chronicle/article/towards-ethics-artificial-intelligence

Vincent, J. (2025). *AI Accountability: How Doing The Right Thing Builds A Better Business*. Forbes. https://www.forbes.com/sites/bruceweinstein/2025/01/29/ai-accountability-how-doing-the-right-thing-builds-a-better-business/

Voigt, P., & von dem Bussche, A. (2017). *The EU General Data Protection Regulation (GDPR): A practical guide*. Springer., DOI: 10.1007/978-3-319-57959-7

Wachter, S., Mittelstadt, B., & Floridi, L. (2017). Why a right to explanation of automated decision-making does not exist in the general data protection regulation. *International Data Privacy Law*, *7*(2), 76–99. DOI: 10.1093/idpl/ipx005

World Economic Forum. (2020a). *The future of jobs report 2020*. Retrieved from https://www.weforum.org/reports/the-future-of-jobs-report-2020

World Economic Forum. (2020b). *A model for ethical AI*. Retrieved from https://www.weforum.org/stories/2020/01/tech-companies-ethics-responsible-ai-microsoft/

World Economic Forum. (2021). *9 ethical AI principles for organizations to follow*. Retrieved from https://www.weforum.org/stories/2021/06/ethical-principles-for-ai/

Zemel, R., Wu, Y., Swersky, K., Pitassi, T., & Dwork, C. (2013). Learning fair representations. *Proceedings of the 30th International Conference on Machine Learning*, 325–333.

Zeng, Z. (2020). *Artificial intelligence, privacy, and ethics: A global perspective*. Springer.

Zuboff, S. (2019). *The age of surveillance capitalism: The fight for a human future at the new frontier of power*. PublicAffairs.

# Chapter 7
# Value Co-Creation and Artificial Intelligence: Towards Responsible Innovation

**Reda Lharti**
https://orcid.org/0009-0001-0880-9264
*UIT Kenitra, Morocco*

## ABSTRACT

*This article examines the ethical challenges posed by the rapid advancement of artificial intelligence across various sectors. While AI brings significant benefits in terms of efficiency and innovation, it also raises serious concerns about bias, accountability, transparency, privacy, and human autonomy. The risk of perpetuating discrimination through biased data, the difficulty of assigning responsibility for harmful outcomes, and the lack of transparency in AI decision-making all contribute to growing ethical dilemmas. Additionally, AI-driven surveillance technologies threaten individual privacy, and the increasing reliance on AI may diminish human judgment in critical areas. The article argues that ethical considerations must be integrated into the development process from the beginning, advocating for regulatory oversight, interdisciplinary cooperation, and active public engagement to ensure AI technologies serve humanity in a fair and responsible manner.*

## INTRODUCTION

In a context marked by accelerated digitalization and the emergence of disruptive technologies, value co-creation has become a central paradigm for understanding contemporary economic and social dynamics. Traditionally perceived as a unilateral product generated by firms, value is now conceptualized as an interactive and col-

DOI: 10.4018/979-8-3373-2597-2.ch007

laborative process, shaped by continuous exchanges among multiple actors (Vargo & Lusch, 2004). This epistemological shift, initiated by the Service-Dominant Logic (S-D Logic), redefines the boundaries between producers and consumers, placing user experience and usage at the heart of value creation (C. K. Prahalad & Ramaswamy, 2004). Artificial intelligence (AI), as a catalytic force, amplifies these interactions while introducing unprecedented challenges in ethics, governance, and inclusion. This chapter explores these tensions through a critical analysis of the theoretical foundations of co-creation, the transformative role of AI, and the imperatives of responsible innovation.

The concept of value co-creation emerged in the early 2000s as a break from traditional marketing and management approaches. According to (C. K. Prahalad & Ramaswamy, 2004), value no longer resides in the product itself but in the customer's lived experience, co-constructed through dynamic interactions. The authors identify four foundational pillars—dialogue, access, transparency, and shared risk management—that transform consumers into active partners. This vision aligns with the Service-Dominant Logic (S-D Logic) of (Vargo & Lusch, 2004, 2008), which posits that value is always determined by its use (value-in-use) and co-produced through the integration of multiple resources. The customer, far from being a passive recipient, mobilizes their skills and knowledge to actualize the firm's value propositions, thereby redefining traditional roles (Grönroos, 2011).

This transition from an exchange-value logic to a use-value logic is accompanied by an ecosystemic approach. (Normann & Ramirez, 1993) conceptualize value constellations, where interdependent networks—firms, customers, suppliers, communities—collaborate within open ecosystems. These dynamic configurations, amplified by digital platforms, transcend organizational boundaries and foster distributed value creation (Payne et al., 2008). Thus, co-creation becomes a multidimensional process, integrating technological, social, and organizational dimensions, where agility and collaboration are key drivers of competitiveness.

Artificial intelligence revolutionizes co-creation mechanisms by automating, personalizing, and scaling interactions. AI tools—chatbots, recommendation systems, predictive models—analyze vast datasets to tailor offerings to individual contexts, thereby enhancing value-in-use (T. Davenport et al., 2020). Recommendation algorithms turn users into implicit "co-designers," whose behaviors continuously shape content propositions (Kumar & Reinartz, 2016). This dynamic personalization aligns perfectly with S-D Logic, where value emerges from the interaction between human and technological resources (Vargo & Lusch, 2008).

AI also facilitates the emergence of intelligent constellations—collaborative ecosystems where actors co-construct innovative solutions at scale. Open innovation platforms leverage algorithms to identify skills or generate code, thereby blurring organizational boundaries (Nambisan et al., 2019). These systems rely on real-time feedback loops, where user data fuels machine learning, creating a virtuous cycle of shared value. However, this automation raises critical questions: Could AI marginalize human agency or replicate structural biases?

The integration of AI into value co-creation is not without ambivalence. On one hand, it democratizes access to innovation—through language tools or adaptive interfaces—making participation more inclusive (Floridi et al., 2018). On the other, it exposes users to significant risks: data exploitation, algorithmic biases, and power asymmetries. (Zuboff, 2019) denounces "surveillance capitalism," where user experiences are monetized without informed consent, turning co-creation into a form of "algorithmic exploitation." Studies on gender and racial biases in automated hiring systems illustrate these pitfalls (Nicolau et al., 2016), reminding us that AI can perpetuate—or even amplify—existing inequalities (Barocas et al., 2021).

These challenges call for hybrid regulation, combining soft law (ethical charters, certifications) and binding legal frameworks. The European Union's AI Act (2024) bans certain practices (mass surveillance) and mandates human oversight for high-risk systems (Floridi et al., 2018). However, as (B. Mittelstadt, 2019) notes, AI ethics cannot be static: it must evolve through inclusive governance involving researchers, citizens, and policymakers.

Faced with these challenges, responsible innovation emerges as an essential framework to reconcile technological progress and human values. Defined by (Owen et al., 2013) as a process "aligned with societal needs," it rests on four pillars: anticipation, reflexivity, inclusion, and responsiveness. Applied to AI, this approach requires anticipating extractive impacts (Crawford, 2021), auditing biases (B. Mittelstadt, 2019), engaging stakeholders (Fukuda-Parr & Gibbons, 2021), and adapting to societal feedback. Initiatives like the Partnership on AI embody this collaborative vision, bringing together firms, NGOs, and researchers to co-construct ethical norms (Hawkins & Mittelstadt, 2023).

Yet tensions persist between disruptive innovation and responsibility. In healthcare, AI improves diagnostics but raises privacy concerns (Topol, 2019). In art, image generators like DALL-E challenge intellectual property norms (Azcárate, 2024). These dilemmas illustrate a central paradox: the more AI advances, the more urgent its ethical alignment becomes. For (Floridi et al., 2018), responsible innovation is not a hindrance but a guide toward AI "for the common good."

This chapter is structured in three parts. The first explores the theoretical foundations of value co-creation, from the seminal work of (Vargo & Lusch, 2004) to (Normann & Ramirez, 1993) value constellations. The second analyzes AI's transformative role, highlighting its potential for personalization, open innovation, and collaborative ecosystems while exposing its ethical limits. Finally, the third section examines the conditions for responsible innovation through regulatory frameworks, inclusive practices, and reflexive governance. In conclusion, this work calls for an ethical co-construction of AI, where technology and humanity coexist in a dynamic of mutual enrichment.

This chapter offers a critical reflection on the opportunities and pitfalls of AI in value co-creation, urging a rethinking of economic models, technological practices, and ethical commitments at the dawn of an era defined by collective intelligence and complex societal challenges.

## METHODOLOGY

This study adopts a narrative review methodology to synthesize and critically interpret existing knowledge on the intersection of artificial intelligence (AI) and value co-creation. Unlike systematic reviews that rely on rigid protocols, the narrative approach allows for an integrative and interpretive exploration of diverse theoretical perspectives, conceptual frameworks, and empirical findings. It emphasizes depth of understanding rather than exhaustive coverage, focusing on how AI technologies reshape co-creation mechanisms, stakeholder interactions, and ethical considerations. Sources were selected based on their conceptual relevance and contribution to debates surrounding Service-Dominant Logic, responsible innovation, and algorithmic ethics. This flexible yet rigorous approach enables the identification of emerging trends, conceptual gaps, and research opportunities, offering a coherent synthesis that bridges fragmented discussions across disciplines. The narrative review thus provides a holistic foundation for theorizing the role of AI in fostering responsible and inclusive value co-creation.

## 1. LITERATURE REVIEW

Value co-creation has emerged as a central paradigm in management sciences, redefining value not as a finished product but as an interactive process among stakeholders. According to (Vargo & Lusch, 2004), the Service-Dominant Logic (S-D Logic) posits that value is always co-created through the exchange of skills and resources, marking a departure from the traditional value-in-exchange logic in favor of value-in-use. This epistemological shift, theorized by (Normann & Ramirez, 1993), extends to value constellations, where interconnected ecosystems (suppliers, customers, partners) collectively generate value, transcending organizational boundaries.

Artificial intelligence (AI) accelerates this dynamic by enabling personalized and scalable interactions. As (C. K. Prahalad & Ramaswamy, 2004) explain, AI facilitates a "value experience" tailored to individual needs, enhancing value-in-use through predictive recommendations or adaptive interfaces. However, its integration into collaborative ecosystems raises major ethical challenges: algorithmic biases, information asymmetries, and the erosion of human agency (Floridi et al., 2018). These risks call for a rethinking of governance models, particularly in intelligent constellations where AI mediates relationships.

Finally, responsible innovation provides a framework to reconcile technological progress and ethics. (Owen et al., 2013) identify four pillars: anticipation, reflexivity, inclusion, and responsiveness. Applied to AI, this approach requires hybrid regulation, combining soft law (ethical guidelines, certifications) and legal constraints (B. Mittelstadt, 2019). Tensions between disruptive innovation and societal responsibility persist, but initiatives such as the co-construction of norms (Jobin et al., 2019) outline pathways for AI that is both competitive and aligned with the common good.

## 1.1. Value co-creation: Definitions and theoretical framework

In an economic landscape profoundly transformed by digitalization, globalization, and hyper connectivity, value co-creation has emerged as a critical theoretical framework for analyzing new forms of interactions among firms, customers, and all stakeholders. This concept breaks with traditional approaches where value was viewed as unilaterally created by firms and then delivered to consumers. Instead, co-creation is founded on the principle that value is jointly produced through continuous and dynamic interactions between businesses and users (C. K. Prahalad & Ramaswamy, 2004).

Introduced in the early 2000s, value co-creation aligns with the Service-Dominant Logic (SDL) developed by (Vargo & Lusch, 2004), which asserts that value is always determined by use (value-in-use) and co-created through the integration of resources from multiple stakeholders. In this perspective, customers are no longer passive recipients but become active contributors to value creation by leveraging their skills, knowledge, and experiences within collaborative environments such as digital platforms, online communities, or service ecosystems (Chandler & Vargo, 2011).

Thus, co-creation emerges as a multidimensional dynamic, incorporating technological, social, and organizational dimensions. It necessitates rethinking roles, processes, and governance logics in value production. Understanding this concept therefore requires an in-depth analysis of its theoretical foundations, practical implementations, and the contextual conditions that foster its emergence and effectiveness in today's complex, interconnected environments.

- **Foundational definition: The age of interaction**

Value co-creation, as conceptualized by (C. K. Prahalad & Ramaswamy, 2004), marks a pivotal shift in understanding firm-consumer relationships. In *The Future of Competition*, the authors introduce co-creation as an interactive, collaborative process where firms and customers jointly build personalized, meaningful experiences. This innovative approach challenges the traditional value-creation paradigm—historically centered on firms as sole producers—by asserting that value now arises from customer experience: "Value is no longer embedded in products or services alone, but in customers' lived experiences" (C. K. Prahalad & Ramaswamy, 2004).

This paradigm rests on four foundational "building blocks": dialogue, access, transparency, and shared risk management. These elements foster an interaction space where consumers become full-fledged partners, actively shaping offerings to their needs, preferences, and usage contexts.

Prahalad and Ramaswamy further advocate "democratizing innovation" by involving consumers early in product/service design. Such engagement generates novel ideas, strengthens customer commitment, and creates sustainable competitive advantage. Thus, in the age of interaction, co-creation transcends marketing strategy to become an organizational philosophy rooted in collaboration, mutual learning, and shared value generation.

- **Service-Dominant Logic (S-D Logic): A foundational framework**

(Vargo & Lusch, 2004, 2008) revolutionized marketing thought with Service-Dominant Logic (S-D Logic), now a cornerstone framework for analyzing value co-creation. Departing from the Goods-Dominant Logic (G-D Logic)—focused on tangible goods production and exchange—S-D Logic positions *service* (the application of competencies for others' benefit) as the core of economic exchange.

(Vargo & Lusch, 2008) argue that value is not firm-created and delivered but inherently collaborative: "Value is always co-created with the customer" (FP6). This process hinges on interactions between firm and customer resources, actualized within the beneficiary's usage context. Their tenth foundational premise (FP10) states that "value is always uniquely and phenomenologically determined by the beneficiary," emphasizing value's subjective, experiential nature.

Here, firms no longer offer finished products but value propositions actualized by customers through their needs, uses, and experiences. The focus shifts to relational processes, knowledge exchange, and resource integration (Vargo & Lusch, 2008). This logic redefines traditional roles: customers become value co-producers, and firms evolve into experience facilitators.

S-D Logic thus provides a robust framework for understanding co-creation in interconnected environments, where producer-consumer boundaries blur in favor of continuous, value-generating interactions.

- **From Value-in-Exchange to Value-in-Use: An epistemological shift**

Value co-creation reflects a fundamental reconceptualization of value—from *value-in-exchange* to *value-in-use*. (Grönroos, 2008, 2011) highlights this epistemological transition, distinguishing between value perceived at the point of sale and that emerging during actual product/service usage.

Under traditional *value-in-exchange* logic, value is seen as inherent to goods/services, measurable at purchase, and solely firm-produced. Conversely, Grönroos' *value-in-use* logic adopts a relational, contextual perspective: "Co-creation occurs when the customer integrates firm resources with their own to create value" (Grönroos, 2011). Thus, value does not precede usage but is constructed through lived experiences via the customer's active resource integration.

This view expands the customer's role from passive recipient to central value creator. Firms, meanwhile, provide resources (often as services) for customers to deploy according to their goals, skills, and contexts.

(Grönroos, 2008) further introduces *service logic*, where value is co-created through interactive service processes rather than delivered as finished entities. This reinforces value's subjective, situational, and co-produced nature, prioritizing *usage* over *transaction*.

- **Value constellations: An ecosystem approach**

Contemporary understanding of value co-creation cannot be confined to a simple dyadic interaction between firms and customers. (Normann & Ramirez, 1993) expanded this perspective by introducing the concept of value constellations, which adopts a systemic and collective approach to value creation. They argue that "firms do not create value alone but design ecosystems where each participant contributes and benefits" (Normann & Ramirez, 1993). This perspective emphasizes interconnected networks—customers, suppliers, partners, communities—that collaborate within an open system to co-produce value.

This model breaks away from (Porter, 1985) linear value chain logic, proposing instead a more fluid and adaptive structure based on stakeholder interdependence. (Payne et al., 2008) further developed this idea, highlighting how service ecosystems enable resource sharing and the co-construction of solutions tailored to users' evolving needs.

Thus, the theoretical framework of value co-creation now rests on three complementary pillars:

1. **Interaction as a value driver**, emphasized by (C. K. Prahalad & Ramaswamy, 2004), who position shared experience as the new locus of value production.

2. **Usage logic (value-in-use)**, advocated by (Vargo & Lusch, 2008), which underscores value's subjectivity and its emergence within usage contexts.
3. **The ecosystem approach**, advanced by (Normann & Ramirez, 1993), which frames co-creation as a distributed process among interdependent actors.

These approaches converge toward a profound redefinition of roles:

Firms are no longer isolated producers but experience facilitators.
Customers become active partners in the valorization process.

This shift in power and responsibility challenges traditional organizational boundaries, demanding a more collaborative vision of strategy and innovation. As (C. Prahalad, 2004) asserts, "Power now belongs to those who know how to engage in dialogue"—a statement reflecting the critical importance of listening, transparency, and co-construction in the experience economy. Here, firms' competitiveness hinges on their ability to embed themselves within agile, value-driven collaborative networks.

## 1.2. Artificial intelligence and value co-creation

The emergence of artificial intelligence (AI) has profoundly transformed value co-creation mechanisms by redefining interactions among firms, customers, and ecosystems. While foundational theories from (C. K. Prahalad & Ramaswamy, 2004) and (Vargo & Lusch, 2008) emphasized dialogue and user experience as central to co-creation, AI introduces an unprecedented technological dimension—automating, amplifying, and personalizing these exchanges. This section explores how AI catalyzes value co-creation while interrogating its ethical and practical limits.

- **AI as an enabler of enhanced interactions**

In value co-creation, artificial intelligence (AI) increasingly serves as a technological facilitator, optimizing the quality, relevance, and fluidity of firm-customer interactions. As (C. K. Prahalad & Ramaswamy, 2004) remind us, interaction is the very foundation of co-creation, enabling the construction of unique, customer-centric experiences. AI thus transcends mere process automation, acting as an intelligent mediator among stakeholders.

AI-powered tools—such as chatbots, virtual assistants, and recommendation systems—can analyze vast volumes of behavioral, contextual, and emotional data in real time. This analytical capacity enables hyper-personalized experiences, continuously adapted to individual preferences. (T. Davenport et al., 2020) note that

"AI transforms customers into active co-designers, capable of shaping offerings through implicit or explicit feedback" (p. 45), reinforcing their participatory role in value creation.

This shift aligns perfectly with (Vargo & Lusch, 2008) Service-Dominant Logic (SDL), where value is co-produced through the interactive use of resources. Here, AI becomes an operant resource—a dynamic asset that interacts, adapts, and generates new co-creation opportunities. By merging technical capabilities with user behaviors, AI fosters contextualized, experiential value, redefining firm-customer relationships through continuous, proactive, and intelligent dialogue—a cornerstone of co-creation in today's digital economy.

- **Personalization and Value-in-Use: AI's contribution**

AI plays a pivotal role in operationalizing value-in-use, as defined by (Grönroos, 2011). Value, he argues, is not embedded in products or services at the point of exchange but emerges through usage, shaped by the user's specific context. AI enables dynamic adaptation to these usage contexts, driving granular, evolving personalization.

Predictive models, computer vision systems, and intelligent recommendation algorithms allow firms to anticipate latent customer needs—sometimes before they are explicitly expressed. This proactive co-creation is exemplified by platforms leveraging crowdsourced data to detect emerging trends and co-design products/services with users in real time.

AI does not replace human agency but augments it. As (Brynjolfsson & McAfee, 2017) assert, "AI doesn't substitute for humans but extends their value-generating capacity by hybridizing human creativity with algorithmic power". This hybridity is evident in creative platforms where AI-assisted tools enhance graphic, musical, or textual design, personalizing experiences and renewing value with each interaction.

In essence, AI shifts the paradigm from standardized production to contextualized usage, where every interaction tailors value to the customer's preferences, skills, and goals.

- **Collaborative ecosystems and AI: Toward intelligent constellations**

AI is reshaping value constellations (Normann & Ramirez, 1993) - dynamic networks of firms, customers, partners, and developers co-producing value within ecosystems. AI makes these ecosystems more fluid, responsive, and interconnected, enabling distributed, intelligent co-creation.

Open innovation platforms harness AI to orchestrate complex networks. Their algorithms identify available skills, recommend collaborations, or even auto-generate code, turning every user into a potential ecosystem contributor. As (Nambisan et al., 2020) observe, “AI dissolves organizational boundaries, enabling large-scale distributed co-creation” by integrating diverse community inputs.

These systems rely on real-time feedback loops: user-generated data is instantly analyzed and fed back into AI models, creating a virtuous cycle of machine learning and shared value. The more users interact, the sharper the algorithms become—and the more relevant the value propositions.

In short, AI transforms ecosystems into intelligent constellations, evolving through continuous learning and collective intelligence. It redefines innovation and collaboration, anchoring co-creation in connectivity, agility, and perpetual adaptation.

- **Ethical challenges and limits of AI in co-creation**

Despite its promise, AI raises critical ethical and sociotechnical challenges. Its efficacy hinges on massive data collection, risking user privacy. (Zuboff, 2019) warns of “surveillance capitalism,” where user experiences are monetized without informed consent, potentially reducing co-creation to “algorithmic exploitation.”

Moreover, AI is prone to bias. Machine learning systems can replicate—or exacerbate—existing stereotypes, as seen in gender- or race-biased hiring algorithms (O’Neil, 2016). Automated decisions based on skewed historical data may skew co-creation toward asymmetry and inequity. (Floridi et al., 2018) thus calls for “algorithmic design ethics,” insisting “AI must empower stakeholders, not manipulate them”.

AI’s proliferation also risks technological divides. Actors lacking skills, resources, or infrastructure may be marginalized, contradicting the inclusive spirit of co-creation (C. Prahalad, 2004; Vargo & Lusch, 2008).

For AI to truly support co-creation, it must be human-centric. As (T. Davenport et al., 2020) summarizes, “AI doesn’t create value alone—it only does so when humans are at the heart of its architecture.” Tomorrow’s co-creation will depend on harmonizing artificial and collective intelligence within ethical frameworks that prioritize diversity, autonomy, and justice in collaborative ecosystems.

## 1.3. Responsible innovation and AI

The rise of artificial intelligence (AI) presents unprecedented challenges in ethics, transparency, and societal impact, making responsible innovation a central imperative. Defined by (Owen et al., 2013) as “a proactive process aimed at aligning innovation with societal values and needs,” responsible innovation in AI involves balancing technological progress, respect for human rights, and the preservation of ecosystems.

- **The Pillars of responsible innovation applied to AI**

Responsible innovation, as conceptualized by (Owen et al., 2013), rests on four fundamental pillars: anticipation, reflexivity, inclusion, and responsiveness. These principles, initially designed to guide scientific and technological practices, take on particular significance when applied to the development of artificial intelligence (AI)—a technology with vast, profound, and still largely uncertain societal impacts.

1. **Anticipation** involves identifying in advance the social, environmental, and economic consequences of AI systems. (Crawford, 2021)) for example, highlights in her book Atlas of AI that this technology relies on invisible extractive practices: mining resources for chip manufacturing, massive energy consumption by data centers, and outsourcing low-cost labeling tasks. This perspective invites a reconsideration of AI's true cost beyond its apparent benefits.
2. **Reflexivity** entails questioning the ethical and epistemic foundations of algorithmic systems. (B. Mittelstadt, 2019) notes that "AI systems often reproduce structural inequalities present in training data", as seen in cases of racial bias in facial recognition technologies. Reflexivity thus demands constant vigilance in the design, validation, and auditing of models.
3. **Inclusion** requires the active integration of all stakeholders in the development process. (The Villani Report, 2018) on "AI for Humanity" proposes citizen consultation mechanisms to co-construct ethical and socially beneficial AI. This approach broadens technological governance by involving not only experts but also civil society actors.
4. **Responsiveness** refers to developers' ability to adapt AI systems to societal feedback and concerns. Microsoft's withdrawal from military projects in 2018 under employee pressure exemplifies this ethical adjustment dynamic (Grant & Weise, 2023).

These principles translate into concrete tools such as Algorithmic Impact Assessments (AIAs), required under the EU's AI regulation (AI Act, 2024), which aim to anticipate and mitigate risks associated with high-impact AI.

- **AI ethics: Between soft law and binding regulation**

The rapid rise of artificial intelligence raises major ethical challenges, from protecting individual freedoms to combating algorithmic discrimination. In response, researchers, NGOs, and international institutions have developed guiding principles to regulate AI development and use.

Among notable contributions, (Floridi et al., 2018) propose five foundational principles for trustworthy AI: transparency, fairness, accountability, respect for human autonomy, and harm prevention. These guidelines have influenced numerous ethical charters, such as UNESCO's Recommendation on the (Ethics of Artificial Intelligence, 2021), which advocates a human rights-based approach. This charter even calls for banning mass surveillance AI, deeming it a direct threat to human dignity and fundamental freedoms.

However, a major limitation of these frameworks is their voluntary nature, which undermines their practical effectiveness. As (Bostrom & Yudkowsky, 2018) argue, "Corporate self-regulation is insufficient: independent oversight is needed to prevent systemic abuses". In other words, while ethical guidelines are normatively essential, they often lack enforcement mechanisms.

Addressing this gap, the European Union took a significant step by adopting its AI Regulation (AI Act, 2024). This law categorizes AI systems by risk level, imposing strict requirements for "high-risk" AI (e.g., in healthcare, security, or justice). Certain practices, such as social scoring and behavior manipulation exploiting psychological vulnerabilities, are explicitly banned.

By combining ethical principles with binding legal frameworks, the EU seeks to establish responsible AI governance. This hybrid approach marks a shift toward proactive regulation, where soft law gradually gives way to enforceable rules with monitoring and sanctions.

- **Tensions between innovation and responsibility**

While AI represents a major advancement for many sectors, it also creates deep tensions between innovation and responsibility. These ethical dilemmas are particularly acute in certain fields where technological promises clash with moral, social, or legal imperatives.

In healthcare, for instance, AI is praised for improving diagnostics, predicting medical risks, and personalizing treatments (citation). Yet, these advances often rely on massive use of sensitive data, raising concerns about informed consent and confidentiality. As (Topol, 2019) notes, "Trust in medical AI depends on the ability to explain algorithmic decisions to patients". The "black box" nature of many algorithms complicates transparency, potentially eroding doctor-patient trust.

In the military domain, tensions are even sharper. The development of autonomous weapon systems—dubbed "killer robots"—has alarmed many scientists. Prominent researchers like Yoshua Bengio and Geoffrey Hinton have publicly condemned such uses, even boycotting autonomous weapons projects as incompatible with human rights and international humanitarian law (Conn, 2017).

The creative arts are not immune to these conflicts. Generative AI like GPT-4 or DALL-E is disrupting traditional cultural production (Hayes, 2023). While enabling innovative assisted creation, it also challenges intellectual property and originality. Artists like (Hayes, 2023) denounce an "algorithmic colonization of creativity," arguing that models trained on existing works without consent undermine creators' rights.

These cases illustrate a central paradox: the more AI advances, the more urgent it becomes to reconcile innovation and ethics. The question is no longer just what AI can do, but what it should do—a crucial consideration for shaping truly responsible innovation.

- **Toward co-constructing responsible AI**

To address growing tensions between technological performance and ethical imperatives, new collaborative approaches seek to embed AI in a more inclusive co-construction process. One key framework is participatory design, which involves end-users from the earliest development phases. According to (Simonsen & Robertson, 2013), this approach "enhances system legitimacy by making users co-designers of the technologies they will use," reducing gaps between intent and real-world application.

Concrete initiatives embody this shift. OpenAI, initially criticized for opacity, has adopted more transparent practices, regularly publishing impact reports on model risks and limitations (Ahmad et al., 2025). Meanwhile, the Partnership on AI brings together tech firms, researchers, and civil society to establish shared ethical guidelines (Camacho et al., 2025). Such collaborations foster collective accountability, preventing AI development from being monopolized by a few dominant players (Askell et al., 2019).

In this vision, responsible innovation becomes a guiding principle. As (Floridi et al., 2018) states, "Responsible innovation is not a brake but a compass for AI serving the common good." It balances technological boldness, ethical caution, and social inclusion—not to slow progress, but to ensure it benefits all stakeholders equitably (Mumford, 2024).

This ethical shift also demands a paradigm change: moving beyond a purely technocentric view toward reflexive, multidisciplinary governance involving citizens, institutions, developers, and users. As (Villani et al., 2018) reminds us, "Innovating responsibly means accepting that progress only matters if it elevates humanity."

Ultimately, the future of AI lies not just in its algorithmic power, but in our collective ability to shape it as a common good—forged through dialogue, transparency, and shared responsibility.

## 2. THE MISSION OF AI IN VALUE CO-CREATION

Artificial Intelligence (AI) has emerged as a transformative force in value co-creation, enabling organizations and stakeholders to collaboratively develop innovative products, services, and experiences. The integration of AI in these processes enhances efficiency, personalization, and inclusivity while presenting ethical and regulatory challenges (Solakis et al., 2024).

### 2.1. Enhancing personalization and user experience

One of AI's most significant contributions to value co-creation lies in its ability to analyze vast datasets to deliver highly personalized experiences (Chandra & Rahman, 2024). Through machine learning and predictive analytics, AI can anticipate consumer needs in real time and tailor offerings based on preferences, behaviors, and usage contexts (citation). This dynamic personalization fosters richer interactions and heightened satisfaction—key drivers of active user engagement (T. H. Davenport & Ronanki, 2018).

Recommendation systems exemplify this potential (Chandra & Rahman, 2024). By continuously learning from user interactions, these algorithms refine content and service relevance, generating unique value with each use (Chen et al., 2025). According to (Kumar & Reinartz, 2016), this adaptive capability strengthens "customer involvement in the value chain, transforming them into co-producers of the experience."

Moreover, conversational agents—such as chatbots and virtual assistants—enhance user experience by providing instant, personalized, and context-aware responses (citation). These automated interfaces act as technological mediators, streamlining exchanges between organizations and their audiences while gathering valuable feedback to iteratively improve AI models. As (Wilson & Daugherty, 2018) note, "AI doesn't just respond—it learns from every interaction, turning customers into active partners in an evolving process."

This bidirectional flow between users and intelligent systems reflects the principles of Service-Dominant Logic (Vargo & Lusch, 2008), where value emerges through usage and is co-created via continuous interactions. In this sense, AI serves as a strategic lever for co-creation, transforming every touchpoint into an opportunity for shared enrichment (Bassano et al., 2020).

Ultimately, by enhancing personalization, responsiveness, and interaction intelligence, AI fosters user-centric value creation within a framework of adaptive, context-driven collaboration (Chandra & Rahman, 2024).

## 2.2. Accelerating innovation and decision-making

Artificial Intelligence (AI) plays a pivotal role in accelerating organizational innovation due to its ability to process and analyze complex datasets at an unprecedented scale (Mirzani, 2024). Deep learning, predictive modeling, and generative AI enable rapid prototyping, facilitating iterative experimentation and faster refinement of new ideas (T. H. Davenport & Ronanki, 2018). This process shortens innovation cycles, improves outcome accuracy, and optimizes resource allocation. As (Cockburn et al., 2018) assert, "AI acts as a discovery catalyst, expanding the exploration of technological possibilities".

In healthcare, for instance, AI-driven drug discovery analyzes millions of molecular structures to identify promising therapeutic compounds, significantly reducing research timelines and costs (Mak & Pichika, 2019). Similarly, in engineering and manufacturing, AI-assisted design tools simulate, test, and optimize complex processes virtually, enhancing system performance and sustainability (Haefner et al., 2021).

Beyond R&D, AI transforms strategic decision-making (citation). By automating insights extraction from big data, AI-powered decision-support systems help managers anticipate market trends, assess consumer behavior, and evaluate competition (Brynjolfsson & McAfee, 2017). These tools enhance organizational agility and enable more responsive, data-driven strategies.

Thus, AI serves as a value co-creation enabler, equipping stakeholders with actionable insights that fuel innovation and facilitate collective, evidence-based decisions.

## 2.3. Promoting inclusive co-creation

Artificial Intelligence (AI) is a powerful tool for making co-creation more inclusive by broadening access to diverse stakeholders. By overcoming traditional barriers related to language, physical abilities, or technical skills, AI enables more equitable participation in value creation (Rane, 2023). Natural Language Processing (NLP) and real-time translation technologies play a key role in this process, enabling seamless multilingual communication—essential for global co-creation projects (Floridi et al., 2018). As (Ghosh & Sanyal, 2021) notes, "Linguistic AI erases cultural and linguistic boundaries, fostering transnational collective intelligence".

Furthermore, AI-powered assistive technologies enhance digital inclusivity. Adaptive interfaces, voice recognition software, and gesture-control systems provide new interaction possibilities for people with disabilities (Setiawan, 2024). These tools not only ensure technical accessibility but also empower users to actively contribute to design processes, expanding the pool of co-creators. According to (Dignum, 2019),

"Ethical AI aims to maximize every individual's capacity to engage in social and economic dynamics".

By integrating diverse perspectives, AI enriches the co-creation ecosystem, encouraging multifaceted and complementary contributions (Lee et al., 2019).This cognitive and cultural pluralism, enabled by intelligent technologies, fuels fairer and more sustainable innovation. When designed inclusively, AI does not merely automate—it democratizes access to collective value creation (Abdalla & Abdalla, 2021).

## 3. ETHICAL CONCERNS AND CHALLENGES OF AI IN CO-CREATION

As artificial intelligence (AI) continues to play an integral role in value co-creation, its implementation presents several ethical concerns that must be carefully considered. While AI enhances efficiency, personalization, and inclusivity, it also raises fundamental questions regarding algorithmic bias, data privacy, transparency, and the displacement of human decision-making (Mehrabi et al., 2021). Addressing these challenges is crucial to ensuring that AI-driven co-creation remains equitable, responsible, and aligned with societal values.

### 3.1. Algorithmic bias and fairness

One of the most pressing ethical challenges in AI-assisted co-creation is algorithmic bias. AI models are often trained on historical datasets that may reflect—or even amplify—preexisting social inequalities. As (Barocas et al., 2019) note, "Algorithmic systems replicate the biases present in their training data, perpetuating invisible forms of discrimination" (p. 17). If left unchecked, these biases can lead to unfair decision-making, directly undermining the dynamics of co-creation (Suresh & Guttag, 2021).

In human resources, for example, automated recruitment tools have been shown to favor certain demographic groups over others, reinforcing structural biases (Binns, 2018). When applied to co-creation, such biases can restrict participation, marginalizing voices that are essential to diverse and meaningful collaboration. This algorithmic exclusion not only threatens fairness but also diminishes the richness of collective innovation (Holten Møller et al., 2020).

To mitigate these risks, robust bias detection and correction mechanisms must be implemented. (B. D. Mittelstadt et al., 2016) emphasize the need for representative datasets, regular AI audits, and fairness criteria in algorithmic design. Additionally, inclusive and interdisciplinary governance is essential to ensuring AI systems align with principles of justice and equity.

## 3.2. Transparency and explainability

Transparency and explainability in AI systems are critical for ethical and inclusive value co-creation. Many AI models, particularly those based on deep learning, operate as "black boxes", making their decision-making processes opaque (Burrell, 2016). This lack of clarity can erode stakeholder trust, as users feel excluded from a process where decisions are neither understandable nor contestable (Dignum, 2019).

In co-creation, this opacity can create information asymmetries, hindering active user participation. As (Ananny & Crawford, 2018) argue, "Without visibility into algorithmic reasoning, it becomes difficult to debate or challenge its outcomes". This lack of transparency limits user engagement and weakens collective accountability.

To address these concerns, Explainable AI (XAI) approaches are essential. These methods aim to make AI decisions interpretable by providing clear, contextual, and accessible justifications (Guidotti et al., 2018). Additionally, implementing transparency protocols—such as documenting decision-making processes and involving stakeholders in system design—can enhance the legitimacy of AI tools.

Ultimately, transparency and explainability are not just technical features but foundational requirements for fostering trust, inclusivity, and accountability in AI-assisted co-creation.

## 3.3. Data privacy and security

AI-driven value co-creation relies on analyzing vast amounts of personal data to deliver personalized and adaptive services (Pousttchi et al., 2019). However, this data dependency raises significant privacy and security concerns (Zuboff, 2019). Risks such as unauthorized access, data breaches, or unethical data use can undermine user trust and lead to harmful consequences (Taddeo & Floridi, 2018).

In co-creation processes—where users actively share sensitive information to improve services—data protection becomes an ethical and strategic imperative. As (Martin & Murphy, 2017) state, "The promise of personalization should never come at the expense of individual privacy". A delicate balance must be struck between data-driven value creation and respect for users' fundamental rights.

To address these challenges, organizations must implement strong data governance frameworks, including encryption, anonymization, and strict access controls (T. Davenport et al., 2020). Compliance with regulations such as the General Data Protection Regulation (GDPR) is also essential in ensuring ethical data collection and processing (Voigt & Von dem Bussche, 2017).

Thus, data privacy and security should not be seen as technical constraints but as pillars of responsible AI. By ensuring transparency and safeguarding user information, businesses can strengthen the legitimacy of their co-creation initiatives and foster long-term trust.

### 3.4. The role of human oversight

The increasing autonomy of AI in decision-making raises concerns about the marginalization of human judgment in co-creation. While AI enhances efficiency and accelerates decision-making, human oversight remains critical to ensuring ethical alignment and contextual understanding (Rahwan, 2018). Complex decisions often involve social, cultural, or moral nuances that algorithmic systems struggle to interpret adequately.

As (Jobin et al., 2019) assert, "Technological autonomy must never replace human moral autonomy". This perspective advocates for a hybrid model where AI assists rather than replaces human judgment. The Human-in-the-Loop (HITL) framework embodies this approach, ensuring continuous human involvement in AI-driven processes—particularly for validation, correction, and interpretation of results (Dignum, 2019).

Such mechanisms enhance accountability and prevent potential ethical pitfalls. They also provide contextual adaptability, which is essential for truly inclusive and responsible co-creation. As (B. Mittelstadt, 2019) notes, AI can only ensure fair decisions if it is "designed to work in synergy with human judgment systems".

Therefore, human oversight is not a barrier to innovation but a foundation for trust and quality in AI systems. By balancing algorithmic autonomy with human control, organizations can foster ethical, context-aware, and human-centered co-creation.

## 4. TOWARDS RESPONSIBLE AI FOR VALUE CO-CREATION

The integration of artificial intelligence (AI) into value co-creation processes offers transformative potential, but its ethical deployment demands a structured, multidisciplinary approach. Addressing challenges such as algorithmic bias, transparency gaps, and human displacement requires frameworks that harmonize technological innovation with societal values. Drawing on scholarly insights and practical initiatives, this section explores four pillars for fostering responsible AI in co-creation: regulatory compliance, algorithmic explainability, human-centered design, and stakeholder education.

## 4.1. Establishing regulations and ethical standards

The implementation of regulations and ethical standards is essential to ensure the responsible development of artificial intelligence (AI), particularly in collaborative contexts like value co-creation. The absence of clear guidelines can lead to major ethical risks, including discriminatory practices, data exploitation, or stakeholder exclusion. As (Jobin et al., 2019) emphasize, ethical principles must serve as a "normative compass" to guide AI applications toward respect for fundamental rights.

International institutions have developed structured frameworks to address these concerns. The OECD Principles on AI (Yeung, 2020) highlight five core values: inclusivity, transparency, robustness, safety, and accountability. These principles aim to ensure that AI systems "promote human well-being" and are deployed with fairness in mind. Similarly, the EU's AI Act (2021) introduces a risk-based regulatory approach, banning certain practices (e.g., mass biometric surveillance) and mandating human oversight for high-risk systems (Floridi et al., 2018).

However, legal norms must be complemented by actionable ethical standards. The IEEE's Ethically Aligned Design initiative, for instance, advocates integrating principles such as autonomy, social justice, and accountability into algorithmic design (Shahriari & Shahriari, 2017). According to (Dignum, 2019), mere "legal compliance" is insufficient—companies must cultivate a culture of responsibility, training technical teams in ethics and establishing internal governance bodies.

Finally, regulations must remain adaptive. As (B. Mittelstadt, 2019) argues, "AI ethics cannot be static" - it must evolve alongside technological advancements, user feedback, and shifting societal contexts. A living, collaborative ethical framework is key to balancing innovation with human values, ensuring AI remains truly responsible.

## 4.2. Developing explainable and auditable algorithms

A major challenge for responsible AI in co-creation is algorithmic opacity. Deep learning systems, while powerful, often function as "black boxes", making their decision-making processes incomprehensible to users—and sometimes even to their own developers (Pasquale, 2015). This lack of explainability undermines transparency, trust, and accountability (Burrell, 2016). As Burrell notes, "Algorithmic opacity is not just technical—it is institutional and intentional", making ethical scrutiny difficult.

To address this issue, Explainable AI (XAI) research has gained momentum (Longo et al., 2024). According to (Doshi-Velez & Kim, 2017), explainability refers to an AI system's ability to make its decisions understandable to humans—a crucial feature in sensitive domains like healthcare, finance, or justice. In co-creation, this allows users to understand, question, and improve AI-driven suggestions (Adadi & Berrada, 2018).

Explainability must be paired with auditability—the ability to verify AI decisions for compliance with ethical and legal standards. Methods like LIME (Local Interpretable Model-agnostic Explanations) (Ribeiro et al., 2016) help interpret AI predictions by analyzing specific cases. Such tools are vital for detecting biases, assessing decision robustness, and ensuring responsible AI governance.

Ultimately, explainable and auditable algorithms are not just a technical necessity but a democratic imperative—empowering all co-creation stakeholders to understand, influence, and oversee AI decision-making. This is fundamental to fostering trust, inclusivity, and effective algorithmic innovation.

## 4.3. Promoting human-centered AI

Human-centered AI prioritizes user autonomy and ethical alignment over pure automation (Ozmen Garibay et al., 2023). Following Norman's principles of user-centered design, stakeholders must be actively involved in AI development to ensure solutions meet real-world needs (Norman, 2013). This participatory approach helps prevent algorithmic biases and enhances solution relevance.

The Human-in-the-Loop (HITL) paradigm embodies this vision by integrating human oversight into AI processes. As (Amershi et al., 2019) highlight, this model ensures that algorithmic decisions remain adjustable by experts, preserving nuance and contextual understanding that machines alone lack. UNESCO's Recommendation on AI Ethics (2021) reinforces this stance, advocating "human control in critical decision-making" to ensure AI supplements—rather than replaces—human judgment (Bean et al., 2025).

However, challenges remain. (Floridi et al., 2018) warns of "moral deskilling", where over-reliance on AI weakens individuals' ethical reasoning. To maintain balance, automation must be paired with human intuition (Jarrahi, 2018). (T. H. Davenport & Ronanki, 2018) illustrate this by showing how AI can provide predictions while allowing humans to refine them based on contextual expertise.

Thus, truly human-centered AI must:

1. Engage users in system design and refinement (Norman, 2013).
2. Maintain human oversight (HITL) to prevent blind automation (Bean et al., 2025).
3. Preserve critical ethical skills by limiting algorithmic dependency (Fukuda-Parr & Gibbons, 2021).

This approach is not merely technical—it reflects a philosophy of complementarity, where AI enhances human capabilities without supplanting them. As (Taddeo & Floridi, 2018) summarizes, "Artificial intelligence should remain a tool serving natural intelligence, not the other way around."

## 4.4. Raising awareness and training co-creation actors

Ethical AI literacy is crucial for all stakeholders in value co-creation. As (B. Mittelstadt, 2019) asserts, "AI ethics cannot be delegated to engineers alone"—a multidisciplinary education blending technology, social sciences, and philosophy is needed to equip participants. This literacy moves beyond a purely instrumental view of AI, incorporating discussions on algorithmic justice, data protection, and systemic biases (Holmes et al., 2022). Without it, co-creation risks reinforcing power imbalances, even under the guise of collaboration.

Structured training programs play a key role in fostering this shared understanding (Jobin et al., 2019). They should target both developers and end-users, emphasizing proactive ethics. (Boddington, 2023) advocates active learning, where participants identify ethical dilemmas before development begins, rather than relying on abstract principles. Meanwhile, internal certifications (Buhmann et al., 2020) can embed these skills in organizational practices, training teams in bias detection, privacy safeguards, and ethical auditing.

Public awareness campaigns further reinforce these efforts by democratizing responsible AI issues. Initiatives like the Montreal Declaration for Responsible AI aim to transform passive users into critical actors who can question AI's societal impacts (Morandín-Ahuerma, 2023). (Taddeo & Floridi, 2018) stress the need to "translate ethical principles into accessible tools", making algorithmic decision-making transparent and understandable.

Decentralized initiatives also play a role. Platforms dedicated to algorithmic fairness (Raji et al., 2020) enable developers and citizens alike to access audit frameworks and bias-testing libraries. Combined with participatory workshops, these tools promote horizontal knowledge-sharing, allowing all stakeholders to shape ethical norms.

Yet challenges persist. The technical complexity of AI often discourages non-experts from engaging (Theodorou & Dignum, 2020). (B. Mittelstadt, 2019) notes that ethical literacy requires more than knowledge transfer—it demands spaces for dialogue where conflicting values can be negotiated. Additionally, disparities in access to education—between corporations and SMEs, or Global North and South—threaten the universality of co-creation efforts (Unwin & Unwin, 2017).

Ultimately, training and awareness-raising go beyond mere knowledge acquisition—they cultivate ethics in action, integrated into every project phase.

As (Floridi et al., 2018) state, "Responsible AI is a process, not a product." This process-driven approach requires continuous commitment, where collective learning and critical vigilance become cornerstones of co-creation.

## CONCLUSION

Value co-creation, redefined by the theories of Service-Dominant Logic (S-D Logic) from (Vargo & Lusch, 2004, 2008) and the foundational work of (C. K. Prahalad & Ramaswamy, 2004), has now emerged as a central paradigm for understanding economic and social dynamics in the age of artificial intelligence (AI). This chapter has shown that value no longer resides in the product or service itself, but in the user experience, co-constructed through continuous interactions between human actors and technological systems. AI, as a catalyst for these interactions, transforms co-creation mechanisms while simultaneously raising ethical and structural challenges that call for responsible innovation.

The theoretical foundations of co-creation, articulated around value-in-use (Grönroos, 2011) and value constellations (Normann & Ramirez, 1993), have been profoundly renewed by AI. It enables dynamic personalization of offerings, accelerates open innovation, and streamlines collaborative ecosystems. Service platforms illustrate this evolution, where users become "implicit co-designers" (Kumar & Reinartz, 2016), their behaviors feeding algorithms in real time to shape increasingly tailored value propositions. AI thus operates as an operant resource (Vargo & Lusch, 2008), integrating data and interactions to generate contextualized value.

However, this transformation is not without ambivalence. As highlighted by (Zuboff, 2019), surveillance capitalism threatens to reduce co-creation to an algorithmic exploitation of data, where user experience is monetized without informed consent. Structural biases, reproduced by systems trained on historically inequitable data (Barocas et al., 2019), risk marginalizing certain stakeholders, contradicting the inclusive ideal championed by (C. K. Prahalad & Ramaswamy, 2004). These challenges demand hybrid regulation, combining soft law and binding legal frameworks, as seen in the European AI Act (2024), which prohibits abusive practices while mandating human oversight for high-risk AI.

Faced with these tensions, responsible innovation emerges as an essential framework. Defined by (Stilgoe et al., 2020) around four pillars—anticipation, reflexivity, inclusion, and responsiveness—it calls for rethinking AI development by integrating ethical and societal considerations from the outset. Anticipating extractivist impacts (Crawford, 2021), auditing algorithmic biases (B. Mittelstadt, 2019), and fostering active citizen participation (Villani Report, 2018) are key levers to align AI with the common good. (Floridi et al., 2018) encapsulates this ideal by stating that "responsible innovation is not a brake, but a guide toward AI serving humanity."

Initiatives like the Partnership on AI embody this collaborative vision, bringing together businesses, researchers, and civil society to co-construct ethical standards. Yet, dilemmas persist, particularly in sensitive sectors like healthcare or art, where AI raises concerns about data confidentiality (Topol, 2019) or intellectual property (Gabriel, 2020). These challenges highlight a central paradox: the more AI advances, the more urgent its ethical alignment becomes (Du & Xie, 2021).

To reconcile technological progress and human values, a co-construction approach is essential. Participatory design (Simonsen & Robertson, 2013) and human-in-the-loop models (Amershi et al., 2019) advocate for human-centered AI, where users are involved from the design phase. Explainable AI (XAI) and decision-making transparency (Guidotti et al., 2018) enhance trust, while frameworks like Algorithmic Impact Assessments (AIAs) mitigate risks. (B. Mittelstadt, 2019) reminds us that "AI ethics cannot be static": it must evolve through inclusive governance, integrating mechanisms for continuous learning and contextual adaptation.

Finally, training and awareness-raising among stakeholders are crucial. A multidisciplinary ethical literacy, spanning technical, social, and philosophical dimensions, helps move beyond an instrumentalist view of AI. As (Floridi et al., 2018) emphasize, "Responsible AI is a process, not a product," requiring collective commitment to turn principles into practice.

In conclusion, this chapter calls for a reconciliation between artificial intelligence and collective intelligence. AI, when designed as a tool for empowerment rather than domination, can amplify human capabilities and enrich co-creation dynamics. Intelligent constellations (Nambisan et al., 2019), fueled by feedback loops and diverse contributions, illustrate this potential. Yet, this future depends on our ability to establish ethical safeguards, promote reflexive governance, and place humans at the heart of innovation.

As (C. Prahalad, 2004) succinctly puts it, "The power now belongs to those who know how to engage in dialogue." In a world marked by complex societal challenges, responsible AI must not be a distant horizon but a co-constructed reality, where technology and humanity coexist in a dynamic of mutual enrichment.

## REFERENCES


Abdalla, M., & Abdalla, M. (2021). *The grey hoodie project : Big tobacco, big tech, and the threat on academic integrity*. 287-297.

Adadi, A., & Berrada, M. (2018). Peeking inside the black-box : A survey on explainable artificial intelligence (XAI). *IEEE Access : Practical Innovations, Open Solutions*, *6*, 52138–52160. DOI: 10.1109/ACCESS.2018.2870052

Ahmad, L., Agarwal, S., Lampe, M., & Mishkin, P. (2025). OpenAI's Approach to External Red Teaming for AI Models and Systems. *arXiv preprint arXiv:2503.16431*.

Amershi, S., Weld, D., Vorvoreanu, M., Fourney, A., Nushi, B., Collisson, P., Suh, J., Iqbal, S., Bennett, P. N., & Inkpen, K. (2019). *Guidelines for human-AI interaction*. 1-13.

Ananny, M., & Crawford, K. (2018). Seeing without knowing : Limitations of the transparency ideal and its application to algorithmic accountability. *new media & society, 20*(3), 973-989.

Askell, A., Brundage, M., & Hadfield, G. (2019). The role of cooperation in responsible AI development. *arXiv preprint arXiv:1907.04534*.

Azcárate, A. L.-V. (2024). *Intermedial semiotics in the age of artificial intelligence. Challenges and opportunities for the arts*. New Techno Humanities.

Barocas, S., Guo, A., Kamar, E., Krones, J., Morris, M. R., Vaughan, J. W., Wadsworth, W. D., & Wallach, H. (2021). *Designing disaggregated evaluations of ai systems : Choices, considerations, and tradeoffs*. 368-378.

Barocas, S., Hardt, M., & Narayanan, A. (2019). *Fairness and Machine Learning*. Fairmlbook. Org.

Bassano, C., Barile, S., Saviano, M., Pietronudo, M. C., & Cosimato, S. (2020). *AI technologies & value co-creation in luxury context*.

Bean, E., Burleigh, C., Haskell, C., Burris-Melville, T., Payne, J., & Pathak, B. (2025). Eavesdropping on UNESCO AI Policy, Leadership, and Ethics. *Journal of Leadership Studies, 18*(4), 98–110. DOI: 10.1002/jls.70007

Binns, R. (2018). *Fairness in machine learning : Lessons from political philosophy*. 149-159.

Boddington, P. (2023). AI ethics. *Singapur: Springer International Publishing*, 48.

Bostrom, N., & Yudkowsky, E. (2018). The ethics of artificial intelligence. In *Artificial intelligence safety and security* (pp. 57–69). Chapman and Hall/CRC. DOI: 10.1201/9781351251389-4

Brynjolfsson, E., & McAfee, A. (2017). *Machine, platform, crowd : Harnessing our digital future* (Vol. 4, Numéro 1). WW Norton & Company New York.

Buhmann, A., Paßmann, J., & Fieseler, C. (2020). Managing algorithmic accountability : Balancing reputational concerns, engagement strategies, and the potential of rational discourse. *Journal of Business Ethics*, *163*(2), 265–280. DOI: 10.1007/s10551-019-04226-4

Burrell, J. (2016). How the machine 'thinks' : Understanding opacity in machine learning algorithms. *Big Data & Society*, *3*(1), 2053951715622512. DOI: 10.1177/2053951715622512

Camacho, M., Perramon, J., Puig-Bosch, X., Dang, V. N., Díaz, O., & Lekadir, K. (2025). Stakeholder engagement : The path to trustworthy AI in healthcare. In *Trustworthy AI in Medical Imaging* (pp. 471–493). Elsevier. DOI: 10.1016/B978-0-44-323761-4.00038-9

Chandler, J. D., & Vargo, S. L. (2011). Contextualization and value-in-context : How context frames exchange. *Marketing Theory*, *11*(1), 35–49. DOI: 10.1177/1470593110393713

Chandra, B., & Rahman, Z. (2024). Artificial intelligence and value co-creation : A review, conceptual framework and directions for future research. *Journal of Service Theory and Practice*, *34*(1), 7–32. DOI: 10.1108/JSTP-03-2023-0097

Chen, J., Zhang, Y., & Liu, Z. (2025). Unlocking the power of algorithmic recommendations : The effect of recommendation characteristics on users' willingness to value co-creation. *Current Psychology (New Brunswick, N.J.)*, *44*(3), 1–19. DOI: 10.1007/s12144-024-07175-y

Cockburn, I. M., Henderson, R., & Stern, S. (2018). *The impact of artificial intelligence on innovation* (Vol. 24449). National bureau of economic research Cambridge, MA, USA.

Conn, A. (2017). An Open Letter to the United Nations Convention on Certain Conventional Weapons. *Future of Life Institute. Available online at:* https://futureoflife.org/2017/08/20/autonomous-weapons-open-letter-2017/*(accessed March 16, 2022).*

Crawford, K. (2021). *The atlas of AI: Power, politics, and the planetary costs of artificial intelligence*. Yale University Press.

Davenport, T., Guha, A., Grewal, D., & Bressgott, T. (2020). How artificial intelligence will change the future of marketing. *Journal of the Academy of Marketing Science*, *48*(1), 24–42. DOI: 10.1007/s11747-019-00696-0

Davenport, T. H., & Ronanki, R. (2018). Artificial intelligence for the real world. *HBR'S 10 MUST*, 67.

Dignum, V. (2019). *Responsible artificial intelligence : How to develop and use AI in a responsible way* (Vol. 2156). Springer. DOI: 10.1007/978-3-030-30371-6

Doshi-Velez, F., & Kim, B. (2017). Towards a rigorous science of interpretable machine learning. *arXiv preprint arXiv:1702.08608.*

Du, S., & Xie, C. (2021). Paradoxes of artificial intelligence in consumer markets : Ethical challenges and opportunities. *Journal of Business Research*, *129*, 961–974. DOI: 10.1016/j.jbusres.2020.08.024

Floridi, L., Cowls, J., Beltrametti, M., Chatila, R., Chazerand, P., Dignum, V., Luetge, C., Madelin, R., Pagallo, U., Rossi, F., Schafer, B., Valcke, P., & Vayena, E. (2018). AI4People—An ethical framework for a good AI society : Opportunities, risks, principles, and recommendations. *Minds and Machines*, *28*(4), 689–707. DOI: 10.1007/s11023-018-9482-5

Fukuda-Parr, S., & Gibbons, E. (2021). Emerging consensus on 'ethical AI' : Human rights critique of stakeholder guidelines. *Global Policy*, *12*(S6), 32–44. DOI: 10.1111/1758-5899.12965

Gabriel, I. (2020). Artificial intelligence, values, and alignment. *Minds and Machines*, *30*(3), 411–437. DOI: 10.1007/s11023-020-09539-2

Ghosh, I., & Sanyal, M. K. (2021). Introspecting predictability of market fear in Indian context during COVID-19 pandemic : An integrated approach of applied predictive modelling and explainable AI. *International Journal of Information Management Data Insights*, *1*(2), 100039. DOI: 10.1016/j.jjimei.2021.100039

Grant, N., & Weise, K. (2023). In AI race, Microsoft and Google choose speed over caution. *The New York Times, 7.*

Grönroos, C. (2008). Adopting a service business logic in relational business-to-business marketing : Value creation, interaction and joint value co-creation. *Otago forum, 2*(9), 269-287.

Grönroos, C. (2011). A service perspective on business relationships : The value creation, interaction and marketing interface. *Industrial Marketing Management*, *40*(2), 240–247. DOI: 10.1016/j.indmarman.2010.06.036

Guidotti, R., Monreale, A., Ruggieri, S., Turini, F., Giannotti, F., & Pedreschi, D. (2018). A survey of methods for explaining black box models. *ACM Computing Surveys*, *51*(5), 1–42. DOI: 10.1145/3236009

Haefner, N., Wincent, J., Parida, V., & Gassmann, O. (2021). Artificial intelligence and innovation management : A review, framework, and research agenda . *Technological Forecasting and Social Change*, *162*, 120392. DOI: 10.1016/j.techfore.2020.120392

Hawkins, W., & Mittelstadt, B. (2023). *The ethical ambiguity of AI data enrichment : Measuring gaps in research ethics norms and practices*. 261-270.

Hayes, C. M. (2023). Generative artificial intelligence and copyright : Both sides of the black box. *Available at SSRN* 4517799. DOI: 10.2139/ssrn.4517799

Holmes, W., Persson, J., Chounta, I.-A., Wasson, B., & Dimitrova, V. (2022). *Artificial intelligence and education : A critical view through the lens of human rights, democracy and the rule of law*. Council of Europe.

Holten Møller, N., Shklovski, I., & Hildebrandt, T. T. (2020). *Shifting concepts of value : Designing algorithmic decision-support systems for public services*. 1-12.

Jarrahi, M. H. (2018). Artificial intelligence and the future of work : Human-AI symbiosis in organizational decision making. *Business Horizons*, *61*(4), 577–586. DOI: 10.1016/j.bushor.2018.03.007

Jobin, A., Ienca, M., & Vayena, E. (2019). The global landscape of AI ethics guidelines. *Nature Machine Intelligence*, *1*(9), 389–399. DOI: 10.1038/s42256-019-0088-2

Kumar, V., & Reinartz, W. (2016). Creating enduring customer value. *Journal of Marketing*, *80*(6), 36–68. DOI: 10.1509/jm.15.0414

Lee, M. K., Kusbit, D., Kahng, A., Kim, J. T., Yuan, X., Chan, A., See, D., Noothigattu, R., Lee, S., & Psomas, A. (2019). WeBuildAI: Participatory framework for algorithmic governance. *Proceedings of the ACM on human-computer interaction*, *3*(CSCW), 1-35.

Longo, L., Brcic, M., Cabitza, F., Choi, J., Confalonieri, R., Del Ser, J., Guidotti, R., Hayashi, Y., Herrera, F., & Holzinger, A. (2024). Explainable Artificial Intelligence (XAI) 2.0 : A manifesto of open challenges and interdisciplinary research directions. *Information Fusion*, *106*, 102301. DOI: 10.1016/j.inffus.2024.102301

Mak, K.-K., & Pichika, M. R. (2019). Artificial intelligence in drug development : Present status and future prospects. *Drug Discovery Today*, *24*(3), 773–780. DOI: 10.1016/j.drudis.2018.11.014

Martin, K. D., & Murphy, P. E. (2017). The role of data privacy in marketing. *Journal of the Academy of Marketing Science*, *45*(2), 135–155. DOI: 10.1007/s11747-016-0495-4

Mehrabi, N., Morstatter, F., Saxena, N., Lerman, K., & Galstyan, A. (2021). A survey on bias and fairness in machine learning. *ACM Computing Surveys*, *54*(6), 1–35. DOI: 10.1145/3457607

Mirzani, Y. (2024). *Artificial intelligence and innovation management : Transformative forces shaping organizational creativity. EPRA International Journal of Multidisciplinary Research*. IJMR.

Mittelstadt, B. (2019). Principles alone cannot guarantee ethical AI. *Nature Machine Intelligence*, *1*(11), 501–507. DOI: 10.1038/s42256-019-0114-4

Mittelstadt, B. D., Allo, P., Taddeo, M., Wachter, S., & Floridi, L. (2016). The ethics of algorithms : Mapping the debate. *Big Data & Society*, *3*(2), 2053951716679679. DOI: 10.1177/2053951716679679

Morandín-Ahuerma, F. (2023). *Montreal declaration for responsible AI: 10 principles and 59 recommendations*.

Mumford, R. (2024). *The Incorporation of an Ethics Framework in the Development of Emerging Information Technology Systems*. Salve Regina University.

Nambisan, S., Lyytinen, K., & Yoo, Y. (2020). *Handbook of digital innovation*. Edward Elgar Publishing. DOI: 10.4337/9781788119986

Nambisan, S., Wright, M., & Feldman, M. (2019). The digital transformation of innovation and entrepreneurship : Progress, challenges and key themes. *Research Policy*, *48*(8), 103773. DOI: 10.1016/j.respol.2019.03.018

Nicolau, M., Perez-Liebana, D., O'Neill, M., & Brabazon, A. (2016). Evolutionary behavior tree approaches for navigating platform games. *IEEE Transactions on Computational Intelligence and AI in Games*, *9*(3), 227–238. DOI: 10.1109/TCIAIG.2016.2543661

Norman, W. (2013). *Stakeholder theory*. International Encyclopedia of Ethics.

Normann, R., & Ramirez, R. (1993). From value chain to value constellation : Designing interactive strategy. *Harvard Business Review*, *71*(4), 65–77.

Owen, R., Stilgoe, J., Macnaghten, P., Gorman, M., Fisher, E., & Guston, D. (2013). A framework for responsible innovation. *Responsible innovation: managing the responsible emergence of science and innovation in society*, 27-50.

Ozmen Garibay, O., Winslow, B., Andolina, S., Antona, M., Bodenschatz, A., Coursaris, C., Falco, G., Fiore, S. M., Garibay, I., Grieman, K., Havens, J. C., Jirotka, M., Kacorri, H., Karwowski, W., Kider, J., Konstan, J., Koon, S., Lopez-Gonzalez, M., Maifeld-Carucci, I., & Xu, W. (2023). Six human-centered artificial intelligence grand challenges. *International Journal of Human-Computer Interaction, 39*(3), 391–437. DOI: 10.1080/10447318.2022.2153320

Pasquale, F. (2015). *The black box society : The secret algorithms that control money and information*. Harvard University Press. DOI: 10.4159/harvard.9780674736061

Payne, A. F., Storbacka, K., & Frow, P. (2008). Managing the co-creation of value. *Journal of the Academy of Marketing Science, 36*(1), 83–96. DOI: 10.1007/s11747-007-0070-0

Porter, M. E. (1985). Technology and competitive advantage. *The Journal of Business Strategy, 5*(3), 60–78. DOI: 10.1108/eb039075

Pousttchi, K., Gleiss, A., Buzzi, B., & Kohlhagen, M. (2019). *Technology impact types for digital transformation. 1*, 487-494.

Prahalad, C. (2004). *The future of competition : Co-creating unique value with customers*. Harvard Business Rress.

Prahalad, C. K., & Ramaswamy, V. (2004). Co-creation experiences : The next practice in value creation. *Journal of Interactive Marketing, 18*(3), 5–14. DOI: 10.1002/dir.20015

Rahwan, I. (2018). Society-in-the-loop : Programming the algorithmic social contract. *Ethics and Information Technology, 20*(1), 5–14. DOI: 10.1007/s10676-017-9430-8

Raji, I. D., Smart, A., White, R. N., Mitchell, M., Gebru, T., Hutchinson, B., Smith-Loud, J., Theron, D., & Barnes, P. (2020). *Closing the AI accountability gap : Defining an end-to-end framework for internal algorithmic auditing*. 33-44.

Rane, N. (2023). ChatGPT and similar generative artificial intelligence (AI) for smart industry : Role, challenges and opportunities for industry 4.0, industry 5.0 and society 5.0. *Challenges and Opportunities for Industry, 4*.

Ribeiro, M. T., Singh, S., & Guestrin, C. (2016). Model-agnostic interpretability of machine learning. *arXiv preprint arXiv:1606.05386*.

Setiawan, A. (2024). Intelligent Systems For Disabilities Technology For A More Inclusive Life. *Available at SSRN* 5062273.

Shahriari, K., & Shahriari, M. (2017). *IEEE standard review—Ethically aligned design : A vision for prioritizing human wellbeing with artificial intelligence and autonomous systems*. 197-201.

Simonsen, J., & Robertson, T. (2013). *Routledge international handbook of participatory design* (Vol. 711). Routledge New York.

Solakis, K., Katsoni, V., Mahmoud, A. B., & Grigoriou, N. (2024). Factors affecting value co-creation through artificial intelligence in tourism : A general literature review. *Journal ofTourism Futures, 10*(1), 116–130. DOI: 10.1108/JTF-06-2021-0157

Stilgoe, J., Owen, R., & Macnaghten, P. (2020). Developing a framework for responsible innovation. In *The ethics of nanotechnology, geoengineering, and clean energy* (pp. 347–359). Routledge. DOI: 10.4324/9781003075028-22

Suresh, H., & Guttag, J. (2021). *A framework for understanding sources of harm throughout the machine learning life cycle*. 1-9.

Taddeo, M., & Floridi, L. (2018). How AI can be a force for good. *Science, 361*(6404), 751–752. DOI: 10.1126/science.aat5991

Theodorou, A., & Dignum, V. (2020). Towards ethical and socio-legal governance in AI. *Nature Machine Intelligence, 2*(1), 10–12. DOI: 10.1038/s42256-019-0136-y

Topol, E. J. (2019). High-performance medicine : The convergence of human and artificial intelligence. *Nature Medicine, 25*(1), 44–56. DOI: 10.1038/s41591-018-0300-7

Unwin, P. T. H., & Unwin, T. (2017). *Reclaiming information and communication technologies for development*. Oxford University Press. DOI: 10.1093/oso/9780198795292.001.0001

Vargo, S. L., & Lusch, R. F. (2004). Evolving to a new dominant logic for marketing. *Journal of Marketing, 68*(1), 1–17. DOI: 10.1509/jmkg.68.1.1.24036

Vargo, S. L., & Lusch, R. F. (2008). Service-dominant logic : Continuing the evolution. *Journal of the Academy of Marketing Science, 36*(1), 1–10. DOI: 10.1007/s11747-007-0069-6

Villani, C., Bonnet, Y., Berthet, C., Levin, F., Schoenauer, M., Cornut, A. C., & Rondepierre, B. (2018). *Donner un sens à l'intelligence artificielle : Pour une stratégie nationale et européenne*. Conseil national du numérique.

Voigt, P., & Von dem Bussche, A. (2017). The eu general data protection regulation (gdpr). *A practical guide, 1st ed., Cham: Springer International Publishing, 10*(3152676), 10-5555.

Wilson, H. J., & Daugherty, P. R. (2018). Collaborative intelligence : Humans and AI are joining forces. *Harvard Business Review*, *96*(4), 114–123.

Yeung, K. (2020). Recommendation of the council on artificial intelligence (OECD). *International legal materials, 59*(1), 27-34.

Zuboff, S. (2019). *Surveillance capitalism and the challenge of collective action. 28*(1), 10-29.

# Chapter 8
# Societal Impact of Ethical Dimensions of AI

**Kashvi Chaturvedi**
https://orcid.org/0009-0005-3192-3400
*Ajeenkya D.Y. Patil University, India*

**Susanta Das**
https://orcid.org/0000-0002-9314-3988
*Ajeenkya D.Y. Patil University, India*

**Rebika Singh**
*Ajeenkya D.Y. Patil University, India*

**Kamaljeet Kaur Siddhu**
*Ajeenkya D.Y. Patil University, India*

**Rakesh Kumar Jain**
https://orcid.org/0000-0002-0216-5437
*Ajeenkya D.Y. Patil University, India*

**Nishakant Ojha**
*Ajeenkya D.Y. Patil University, India*

## ABSTRACT

*The ethical, legal, & sociological consequences of AI are examined in this study, with focus on topics like regulatory frameworks, AI literacy, and AI powered social robots. AI is being incorporated into many sectors of the economy, transforming domains like healthcare, and education. In the context of eldercare, AI-powered social robots provide advantages like company and help with everyday chores, but they also present serious privacy, emotional health, and dependency issues. The study looks at a number of ethical frameworks, like European Union AI Act. EMMA approach, which combines moral and decision-making topics, is also covered. Intellectual property rights, liability, and the development of precise regulatory standards to handle AI's international character are important legal issues. According to the study's findings, ethical standards that strike a balance between innovation and human dignity, openness, and justice must direct AI's social integration. Robust legal frameworks and interdisciplinary cooperation is required so that all the advantages of AI are fully experienced.*

DOI: 10.4018/979-8-3373-2597-2.ch008

## 1. INTRODUCTION

AI research was first centered on technological advancement, but it is currently gaining greater attention for its effects on day-to-day living. AI is increasingly being incorporated into consumer goods and services, much how the internet went from being a technological advancement to becoming a necessary component of daily life. Common household goods, automobiles, and professional services like financial counseling, legal advice, and medical diagnostics increasingly incorporate it (Sanjeetha & Abeygunawardhana, 2024).

The employment market is changing as a result of AI's development. AI is predicted to replace many current professions, especially those that need repetitive work, even if it will also generate new employment prospects. Around 47% of American employment and 15% of work hours worldwide might be automated by 2030, as mentioned in the paper by Ng et al. (2021). In the digital age, AI literacy is just as important as traditional literacy since people need to acquire AI-related skills to stay competitive (Ng et al., 2021).

AI literacy is the ability to comprehend and use AI technologies in the job, in the classroom, and in daily life. At first, AI instruction was restricted to computer science courses at the university level that required sophisticated programming knowledge. But now that AI tools and technology are more widely available, teachers are implementing AI literacy in K–12 classrooms (Ng et al., 2021).

Ethics has always been an essential component of human society. The necessity for inclusive, secure, and resilient urban settings is emphasized by the Sustainable Development Goals (SDGs), especially SDG-11 (sustainable cities and communities). Instead of being a disjointed concept, ethics ought to be a universal one that aids in the sustainability and equity of communities (Pastor-Escuredo et al., 2022).

The discipline of AI ethics, which focuses on moral standards, rules, and techniques for making sure AI acts responsibly, has arisen in response to these issues. The two main facets of AI ethics are (1) comprehending the moral principles that direct AI research and (2) developing AI systems that obey to these principles. Even though there has been a lot of discussion on AI ethics, the field is still in its infancy, and the research that has been done so far has tended to concentrate on particular issues like explainability, bias, privacy, and security (Huang et al., 2022).

Language models, particularly OpenAI's GPT series, have drawn a lot of interest among AI technologies. The most recent version of ChatGPT is being utilized for a variety of tasks, from creating original content to sending emails (Ghandour et al., 2024).

By acting as social catalysts, AI-powered Socially Assistive Robots (SARs) can promote interaction between senior citizens and their peers. It has been demonstrated that certain models, such as the Paro robot seal, help older people socialize better and experience less stress (Vozna & Costantini, 2025).

The paper studies various models like those from IBM (International Business Machines Corporation), UNESCO (United Nations Educational, Scientific and Cultural Organization), OECD (United Nations Educational, Scientific and Cultural Organization) and the European Union AI Act. There is also mention of Ethical Management of AI (EMMA) which blend three essential viewpoints: Decision-making by managers, The moral ramifications of using AI and Businesses handling AI. It also discusses how well-known ethical framework is used to investigate 5 areas of ChatGPT, which are Information rights and obligations, Property rights and responsibilities, Control and accountability, System quality, and Quality of life. AI literacy, how teachers can impart AI literacy in an effective manner, evaluating AI literacy of students and moral issues that are raised by AI literacy are few other important topics. However, there are privacy and security issues because SARs and ChatGPT gather a lot of private information, such as home and health records. Based on philosophical underpinnings and the unique characteristics of AI, the study suggests ethical guidelines for AI designers. It also offers methods for incorporating ethics into the development and application of AI, guaranteeing that these systems are in line with society's larger moral obligations. The chapter concludes with presenting bibliometric analysis for analyzing the trend on the chapter theme. Figure 1 presents a visual depiction explaining the flow and subsections of the chapter.

*Figure 1. Flowchart for the Chapter.*

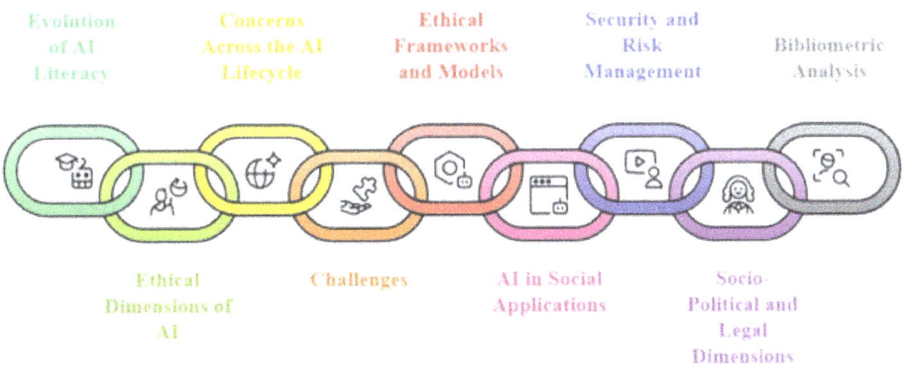

## 2. LITERATURE REVIEW

AI has revolutionized our understanding of both natural and artificial intelligence. With the goal of mimicking or outperforming human cognitive abilities, AI is now widely used in a variety of industries, including healthcare, education, and transportation. These systems can now perform better than humans in areas like pattern recognition, decision-making, and language understanding because of technologies like machine learning, deep learning, and natural language processing (Sanjeetha & Abeygunawardhana, 2024).

Business intelligence makes extensive use of AI, which enhances decision-making and automates sectors including manufacturing, services, and agriculture. Big data, the Internet of Things (IoT), and AI are transforming industries and producing better goods and services. Self-driving cars, smart appliances, and automated medical equipment are just a few examples of the increasingly prevalent AI-driven smart products (Sanjeetha & Abeygunawardhana, 2024).

Research on AI's wider societal effects is still lacking, especially with regard to smart devices and services. While research emphasizes AI's accuracy and efficiency, it frequently ignores its ethical ramifications, including bias, privacy issues, and job loss. To guarantee the appropriate use of AI, more precise rules and guidelines are also required (Sanjeetha & Abeygunawardhana, 2024).

### 2.1. Digitalization and Ethics

AI frequently develops more quickly than moral laws can keep up. Although the Sustainable Development Goals (SDGs) of the UN offer a comprehensive framework for world advancement, digitalization has not yet been completely included into these programs. The pace of efforts to leverage technology and data for social good has slowed, especially in private-public organizations. AI development lacks appropriate monitoring in the absence of ethical principles. Despite the fact that AI has transformed many businesses, a well-organized ethical shift has not yet resulted. To ensure responsible and sustainable digitization, a deeper ethical base is required (Pastor-Escuredo et al., 2022).

The first step in addressing the ethical issues raised by AI is to recognize and comprehend the possible risks connected to its creation and application. This knowledge can subsequently be used to establish ethical AI rules and principles, which will result in the production of morally sound systems. Concerns including lack of transparency, privacy violations, accountability, bias and discrimination, safety and security threats, and abuse possibilities are only a few of the ethical challenges surrounding AI. The ethical concerns, like lack of transparency, privacy, violations etc., are some of the topics, which also suggests a novel classification that divides them into three levels: environmental, societal, and individual (Huang et al., 2022).

A. *Individual Level*: This covers moral dilemmas that have an immediate impact on people, such their rights and welfare. Here, privacy, data security, and the effect of AI on individual liberty are issues (Huang et al., 2022).
B. *Level of Society*: This level discusses the wider effects of AI on society at large. Concerns include the loss of jobs as a result of automation, the availability of AI technologies, and the possibility that AI would worsen prejudices and social injustices (Huang et al., 2022).
C. *Environmental Level*: This component considers how AI affects the environment, including how AI technologies use energy, produce pollution, and consume resources (Huang et al., 2022).

OpenAI created ChatGPT, a potent AI language model that can produce text for a variety of uses that looks human. It employs a transformer-based methodology with two main stages:

I. Pre-training is the process of learning grammar, facts, and reasoning from a large amount of online data (Ghandour et al., 2024).
II. Fine-tuning: Improving the model's accuracy and conformity to ethical standards by refining it on a particular dataset (Ghandour et al., 2024).

## 2.2. Concerns and Challenges

Industry 5.0 is one new trend that stresses collaboration between humans and AI rather than only automation. Nevertheless, there aren't many precise rules for handling AI morally in this changing environment (Brendel et al., 2021).

The development of AI is being tackled from a broad perspective, considering business, science, and the need to increase public trust. Developing an ecosystem that puts human values, rights, and societal needs first is a major problem in order to guarantee that AI helps people without endangering them. Given how AI is changing sectors including manufacturing, banking, healthcare, security, education, and mobility, trust is crucial for productive human-AI collaboration. Work dynamics are also impacted by this change, necessitating new approaches to human-AI collaboration. The proper regulatory framework is required to make AI systems reliable. Conventional laws deal with harm produced by humans, while AI creates new problems like hate speech, internet disinformation, and worries about AI's increasing autonomy. Whether to build dynamic legal systems that foresee future hazards or modify current laws to accommodate AI is a topic of debate among policymakers.

Ex ante impact evaluations, which analyze possible risks prior to the deployment of AI systems rather than evaluating their consequences after implementation, are one important suggestion (Boni, 2021).

### 2.2.1. Concerns at Different Levels

(i) *Individual Level Ethical Concerns*: AI has an impact on human dignity, autonomy, privacy, and personal safety. Safety concerns could include mishaps by autonomous systems, like self-driving cars, privacy concern arises due to need of enormous volumes of personal data by AI, and by making decisions for people, AI can also undermine human autonomy by possibly reducing the capacity for autonomous thought. Furthermore, technology like deadly autonomous weapons have the potential to violate human dignity, which is a fundamental human right (Huang et al., 2022).

(ii) *Moral Concerns at the Community Level*: Concerns regarding justice, accountability, transparency, surveillance, democracy, employment displacement, and interpersonal connections are brought up by AI on a larger societal scale. AI systems have the potential to continue prejudice and discrimination. To guarantee appropriate governance, responsibility and accountability for AI decisions must be precisely specified. Because AI is a "black-box," it is challenging to comprehend how judgments are made, which raises concerns about transparency. AI-powered mass surveillance also jeopardizes civil liberties and privacy (Huang et al., 2022).

(iii) *Moral Concerns in the Environmental Field*: The creation and application of AI use a lot of natural resources, which raises environmental issues. Rare materials are depleted during the manufacturing of AI-related technology, such as chips and sensors, and their disposal causes pollution. AI systems use a lot of energy since they need a lot of processing power. AI must integrate with global sustainability goals by striking a balance between environmental responsibility and technological advancement (Huang et al., 2022).

### 2.2.2. Concerns based on AI Lifecycle

A. *Business Analysis*: Establishing just and moral objectives for the use of AI.

B. Data Engineering: Ensuring impartial and privacy-conscious data gathering, labeling, and processing.

C. *ML Modeling*: Ensuring decision-making transparency and preventing bias in AI models.

D. *Model Deployment*: Creating guidelines for appropriate use prior to the deployment of AI systems.
E. *Operation & Monitoring*: Constantly evaluating the effects of AI and making necessary adjustments to avoid harm.

### 2.2.3. Ethical and Legal Difficulties

ChatGPT can reflect biases from its training data and does not automatically guarantee fairness, despite its remarkable speed and processing capacity. Even so, it has the capacity to promote constructive social change when applied appropriately (Ghandour et al., 2024).

There is now more pressure to change rules and regulations to better govern AI technology as a result of ChatGPT's emergence. Among the main moral issues are:

a) *Concerns about privacy*: ChatGPT lacks built-in safeguards for personal information, which is against privacy regulations such as the General Data Protection Regulation (GDPR) (Ghandour et al., 2024).
b) *Copyright issues*: Because ChatGPT is trained on a variety of online information, such as blogs, poetry, and court documents, it may produce responses that contain content protected by copyright, which could result in intellectual property problems (Ghandour et al., 2024).
c) *Accountability and control*: Because ChatGPT generates its responses automatically, it is challenging to identify the information's originating sources or properly govern its behavior (Ghandour et al., 2024).
d) *System transparency*: It can be difficult to understand ChatGPT's decision-making process, which raises questions about explainability and trust (Ghandour et al., 2024).
e) *Impact on society*: Careful consideration must be given to how AI-generated content influences people's decisions, employment, and general well-being (Ghandour et al., 2024).

## 2.3. Building an Ethical Framework

The ethical issues surrounding AI's application have drawn a lot of attention as it grows more pervasive in our daily lives. To guarantee AI is used ethically, a number of organizations—including governments, businesses, and academic institutions—have created rules.

In 2019, a significant study examined 84 ethical AI guidelines from around the world and discovered recurring themes. Since then, though, numerous new guidelines have been published. The majority of the 146 AI ethics standards that have been recognized as of 2021 have been released within the previous five years (Huang et al., 2022). Eleven important ethical principles surfaced from the analysis of these guidelines and it is the responsibility of designers to accomplish all these through their application:

i. *Transparency*: In order for people to comprehend how decisions are made, AI systems should be comprehensible and explicable.
ii. *Justice and Fairness*: AI shouldn't treat people or groups differently.
iii. *Accountability and Responsibility*: Those who develop and apply AI must take responsibility for its results.
iv. *Nonmaleficence*: AI shouldn't endanger or hurt humans.
v. *Privacy*: AI needs to abide with privacy regulations and safeguard personal information.
vi. *Beneficence*: AI ought to help people and make a constructive contribution to society.
vii. *Freedom & Autonomy*: AI shouldn't take away people's freedom to choose for themselves.
viii. *Solidarity*: AI should foster societal unity rather than sow discord.
ix. *Sustainability*: Development of AI should take the environment into consideration.
x. *Trust*: For AI to be accepted by the general population, it must be dependable and trustworthy.
xi. *Dignity*: AI ought to uphold morals and human rights.

In order to assist enterprises in operationalizing ethical considerations in their decision-making processes, especially with reference to AI, the EMMA framework is introduced. It highlights how businesses must modify the framework in accordance with their ethical principles, environmental considerations, and decision-making processes. Three stages of decision-making are distinguished by the framework: strategic, tactical, and operational. While tactical decisions convert this strategy into workable plans, strategic decisions concentrate on the organization's long-term goals and moral principles. Operational decisions handle ethical dilemmas that may come up during day-to-day operations and require putting these strategies into action. The framework uses a variety of ethical streams, such as meta-ethics, normative ethics, applied ethics, and descriptive ethics, to produce ethical considerations. To evaluate the ways in which external elements impact organizational success and decision-making, the framework also includes a PESTEL study (Political, Economic, Social, Technological, Environmental, and Legal) (Brendel et al., 2021).

The EMMA framework is a result of this integrated approach, which guarantees that ethical standards, decision-making, and the larger organizational environment all incorporate AI ethics. Businesses may preserve their competitiveness and promote ethical AI development that benefits society and businesses alike by doing this (Brendel et al., 2021).

Establishing unambiguous ethical standards is essential to ensuring the appropriate and humane application of AI technology as it develops. Additional laws like the Machinery Directive (2006/42/EC) and ISO 13482:2014 address certain organizational and technical measures needed for data security, even if GDPR provide the framework for data protection in the EU (Vozna & Costantini, 2025).

With a focus on human dignity, autonomy, and the avoidance of prejudice, UNESCO has been instrumental in establishing ethical criteria for robotics. Their rules place a strong emphasis on the value of responsibility, transparency, and user permission in AI applications. Additionally, UNESCO supports public education about robotics, legal accountability for abuse, and defined roles in AI development. Additionally, the World Health Organization (WHO) has released ethical standards for AI in healthcare that place a strong emphasis on inclusivity, responsibility, and openness. These rules guarantee that AI-powered solutions, such social robots for senior care, respect the highest moral principles and encourage equitable access for people from all socioeconomic backgrounds (Vozna & Costantini, 2025).

AI must respect human autonomy, integrity, and fundamental rights, according to the European Commission and European Parliament. The AI Act, a legislative framework that divides AI systems into four risk categories—unacceptable, high, limited, and minimal risk—which was introduced by the European Commission in 2021. In order to ensure that high-risk AI applications—like those in healthcare and law enforcement—meet legal, ethical, and technical criteria, the Act requires tougher regulations. The AI Act, officially known as Regulation (EU) 2024/1689, encourages innovation through regulatory sandboxes and fortifies the EU's commitment to responsible AI use (Vozna & Costantini, 2025).

A framework that categorizes AI systems according to their capacity for self-learning and their influence on humans was created to assist corporations in managing AI ethics:

a) Low human effect, low self-learning: Data analysis tools (lower ethical risk) are one example.
b) AI chatbots influencing customer service is an example of a medium-self-learning, medium-human impact (moderate ethical risk) (Brendel et al., 2021).
c) AI in healthcare that makes treatment decisions is an example of high self-learning, high human impact (high ethical risk) (Brendel et al., 2021).
d) Stricter ethical control is necessary for AI systems in the higher-impact categories since errors might have grave repercussions (Brendel et al., 2021).

## 2.4. Ethical Management of AI in Organizations

Organizations must assess AI projects according to their possible effects on individuals and society in order to make sure AI technology complies with ethical norms. Businesses can divide AI systems into three tiers using an AI positioning matrix: low self-learning and low human impact (low ethical issues), medium self-learning and medium impact (moderate ethical concerns), and high self-learning and high impact (high ethical problems). The greater the AI's capacity to learn and impact human lives, the greater the need for ethical supervision.

A three-part strategy is required to manage AI ethics successfully. First, ethical considerations should be incorporated into managerial choices, particularly in high-impact AI programs. Second, in order to make ethical decisions and ensure that AI development is in line with moral principles, organizations require a defined ethical reference framework. This framework ought to be adaptable while still being sufficiently detailed to evaluate new goods, services, and internal procedures. Third, environmental factors, such as stakeholder interests, social effects, and political environments, must be taken into account while making ethical decisions. Employers should give staff members and decision-makers the resources and direction they need to handle these moral dilemmas (Brendel et al., 2021).

## 2.5. Ethical Obligations of Designers

By analyzing various ethical theories, such as deontology, utilitarianism, social contract theory, and care ethics, it highlights the need of comprehending moral obligations towards both individuals and society (McGraw, 2024).

Deontology emphasizes the inherent rightness or wrongness of deeds based on upholding moral obligations rather than their effects. The consequentialist theory of utilitarianism seeks to maximize the pleasure of the largest number of individuals by evaluating actions according to their results. Utilitarian design concepts in AI encourage developers to build systems that yield positive results while also taking into account potential drawbacks, including algorithmic biases. According to the social contract idea, social agreements give rise to ethical obligations. According to this theory, moral obligations might change depending on cultural norms and values, hence it's critical to comprehend the particular social environment in which AI systems are used. Care ethics places a strong emphasis on compassion, empathy, and the value of human connections in moral decision-making. The core of this idea is Kant's categorical imperative, which emphasizes that deeds must respect human autonomy and be universally applicable (McGraw, 2024). Along with these designers should uphold their ethical duty of making sure that AI is created in a fashion that upholds core principles and advances humanity. These obligations tackle the particular difficulties that artificial intelligence poses and are founded on established ethical standards (McGraw, 2024).

i. *(Non-Maleficence) Do No Harm*: AI should prioritize safety, security, and dependability which requires designers to carry out risk analyses and put protections in place to avoid unforeseen consequences (McGraw, 2024).
ii. *Social Utility (Benefit Society)*: AI should be created with the goal of enhancing society by tackling significant problems and reducing adverse consequences like social inequality and employment loss (McGraw, 2024).
iii. *Honor User Autonomy*: People should be able to make educated decisions and have clear information about how AI impacts their life (McGraw, 2024).
iv. *Assure Equitable Treatment*: AI ought to be devoid of prejudice and discrimination. In order to guarantee that AI serves everyone and not just a chosen few, developers must actively seek to eradicate biased outcomes (McGraw, 2024).
v. *Be Open and Responsible*: AI decision-making ought to be transparent and intelligible. Developers must assume accountability for AI's actions, guaranteeing transparent supervision and procedures for resolving any problems (McGraw, 2024).

## 2.6. Aligning Ethical Principles with SDGs

The study emphasizes how ethics, digitalization, and the United Nations Sustainable Development Goals (UNSDGs) are related (Pastor-Escuredo et al., 2022):

a) *SDG-11 (Sustainable Cities and Communities)*: To tackle urban issues including infrastructure, transit, and pollution, cities need to be resilient, inclusive, and technology-driven. AI can be used to monitor and enhance urban conditions.
b) *SDG-10 (Reduced Inequalities)*: By guaranteeing equitable access to technology and services, digital tools help advance social and economic equality.
c) *SDG-8 (Decent Work and Economic Growth)*: Automation driven by AI and digital workspaces should complement human labor rather than replace it. Remote work options and the metaverse may promote economic expansion while maintaining decent working conditions.
d) *SDG-12 (Responsible Consumption and Production)*: Through data-driven innovations, digitalization may increase the effectiveness of the supply chain, cut waste, and encourage sustainable consumption.

## 2.7. Methods to Evaluate Ethical AI

The goal of ethical methods is to develop AI systems that are capable of moral decision-making by using accepted ethical ideas. There is discussion of three primary ethical theories:

a. The study of virtue ethics focuses on an AI system's moral nature and makes sure it exhibits qualities like kindness and honesty (Huang et al., 2022).
b. Deontological ethics: Makes sure AI behaves in line with established ethical standards by emphasizing adherence to moral laws or obligations (Huang et al., 2022).
c. Consequentialist ethics seeks to maximize total value by evaluating AI activities according to their results (Huang et al., 2022).

Three approaches are investigated for integrating ethical reasoning into AI:

a. *Top-down methods*: AI abides by ethical guidelines that are expressly programmed, much as Asimov's laws of robotics (Huang et al., 2022).
b. *Bottom-up methods*: Just as humans acquire morality via experience and observation, AI also learns ethical conduct (Huang et al., 2022).
c. *Combination of two methods*: Combining top-down and bottom-up strategies to strike a balance between rigid ethical standards and flexibility is known as a hybrid strategy (Huang et al., 2022).

Every strategy has advantages and disadvantages. Although they provide predictability, top-down approaches could not be flexible. While bottom-up methods allow AI to learn and adapt, they might not be transparent. The goal of hybrid approaches is to incorporate the best features of each (Huang et al., 2022).

In order to avoid ethical dilemmas, technological methods concentrate on enhancing AI systems. This comprises:

i. Increasing the transparency and comprehensibility of AI decisions is known as explainable AI (XAI) (Huang et al., 2022).
ii. Reducing bias in AI decision-making is known as "fair AI." (Huang et al., 2022).
iii. AI privacy preservation: Making sure AI protects user privacy (Huang et al., 2022).
iv. Creating systems to keep AI developers accountable for their actions is known as accountability in AI (Huang et al., 2022).

Although this field of study is still in its infancy, groups like AAAI and ACM are aggressively pursuing AI accountability, transparency, and justice. The goal of conferences and research projects is to guarantee that AI benefits humanity in a morally and responsibly responsible manner (Huang et al., 2022).

The three primary methods for evaluating AI ethics—testing, verification, and standards—are examined in this section (Huang et al., 2022).

Evaluating AI Ethics:

A. *The Moral Turing Test (MTT)*: It assesses the moral thinking of an AI system by determining if a human interrogator can discern between the ethical decisions made by the AI and those made by a human. The AI is regarded as a moral agent if its moral judgments are identical to those of a human. If the AI is not found to be less moral than the human subject, it passes (Huang et al., 2022).
B. *Expert and Nonexpert Tests*: Both ethical professionals and regular people (non-experts) can evaluate AI's moral judgments. Nonexpert tests depend on public views of morality, whereas expert tests employ professional ethical standards (Huang et al., 2022).

Verification entails demonstrating that an AI system continuously complies with moral guidelines. This procedure consists of:

a. *Formal Verification*: This approach uses thorough testing to guarantee that an AI system satisfies established ethical norms. A counterexample is given to show where the AI went wrong if it does not match these requirements (Huang et al., 2022).
b. *Design Verification*: This method guarantees that AI ethics are in line with the desired ethical objectives. It aids in avoiding problems like dishonesty and faulty moral judgment (Huang et al., 2022).

Numerous groups have developed industry standards and ethical principles for AI, such as:

A. Codes of Conduct for Professionals:
    a) In 2014, the Australian Computer Society (ACS) created a code of conduct that outlines the moral obligations of technology workers (Huang et al., 2022).
    b) To help with moral decision-making in computing, the Association for Computing Machinery (ACM) revised its Code of Ethics in 2018 (Huang et al., 2022).
B. Global Guidelines:

a) The IEEE P7000 series, which addresses algorithmic bias, data privacy, and responsible AI development, was created by the IEEE Global Initiative on AI Ethics (Huang et al., 2022).
b) A collaborative effort by international organizations, ISO/IEC JTC 1/SC 42 has been addressing ethical, trustworthiness, and governance issues related to AI (Huang et al., 2022).

There is still a discrepancy between theory and reality even if these principles offer ethical guidelines for AI. Due to budget constraints, smaller organizations find it difficult to embrace ethical frameworks, even though major corporations such as Microsoft and IBM have done so. To create workable, legally binding ethical norms for AI in all sectors, further work is required (Huang et al., 2022).

Parson et. Al (2019) in his report summarizes the preliminary results of eight workgroups that worked together at First International Summer Institute on Artificial Intelligence and Society (SAIS) hosted by Alberta Machine Intelligence Institute (Amii) in Edmonton on several ethical and useful applications of AI. An overview of each group's objectives and results is provided below:

i. *Mobility Systems and Embedded Values*: This group studied the effects of AI-powered navigation systems on society, including the rise in traffic in residential neighborhoods. As these systems grow increasingly linked, they emphasized the necessity of taking into account a number of principles, including as safety and fair access to mobility (Parson et al., 2019).
ii. *Meaningful Human Control*: The significance of human participation in high-stakes decision-making, like that in medicine and weapons, was the main focus of this group. They investigated the advantages and disadvantages of retaining human control and are creating standards for applying suitable degrees of control in various situations (Parson et al., 2019).
iii. *The goal of the group AI*: Without Math is to develop non-technical teaching resources on AI principles for judges and journalists, among other audiences. They are creating materials that, without the use of formal mathematical terminology, explain important concepts in AI and machine learning (Parson et al., 2019).
iv. *Principles of AI Policy Analyst Design*: This team looked into how AI might complement or take the role of policy analysts. They argued for a model where AI augments human capacities, concluding that although AI might handle data synthesis, human interaction is still essential for effective policy analysis (Parson et al., 2019).
v. *Evaluation Instrument for AI Products' Ethical Effects*: A framework for operationalizing ethical standards for AI initiatives was created by this group. They

developed a grading system to assess compliance with these guidelines, highlighting the necessity of continuing conversations regarding moral ramifications (Parson et al., 2019).

vi. *Criticism of EU AI rules*: This team evaluated the EU's ethical rules for AI, namely a developer-guideline checklist. They contended that checklists have the potential to oversimplify difficult ethical issues and encourage "ethics-washing" as opposed to sincere ethical participation (Parson et al., 2019).

vii. *AI and Agency*: This group investigated the idea of agency in relation to AI, asking whether it applies to both humans and AI and how it is defined. They talked about how these concepts affect moral responsibility and accountability (Parson et al., 2019).

viii. *AI as a Tool for Social Progress*: This group compared dystopian perspectives with the potential of AI to propel social and political advancement. They argued for AI systems that put user values and well-being first, identifying the technical and sociopolitical prerequisites required for AI to advance liberty and equality (Parson et al., 2019).

## 2.8. AI Literacy and Awareness

The idea of literacy in general, which expanded from digital literacy in the 1970s to include knowledge of AI technology today, is the foundation of AI literacy. Few primary components of AI literacy were discovered by researchers:

A. *Know and Understand AI*: This entails teaching students the essential ideas and abilities of AI without requiring any prior expertise. It places a strong emphasis on comprehending the technologies underlying AI applications (Ng et al., 2021).

B. *Use and Apply AI*: Students should be able to apply AI principles in a variety of settings. This entails being aware of moral issues and applying AI sensibly (Ng et al., 2021).

C. *Analyze and Develop AI*: Higher-order thinking abilities are also a part of AI literacy, which helps people evaluate AI technology critically and work with AI systems in productive ways (Ng et al., 2021).

The key to overcoming mistrust, which breeds uncertainty and uneasiness, is knowledge and training. According to a 2020 survey, 86% of participants supported AI training initiatives, and 90% of respondents said that the best approach to assist people in adapting was to enhance their AI-related abilities. Companies also underlined the value of AI digital literacy, which ought to be taught in schools to improve flexibility. Understanding AI's effects on human psychology and behavior should be part of any awareness-raising campaign. AI can cause problems including psychological dependence, deceit, etc. while also offering advantages, such as enhancing human cooperation and communication through interaction support. In order to help individuals navigate AI's influence in daily life, digital literacy development should cover both the advantages and disadvantages of AI (Boni, 2021).

## 2.9. Educators' Role

By using relevant learning materials, pedagogical strategies, and subject matter, educators can significantly contribute to the development of AI literacy. The framework of Technological, Pedagogical, and Content Knowledge (TPACK) is utilized to comprehend the appropriate integration of AI literacy into courses. Teaching AI principles requires the use of age-appropriate hardware and software. For different educational levels, different teaching strategies are advised. While secondary students should work together on projects, younger pupils are encouraged to use playful and inquiry-based methods. Students at universities are able to apply their knowledge to real-world problems and take on increasingly challenging assignments (Ng et al., 2021). For teachers to successfully integrate AI literacy into their classrooms, teacher training programs must provide them with the necessary AI understanding and instructional techniques. Researchers assess students' AI literacy using both quantitative and qualitative techniques (Ng et al., 2021).

## 2.10. Information Rights and Obligations

The threat to privacy posed by AI models such as ChatGPT is one of the most significant ethical issues. Based on the data they process, these algorithms are able to deduce private information about specific people. Furthermore, there is worry that improper usage or illegal access to data could pose major security threats (Ghandour et al., 2024).

A. *Information Rights and Responsibilities*: Individuals ought to be in charge of their data, including being aware of what information is gathered, how it is put to use, and who can access it along with the option to seek deletion, restrict usage, or fix errors. However, organizations also have obligations, such as being open and honest about their data usage, utilizing robust security methods to protect data, and preserving privacy rights and utilizing data solely for appropriate purposes (Ghandour et al., 2024).
B. *AI's Effect on Privacy*: Large volumes of personal data are processed by AI-powered systems, such as chatbots used in customer support. But a lot of individuals don't know how much information is gathered or how their online activity is examined. Because AI may uncover personal information about people that

they never meant to disclose, this could pose privacy problems (Ghandour et al., 2024).

C. *The Need for More Strict Privacy Laws*: As AI develops, privacy issues grow more complicated, necessitating strict data protection regulations. Lawmakers must amend existing legislation to address such violations and establish precise guidelines for the gathering and use of data. Many nations currently have laws in effect, including The European Union's GDPR, California, USA's CCPA, Singapore's PDPA, The UK's Data Protection Act of 2018, South Africa's POPIA. Respecting these rules helps businesses stay out of trouble with the law, protects users, and increases stakeholder and customer trust (Ghandour et al., 2024).

## 2.11. System Quality: Ethical Considerations and Management Strategies for ChatGPT

*A. Misinformation and System Failure*: AI models can occasionally produce inaccurate or deceptive data—a condition referred to as "hallucination." This can be particularly risky when AI gives inaccurate financial, legal, or medical advice, which could be harmful. To cut down on errors, businesses can use performance monitoring, quality assurance, and thorough testing. Industries with stringent safety regulations, such as healthcare and aircraft, can teach us a lot (Ghandour et al., 2024).

*B. Explainability and Transparency*: ChatGPT and other AI systems are frequently referred to as "black boxes" due to the complexity of their decision-making processes, which make them difficult for users to understand. This lack of openness can cause problems with accountability and mistrust, particularly when AI influences significant choices. Explainable AI (XAI) techniques can assist make sense of how AI models operate. Businesses should outline the limitations of AI and describe the process by which results are produced (Ghandour et al., 2024).

*C. Discrimination and Bias*: Large volumes of internet data, which may carry unspoken biases, are used to train AI algorithms. AI may therefore inadvertently perpetuate discrimination, producing unjust results. Employ techniques for detecting and mitigating prejudice. Use data de-biasing and fairness-aware machine learning methods. Check AI systems frequently for unjust biases and make the necessary corrections (Ghandour et al., 2024).

## 2.12. The Unique Concerns of AI

AI technology presents special ethical problems that call for increased responsibility from its creators. In contrast to previous technologies, AI systems are sophisticated and autonomous, which might have unanticipated consequences that have a big influence on society. Transparency and intricacy are major issues. It may be challenging to identify biases or unexpected consequences since even the people who developed AI may not fully comprehend how these systems operate. Autonomy is another significant problem because AI is capable of making decisions on its own, which could have disastrous repercussions. The unpredictable nature of AI can sometimes perpetuate prejudices or result in unanticipated harm, as in the case of social media algorithms spreading false information or predictive policing. Proactive safety measures and continuous monitoring are required to address these threats. Transparency, accountability, equity, and inclusivity must be given top priority by developers and legislators to guarantee AI is created responsibly. This entails preserving human values by making AI systems intelligible, disseminating their advantages broadly, and incorporating a range of viewpoints in decision-making (McGraw, 2024).

## 2.13. AI-driven Social Robots

Social robots with AI capabilities can become crucial to elder care because they could help with everyday duties, offer emotional support, and stimulate cognitive processes. Research indicates that these robots can be useful in healthcare settings, despite the fact that older persons have varied opinions about them. These AI-powered robots can customize interactions and adjust to the needs of people. They can aid in stroke recovery, mental health treatment, anxiety reduction, Alzheimer's disease companionship, and the development of communication skills in autistic people. They also assist medical personnel in providing remote care, collecting patient data, and educating individuals about their health (Vozna & Costantini, 2025).

Designing successful technology requires an understanding of how older individuals view social robots. According to preliminary research, elderly people frequently had an unfavourable opinion of robots because they were too mechanical, particularly in sensitive situations. Long-term use, however, showed more positive attitudes, with older persons using robots for games, casual discussions, and daily updates (Vozna & Costantini, 2025).

When incorporating robots into elder care, caregivers' viewpoints are equally crucial. According to studies, social robots can improve daily tasks and the care experience for older individuals, strengthening bonds between them and their caretakers. A thorough comprehension of the relationships between elderly people, caretakers, and robots is necessary for successful integration (Vozna & Costantini, 2025).

Studies have looked into how social robots, such as Pepper, might offer individualized interactions by picking up on users' moods and routines. Acceptance is greatly influenced both the user's and the robot's personalities, but further study is required to completely comprehend this relationship, particularly for elderly persons who live independently (Vozna & Costantini, 2025).

Privacy, autonomy, and mental health should all be taken into account when using AI-powered social robots in elder care. Technology has to be used to enhance instead of replace, older people's autonomy in making decisions about their own care. It takes cooperation between various disciplines, including engineering, psychology, and philosophy, to design robots that respect their choices (Vozna & Costantini, 2025).

One of the main worries is that robots may restrict rather than increase independence if they are not properly designed. According to research, assistive technology is necessary to enable seniors to continue living their lives as they see fit while getting assistance. It's also important to think about how these robots will affect people emotionally. Many elderly people may grow emotionally attached to robots because they perceive them as friends. Although this can lessen loneliness, it also begs the question of whether these exchanges actually satisfy social needs or lead to inflated expectations. Companion robots can be comforting for elderly people and dementia patients, but an excessive dependence on them may limit human interaction, which is important for mental health (Vozna & Costantini, 2025).

## 2.14. Security

An ethical framework that focuses on five main areas has been developed to address legal difficulties:

A. *Information rights and responsibilities*: Safeguarding user information and maintaining privacy (Ghandour et al., 2024).
B. *Property rights and responsibilities*: Handling issues related to intellectual property and copyright (Ghandour et al., 2024).
C. *Control and accountability*: Taking ownership of AI choices and their negative effects (Ghandour et al., 2024).
D. *System quality*: Making sure AI results are accurate, transparent, and equitable (Ghandour et al., 2024).
E. *Quality of life*: Assessing the impact of AI on individuals' health and society at large (Ghandour et al., 2024).

Significant privacy and data security issues are brought up by the growing usage of AI in social robots, especially in healthcare contexts where these robots gather and handle sensitive personal data. Although there are basic privacy rules, there aren't many frameworks designed specifically for robots to systematically identify and reduce threats (Vozna & Costantini, 2025).

When implementing robots in elder care, transparency is essential. To control expectations and prevent misconceptions, older individuals should be made aware of the capabilities and limitations of these robots. A key component of the moral application of AI-powered healthcare robots is respect for human dignity. Personal data is protected and responsible AI deployment is ensured by legal frameworks like the General Data Protection Regulation (GDPR) and safety regulations like the Machinery Directive. International safety standards like ISO 13482 offer recommendations for creating personal care robots that put user comfort and safety first, while GDPR places a strong emphasis on data protection, transparency, and the right to consent (Vozna & Costantini, 2025).

A flexible regulatory architecture is essential given the speed at which AI is developing. According to the "risk pyramid" strategy put forth by the European Commission, high-risk AI systems must undergo stringent safety and ethical inspections before being released onto the market (Boni, 2021).

## 2.15. Ex ante impact assessments

It takes both technological and ethical factors to build reliable human-machine collaboration. To guarantee that AI systems operate responsibly and satisfy user demands, developers must incorporate ethical standards into the technology. Preventing bias and discrimination in AI, which frequently results from subpar datasets, is a significant challenge. AI systems may inadvertently perpetuate discrimination if they are trained on biased data. Strict data selection criteria and European data usage guidelines are required to solve this, as stated in the proposed Data Governance Act and Data Act. Users should also have the means to manage their own data (Boni, 2021).

Verifying that training datasets adhere to privacy and fairness requirements is necessary to guarantee that high-risk AI systems are in line with ethical principles and fundamental rights. Both users and providers of AI have obligations: users must supervise AI operations and finish data protection assessments, while providers must register AI systems, guarantee quality, control risks, and notify authorities (Boni, 2021).

Experts stressed the significance of self-assessed risk assessments prior to AI implementation during talks on AI regulation. The implementation of a required Human Rights Impact Assessment for AI systems is one important suggestion, which would call for disclosure, outside evaluations, and interaction with impacted populations. Additional alternatives like Data Hygiene Certification and ethical Technology Assessments have been suggested, despite the fact that this compliance strategy may need a lot of resources (Boni, 2021).

In the end, building trust in AI requires the implementation of strict pre-deployment (ex-ante) protocols. The foundation for moral and open human–AI cooperation will be strengthened as more businesses and individuals take part in these evaluations (Boni, 2021).

The ethical, social, and legal implications of artificial intelligence (AI) were investigated in this study using a qualitative research methodology. The study sought to gather a range of viewpoints on the effects of AI by conducting in-depth interviews with specialists including legal scholars, policymakers, technology ethicists, and AI engineers. To guarantee a comprehensive grasp of the topics, participants were carefully chosen, and their anonymity was protected by adhering to ethical standards (Wang et al., 2023).

Semi-structured interviews were used to gather the data, giving participants a chance to express their opinions while discussing important subjects including privacy, bias, accountability, and regulation. To find important patterns and themes, the interviews were transcribed and subjected to thematic analysis software (Wang et al., 2023).

A. Sociopolitical Dimension
   i. *Privacy and Data Governance*: As AI continues to blur the boundaries between data utility and privacy infringement, participants underlined the importance of robust data protection measures (Wang et al., 2023).
   ii. *Bias and Discrimination*: Fairness and inclusivity are essential in the creation of AI systems because they can reinforce societal biases if they are trained on uneven data (Wang et al., 2023).
   iii. *AI and Employment*: Although AI increases productivity, it also jeopardizes employment stability, underscoring the need for workforce retraining (Wang et al., 2023).
   iv. *Digital Divide*: Social inequality is exacerbated by limited access to AI technology, especially for people without internet or digital literacy (Wang et al., 2023).
   v. *AI in Governance*: Although AI has the potential to improve government operations, it also brings up issues with accountability and monitoring (Wang et al., 2023).

B. Legal Dimension
   i. *Intellectual Property Rights*: Current copyright and patent regulations make it difficult to identify who owns AI-generated work (Wang et al., 2023).
   ii. *Liability and Accountability*: Determining who is responsible for AI mistakes is difficult and calls for more precise legal regulations (Wang et al., 2023).

iii. *Regulatory Frameworks*: Coherent frameworks are required since current AI policies are disjointed and insufficiently address AI's worldwide impact (Wang et al., 2023).
iv. *AI Ethics and Law Integration*: In order to balance innovation and public safety, legal institutions must change to incorporate ethical issues (Wang et al., 2023).
v. *AI and Human Rights*: AI should uphold and strengthen human rights rather than jeopardize non-discrimination, freedom of speech, or privacy (Wang et al., 2023).

## 3. BIBLIOMETRIC ANALYSIS

The following section presents bibliometric analysis on the topic "Societal Impact of Ethical Dimensions of AI". This analysis presents and helps in understanding the research trends and focus that have been received by the topic over past 20 years. The data for the analysis is taken from Lens for the years 2000-2025. The first 4 images i.e. Figure 2 to Figure 5 are made using VOSViewer that illustrates the grouping (or clustering) of different keywords, explain the relation between AI and Humans, timeline chart and density chart respectively. Figure 6, Figure 7 and Figure 9 are made using Python code for the same data as extracted earlier explaining the growth in research, important keywords and contribution of different countries. Figure 8, a tree map, is made using Origin Labs depicting Publishers' contributions.

Based on information taken from Lens and displayed using VOSViewer, Figure 2 shows a thorough bibliometric network clustering map that depicts the intellectual framework of research on the societal impact of ethical aspects of artificial intelligence (AI). The map shows several connected clusters that correspond to different but related theme topics. The "AI ethics," "algorithmic bias," and "responsible AI" clusters are the biggest and densest, suggesting that these are the main areas of focus for research. Secondary clusters highlight "privacy," "autonomy," "fairness," "transparency," and "governance," which reflects the growing range of conversations on social justice and moral responsibility in AI systems.

Strong connections between these clusters imply that academics are increasingly taking a multidisciplinary approach to AI ethics, combining legal, philosophical, and technical knowledge. The connection of writers and institutions across geographies further demonstrates a significant level of international collaboration in the visualization. Overall, the network clustering shows that ethical AI research has developed from dispersed endeavors into a coherent and interconnected body of knowledge, highlighting a common worldwide commitment to comprehending and reducing the societal effects of developing AI technologies.

*Figure 2. Bibliometric Analysis on Societal Impact of Ethical AI with data from Lens and made using VOSViewer: Network Clustering (Lens, 2025; Van Eck & Waltman, 2022).*

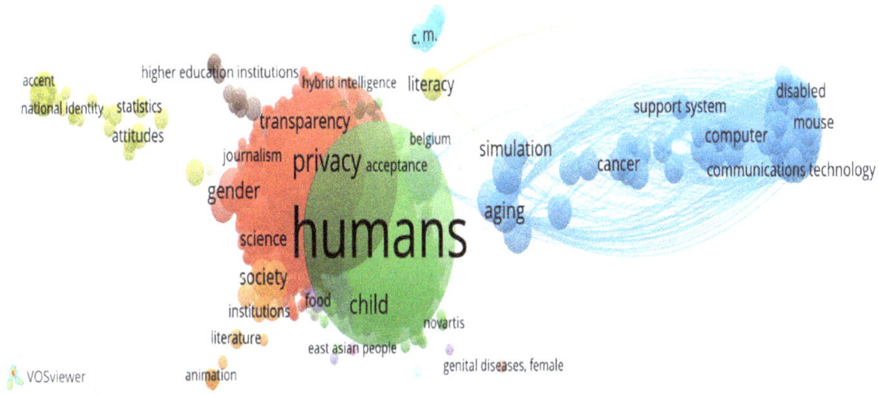


A magnified image of a significant portion of the bibliometric network is provided in Figure 3, which focuses on the connection between human-centered concerns and artificial intelligence. Sub-clusters where nodes like "human–AI interaction," "trust," "empathy," "autonomy," "accountability," and "decision-making" are highly related are isolated by this depiction. These nodes' popularity indicates that academics are becoming increasingly interested in how AI technologies affect social interactions, ethical responsibility, and human behavior. The close relationship between "ethics" and "human values" indicates that researchers are still working to include moral and cultural considerations into the development and application of AI. Furthermore, the clustering pattern shows that ethical AI is now seen as a humanistic endeavor focused on moral agency and social well-being rather than just algorithmic fairness. The field is heading toward more integrative frameworks, where human-centered AI is viewed as an interdisciplinary challenge involving technology, psychology, sociology, and philosophy, as this targeted visualization demonstrates. It highlights a paradigm change from the technological determinism of previous decades to a value-driven, participative approach that stresses the co-evolution of intelligent systems and humans.

*Figure 3. Focus on Relation between Artificial Intelligence and Humans: Highlighted Part of Network Clustering (Lens, 2025; Van Eck & Waltman, 2022).*

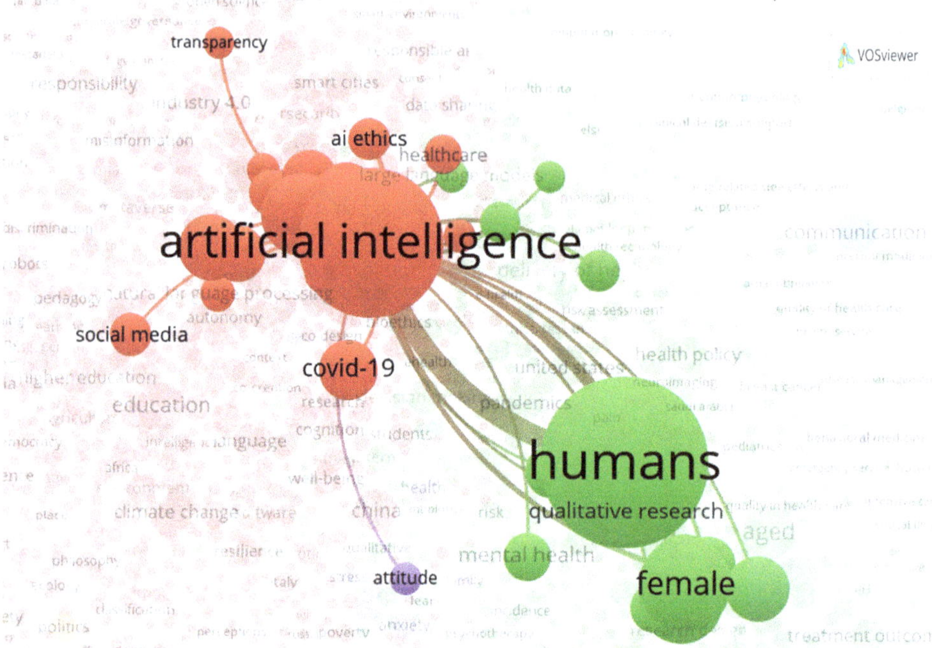


According to VOSViewer's timeline visualization, Figure 4 shows the temporal evolution of research issues within the societal and ethical dimensions of AI. The data shows a dynamic evolution of scholarly concentration between 2000 and 2025. Keywords like “machine ethics,” “moral reasoning,” and “philosophy of AI” dominated writings in the early 2000s that mostly dealt with philosophical and theoretical topics. The literature shifted toward applied ethics between 2010 and 2015, focusing on subjects like “data privacy,” “autonomous decision-making,” and “robot ethics.” Studies addressing “algorithmic bias,” “transparency,” and “governance frameworks” have significantly increased since 2016, coinciding with the growth of deep learning and large-scale data analytics. The years 2020–2025 show a shift toward practical frameworks like “explainable AI,” “AI for social good,” and “ethical governance models,” indicating a developing field that is more concerned with practical application and accountability. The timeline shows how the conversation has shifted from abstract moral analysis to concrete ways to guarantee that AI systems respect sustainability, equity, and human rights. This course reflects the general tendency in society to formalize AI ethics in technological governance, education, and policy.

*Figure 4. Bibliometric Analysis on Societal Impact of Ethical AI with data from Lens and made using VOSViewer: Timeline Chart (Lens, 2025; Van Eck & Waltman, 2022).*

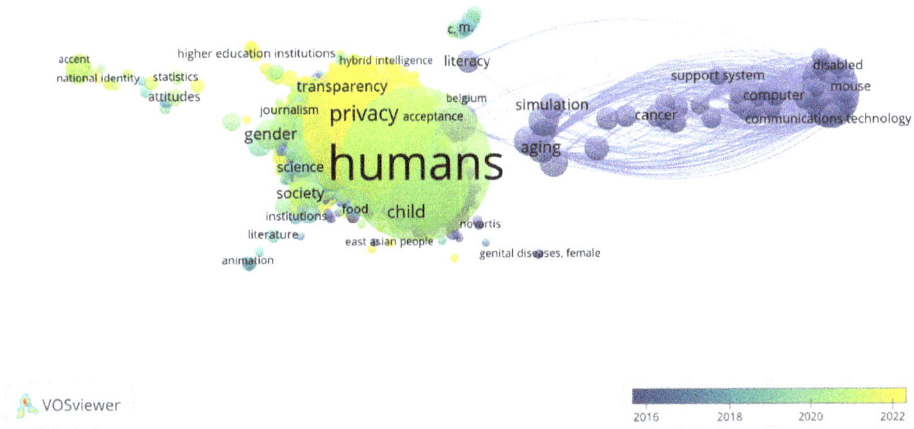


A density visualization produced by VOSViewer is shown in Figure 5, which sheds light on the predominance and intensity of research themes related to the ethical and societal aspects of AI. The density map depicts areas of concentrated intellectual activity using color gradients. Frequently common terms like "AI ethics," "society," "accountability," and "machine learning" are represented by bright yellow and red zones, signifying well-known and frequently referenced research issues. The surrounding green and blue patches, on the other hand, stand for new or understudied topics like "sustainability," "human rights," and "AI regulation." The density distribution illustrates how the field has expanded and deepened over time, covering a variety of ethical issues. Additionally, this image shows how new topics like ecological effect, inclusion, and global justice are becoming more prominent while more conventional ones like bias and fairness continue to be crucial. The shift from reactive ethical reactions to proactive, systemic approaches that incorporate ethical design principles into AI creation is reflected in this diversification. As a result, the density map provides a visual representation of the development and diversity of research interests in AI ethics, demonstrating how the academic community is working together to create a multifaceted understanding of the influence of AI on society.

*Figure 5. Bibliometric Analysis on Societal Impact of Ethical AI with data from Lens and made using VOSViewer: Density Map (Lens, 2025; Van Eck & Waltman, 2022).*

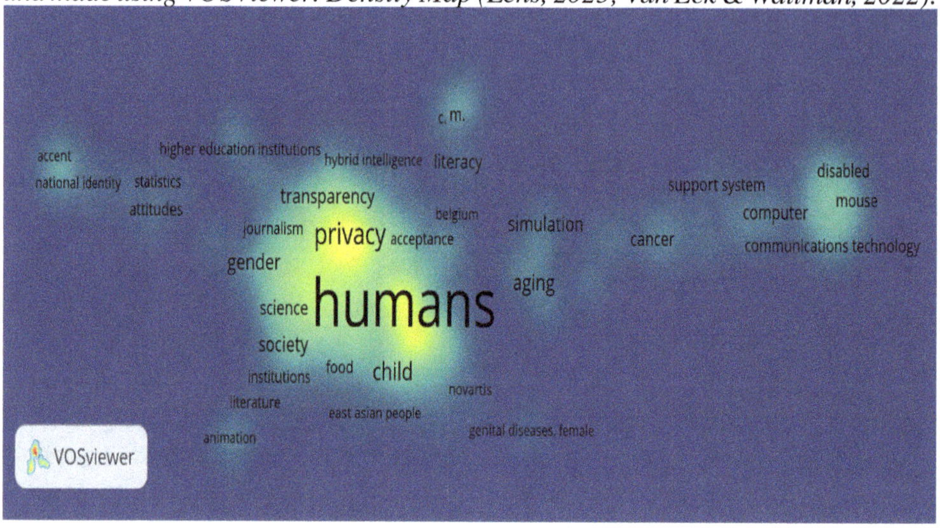


A longitudinal examination of publication trends from 2000 to 2025 is shown in Figure 6, which was produced using data visualization tools based on Python. The graph shows a distinct and consistent upward trend in the quantity of academic publications on the moral and social implications of AI. The paucity of research in the early 2000s was indicative of a developing understanding of the ethical issues of AI. However, as AI applications became more commonplace around 2010, publication activity started to slightly increase. Following 2016, there is a noticeable rise that coincides with international regulatory initiatives, ethical AI frameworks, and prominent discussions on algorithmic bias, data privacy, and accountability. Between 2019 and 2024, when large-scale AI systems proliferated and international regulation negotiations like the EU's AI Act took place, there was the most noticeable rise. This increase shows that ethical AI is now a mainstream research priority rather than a specialized academic topic. As academics from computer science, philosophy, sociology, and law come together to discuss the ethical issues raised by AI, the steady expansion also demonstrates the rise in multidisciplinary cooperation. Overall, this graphic highlights how society and academia are becoming more involved with AI ethics in reaction to the quick development of technology.

*Figure 6. Line Chart Depicting Number of Publications from year 2000-2025 (Lens Data)*

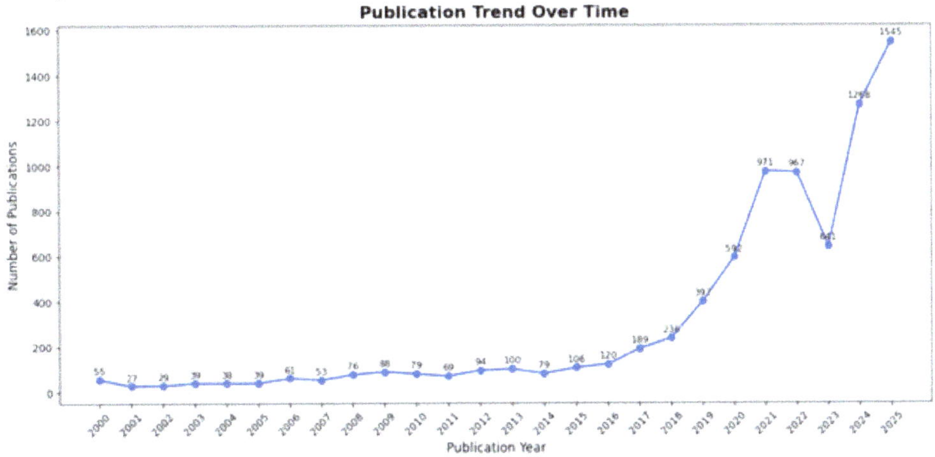


A Python-generated word cloud that graphically summarizes the frequency and prominence of recurrent phrases in the bibliometric dataset is shown in Figure 7. The literature's dominating conceptual anchors are shown by larger terms, which indicate higher occurrence rates. Key phrases like “ethics,” “AI,” “society,” “governance,” “bias,” and “accountability” take up the most space, indicating how important they are to the current conversation. Words like “fairness,” “trust,” “transparency,” “autonomy,” and “justice” are frequently used, suggesting a deeper examination of ethical and societal issues in the application of AI. Emerging concepts like “sustainability,” “regulation,” and “human-centered AI” are also included in the graphic, indicating a move toward inclusive and comprehensive ethical frameworks. The variety of keywords highlights the field's interdisciplinary nature, incorporating ideas from public policy, technology studies, ethics, and human-computer interaction. The word cloud's general structure highlights the field's transition toward practical and context-sensitive solutions by reflecting a thematic balance between normative ethics (principles, values) and applied ethics (governance, accountability). The graphic illustrates how academic attention has shifted from theoretical moral discussions to the real-world challenges of guaranteeing fair and open AI development and application.

*Figure 7. Word Cloud of Some Important Keywords (Lens, 2025)*

A tree map created with OriginLabs is shown in Figure 8, which shows the relative contributions of various publications to the body of research on the ethical and societal ramifications of artificial intelligence. A publisher's proportionate output is represented by each segment, with larger blocks denoting higher publication quantities. Due to the topic's cross-disciplinary appeal spanning engineering, computer science, social sciences, and humanities journals, the graphic shows that large academic publishers like Springer, Elsevier, IEEE, and Taylor & Francis dominate the area. Significantly, open-access publishers like MDPI and Frontiers are becoming more prevalent, indicating a trend toward increased accessibility and worldwide distribution of ethical AI research. The field's growth into a number of academic fields, such as policy studies, education, and health informatics, is also reflected in the range of publishers. This distribution demonstrates how conversations around AI ethics have crossed traditional disciplinary lines and developed into a worldwide activity. A more inclusive discussion on how AI impacts various cultures and cultural contexts is made possible by the addition of regional and specialist publications, which also signifies the democratization of research distribution. Overall, the tree map illustrates a publication environment that is becoming more diverse, open, and cooperative, supporting the democratization of information in AI ethics.

*Figure 8. Tree Map Depicting Publishers' Contribution in Research (Lens, 2025; OriginLabs, 2025).*

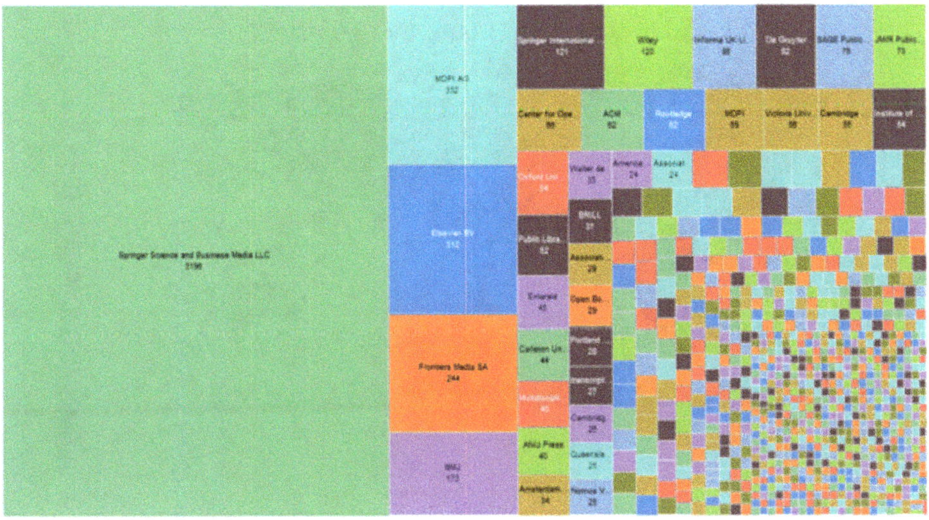


Using a Python-generated visualization based on Lens data from 2000 to 2025, Figure 9 shows the regional distribution of academic contributions to the subject of ethical AI. According to the data, the countries with the highest publishing output include the United States, the United Kingdom, China, Germany, and Canada, indicating their significant institutional investment in AI ethical research. Together, European countries show high levels of participation, which is consistent with EU-wide programs that prioritize human-centered and reliable AI. The involvement of nations like South Korea, Japan, Australia, India, and others shows that the global conversation is expanding, with Asia-Pacific regions becoming major participants in recent years. The increasing contributions from developing countries point to a greater understanding of the socio-cultural and economic aspects of AI ethics outside of Western contexts. The globalization of AI ethics and its incorporation into regional policy frameworks, including local modifications of fairness, accountability, and transparency, are highlighted by this geographical diversity. The graphic shows qualitative diversity in ethical concerns across locations in addition to quantitative variations in publication output. When taken as a whole, this trend represents a move toward international cooperation, enhancing the global ethical AI ecosystem by addressing global issues like bias, privacy, and governance through common frameworks and regional interpretations.

*Figure 9. Contribution of Different Countries in Research (Lens, 2025)*

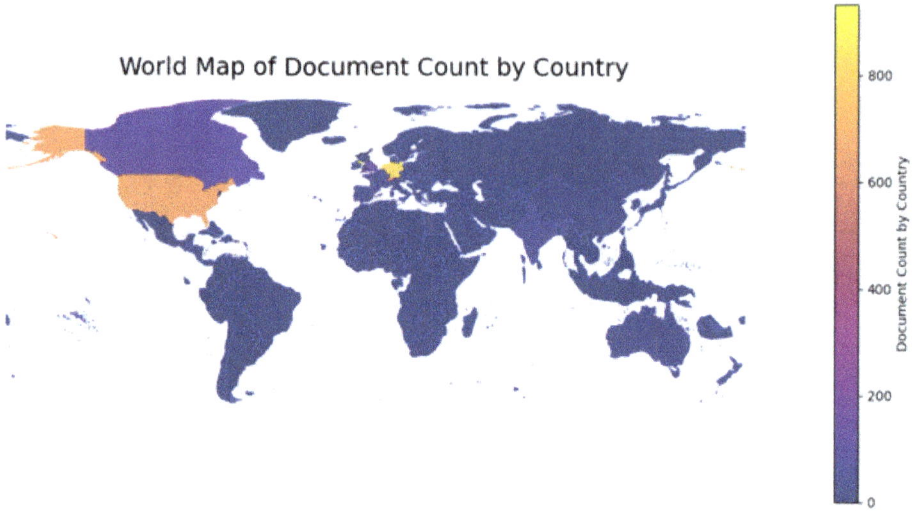


## 4. CONCLUSION

From healthcare to education and government, artificial intelligence is changing many industries. While there are many advantages, there are also moral, societal, and legal issues. This study draws attention to the important issues surrounding the creation and application of AI, especially the moral conundrums pertaining to algorithmic bias, data privacy, accountability, and human autonomy. It is crucial to make sure AI complies with moral standards and societal norms as it continues to permeate daily life.

The application of AI-powered social robots in eldercare has been noted as one of the main issues. These robots help with everyday chores and offer companionship, but they also pose serious privacy and emotional health concerns. The study emphasizes how crucial it is to build AI systems that empower people rather than reducing human interaction or fostering reliance. As these robots gather sensitive personal data, privacy concerns are especially urgent, calling for strong data protection regulations and moral protections.

Another important area of attention is AI literacy. People of all ages need to acquire the abilities to comprehend, engage with, and evaluate AI-driven technologies as they become more pervasive in society. To ensure that people are prepared to interact with AI responsibly, effective AI education must be conducted at many levels, from early education to professional training.

In order to determine best practices for the creation of ethical AI, the study also assesses a number of international ethical frameworks, such as those from the OECD, UNESCO, and the European Union AI Act. A method that combines ethical responsibility with managerial decision-making is the Ethical Management of AI (EMMA) model. The study emphasizes the necessity of precise legal frameworks, particularly with relation to accountability, intellectual property rights, and the worldwide regulatory issues surrounding AI.

The chapter also includes bibliometric analysis explaining various important parts on the chapter topic. It includes figures made from VOSViewer, OriginLabs, and Python code with data from Lens. The analysis presents and explains the contribution from various countries and publishers, important keywords, growth in research and relation between various keywords.

The study's main conclusion is that ethical foresight should guide the development of AI, striking a balance between societal benefit and technological progress. To guarantee that AI benefits humans fairly, the study suggests stricter laws, interdisciplinary cooperation, and open AI governance. In the end, AI should improve human lives while upholding privacy, dignity, and fundamental rights.

## REFERENCES


Boni, M. (2021). The ethical dimension of human–artificial intelligence collaboration. *European View*, *20*(2), 182–190. DOI: 10.1177/17816858211059249

Brendel, A. B., Mirbabaie, M., Lembcke, T., & Hofeditz, L. (2021). Ethical management of Artificial Intelligence. *Sustainability (Basel)*, *13*(4), 1974. Advance online publication. DOI: 10.3390/su13041974

Ghandour, A., Woodford, B. J., & Abusaimeh, H. (2024). Ethical Considerations in the use of ChatGPT: An exploration through the lens of five moral dimensions. *IEEE Access : Practical Innovations, Open Solutions*, *12*, 60682–60693. DOI: 10.1109/ACCESS.2024.3394243

Huang, C., Zhang, Z., Mao, B., & Yao, X. (2022). An Overview of Artificial intelligence ethics. *IEEE Transactions on Artificial Intelligence*, *4*(4), 799–819. DOI: 10.1109/TAI.2022.3194503

Lens. (2025). *Lens*. Retrieved October. https://www.lens.org/lens/

McGraw, D. K. (2024). Ethical responsibility in the design of artificial intelligence (AI) systems. *International Journal of Responsibility*, *7*(1). Advance online publication. DOI: 10.62365/2576-0955.1114

Ng, D. T. K., Leung, J. K. L., Chu, S. K. W., & Qiao, M. S. (2021). Conceptualizing AI literacy: An exploratory review. *Computers and Education Artificial Intelligence*, *2*, 100041. Advance online publication. DOI: 10.1016/j.caeai.2021.100041

*OriginLab - Origin and OriginPro - data analysis and graphing software*. Retrieved October, 2025, from https://www.originlab.com/

Parson, E. T. A., Fyshe, A., & Lizotte, D. (2019). Artificial Intelligence's Societal Impacts, Governance, and Ethics. *UCLA School of Law, Public Law Research Paper*.

Pastor-Escuredo, D., Treleaven, P., & Vinuesa, R. (2022). An ethical framework for artificial intelligence and sustainable cities. *AI*, *3*(4), 961–974. DOI: 10.3390/ai3040057

Sanjeetha, M. B. F., & Abeygunawardhana, P. K. (2024). AI in Smart Devices and Services: A Systematic Review of Social Impacts and Ethical Implications. *TechRxiv*. DOI: DOI: 10.36227/techrxiv.172671793.34847626/v1

Van Eck, N. J., & Waltman, L. (2022). *VOSviewer* manual (Version 1.6.18). Leiden University, CWTS. Retrieved from https://www.vosviewer.com/documentation/Manual_VOSviewer_1.6.18.pdf

Vozna, A., & Costantini, S. (2025). Ethical, legal, and societal dimensions of AI-Driven social robots in elderly healthcare. *Intelligenza Artificiale*, *19*(1), 30–40. DOI: 10.1177/17248035241310192

Wang, J., Mao, W., & Wenjie, W. (2023). The Ethics of Artificial Intelligence: Sociopolitical and Legal Dimensions. *Interdisciplinary Studies in Society, Law, and Politics*, *2*(2), 27–32. DOI: 10.61838/kman.isslp.2.2.6

## ADDITIONAL READING

Giorgianni, G., Ali, M. N., Jebbor, I., Wafik, H. M. A., Benmamoun, Z., Khlie, K., Rabby, S. K., Md, A. H., & Mahbub, S. (2024). Unveiling The Potential Of Ai: Impacts On Industries And Ethical Considerations. *Educational Administration: Theory and Practice*, *30*(3), 2654–2661. DOI: 10.53555/kuey.v30i3.7484

Kazim, E., & Koshiyama, A. S. (2021). A high-level overview of AI ethics. *Patterns (New York, N.Y.)*, *2*(9), 2666–3899. DOI: 10.1016/j.patter.2021.100314

Li, F., Ruijs, N., & Lu, Y. (2022). Ethics & AI: A Systematic Review on Ethical Concerns and Related Strategies for Designing with AI in Healthcare. *AI*, *4*(1), 28–53. DOI: 10.3390/ai4010003

McLaren, B. M. (2003). Extensionally defining principles and cases in ethics: An AI model. *Artificial Intelligence*, *150*(1-2), 145–181. DOI: 10.1016/S0004-3702(03)00135-8

Morley, J., Machado, C. C., Burr, C., Cowls, J., Joshi, I., Taddeo, M., & Floridi, L. (2020). The ethics of AI in health care: A mapping review. *Social Science & Medicine*, *260*, 113172. Advance online publication. DOI: 10.1016/j.socscimed.2020.113172

Yu, H., Shen, Z., Miao, C., Leung, C., Lesser, V. R., & Yang, Q. (2018). Building Ethics into Artificial Intelligence. *arXiv (Cornell University)*. https://doi.org//arxiv.1812.02953DOI: 10.24963/ijcai.2018/779

## KEY TERMS AND DEFINITIONS

**AI Literacy:** It is the ability to understand, use, and evaluate artificial intelligence (AI) technologies.

**EMMA:** It is a framework focusing on managing the ethical aspects of AI development and deployment, encompassing managerial decision-making, ethical considerations, and environmental dimensions.

**Ethical framework:** It is a set of principles that guide the development and use of AI in a responsible way.

**Ethical Principles:** It includes fairness, privacy, accountability, transparency, and explainability.

**Ethics:** It is a framework that helps manage the risks of AI while also using its benefits.

**Information Rights:** It encompasses the legal claims and duties related to the communication, collection, access, use, and control of information, including rights like privacy, intellectual property, free expression, and access to information.

**Legal challenges:** AI presents significant legal challenges, including data privacy concerns, potential for bias and discrimination, intellectual property issues related to AI-generated content.

**SAR:** A type of robot designed to aid users through social interaction and engagement rather than physical tasks, providing emotional, cognitive, and social support for specific goals like behavior change, learning, or therapy.

**Security:** AI security helps protect AI systems from cyberattacks, data breaches, and other threats

**UNSDG:** It is a group that works to improve the effectiveness of UN development activities at the country level, focusing on the 17 Sustainable Development Goals (SDGs).

Chapter 9

# AI in Education: Ethical Dilemmas and Opportunities for Equity

**Mustafa Kayyali**
https://orcid.org/0000-0003-3300-262X
*Maaref University of Applied Sciences, Syria*

## ABSTRACT

*The integration of artificial intelligence into education systems holds transformative potential—customized learning, predictive analytics, and intelligent tutoring are reshaping how knowledge is accessed and delivered. Yet beneath this promise lie complex ethical dilemmas. From algorithmic bias that reinforces existing inequalities to opaque decision-making systems that challenge human agency, AI introduces risks that disproportionately affect already marginalized students. This chapter critically examines the ethical dimensions of AI in educational contexts, highlighting tensions between innovation and equity. Drawing on global case studies and policy analyses, it explores how AI can both hinder and help the pursuit of just and inclusive education. Ultimately, it calls for a human-centered, equity-driven framework for designing and deploying AI in education—one that resists technocentrism and embraces accountability, transparency, and ethical foresight.*

## INTRODUCTION

The integration of artificial intelligence into educational systems represents one of the most significant paradigm shifts in the history of teaching and learning. No longer confined to science fiction or tech conferences, AI has found its way into classrooms, universities, assessment platforms, administrative workflows, and even into the private lives of students and educators. Algorithms are recommending per-

DOI: 10.4018/979-8-3373-2597-2.ch009

sonalized learning paths, grading essays, identifying students at risk of dropping out, and analyzing vast datasets to shape policies (Elwarraki et al, 2023). At first glance, these developments seem like unambiguous steps forward—more efficiency, more access, more personalization. But as history has often shown, technological advancement rarely arrives without ethical and social implications, especially in domains as human-centered as education.

The allure of AI in education lies in its promise to individualize learning and democratize access to high-quality instruction (Bulathwela et al, 2024). For under-resourced schools and overstretched teachers, AI tools offer relief and potential. Adaptive learning platforms can tailor lessons to the pace and needs of each student. Predictive analytics can help identify gaps in comprehension or flag mental health concerns. Language models can assist non-native speakers in expressing themselves more fluently. In an ideal world, AI would act as a tireless assistant, lifting burdens and extending the reach of educators. Yet this vision remains deeply complicated by the realities of algorithmic bias, surveillance, opacity, and inequitable access to digital infrastructure.

Ethical concerns surrounding AI in education are not peripheral—they are central. These technologies are shaped by datasets that often carry historical and structural biases (Dieterle et al, 2024). For instance, if a predictive model is trained on past performance data from schools with racially or economically stratified outcomes, its recommendations may reproduce and legitimize those same inequalities. Moreover, the increasing reliance on AI for decision-making introduces questions of accountability. Who is responsible when an AI-powered admissions tool disadvantages a qualified applicant? How can a student contest a grade issued by a machine learning model they do not understand? Can we teach students to be critical thinkers in systems governed by opaque algorithms?

At the same time, the digital divide continues to grow. AI-based tools are often marketed as universal solutions, yet their effectiveness depends heavily on infrastructure—stable internet, compatible devices, and digital literacy—all of which are unevenly distributed across and within countries (Mutambara, 2025). This inequality creates a risk that AI will deepen existing educational disparities rather than alleviate them. The most privileged schools and students benefit from the latest innovations, while others are left further behind. The danger is not only technological exclusion but also pedagogical homogenization, where the richness of human interaction is replaced by automated learning loops optimized for efficiency, not depth.

Another layer of complexity emerges from surveillance and data ethics. In the name of personalization and safety, educational institutions are increasingly collecting granular data about student behavior—clicks, keystrokes, attention spans, and even facial expressions. Such data is used to train AI models, but it also opens the door to invasive monitoring and commodification of student information. In some cases, these systems blur the line between education and control, transforming learners into data points to be analyzed, predicted, and managed.

In this chapter, we delve into these multifaceted ethical dilemmas, not to reject the role of AI in education but to advocate for its responsible, equitable, and humane integration. Drawing on case studies, philosophical frameworks, and global policy debates, we explore how AI can both challenge and reinforce inequality, how it can either undermine or uplift student agency, and how ethical design and governance can steer educational AI toward more just outcomes. The aim is not to provide definitive answers, but to illuminate the critical questions we must ask—and keep asking—as we reshape the future of education through intelligent systems.

Ultimately, the question is not whether AI will play a role in education, but what kind of role we will allow it to play—and for whom. In addressing this, we must confront a deeper issue: do we view education as a space for cultivating human flourishing and democratic participation, or as a system to be optimized for performance and compliance? How we answer this will determine the ethical trajectory of AI in education, and the kind of world our learners will inherit.

## Reimagining Education Through AI: Promise or Peril?

There is something both intoxicating and disquieting about the idea of machines "teaching" us. For centuries, education has been considered the most profoundly human of endeavors—a messy, iterative dance of minds, emotions, and contexts. Yet today, in lecture halls and learning apps, across dashboards and dashboards within dashboards, algorithms are beginning to do what only human educators once did. They listen, recommend, evaluate, and respond. They adjust lessons, personalize feedback, monitor engagement, and sometimes, in an eerie inversion, they do all this while remaining largely unseen. It's tempting to declare this moment as a new dawn. But somewhere between the excitement of innovation and the unsettling murmur of automation, a deeper question begins to rise: are we reimagining education to better serve learners, or are we reengineering learners to better suit machines?

The promises are loud and convincing. AI can deliver personalized education at scale, we're told—tailoring lessons to each student's pace, recognizing learning styles, flagging areas of confusion in real time, and even predicting who might struggle long before the human teacher notices (Maghsudi et al, 2021). That vision dazzles. In overcrowded classrooms, where a single teacher may juggle the needs of thirty or more students, the notion that a tireless digital assistant could bridge the gaps feels like a kind of salvation. No more one-size-fits-all instruction. No more waiting for a struggling child to fall so far behind that intervention comes too late. And in some places, these dreams are no longer hypothetical.

Take the learning platforms that analyze millions of student responses to adjust future quizzes. Or the facial recognition systems that claim to track attention levels during online classes. Or AI tutors—some of which now converse fluidly in natural language—capable of answering questions, breaking down complex concepts, and even detecting emotional cues. The logic is seductive: If education is about transmitting knowledge, then surely intelligent systems that process more data, faster, and with more consistency than any human possibly could, must be superior. But that logic carries a trap.

Because education isn't merely the transmission of information. It is the formation of thinking, feeling beings. It is trust built over time. It is the awkward silence that stretches between a question and an answer. The sighs, the resistance, the sudden spark of insight that comes not from being told, but from struggling to figure something out. Machines don't experience that (Hanemaayer, 2021). They don't know what it feels like to be nine years old, confused by long division and embarrassed to ask for help. They don't recognize the nuances in a student's gaze, the quiet fatigue of hunger or trauma or sleeplessness. And in this way, they cannot fully teach.

Still, we keep handing over more of the pedagogical terrain to machines. Not because they are better teachers—but because they are cheaper, faster, and unrelenting. The fear lurking beneath the promise is this: once efficiency becomes the goal, empathy becomes an afterthought. The peril isn't in the machines themselves. It's in the way we start designing education around what machines can do best, rather than around what humans need most.

There's another wrinkle here. Personalization, for instance, is often touted as AI's greatest gift to learning (Vallabhaneni et al, 2024). And on the surface, who could argue with that? Every learner is unique—so why shouldn't their educational path reflect that individuality? But beneath the surface lies a quieter tension: personalization requires surveillance. In order to customize a learning journey, AI must collect and analyze enormous volumes of data. Clicks, scrolls, response times, hesitation gaps, facial expressions—every digital move becomes a data point, every behavior a signal to be read. Over time, the student becomes less a person and more a profile: dynamic, yes, but ultimately reduced to patterns that the system can learn from and optimize.

The irony is biting. In attempting to individualize education, we may be commodifying individuality itself. And yet, we cannot ignore the real, powerful potential of AI to address structural inequities (Lin & Chen, 2022). In contexts where access to qualified teachers is limited—or where educational resources are underfunded, fragmented, or politically volatile—AI systems can fill gaps. They can translate content, read aloud to visually impaired learners, simulate laboratory experiments for students who lack physical equipment. In these scenarios, the arrival of AI is not just a novelty; it can be a lifeline. But even lifelines must be held with care. Because technologies, especially those as powerful and opaque as AI, do not exist in a vacuum. They inherit the biases of their creators, the assumptions of their training data, and the limitations of their design parameters. When AI is deployed in education, those biases don't just stay in the code—they manifest in real outcomes: in who gets flagged for support, in which dialects are "understood" by speech recognition systems, in how success is defined, measured, and rewarded.

A sobering case from the United Kingdom exemplifies this risk. During the COVID-19 pandemic, when traditional exams were canceled, an algorithm was used to predict students' final grades (Gaftandzhieva et al, 2022). The results were catastrophic. Students from affluent schools were disproportionately awarded higher marks, while those from disadvantaged backgrounds—despite strong teacher assessments—were downgraded. The algorithm had no malicious intent. It was simply trained on historical data, which itself reflected decades of systemic inequity. The outrage that followed was not just about the grades—it was about the cold, faceless injustice of being misjudged by a machine. And yet, as unsettling as that story is, it's not isolated. In the United States, predictive models are being used to identify students at risk of dropping out. In theory, this allows early intervention. But what happens when the model begins to shape how teachers treat those students? What happens when expectations shift based on a prediction rather than a person's lived potential? There's a fine line between using AI to illuminate invisible struggles and allowing it to become a self-fulfilling prophecy. All of this brings us to a kind of ethical crossroads.

To reimagine education with AI is not to simply digitize the classroom, but to rethink its values, its priorities, its very soul (Hossain & Islam, 2024). Will we design systems that center the learner, or ones that serve institutional convenience? Will we use AI to empower teachers—or to monitor, rank, and eventually replace them? Will we preserve the unpredictable, messy, deeply human essence of learning—or smooth it into something measurable, manageable, and ultimately less alive?

Some voices argue that resistance to AI in education is merely technophobia—a clinging to outdated norms in the face of inevitable progress (Hutson & Plate, 2024). But that's a misreading of the conversation. The skepticism isn't about fearing machines. It's about fearing what we might lose in our rush to adopt them: nuance, patience, humility, and the capacity for wonder. Education isn't just about outcomes. It's about the unpredictable journey toward becoming. And here's the thing: not everything that counts can be optimized. Not every teaching moment can be logged. The warmth of a teacher pausing to ask how you're doing. The sideways glance that tells a student you believe in them. The spontaneity of a lesson derailed by an unexpected question that turns into an unforgettable discussion. These are not bugs in the system. They are the system. And they're the pieces that no AI, no matter how advanced, can truly replicate. But perhaps the most unsettling risk is subtler still—the slow normalization of algorithmic thinking itself. As more and more of our educational tools become powered by predictive systems, we may begin to internalize their logic. Efficiency over empathy. Optimization over intuition. Risk-avoidance over curiosity. That's not just a shift in technology. It's a shift in mindset.

Still, there's room for something better. AI in education need not be either savior or saboteur. It can be a partner—humble, transparent, and human-guided (Chen et al, 2024). But this requires intention. It requires design choices that prioritize dignity over data, that bake ethics into the architecture, not as an afterthought but as a foundation. It also demands dialogue—a willingness to include educators, students, parents, and marginalized communities in the conversation about how these tools are built and used.

To reimagine education through AI is not to chase the latest gadget (Jackaria et al, 2024). It is to reassert the fundamental question: what is education for? If our answer is merely efficiency, we've already surrendered something vital. But if our answer is transformation—of the self, the community, the world—then AI must be our servant, not our master. And so, we sit in this moment of promise and peril, trying to hold both with honesty. It's not easy. The allure of innovation is strong, and the discomfort of uncertainty is real. But if education is about preparing the next generation to live wisely, justly, and fully in a complex world, then the way we use AI must reflect those same commitments. Maybe that's the most important lesson of all.

## Unpacking Ethical Dilemmas in Algorithmic Pedagogies

There's something unsettling about an algorithm making decisions in the classroom. It creeps in quietly, at first—a nudge here, a recommended video there, a predictive quiz score sent to a teacher's dashboard. Then it begins to settle in more deeply, not just offering assistance but reshaping the very contours of learning itself. What began as a tool starts to become a filter, a gatekeeper, sometimes even an arbiter of who succeeds and who does not. And the scariest part? Much of this happens invisibly, beneath the surface of what most students and teachers are even aware of. The ethical tensions at the heart of algorithmic pedagogies aren't about fear of technology—they're about a slow erosion of human judgment and agency in spaces that should be deeply human.

To speak of algorithms in education is, in many ways, to speak of power—but disguised in the language of optimization (Blikstein & Blikstein, 2021). We don't question a grading algorithm because it seems neutral, mathematical, rational. But neutrality, as history has repeatedly shown, is often a carefully curated illusion. Algorithms, after all, are built by people. They are trained on data collected in particular contexts, from particular populations, reflecting the assumptions and inequalities embedded in those societies. If the past was biased, and that past becomes the training data, then the future, too, becomes biased—only this time with the seductive authority of "objectivity."

One can imagine, for instance, a system that predicts which students are likely to fail a math course. It might flag students who historically performed poorly on similar assignments, or who spent less time on the digital learning platform, or who missed a certain number of questions in a quiz. To an administrator, such insights might look like progress—early intervention, data-driven decision-making, better outcomes. But what if the underlying data was shaped by structural disadvantage? What if the system is more likely to flag students from low-income backgrounds because those students are more likely to have inconsistent internet access, less time to study after school, or learning gaps from under-resourced primary education? The algorithm doesn't see these nuances. It sees patterns. And when we mistake patterns for truth, we risk coding injustice into the very core of our pedagogical processes.

The ethical dilemma here isn't just that bias exists—bias exists in every human system (Peters, 2023). The problem is that algorithmic bias is harder to detect, harder to challenge, and harder to escape. Students can't question an algorithm the way they might question a teacher's grade. Teachers may not even know how the algorithm reached its conclusion. The entire process becomes opaque, inscrutable—a black box with profound authority. And as a result, the pedagogical relationship subtly shifts. No longer is learning a negotiation between student and teacher, between effort and feedback. It becomes a performance for a machine, judged by criteria that remain largely invisible.

*Figure 1 — Ethical Risk Landscape in AI-Enabled Education*

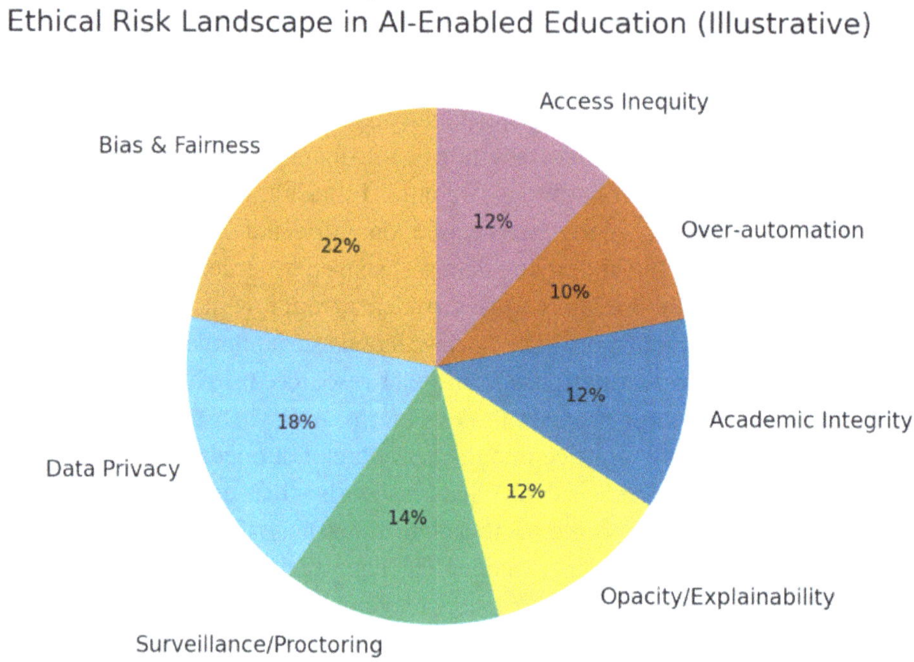


That invisibility breeds a quiet kind of helplessness. A student might ask why they didn't receive a recommendation for the advanced course. A teacher might wonder why a particular student was flagged as "at risk" despite showing promise in class (Jacob Kirksey & Elefante, 2024). And often, the answer is: "That's what the system says." There's no appeal process for a statistical inference. There's no conversation with a neural network. And the ethical weight of those decisions—the human cost—gets lost in the technicality.

It's tempting to believe that more data equals better decisions. That if we just feed the system enough variables, tweak enough parameters, or refine the model with enough granularity, we'll eventually arrive at fairness. But fairness is not a math problem. It's a human value. And when we outsource our ethical responsibilities to systems optimized for efficiency, we inevitably lose something—context, nuance, empathy. Algorithms are good at averages, but education is about exceptions. It's about that student who defies the odds, who surprises you, who struggles in silence for weeks before suddenly blooming. No predictive model can account for that spark. And the ethical landscape gets even murkier when commercial interests enter the equation (Kayyali, 2025 A). Many of the algorithmic tools used in classrooms today are developed by private companies, not public institutions. These companies—often motivated more by market share than by educational equity—hold the keys to the algorithms, the training data, the proprietary models (Kayyali, 2022). This raises urgent questions. Who owns the data generated by students' interactions? Who profits from their learning patterns? And what happens when educational software doubles as a surveillance apparatus, tracking attention, keystrokes, facial expressions, even emotional states—all in the name of "engagement"?

It's worth pausing here. Imagine being a teenager whose every eye movement is tracked during a virtual lesson, your attention scored in real time, your emotional state flagged by facial recognition software. What kind of learning environment does that create? One built on trust and curiosity? Or one shaped by performance anxiety, control, and quiet fear? When pedagogy becomes surveillance, we risk turning students into data subjects first, learners second.

There's also the matter of consent—real, informed, meaningful consent. Most students, especially younger ones, do not fully understand what data is being collected about them, how it's being used, or who has access to it (Szymkowiak et al, 2021). Parents may sign off on terms of service they barely read. Teachers may rely on platforms without knowing what lies under the hood. And in this fog of unawareness, profound ethical breaches can occur—not out of malice, but out of complacency. The ethics of algorithmic pedagogy demand not just good intentions, but informed participation. And it isn't only the students who are affected. Teachers, too, find themselves redefined by algorithmic systems. Some platforms "coach" educators by analyzing classroom behavior, suggesting instructional adjustments, or ranking teacher performance based on student metrics. On one hand, this could be empowering—offering new insights into pedagogy, highlighting blind spots, fostering professional growth. On the other, it can feel like micromanagement at a granular, almost molecular level. Teachers become data points in their own classrooms. Their creative autonomy, their pedagogical instincts, their idiosyncratic magic—all slowly flattened by dashboards and benchmarks.

Even the very idea of success becomes suspect in such systems. What is the algorithm optimizing for? Test scores? Completion rates? Engagement time? These metrics, while measurable, are not synonymous with learning. A student might spend more time on a platform not because they are engaged, but because they are confused. A high test score might reflect rote memorization, not deep understanding. When algorithms define success, the goals of education begin to warp. And once warped, they shape behaviors—students game the system, teachers teach to the algorithm, institutions chase data points that look good on reports but tell us little about actual learning.

This creeping instrumentalization—the reduction of complex educational experiences into trackable outcomes—is not just a technical issue (Torabian, 2022). It's a philosophical one. It challenges the very nature of what we believe learning should be. Should students be shaped into optimized outputs, calibrated for efficiency and predictability? Or should education remain a space for ambiguity, for contradiction, for the slow, nonlinear unfolding of thought?

There's no easy answer. But perhaps the ethical starting point is humility. A recognition that no system, no matter how advanced, can fully replace the judgment, empathy, and unpredictability of human relationships. That education is not a process to be controlled but a dialogue to be nurtured. Algorithms can assist, illuminate, support. But they must not dominate, dictate, or dehumanize (Schultz et al, 2024). To get there, we need new models of transparency—algorithms whose workings are open to scrutiny, whose decisions can be explained and challenged. We need participatory design—systems built not just for schools but with schools, incorporating the voices of students, teachers, parents, and marginalized communities. And we need ethical frameworks that aren't bolted on after the fact, but embedded from the ground up—ethical by design, not by compliance.

This isn't about rejecting technology. It's about reclaiming agency. About asking hard questions not just because we can, but because we must. Because once we hand the keys of education to systems we do not fully understand, it becomes harder to notice when something goes wrong. And by the time we do, the damage may already be coded in. The ethical dilemmas of algorithmic pedagogies are not abstract puzzles to be solved. They are living questions, threaded into the everyday realities of classrooms around the world. They demand vigilance, reflection, and above all, a refusal to let convenience replace conscience. If we can hold on to that—if we can insist that education remain a profoundly human endeavor—then maybe, just maybe, the algorithms can serve us, rather than the other way around.

## Data, Privacy, and Surveillance in the Learning Ecosystem

Something has shifted in the classroom—not in a sudden, theatrical way, but like a slow fog that settles so gradually no one notices it until everything feels just slightly off. Walls that once absorbed chalk dust and the hum of whispered ideas now reflect the cold glow of screens. Children log in. Teachers scroll dashboards. Every movement—every hesitation, click, scroll, question, even silence—is translated into data, silently sucked into servers that sit nowhere near the school and belong to no one in the room. The blackboard is gone. In its place: an invisible apparatus humming quietly behind the screen, watching, recording, analyzing. This isn't education as we used to know it. It's something else. Something that asks—not always kindly—what it means to learn in a world that treats students as sources of data before recognizing them as human beings.

At the heart of this transformation lies a double-edged promise: that data, when collected and processed intelligently, can make education more responsive, more personalized, even more just (Khan et al, 2024). Who wouldn't want that? Adaptive technologies that tune themselves to a student's unique rhythm. Early-warning systems that detect a struggle before it becomes a failure. Platforms that map a learner's progress in fine-grained detail, offering pathways tailored to each mind. It's difficult to argue with these dreams. But dreams, especially digital ones, tend to omit the fine print. What's the cost of this precision? Whose eyes are behind the interface? Who controls what is seen—and more importantly, what is not?

Most students today begin leaving data trails long before they're capable of consent. From their earliest interactions with educational software—some introduced as early as preschool—they generate patterns that are logged, stored, cross-referenced, and sometimes sold. Platforms track not just right or wrong answers, but how long a child hovers over a question, the way their eyes move across the screen, the time they log in and out. These are not just academic datapoints. They're behavioral signatures. And slowly, those signatures begin to form profiles—sometimes benign, sometimes predictive, always persistent.

What does it mean for a child to grow up inside an ecosystem where their every educational breath is monitored? It's not just a technical question. It's existential. Because the accumulation of data doesn't just describe the learner—it constructs them. If a student is told, through a dashboard or recommendation, that they are "below average," that narrative can become internalized. Invisible predictions begin to shape visible behaviors. A learner might stop trying in math not because they can't improve, but because the system never expected them to (Schoenfeld, 2022). Privacy violations in education are not always breaches or hacks. Sometimes, they are quiet acts of epistemic violence—moments where a machine's summary of a child becomes more real than the child's own story. And then there's the issue of surveillance—not as a metaphor, but as a literal condition. Surveillance has long existed in schools, of course. Bells that regulate time, attendance registers, hall monitors, disciplinary logs—all tools of order. But algorithmic surveillance is different. It's automated. It's scaled. It's ambient. It doesn't blink or forget. Learning management systems now record every interaction. Proctoring software uses webcams to monitor students' faces for signs of "suspicious" behavior. Some systems analyze typing patterns to detect cheating. Others employ facial recognition to verify identity. In one case, a student of color was repeatedly flagged by a proctoring algorithm that failed to recognize darker skin tones, forcing her to shine a bright lamp on her face just to take a test.

These aren't just technical glitches. They are reminders that the machines watching us weren't made with all of us in mind. And that's the ethical chasm: when surveillance technologies enter education, they bring with them not just the logic of accountability, but the architecture of control. Students are no longer just learning—they're being watched learning. And being watched changes how people behave.

*Table 1. Ethical Risk Mitigation Matrix*

| Risk | How it Manifests | Potential Harm | Mitigations (Illustrative) | Priority |
|---|---|---|---|---|
| **Bias & Fairness** | Unequal recommendations/ grades across subgroups | Entrenched inequities; mistrust | Bias audits; representative data; human override; continuous monitoring | High |
| **Data Privacy** | Over-collection; unclear consent; long retention | Exposure, re-identification, loss of trust | Data minimization; consent dashboards; deletion by default; DPIAs | High |
| **Surveillance/ Proctoring** | Face/keystroke monitoring; attention scoring | Chilling effects; disparate impacts | Proportional use; alternatives; impact assessments; opt-out paths | High |
| **Opacity/ Explainability** | Black-box scoring with no recourse | Due-process concerns; learned helplessness | Explainable models; appeal workflows; documentation | Medium-High |
| **Academic Integrity** | Undisclosed AI use; contract cheating | Assessment validity erosion | Assessment redesign; disclosure norms; AI literacy | Medium |
| **Over automation** | Replacing human judgment with dashboards | DE professionalization; brittle systems | Human-in-the-loop; escalation rules; teacher agency | Medium |
| **Access Inequity** | Tools require bandwidth/ devices not available to all | Widened achievement gaps | Offline-first design; device sharing plans; universal access funds | High |

But perhaps what's most insidious isn't even the overt surveillance. It's the normalization of it. The quiet acceptance that to be a good student is to be a measurable one. That if you can't be tracked, you don't exist. That privacy is a luxury education can no longer afford. We don't question these systems often enough (Komljenovic, 2022). Perhaps because they're cloaked in good intentions. Perhaps because they promise safety, or because they offer administrators tidy reports and colorful graphs. Perhaps because questioning them feels like questioning progress itself. But here's the thing: when surveillance becomes seamless, resistance becomes radical.

What happens to learning when curiosity becomes a liability? When every question asked is logged, and every moment of inattention is interpreted as disengagement? There's a profound difference between observation and monitoring. Observation sees the learner as a subject to be understood. Monitoring treats them as a risk to be managed (Andersson & Granberg, 2022). In monitored classrooms, mistakes become metrics. Struggles become signals. The act of learning—a fundamentally vulnerable, often messy process—gets flattened into a dataset, optimized and judged before it has a chance to unfold.

There's a particularly eerie irony in the use of artificial intelligence to gauge "engagement." Some platforms claim to detect boredom through facial expression analysis, issuing prompts or reminders when students look distracted. But what does boredom look like, really? A wandering gaze? A head resting on a hand? For whom? In what cultural context? And who decided that distraction is always a problem? Daydreaming, after all, has birthed revolutions of thought. Many of the world's great ideas arrived not through focused attention, but through idle, drifting minds allowed to roam. By pathologizing inattention, these systems risk stripping education of its serendipity.

The question of who owns educational data is another rabbit hole, deep and largely uncharted. In many countries, student data is not clearly protected by law. Companies offering free software to schools often monetize the information they collect—either directly, through advertising models, or indirectly, by refining their algorithms with user data (Wixom et al, 2023). And while schools may enter into agreements, students rarely do. Their identities, behaviors, struggles, and growth become commodities, used to fine-tune products or predict markets. Even anonymized data, when cross-referenced with other datasets, can often be re-identified. The digital body of a learner is far more exposed than their physical one. And once data is collected, it rarely disappears. A student's academic journey—from elementary arithmetic errors to college application essays—can now live forever in cloud storage, susceptible to breaches, repurposing, or reinterpretation (Peralta, 2024). That permanence raises ethical alarms. What happens when a misinterpreted behavioral pattern from middle school follows a student into their adult life? What rights do learners have to delete, contest, or revise their digital pasts?

There's also the false comfort of consent. Schools may ask parents to sign data-sharing agreements, but these are often buried in jargon, presented as non-negotiable, or bundled with essential services. The idea that families freely choose these systems is, more often than not, a polite fiction. True consent requires transparency, clarity, and meaningful alternatives. In many educational settings, none of these exist.

The dilemma deepens when one considers the global disparity in how data rights are protected. In affluent regions with strong regulatory frameworks, students may enjoy some semblance of control. But in under-resourced or marginalized communities—often the same communities targeted by philanthropic tech interventions—students become test subjects for systems they had no role in designing and limited means to resist. The ethics of data in education are inseparable from broader questions of justice, equity, and power. And yet, amid all this, the possibility of a different path still lingers. Technology need not violate privacy to support learning. Data can be used ethically—when it is collected minimally, stored securely, interpreted contextually, and governed transparently (Ige et al, 2022). But this requires a seismic shift in mindset. From extractive to relational. From top-down to participatory. From convenience to consent.

One radical idea: educational institutions could treat student data the way libraries treat books—not as assets to be hoarded or exploited, but as borrowings to be returned, respected, and cared for. Another: students could have data diaries—personal logs where they choose what to share, when, and with whom. A digital portfolio built not by a platform, but by the learner, shaped by their own reflections and aspirations. These are not utopian fantasies. They are ethical necessities. It's time we stop pretending that privacy and learning exist on opposite ends of a spectrum. They don't. A student who feels watched will not take intellectual risks. A teacher who fears being judged by an algorithm will not innovate. Real learning—messy, vulnerable, transformative learning—requires trust. And trust cannot grow in shadows.

In the end, the question is not whether data belongs in education. It already does. The real question is: can we wield it without doing harm? Can we protect the sacredness of the learning space from the creep of commodification and control? Can we build systems that listen without eavesdropping, that support without surveilling, that empower without extracting? If the answer is yes, it will not be because we followed the path of least resistance. It will be because we chose, deliberately, to put human dignity at the center of our technological decisions. And in that choice—quiet, difficult, sometimes unpopular—we may yet reclaim education as a space of freedom, not fear.

## AI and Educational Equity: Bridging or Widening the Gap?

The word "equity" sits heavy in the mouth, especially in the world of education, where it's been passed around like a mantra, printed in mission statements, chanted at conferences, etched into policy documents. And yet, for all its rhetorical weight, it remains one of those ideas that seems always just out of reach in practice—tangible enough to name, elusive enough to avoid measuring. Now, in the age of algorithms and machine intelligence, it's being invoked yet again, this time in the context of "AI for good." But what happens when machines, born from the data of an unequal world, are tasked with building something fairer? Can the very tools that emerged from systems marked by structural injustice be trusted to heal those same wounds—or do they risk cutting deeper?

It's easy to get swept up in the optimism. Artificial intelligence, after all, doesn't tire, doesn't discriminate intentionally, doesn't forget (Abercrombie, 2022). It can deliver instant feedback to a student in a rural village with no access to a math tutor. It can translate lessons in real time for refugees learning in a second—or third—language. It can scaffold learning experiences for students with disabilities, offering interfaces that adapt to different sensory, cognitive, or motor needs. It can analyze performance trends across districts and flag disparities that might have taken years to surface otherwise. In short, it has the potential to see what we've failed to see, and fix what we've long failed to fix. That's the promise. But the reality? That's much more complicated.

Equity, at its core, demands not sameness, but justice. And justice, as any educator will tell you, is messy, context-bound, and resistant to shortcuts. It asks us to look not just at outcomes, but at histories—at the layered, inherited disadvantages that shape who has access to what, and why. AI, unless deliberately designed otherwise, does not do this well. It is, by nature, a system of patterns (Haugeland et al, 2023). It thrives on probability, correlation, and abstraction. But equity, true equity, is about the specifics—the exceptions, the overlooked, the outliers who don't fit the model but still matter deeply. And that's where the tension begins.

Take, for instance, adaptive learning platforms. These systems, now widely used in both formal and informal education, are meant to personalize the experience for each learner. On paper, that's beautiful. A struggling student gets more support. A gifted learner moves ahead. The middle finds its rhythm. But what often goes unnoticed is that these platforms are only as "intelligent" as the data they're trained on. And that data—surprise—is rarely neutral.

If a model is trained primarily on students from affluent backgrounds, with stable internet, quiet homes, and abundant parental support, then it learns what success looks like in that environment. Introduce a student from a low-income neighborhood, grappling with housing insecurity or food scarcity, and the system may not recognize their patterns. It may flag them as disengaged, slow, erratic. But it's not their learning that's broken—it's the system's lens that's narrow. And once flagged, once labeled by the algorithm, those students may begin receiving "remedial" content, narrowing their exposure to complex material, shrinking their educational horizons under the guise of support.

*Figure 2. Distribution of Model Errors by Socioeconomic Context*

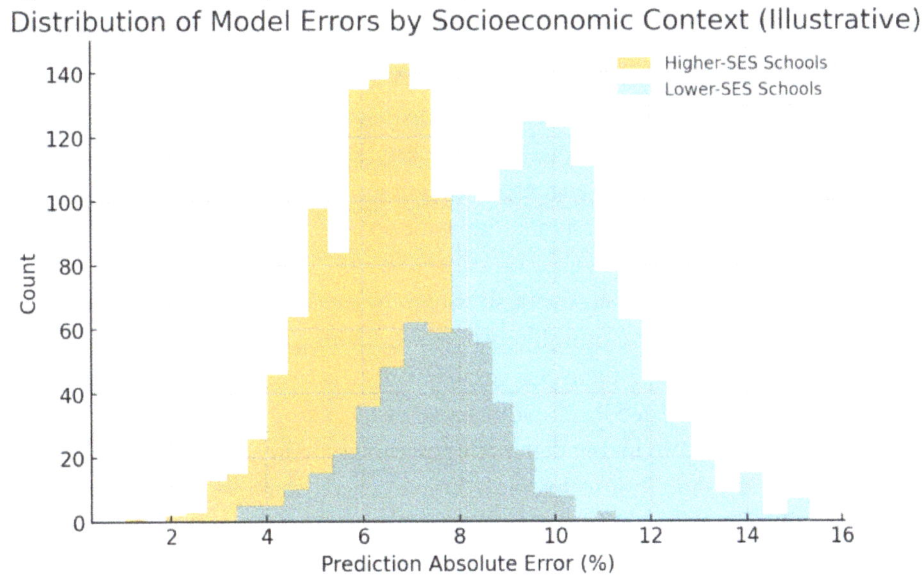


The machine, in trying to help, ends up boxing them in. That's one of the more subtle dangers of AI in education: not outright exclusion, but algorithmic benevolence that smothers possibility (Coeckelbergh, 2024). Systems that try to "optimize"

student pathways can easily reinforce low expectations. If the model predicts you won't do well in science, it may stop recommending advanced science courses. It doesn't mean to discourage. It just does the math. But in doing so, it quietly performs a kind of digital tracking—an echo of the very practices equity-minded educators have spent decades trying to dismantle (Kayyali, 2024). And then there's the infrastructure gap, which undergirds everything.

AI may be scalable, but it is not universally accessible. To benefit from AI-enhanced learning tools, students need reliable internet, up-to-date devices, digital literacy, and environments conducive to learning (Wong, 2024). Many do not. The global digital divide is not closing fast enough, and within nations, disparities often fall along lines of race, income, disability, and geography. So even if the software is free, the conditions to use it are not. What good is an AI tutor if the power cuts out every two hours? What good is real-time feedback if a child shares a single device with three siblings?

These aren't hypothetical scenarios. They're daily realities for millions. And unless intentionally addressed, AI will simply mirror the geography of privilege—enhancing opportunities for those already ahead, and leaving behind those who were always running uphill. But it's not just about who has access to AI. It's also about who builds it—and for whom. The edtech landscape is still dominated by developers from the Global North, often designing for imagined "universal" learners that look suspiciously like themselves (Henry, 2024). This leads to tools that reflect particular pedagogical assumptions, cultural norms, and linguistic frameworks—ones that may not translate across borders, or even neighborhoods. A speech recognition system that performs well with American English may stumble over African-American Vernacular English, or Indian-accented English, or the clipped tones of a shy child. When the system fails to understand, it doesn't apologize. It just fails silently—and the learner pays the price.

True equity in AI-enhanced education would mean systems that are not only technically accessible, but culturally responsive (Lata, 2024). That means building models trained on diverse datasets, yes—but also involving communities in the design process, inviting feedback from those at the margins, resisting the urge to generalize. It means recognizing that intelligence takes many forms, and that success is not a universal shape. It also means asking uncomfortable questions about data ownership, algorithmic transparency, and the silent influence of venture capital on what gets built, and why.

There's an especially troubling paradox at play here: the very populations that AI could most benefit—historically underserved students—are also the most likely to be harmed by it. Why? Because they're more likely to be subjected to untested technologies, to platforms that haven't been audited for bias, to environments where human oversight is stretched thin. In these contexts, AI becomes a substitute for adequate staffing, rather than a supplement to good teaching. A triage tool, not a transformative one.

Think of overcrowded urban schools where teachers juggle classrooms of forty or more. Administrators might turn to AI systems to flag at-risk students, automate grading, or monitor attendance patterns (Embarak & Hawarna, 2024). But when those systems misfire, who will intervene? When a child is wrongly labeled as disengaged, or when a subtle learning disability is mistaken for laziness by the algorithm, who will notice? The assumption that AI will lighten the load often leads to reduced human attention. And in the absence of that attention, small errors become compounding injustices.

That's not to say AI has no place in the pursuit of educational equity. On the contrary, it can be a powerful ally—if deployed with humility, caution, and a deep commitment to justice. AI can amplify human insight, uncover hidden patterns, spotlight inequities, and facilitate personalized learning at scales previously unimaginable (Costa, 2023). But it can't—and shouldn't—replace the human relationships that sit at the heart of transformative education.

Some hopeful models exist. In regions where AI has been used to support multilingual instruction, for instance, it has helped bridge linguistic divides that once excluded entire communities. In inclusive education settings, AI tools that adapt to sensory and motor differences have empowered students with disabilities to participate more fully (Chopra et al, 2024). In some cases, AI-driven data analysis has illuminated funding inequities, pushing policymakers to allocate resources more fairly. These examples matter. They show what's possible when technology serves equity, not the other way around. But for every success story, there are cautionary tales. Platforms rolled out without community consultation. Algorithms used to rank teachers or students without context. Disciplinary systems that flag "problem behaviors" based on biased data. Every such instance is a reminder that equity is not an inevitable outcome of technology. It is a choice—a series of choices, actually—about how systems are built, governed, and used.

Ultimately, the question isn't whether AI will bridge or widen educational gaps. The question is whether we're willing to do the hard, unglamorous work of designing it for justice. That means resisting the fetish for scale, the worship of efficiency, the impulse to automate without reflection. It means privileging care over convenience, process over product, inclusion over innovation. It means listening—to students, to teachers, to communities whose voices are often sidelined in these conversations. Equity isn't a feature you can toggle on or off in a platform. It's a principle that must inform every line of code, every design decision, every policy implementation (LeBlanc, 2021). And that requires more than tech literacy. It requires moral imagination. Can AI help create a world where every child—regardless of zip code, skin tone, native language, or neurological wiring—has access to meaningful, empowering education? Perhaps. But only if we stop mistaking neutrality for fairness. Only if we remember that the future of learning isn't just about what machines can do, but about what kind of humans we want to be.

## Towards a Just and Inclusive AI-Driven Educational Future

It's tempting to look to the future as though it's a train we're either aboard or left behind by. That somehow, the tracks have already been laid, the direction set, the destination fixed. But technology—contrary to the myths we like to tell about it—is not inevitable. AI is not gravity. It does not fall upon us. It is designed, curated, implemented (Crawford, 2021). Which means it can be changed. Rerouted. Reimagined. And when it comes to education—a realm so deeply entangled with dreams, identities, aspirations—the question isn't just how we use AI, but who decides what it's used for, and whose vision of "progress" it ends up serving.

To speak of a just and inclusive AI-driven educational future is to wrestle with contradiction. The kind of contradiction that doesn't resolve neatly, doesn't fold into bullet points or product demos. It's the kind that sits like a knot in the stomach: the awe of what's possible tangled with the unease of who's being forgotten. Because for all the noise around innovation, the deeper work lies in the silences—listening for the students who aren't heard by the algorithm, the teachers overwhelmed by dashboards, the communities excluded from the design table altogether.

*Figure 3. AI Use vs. Ethical Guardrail Adoption in Education*

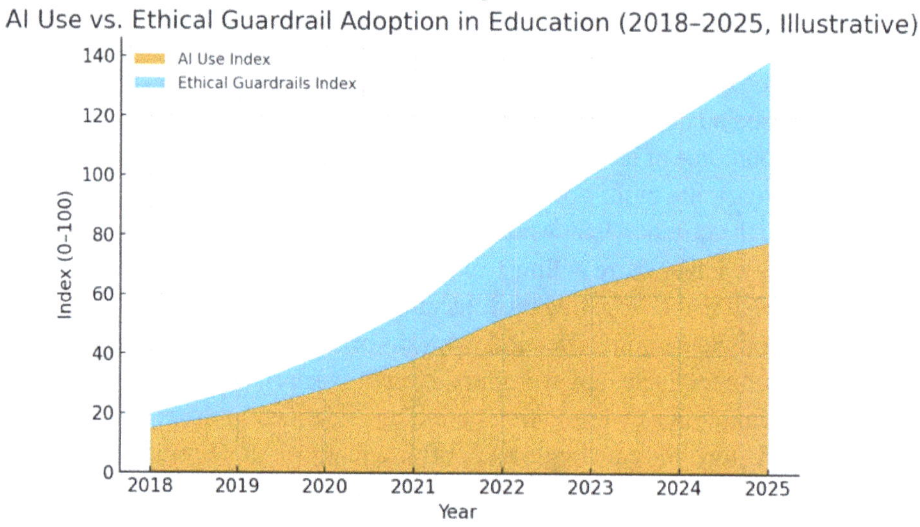

In truth, what's often missing from the grand narratives of AI is a sense of humility. A willingness to admit that smart doesn't always mean wise. That being able to predict a student's next test score doesn't mean you understand who they are, or what they carry (Willingham, 2021). That even the most finely-tuned model can misread pain as distraction, curiosity as misbehavior, silence as disengagement. Machines can scan data, but they can't feel the tremble in a voice. They don't know how poverty smells on a child's clothes. They don't notice when a student comes in hungry, not for knowledge, but for food. Which is why any future that claims to be just must begin by re-centering the human. Not in a nostalgic way, as though yearning for chalk and blackboards. But in an ethical way—a future where technology bends toward the needs of people, not the other way around (Kayyali, 2025 B). That means rejecting the fantasy of neutrality. Algorithms aren't neutral. They are made by people, trained on histories, steeped in assumptions. And so, the first act of building a more inclusive future must be to bring those assumptions into the light.

Who defines what counts as "success" in AI-driven education? Is it completion rates? Exam scores? Time-on-task? Or is it something less quantifiable—resilience, empathy, the courage to ask strange questions? And who gets to decide which metrics matter? If we're not careful, we'll end up encoding old inequities into new systems, dressing them in sleek interfaces and calling it progress.

There's a need to push back against the cult of efficiency. Education is not a factory. Students are not raw inputs to be processed. The most transformative learning often resists streamlining. It happens in meandering conversations, in the teacher who lingers after class, in the moment when a student's worldview cracks open just a little (Alarifi, 2023). If AI is to serve education, it must be designed not just to optimize—but to honor complexity.

That means systems that are transparent. Not black boxes that spit out grades or risk scores, but tools that explain their reasoning, allow for contestation, invite dialogue. Students should have the right to know how they're being evaluated, and by whom—or by what. Teachers should be able to override algorithmic recommendations when their judgment says otherwise (Guo et al, 2022). Parents should be able to ask questions without needing a computer science degree to understand the answers. And if we're serious about inclusion, we need to talk about voice. Whose stories shape the development of AI in education? Too often, the conversation is dominated by technologists, investors, and policymakers—people who, however well-intentioned, are often far removed from the everyday realities of classrooms. Students, especially those from marginalized backgrounds, are rarely invited to imagine the future of the tools that will shape them. But inclusion isn't just about representation. It's about participation. It's about co-creation. Designing with, not for.

That shift—from user to shaper—could be transformative. Imagine curricula that help students critically interrogate the AI systems they interact with. Imagine a generation not just digitally literate, but algorithmically aware—able to understand, challenge, and even build their own models. Not passive consumers of AI, but active authors of its trajectory. Equity in this context means more than access to devices. It means access to power (LaForce & Bright, 2022). The power to shape how learning technologies evolve. The power to resist their misuse. The power to demand systems that reflect one's culture, language, pace, and dreams. A just future requires us to confront not only the design of AI, but the distribution of agency. And then there's the matter of care—a word too rarely spoken in conversations about tech. But care is central. Systems that don't care—about context, about harm, about the dignity of those they touch—will fail, even if they "work." A just future must be built with care as both method and goal. That means listening deeply, designing slowly, iterating with humility. It means recognizing that not all problems are best solved by technology, and that sometimes, the most ethical choice is to leave a tool unused.

One might imagine a future where AI acts not as overseer, but as quiet companion. Where its intelligence is shaped by values, not just data. Where it supports relationships rather than replacing them. Where students are not scored and sorted, but seen and supported. Where AI doesn't dictate the rhythm of the classroom, but responds to it—lightly, attentively, respectfully (Koehl, 2025). But let's not pretend this vision will materialize on goodwill alone. It requires policy. It requires regulation. It requires rigorous oversight and public investment. Ethical intentions are not enough. There must be safeguards—against bias, against surveillance, against the entanglement of education with profit motives that undermine learning for the sake of learning. There must be robust standards for data privacy, meaningful consent, algorithmic transparency. There must be room for refusal—for students and educators to say no, to opt out, to choose another way.

It also requires memory. A future worth building is one that remembers the past—not as nostalgia, but as grounding. We must remember the harms already done in the name of educational "reform." The standardized testing regimes that gutted creativity. The tracking systems that sorted students by race and class. The edtech solutions that promised miracles and delivered metrics. AI cannot be allowed to become the next iteration of this pattern: the shiny object that obscures structural rot (Ochoa, 2024). Instead, we might hold onto the idea that education—real education—is a long conversation. One that spans generations. One that doesn't always seek answers, but dares to ask better questions. And perhaps the role of AI in that conversation is not to speak louder, but to listen better. To catch what might otherwise be missed. To offer possibilities without closing doors.

So, what does that look like, in practice? Maybe it means schools investing not only in tools, but in time—time for teachers to explore and evaluate AI systems, time for students to reflect on their experiences with them. Maybe it means building community review boards for educational technologies, where parents, teachers, and students collectively assess whether a tool aligns with their values. Maybe it means funding open-source alternatives to corporate platforms, so that educational infrastructure isn't controlled by a handful of tech giants. Maybe it means resisting the urge to automate at all costs, and remembering that some things—some of the most important things—are best done slowly, face-to-face, together. A just and inclusive AI-driven future will not be defined by the sophistication of our models, but by the moral clarity of our choices (Sargiotis, 2024). Will we prioritize equity over convenience? Care over control? Humanity over hype? These are not rhetorical questions. They are live wires in the systems we build and the cultures we shape. And the answers won't come from algorithms. They will come from us—from the choices we make now, in this hinge moment, before the dust settles, before the infrastructure hardens, before the narratives calcify. The future is not a given. It is a project. A commitment. A promise we make to the next generation—that education, whatever shape it takes, will remain a space of possibility, not prediction. A space of freedom, not sorting. A space where every learner is more than a datapoint. We owe them that.

## CONCLUSION

The deployment of artificial intelligence in education presents both an extraordinary opportunity and a profound ethical test. As we stand at the intersection of technological innovation and human development, we are called to make choices—not just about tools and systems, but about values, priorities, and the kind of society we aspire to build. AI, when thoughtfully designed and equitably implemented, has the potential to personalize learning, streamline administration, and democratize access to knowledge. Yet without intentional ethical guardrails, it may also entrench inequalities, obscure decision-making, and transform learners into passive subjects of algorithmic governance.

This chapter has explored the ethical dilemmas inherent in educational AI, from algorithmic bias and digital inequity to data privacy and the erosion of human agency. These challenges are not hypothetical—they are real, urgent, and already affecting students and teachers across the globe. Algorithms trained on biased data can reinforce exclusion; facial recognition systems can misidentify students of color; and over-reliance on automation can deskill educators while narrowing pedagogical diversity. These are not merely technical problems to be patched—they are ethical tensions that require critical reflection, inclusive dialogue, and systemic action.

At the same time, we must recognize the opportunities that AI brings—not as silver bullets, but as tools that, when aligned with educational justice, can help us address long-standing challenges. For students with disabilities, AI can offer accessibility solutions once thought impossible. For educators, intelligent platforms can help reduce administrative overload, freeing up time for meaningful engagement. For policymakers, data-driven insights can inform more responsive and targeted interventions. The promise of AI is real—but it must be pursued without sacrificing the human essence of education.

The way forward requires more than ethical guidelines or abstract principles. It demands participatory governance, where students, teachers, parents, technologists, and policymakers co-create the future of AI in education. It calls for transparency in algorithms, equity in access, and accountability in outcomes. It also demands an ongoing commitment to examining power dynamics—who builds the technology, who benefits from it, and who bears the risks. Education cannot be reduced to a series of metrics and outputs; it is a deeply human endeavor, shaped by culture, emotion, creativity, and context.

If we are to harness AI for the public good, we must center justice, not just efficiency; humanity, not just innovation. This means rejecting the logic of inevitability and embracing the possibility of shaping AI systems that reflect our highest ideals rather than our lowest biases. Ethical AI in education is not a destination but a continuous process—a commitment to reflection, revision, and resistance against the reproduction of systemic harm.

In the end, the future of AI in education is a mirror to the future of education itself. Will we allow machines to define what learning means, or will we define how machines should support human learning in all its diversity and complexity? The answer to this question will shape not only classrooms and curricula, but the fabric of our societies. Let us ensure that our answer is rooted in equity, dignity, and the unwavering belief in the transformative power of education for all.

## REFERENCES


Abercrombie, C. (2022). *What You Don't Know: AI's Unseen Influence on Your Life and How to Take Back Control*. Post Hill Press.

Alarifi, I. M. (2023). *Learning Limitless Knowledge: Transformative Learning Pathway to Unlocking and Harnessing the Endless Power of Knowledge*. Balboa Press.

Andersson, C., & Granberg, C. (2022, March). From passenger to pilot–using formative assessment to support students with intellectual disabilities to become self-regulated learners. [). Frontiers Media SA.]. *Frontiers in Education, 7*, 845208. DOI: 10.3389/feduc.2022.845208

Blikstein, P., & Blikstein, I. (2021). Do educational technologies have politics? A semiotic analysis of the discourse of educational technologies and artificial intelligence in education. Algorithmic rights and protections for children.

Bulathwela, S., Pérez-Ortiz, M., Holloway, C., Cukurova, M., & Shawe-Taylor, J. (2024). Artificial intelligence alone will not democratise education: On educational inequality, techno-solutionism and inclusive tools. *Sustainability (Basel), 16*(2), 781. DOI: 10.3390/su16020781

Chen, L., Chang, L., Lin, H., Tu, J., Zhou, Y., & Han, Y. (2024). Savior or saboteur? A nationwide study on digital economy and depression in China. *Journal of Affective Disorders, 365*, 578–586. DOI: 10.1016/j.jad.2024.08.147 PMID: 39187198

Chopra, A., Patel, H., Rajput, D. S., & Bansal, N. (2024). Empowering Inclusive Education: Leveraging AI-ML and Innovative Tech Stacks to Support Students with Learning Disabilities in Higher Education. In *Applied Assistive Technologies and Informatics for Students with Disabilities* (pp. 255–275). Springer Nature Singapore. DOI: 10.1007/978-981-97-0914-4_15

Coeckelbergh, M. (2024). *Why AI undermines democracy and what to do about it*. John Wiley & Sons.

Costa, S. (2023). Artificial intelligence in education: Challenges and opportunities in learning.

Crawford, K. (2021). *The atlas of AI: Power, politics, and the planetary costs of artificial intelligence*. Yale University Press.

Dieterle, E., Dede, C., & Walker, M. (2024). The cyclical ethical effects of using artificial intelligence in education. *AI & Society, 39*(2), 633–643. DOI: 10.1007/s00146-022-01497-w PMID: 36185064

Elwarraki, O., Aammou, S., Jdidou, Y., & Lahiassi, J. (2023). Toward the future of personalized learning: Emerging trends and challenges in recommendation systems. ICERI2023 Proceedings, 7226-7235.

Embarak, O. H., & Hawarna, S. (2024). Automated AI-driven system for early detection of at-risk students. *Procedia Computer Science*, *231*, 151–160. DOI: 10.1016/j.procs.2023.12.187

Gaftandzhieva, S., Talukder, A., Gohain, N., Hussain, S., Theodorou, P., Salal, Y. K., & Doneva, R. (2022). Exploring online activities to predict the final grade of student. *Mathematics*, *10*(20), 3758. DOI: 10.3390/math10203758

Guo, L., Flathmann, C., Anaraky, R., McNeese, N., & Knijnenburg, B. (2022, June). The effect of recommendation source and justification on professional development recommendations for high school teachers. In *Proceedings of the 33rd ACM conference on hypertext and social media* (pp. 175-185). DOI: 10.1145/3511095.3531280

Hanemaayer, A. (2021). Don't touch my stuff: Historicising resistance to AI and algorithmic computer technologies in medicine. *Interdisciplinary Science Reviews*, *46*(1-2), 126–137. DOI: 10.1080/03080188.2020.1840222

Haugeland, J., Craver, C. F., & Klein, C. (Eds.). (2023). *Mind Design III: Philosophy, Psychology, and Artificial Intelligence*. MIT press.

Henry, J. V. (2024). Theorizing 'The Gap'twenty years later: Global development, design, and speculative ethics in edtech research. *Postdigital Science and Education*, *6*(1), 231–258. DOI: 10.1007/s42438-023-00429-1

Hossain, M. E., & Islam, M. A. (2024). AI and the Future of Education: Philosophical Questions about the Role of Artificial Intelligence in the Classroom. *International Journal of Research and Innovation in Social Science*, *8*(IIIS, 3s), 5541–5547. DOI: 10.47772/IJRISS.2024.803419S

Hutson, J., & Plate, D. (2024). The Algorithm of Fear: Unpacking Prejudice Against AI and the Mistrust of Technology. Journal of Innovation and Technology, 2024.

Ige, A. B., Chukwurah, N., Idemudia, C., & Adebayo, V. I. (2022). Ethical considerations in data governance: Balancing privacy, security, and transparency in data management. Journal of Ethical Data Practices.[Year unavailable].[Epub ahead of print].

Jackaria, P. M., Hajan, B. H., Mastul, A. R. H., & Sali, F. Z. (2024). Generation AI in a Reimagined Classroom: Challenges, Opportunities and Implications to Education. Exploring Youth Studies in the Age of AI, 174-185.

Jacob Kirksey, J., & Elefante, J. (2024). Familiar faces in high school: How having the same peers from year-to-year links to student absenteeism. [JESPAR]. *Journal of Education for Students Placed at Risk*, *29*(1), 23–42. DOI: 10.1080/10824669.2022.2102982

Kayyali, M. (2022). Equity, Equality, Diversity, and Inclusion as Key Performance Indicators in Higher Education. *Online Submission*, *3*(5), 1–3.

Kayyali, M. (2024). Future directions towards sustainable educational equity: Policies and practices. In *Exploring Educational Equity at the Intersection of Policy and Practice* (pp. 81–100). IGI Global. DOI: 10.4018/979-8-3693-1614-6.ch005

Kayyali, M. (2025a). Ethical Implications of Generative AI in Education: Privacy, Bias, and Integrity. In Transformative AI Practices for Personalized Learning Strategies (pp. 185-218). IGI Global Scientific Publishing.

Kayyali, M. (2025b). Ethical Considerations in Higher Education: Accountability Navigating Complexities. In Accountability in Higher Education: Navigating Current Issues and Trends (pp. 145-184). IGI Global Scientific Publishing.

Khan, A. A., Hussain, T., Al Siyabi, M., Al Shibli, A., Hussain, S., & Pillai, J. R. (2024). Challenges and Prospects of Cutting-Edge Technology in Education. *Studies on Education, Science, and Technology*, *2024*, 104.

Koehl, S. (2025). Designing AI Software for Large Classroom Engagement (Doctoral dissertation).

Komljenovic, J. (2022). The future of value in digitalised higher education: Why data privacy should not be our biggest concern. *Higher Education*, *83*(1), 119–135. DOI: 10.1007/s10734-020-00639-7 PMID: 33230347

LaForce, S., & Bright, D. (2022). Are we there yet? The developing state of mobile access equity. *Assistive Technology*, *34*(6), 628–636. DOI: 10.1080/10400435.2021.1926372 PMID: 34048330

Lata, P. (2024). Towards Equitable Learning: Exploring Artificial Intelligence in Inclusive Education.

LeBlanc, P. (2021). *Students first: Equity, access, and opportunity in higher education*. Harvard Education Press.

Lin, T. A., & Chen, P. H. C. (2022). Artificial intelligence in a structurally unjust society. Feminist Philosophy Quarterly, 8(3/4).

Maghsudi, S., Lan, A., Xu, J., & van Der Schaar, M. (2021). Personalized education in the artificial intelligence era: What to expect next. *IEEE Signal Processing Magazine*, *38*(3), 37–50. DOI: 10.1109/MSP.2021.3055032

Mutambara, A. G. (2025). *Artificial intelligence: A driver of inclusive development and shared prosperity for the Global South*. CRC Press. DOI: 10.1201/9781003511014

Ochoa, K. S. (2024). Can Artificial Intelligence Mark the Next Architectural Revolution? Design Exploration in the Realm of Generative Algorithms and Search Engines. Decoding Cultural Heritage: A Critical Dissection and Taxonomy of Human Creativity Through Digital Tools, 3.

Peralta, A. (2024). Best Practices for Managing Privacy and Security of Cloud-Based Student Data in Primary and Secondary Education.

Peters, U. (2023). Extended implicit bias: When the metaphysics and ethics of implicit bias collide. *Erkenntnis*, *88*(8), 3457–3478. DOI: 10.1007/s10670-021-00511-9 PMID: 35035000

Sargiotis, D. (2024). Fostering Ethical and Inclusive AI: A Human-Centric Paradigm for Social Impact. Available at *SSRN* 4879372. DOI: 10.2139/ssrn.4879372

Schoenfeld, A. H. (2022). Why are learning and teaching mathematics so difficult? In *Handbook of cognitive mathematics* (pp. 1–35). Springer International Publishing.

Schultz, M. D., Clegg, M., Hofstetter, R., & Seele, P. (2024). Algorithms and dehumanization: A definition and avoidance model. *AI & Society*, •••, 1–21.

Szymkowiak, A., Melović, B., Dabić, M., Jeganathan, K., & Kundi, G. S. (2021). Information technology and Gen Z: The role of teachers, the internet, and technology in the education of young people. *Technology in Society*, *65*, 101565. DOI: 10.1016/j.techsoc.2021.101565

Torabian, J. E. (2022). Education and Culture: The Instrumentalisation. In *Wealth, Values, Culture & Education: Reviving the essentials for equality & sustainability* (pp. 91–135). Springer International Publishing. DOI: 10.1007/978-3-030-92893-3_4

Vallabhaneni, A. S., Perla, A., Regalla, R. R., & Kumari, N. (2024). The Power of Personalization: AI-Driven Recommendations. In Minds Unveiled (pp. 111-127). Productivity Press.

Willingham, D. T. (2021). *Why don't students like school?: A cognitive scientist answers questions about how the mind works and what it means for the classroom*. John Wiley & Sons.

Wixom, B. H., Beath, C. M., & Owens, L. (2023). *Data is everybody's business: The fundamentals of data monetization*. Mit Press. DOI: 10.7551/mitpress/14867.001.0001

Wong, K. K. (2024, June). Blended Learning and AI: Enhancing Teaching and Learning in Higher Education. In International Conference on Blended Learning (pp. 39-61). Singapore: Springer Nature Singapore. DOI: 10.1007/978-981-97-4442-8_4

## KEY TERMS AND DEFINITIONS

**Algorithmic Bias:** systematic errors in AI outputs that unfairly affect certain groups due to skewed data or design.

**Explainability:** the degree to which an AI system's reasoning can be understood and challenged by humans.

**Data Minimization:** collecting only the least amount of student data needed for a clearly stated educational purpose.

**Ambient Surveillance:** continuous background monitoring of learners via platforms, proctoring, or device sensors.

**Human-in-the-Loop:** workflows where educators can review, override, and correct AI recommendations or decisions.

**Equity-by-Design:** building policies, tools, and assessments from the ground up to prevent disparate impacts.

**Guardrails:** practical governance (audits, appeals, consent, access funds) that constrain AI use toward safe, fair outcomes.

# Chapter 10
# Ethical Imaginaries of Geospatial Artificial Intelligence:
## A Sociotechnical Approach and Open Spatial Data Global Policy

**Federico Monaco**
https://orcid.org/0000-0001-8908-7365
*Mars Planet, Italy*

## ABSTRACT

*The convergence between Geography and AI promises innovative applications and digital services for many fields, such as epidemiology and public health, marketing, environmental monitoring and disaster management, urban planning, and navigation. In terms of ethics, such emerging technology presents several issues at different scales. Bias and ethical issues are already part of the debate inside the GeoAI community, mainly about transparency, justice & fairness, non-maleficence, responsibility and privacy. Understanding the potential and the risks of GeoAI might provide guidance and counter measures for the industry, policy makers, and communities. A transparent and standardised open data global policy might provide a sustainable and global solution in the long term. It is here recommended to consider both technical and social aspects in order to frame ethical issues by a sociotechnical approach and intervene in the emerging coproduction of the sociotechnical imaginary of GeoAI.*

DOI: 10.4018/979-8-3373-2597-2.ch010

## INTRODUCTION

Geospatial Artificial Intelligence (GeoAI) represents a transformative intersection of research in full growth for the geographical domain of the scientific community and the location services market, and sits at the intersection of artificial intelligence, geospatial big data, and high-performance computing to address compute-intensive spatial problems and enable large-scale geospatial intelligence (Li & Hsu, 2018; Li et al., 2020; Zhu et al., 2017). GeoAI community promises to uncover patterns and insights about spatial conditions and phenomena represented in geo datasets. The automation of analysis of complex geospatial data can enhance the accuracy and efficiency of geographic information systems in order to address the challenges of progress in other domains. Applications for Earth Observation (EO) and geospatial data-related tasks, become possible with the availability of multispectral, hyper-spectral and multi-source imagery.

The peculiarity of GeoAI is spatial data, as it features a whole framework of research on ethical issues. The creation of more intelligent geographic information as well as methods, systems, and services for a variety of downstream tasks (Janowicz et al., 2020) requires a deep understanding of the risks and issues in every operational and methodological phase with present and long term consequences for individuals and communities. A technology able to "mimic the processes of human perception, spatial reasoning, and discovery about geographical phenomena and dynamics" (Gao, 2021) should include an analytical framework on how such processes should be imitated and simulated given ethical concerns and risk of bias.

Geospatial big data play a fundamental role in the study of phenomena such as mapping, environmental risk analysis and monitoring, urban planning, disaster response, public health and the management of assets and infrastructures with predictive strategies. Thanks to the availability of datasets, the capability of GeoAI empowers research and industry to advance in fields like Planetary Science, Remote Sensing, Navigation and Disaster Risk Management. GeoAI applications may simulate realistic scenarios of disaster, and identify emerging trends in climate change (Raj et al., 2025), and enable swift disaster assessment (Adhikari, 2024). Appropriate processing chains are able to convert the pixel values to information useful for decision makers (Tuia et al., 2024), while computational processes can provide intuitive user interfaces for a wider access and use AI-enabled EO services. Data about the conditions of natural resources and population are collected from satellites and other different sources including humans and communities. All above makes explicit the importance of data and how these should be carefully and responsibly managed in order to prevent misuse and abuse.

At global level a fulfilling and optimistic imaginary is still missing, while there are several clear challenges at play for GEOAI (Xing & Sieber, 2023). The risk of ethical issues and bias represents a strong limitation to the current state of the research and applications, and might affect in the future users, communities and people.

As GeoAI represents an opportunity to delegate to machines and enhance the workflow analysis of maps and data about places and spatial phenomena to perform highly reliable results as significant patterns, how then does GeoAI might be harmful and negatively affect our existences?

According to Janowicz (2023), a philosophical foundation should be able to position GeoAI in terms of sustainability, diversity, and possible bias. For instance, uncertainty in geographical information itself, such as location distributed phenomena, may weaken the management, evaluation and decision making outcome about people (Zhang & Goodchild, 2002).

Several guidelines on ethical AI have included five main ethical principles: transparency, justice & fairness, non-maleficence, responsibility and privacy (Jobin et al., 2019). In the case of AI applied to Geography many peculiar issues are at work, given that "a unique type of information in that knowledge of one person's location reveals highly sensitive information about nearby individuals or groups" (McKenzie et al., 2023). The integration of artificial intelligence in geosciences can deliver more accurate estimations, efficient predictions, and innovative solutions. By leveraging algorithms and machine learning, AI empowers geoscientists and other scientific communities to work in innovative ways with complex and big geodata sources (Boutayeb et al., 2025). However, innovation implies in this case the analysis of personal location and movement data and the consequent risks of bias generating and possible misuse of information provided. As the societal implications of progress in geosciences can empower stakeholders and policymakers to optimize resource management, design effective strategies, and generate progress for the benefit of the public, at the same time there is an increase of risk about privacy and ethical issues that should be tackled by a sociotechnical approach able to work also on the shared vision of progress about such technology. An intervention from international organizations and national agencies should be able to regulate and enforce the rights of citizens about their positioning and privacy data, while an open data policy could prevent non-transparent use and exclusion of the stakeholders.

## THE EMERGENCE OF GeoAI

A brief history of how computation was sold to the geosciences can illustrate both the technical progress and the ethical risks in this field. Given its interdisciplinary nature, "Geography has the natural advantage of embracing new theories, methods, and tools from other disciplines" (Li & Hsu, 2022) and deserves a keen analysis of how the introduction of AI can affect the discipline and the stakeholders. A first attempt to integrate Geography with technology was documented in 1970 about a computer simulation of the urban growth in the Detroit area (Tobler, 1970). In 1983 the term "Automated Geography" was adopted to mean the application of computational methods and techniques "to portray spatial properties, to explain geographic phenomena, and to solve geographic problems" (Dobson, 1983, p.136). At that time, the debate on the ethical practice was about simple good behaviour and adherence to standards of Geography (Crampton, 1995), while the commitment of part of the geographical scientific community brought to the adoption of the term "Geographical Information Science", or GIScience (Goodchild, 1992) with further claiming to the involvement of the public and the need for standards (Goodchild, 2009). A primer publication named "Artificial Intelligence in Geography" was published in 1997 and included the first AI techniques and methods that could be adopted for geographical practitioners (Openshaw & Openshaw, 1997). The introductions of new ontologies such as expert systems, neural networks, fuzzy systems, and evolutional computation, brought several discussion and criticism in the community about the use and concerns on new theoretical and epistemological terms (Couclelis, 1998). A transient definition was then "CyberGIS", referring to a 5 year research program about software for the scientific community with the following characteristics: high-performance and scalable, open and distributed, collaborative, service-oriented, user-centric, and community-driven (Wang et al., 2013).

In recent years, AI applications, such as image processing, smart sensors, and intelligent inversion, have garnered attention from the geosciences community, envisioning a transition of geosciences from qualitative to quantitative analysis.

It was only at the 2017 ACM SIGSPATIAL conference that the term "GeoAI" was introduced to the community (Mao et al., 2017.; Lunga et al., 2021), while defined as "a new trans disciplinary research area that exploits and develops AI for location-based analytics using geospatial big data" (Li & Hsu, 2022). The same term was then adopted to include Machine Learning (ML) and Deep Learning (DL) (Choi, 2023).

The first consequence was the need for a geospatial data management about EO, geographic information system/building information modeling (GIS/BIM) integration, and 3D/4D city planning (Breunig et al., 2020), while the new visions for a geospatial infrastructure began to include open access, sharing, engagement, and the integration of emerging technologies like Big Data, AI, and data science (Li et al., 2020). As scientific production, special issues have then been published in prominent journals, such as the International Journal of Geographic Information Science, GeoJournal, Geoinformatica, Geographies, or Applied Sciences, covering spatially explicit models, and applications in urban analytics, image classification and land cover mapping, crowd-sourced image and text analysis, natural resource management and infrastructure monitoring (Hochmair et al., 2025). Nowadays, an ecological approach, given social, technical and environmental challenges, is making rethink GIScience in terms of education and training (Frazier et al., 2025), because GeoAI is not just a tool, but a potential revolution in geospatial data analysis. It is rapidly becoming a powerful disruptive force, able to provide unprecedented quantitative and qualitative insights into complex environmental and societal challenges. From mapping and modeling land cover changes to mapping flood-risk areas, GeoAI is set to transform geospatial decision-making, harnessing the power of ML and DL for more effective solutions.

## WHAT IS INSIDE GeoAI MODELS

GeoAI models can analyse more efficiently massive geospatial datasets, allowing experts to automate complex tasks, such as feature engineering, improve predictions, and uncover patterns and trends that would be difficult to detect by other means. GeoAI pattern discovery is also valuable for solving real-world problems such as urban planning, forest conservation, disaster risk management, and climate change adaptation; for example, it provides innovative solutions for the preservation of land and forests detecting fires (Purbahapsari & Batoarung, 2022).

EO with the integration of ML/DL techniques are drivers of innovation in Earth System Observation and Prediction (ESOP) for weather and climate, enhancing simulation and prediction capabilities of the Earth's complex dynamics.

However, despite these benefits, GeoAI has its challenges given strong uncertainty and the need to intervene to train models, often transferring bias, and generating or replicating ethical issues.

The complexity of ML and DL models often results in a lack of transparency due to their “black box” nature: you use it, but can't access how it works. It is a condition that makes it difficult for the users to trust the results, as there is no way to check and control the whole workflow. Another issue comes from data quality, which can significantly impact the performance of GeoAI models, leading to potential biases, or inaccuracies in predictions. Furthermore, spatial autocorrelation and spatial heterogeneity are not readily incorporated into GeoAI, limiting its ability to fully capture the underlying spatial dynamics of geospatial data. It is still up to the human experts to interpret these results requiring expensive and specialized knowledge, which may limit the accessibility to GeoAI services for broader audiences.

Such limitations in current GeoAI research are anyway an advantage in terms of needs of innovation; new methods and qualitative solutions are required, engaging theories in Geography, enhancing model explainability, quantifying uncertainty, and improving model generalizability (Hu et al., 2024). When looking at the whole workflow starting from data collection, a main issue is about the need for correctness

(Hochmair et al., 2024). Other limitations come from GeoAI models that are largely inflexible and still sensitive to human made labelling, clustering and classification, although automated data labelling is yet to be a fruitful field of research (Albrecht et al., 2021). Generalization is another challenge, as it is difficult for a model to generalize from one source of images to another, especially when they have different spectral features. Given the adoption of multi-source methods, heterogeneity in image quality from different data sources brings in further issues. To end the list of the many issues at hand, current GeoAI models lack higher level reasoning to help the experts interpret the results (Hu et al., 2024).

GeoAI derives from Geography as emerging discipline, so the question we want to answer now is: how does AI models affect and influence Geospatial Science and what are the consequences for Geography?

According to a recent review, applications can be grouped in four groups: mobility, natural resource and disaster management, cartography and mapping, and geospatial data handling and curation (Hochmair et al. 2025). It is also important to consider the peculiar methods of GeoAI, because AI techniques and methods include expert and fuzzy systems, neural networks, and evolutional computation. GeoAI applies data mining, machine learning, and other traditional AI methods that have been around for many years. The integration and development of AI and ML has become crucial in analysing big geospatial data. It is important to understand what are the challenges and advances in spatial and spatio-temporal ML methods and the software and infrastructures to support them.

Pre-trained foundation models (PFMs) used nowadays in big data analysis are successful in fields such as Natural Language Processing and Computer Vision; because they are typically trained in a task-agnostic manner, their ability to handle geospatial data and related tasks is still limited (Huang et al., 2023) as PFMs lack the inherent capability to fully comprehend and address geospatial challenges (Wang et al., 2024). Geospatial tasks often require substantial prior knowledge, making the integration of GIS theories and spatial knowledge into the hierarchical features of deep learning a key challenge for the future development of GeoAI. A research program to bypass the limitations of PFMs has been the development of geo-foundation models (GeoFMs), that must be specifically tailored to geoscience questions (Hochmair et al., 2025), broadening the variety of data and processes used, but becoming more flexible only by human labelling and model explainability. Other issues may come from the scarce availability of data, and lack of useful assets in terms of frequency of data and events available to train models. As models learn from the frequency of data, the intensivity of data about peculiar geographic areas like main cities might generate inequalities and discriminacy considering that critical phenomena can happen less frequently, like pandemics in rural areas for instance. Another solution might consist in adopting similarity models, which can be trained on the uniqueness of sample data (Zhu & Turner, 2022), but might hide bias and ethical issues as well because there is no way to make comparisons or work at a significant statistical stage. There are also positive insights represented by Deep Machine Learning techniques, such as the Convolutional Neural Network (CNN), that have become of interest in AI research and contributed to the growth of the GeoAI community (Li & Hsu, 2022), making methods scalable in processing geospatial big data (Hochmair et al., 2025). To provide an example here, Deep Machine Learning techniques are applied to detect traffic anomaly, image fusion, and semantic segmentation.

Deep Learning is transforming the data analytics paradigm in two ways. Some deep learning models, like CNNs, can extract automatically prominent features from the data to help differentiate object classes so that accurate predictions can be made; other deep learning models are able to run local convolution operations into the learning process. Technically speaking, such models are able to break down the numerous interdependencies residing in the global computation of traditional neural network models.

A proper attention must be addressed to how ML and AI are applied to spatial and spatio-temporal data along the entire modelling pipeline (Arundel et al., 2023; Mai et al., 2025) in order to guarantee the reliable accuracy and uncertainty assessment, model interpretation, and transferability (Prasad et al., 2024). In fact, bias and unfairness can be part of the whole machine learning cycle (Tursunalieva et al., 2024) bringing in further reworking with fairness metrics, featured in tools, datasets, and the need for bias mitigation techniques (Pagano et al., 2023). Possible organizational principles to create a taxonomy can be, for instance, considering issues caused by data generation, data collection, or institutional bias (Ntoutsi et al., 2020), or group types of bias based on how they interact with the data, the algorithm, or the user (Mehrabi et al., 2021), or even sources of harm in the forms of bias as negative consequences at the different stages in the ML life cycle for organizational structure, from data collection to deployment (Suresh & Guttag, 2021).

As previously mentioned, a multi-source methodology is growing interest in the community, combining different datasets from different sources and technologies. This represents on one side a systematic challenge, while on the other enhances the risks and reduces the transparency of the whole workflow (Bordogna & Fugazza, 2023). A multi source methodology can include a human centric approach with humans-as-sensors (Goodchild, 2007), and GeoAI mixed-expertise with human collaboration is obviously at higher risk for the possible bias and ethical issues about such methods (Ye et al., 2025).

## ETHICAL ISSUES AND THE GEOAI SCIENTIFIC COMMUNITY

The academic literature on responsible AI often emphasizes general principles, with comparatively limited focus on the geospatial domain. Notably, there is a lack of a comprehensive, transnational framework addressing the intersection of geospatial data, ethics, and governance. Scholars have called for frameworks that foster ethical awareness, education, and global best practices to guide the responsible adoption of AI in spatial contexts (Kitchin, 2016; Goodchild, 2020; UN-GGIM, 2020).

Such approach should include at least four goals to maximise the consideration of ethical issues: to scrutinize the relationship between geospatial data and ethics, to foster ethical awareness and discussion, to educate and inform, and to provide guidance and best practices at global level (Sahithya et al., 2024). It is worth remarking that the integration of AI and Geography, with the availability of machine readable data, rework the concept of the geographer himself and its role in the discipline. Another crucial step should be about the interpretability of models adopted by GeoAI for AI-informed decisions, as, at the end, automation and interpretation of data affect the human role of decision maker, bringing in several ethical issues in the geospatial domain itself and its possible advancements (Janowicz, et al., 2020).

The fast evolution of the field and the several applications work against an ultimate definition of GEOAI, that should also include possible misuse and the risks. A possible solution could be the development of transparent and unbiased "robust GeoAI models" (Li & Arundel, 2022), models that could be considered critical for the community and the public at large. Such models should be extensive in putting at safe values and privacy for large communities and the heterogeneity of identities, while should be transparent through the whole framework. When considering an ethical relativist approach, different societies have their own customs that we must acknowledge with the rules of conduct based on acceptable and unacceptable behaviour (Siau & Wang, 2020) to identify the ethically correct actions (Anderson & Anderson, 2011) for the development of robust GeoAI models.

According to Kang, Gao and Roth (2024), five potential ethical challenges need to be addressed in the integration of GeoAI for Cartography (commodification, responsibility, privacy, bias, and then this three together, transparency, explainability, and provenance), naming "Ethics of GeoAI for cartography" a set of possible rules of conduct for researchers, professionals, and policymakers regarding acceptable and unacceptable application of GeoAI methods for cartographic design. This would become possible with the "careful examination of the responsible dimensions" (Ghamisi, et al., 2024) into the process of integration between EO and AI. Such dimensions have to do as much with methods as data. For instance, geographical data analysed by ethically unsupervised models can generate technical or management risks of unauthorized tracking of individual movements and potentials for misuse of location-based information. As stated in the "Geoprivacy Manifesto" "ubiquitous positioning devices and localizing APIs make information about an individual's location much easier to capture than other kinds of personally identifiable information" (Keßler & McKenzie, 2018). Multi-source models put at higher risk privacy as they are the reliance on human location trajectories and the geographically tagged social media data (Dilmaghani et al., 2019), with the result of several behavioral data science privacy issues (Saura et al., 2022). It is important to remark that datasets from Geospatial information are tied to further spatial contexts that profile people; moreover, the combination of geospatial big data with anthropometric attributes like facial and gait features bring further concerns on privacy and security risks associated with GeoAI; it becomes easier to know where you are and who you are. The scientific community is aware of the need to adopt countermeasures like geomasking techniques (Lorestani et al., 2024), and advocate for the development of privacy-preserving and secure GeoAI foundation models (Rao et al., 2023) to ensure much responsible and equitable adoption and use of spatially explicit models (Janowicz et al., 2019), while further work is expected to develop shared ethical geospatial practices (Sahithya, et al., 2024). The awaited EU AI Act should define the ground about clear legal requirements and make bias mitigation obligatory in the field (Matuszczyk et al., 2025) at least in Europe.

## SOLUTIONS FOR A RESPONSIBLE GeoAI

A responsible GeoAI approach must entitle stakeholding communities to be capable to understand, design and develop spatial methods, processes, algorithms, and systems to discover spatial patterns that reduce adverse impacts and risks on the communities themselves (Gupta et al., 2023). A comprehensive and actionable framework is needed towards a much responsible, sustainable, open and optimised AI adoption in the geospatial domain (Marasinghe et al., 2024).

In order to tailor GeoAI according to the specific needs, priorities, and contexts, a participatory approach of diverse stakeholders should include local communities and domain specialists to co-design solutions and operate with high-quality and context appropriate data to increase AI explainability and AI interpretability (Mai et al., 2025). Data shouldn't compromise the performance and reliability of GeoAI systems, but respect strict protocols for the acquisition, types, quality, and management of data (Salman et al., 2024) also to overcome "the problem of many hands" (Timmermans et al., 2010) of responsibility in technology.

There are new methods, techniques, and tools to provide privacy-preserving techniques for the so-called Location-Based Services (LBSs) (Jiang et al., 2021). These make possible obfuscating, anonymizing, encrypting, and protecting location information by differential privacy techniques and offloading modes (Zhang et al., 2024). Data privacy techniques concern either protection mechanisms, or privacy attacks (McKenzie et al., 2023): protection mechanisms deal about privacy-preservation, anonymization, and obfuscation, while privacy attacks concern re-identification, de-anonymization, and privacy exposure. Approaches about obfuscation & anonymization include the addition of noise either randomly or following some structured probability distributions in order to "geomask" (Arbia et al., 2022), the location data and metadata; for instance, geomasking has made possible during the Covid-19 outbreak to "mask" the real address of positive cases while preserving the local spatial cluster patterns (Houfaf-Khoufaf et al., 2021) and had only been effective because of the adoption of an open data policy at global level.

Anonymization is another interesting solution as it preserves individual privacy, while actions and locations can be still tracked. Spatial k-anonymity can ensure an individual level of anonymity by reporting locations as datasets of "k" potential locations instead of the actual one; this means that larger values of "k" increase spatial anonymity, but in the meanwhile they decrease the utility of the location information (McKenzie & Zhang, 2023). Technically, trajectories can be easily detected and there are possible solutions to provide differential masking and synthetic data generation to protect users' privacy (Miranda-Pascual et al., 2023).

A growing market is the one represented by synthetic data; their generation and use is an alternative to obfuscating or anonymizing real location data. Synthetic data don't provide the real data at all, but generate data sets with similar properties of the original ones, according to parts or the whole information needed. ML and other data synthesis methods can be used to generate synthetic data (Nikolenko, 2021); for instance, synthetic data can make possible to understand crowd dynamics pre-training a crowd counter on the synthetic data, then fine tuning it using the real data and choosing a crowd counting method (Wang et al., 2019).

Cryptographic techniques can also be adopted to hide the data instead of distorting or creating dummy datasets, as description keys are generated and shared only with the direct stakeholders, so that part of the utility data stay public, but are of no interest as the spatial coordinates are encrypted (Sun et al., 2019). An encryption method called homomorphic encryption is used especially in Health research and Demography (Munjal & Bhatia, 2022) and allows the analysis of encrypted data and the extraction of patterns without the need to decrypt it. Such analysis can result in the extraction of patterns and insights without having access to the original unencrypted private information. On the other side of Data privacy techniques, there are methods about re-identification and privacy attacks to seize protected or scrambled location data. Such methods use external datasets, IP, geosocial tagging, and co-location techniques, while on the side of statistical techniques, inference and probabilistic modelling make possible by emerging GeoAI techniques to re-identify records in datasets, putting privacy and other ethical issues at risk. All above are the main solutions that can provide a reduction of the accuracy of information and, therefore, guarantee tailored degrees of geospatial data anonymity. It is important to remark that a pivotal challenge will consist in the development of performative GeoAI with effective privacy-preserving techniques while maintaining the quality of geospatial predictions (Mai et al., 2025).

## A FOCUS ON PRIVACY IN GeoAI

Location-based services and application areas of GeoAI can affect privacy of individuals and groups in diverse ways. Insights on how technologies can harm people privacy in terms of fraudulent access and use of personal and sensible data can be classified according to: advertising, health, security & surveillance (McKenzie et al., 2023). GeoAI services might rely also on personal or sensitive spatial data. Datasets including individual mobility patterns, residential locations, geo-tagged images, and land use, can directly or indirectly expose personally identifiable information. When AI is integrated with geospatial data, there is a high risk of privacy breaches, particularly when datasets are combined, re-identified, or released for public use (Mai et al., 2025). In the case of privacy risks from multi-modal analysis the technology makes available are information from tracking activities of users on social media that must be included in order to track disasters and provide first responding actions (Zhou et al., 2022; Hanny & Resch, 2025). Specht (2020), in a work wisely written during the pandemic, considered what mapping and tracking has to do with crisis situations: the peculiar and risky use of data happen when people are more vulnerable.

The debate is not easy when it comes to Public Health and Privacy. We should take in account that, in the case of tracking diseases like during Covid-19, the not attendancy of enforcing privacy and overlooking privacy issues have contained the infection and reduced the contagion risks (Afroogh et al., 2022). Furthermore, there is a problem of digital divide about possible re-use of personal data, given the role played by tech companies in the public health domain and their power dynamics with the government and public research sector. The result of digital epidemiology tools that undermine equity, fairness, public trust, just distribution of benefits, autonomy, and minimization of group harm is a concrete risk (Ferretti & Vayena, 2022).

Computational uses of personal data and spatial services need further insight and solutions in every field. Possible issues about privacy identify some of the leading concerns that should be addressed by communities and by the governments in collaboration with the industry. McKenzie, Zhang, and Gambs (2023) have outlined three ways in which privacy within GeoAI can be improved: privacy by design, spatial privacy, and enhancing regulations. At the same time, three emerging privacy topics could play further particular interest to the GeoAI community: fake geospatial data. publicly accessible and integrated tools, policy development.

## GLOBAL STAKEHOLDERS IN GEOAI

Several governments, organizations, research centers and companies are involved in GEOAI technology development and different players tackle ethical issues in different ways on different levels. For instance, SpaceX and Starlink include privacy and the implications of surveillance, while Google Earth Engine advocates for the responsible use of satellite data protecting private lands and areas of conflicts. Others, like Esri, a leading provider of GIS (Geographic Information Systems) software, supports ethical decision-making by developing tools that emphasize transparency, inclusivity, and accountability, like Planet Labs, a global enterprise that emphasizes also transparency in data sharing and collaborates with local governments to provide satellite imagery for humanitarian causes.

Space agencies have a different approach. The European Space Agency (ESA) prioritizes environmental protection, sustainability, and international collaboration to ensure that data is used for public good, minimizing risks related to surveillance and exploitation. National Aeronautics and Space Administration (NASA) emphasizes transparency, collaboration with international agencies, and the responsible use of its data. They also focus on ensuring that their technologies benefit underserved communities and aid in global challenges like climate change. At European level, implementing approaches provided by the General Data Protection Regulation (GDPR) are intended to promote a mixture of ethical, legal and technical principles: lawfulness, fairness and transparency; purpose limitation; data minimization; accuracy; storage limitation; integrity and confidentiality; accountability.

An international organization that represents the Neogeography Community and is aware of the impact of new technologies related to spatial studies and applications is ISDE, International Society of Digital Earth based on the concept to construct a 3D digital representation of Earth. Such a project aims to develop a comprehensive monitoring, simulation, and prediction system to analyse and better comprehend the intricate interactions between natural phenomena and human activities.

The concept of Digital Twins (DTs) of the Earth has emerged as a revolutionary approach to address climate resilience, disaster risk management, and sustainable development through highly detailed, high fidelity digital replicas of the Earth system (Hazeleger et al., 2024).

The Digital Earth Initiative (DEI) is a long term global effort to create comprehensive, interactive digital representations of the Earth, integrating satellite data,

GIS, and other Earth observation tools to provide a more detailed, real-time view of the planet. This initiative was first proposed by Goddard Space Flight Center from NASA, and has since been taken up by various international space agencies, including ESA and China's National Space Administration (CNSA). About privacy and surveillance issues there is an approach that advocates for international privacy standards for satellite data, making geospatial data freely available, but respecting politically sensitive regions, identities and freedom. Also data accessibility and equity is an issue, while the goal is to democratize access to geospatial data by open-source platforms, like the Global Earth Observation System of Systems (GEOSS), making satellite data available to researchers, governments, and the public. The initiative promotes inclusive data collection and emphasizes the use of diverse balanced datasets that accurately represent different geographic areas and communities, taking account of socio-economic factors, environmental justice, and indigenous knowledge.

Such scenarios should feature how technologies might concern society and how society with services, policies and issues does drive such technologies as well. The evolution of visions for geospatial technology can help to identify and simulate possible ethical issues and emphasize the necessity of a global framework for geospatial infrastructure to address challenges such as Big Data, Machine Learning and Artificial Intelligence. Thinking of converging to a unique infrastructure and standard is unthinkable at the moment, while risks and consequences of the present and future ethical issues for location services boosted by AI -a emerging and edge cutting technology- must become part of a common and global strategy for the sake of democracy and the global stability. The possible applications and digital services involve various disciplines in reflecting on the opportunities and challenges of governance, as GeoAI spans across multiple aspects and dimensions of control, coordination, incentives and trust. Proper stress should be put on GeoAI by diverse ranges of perspectives and approaches with a “good governance” approach (van Dijck et al., 2025). Given that policies and rules hardly follow the pace of innovation of technology processes and achievements, ethical issues become often and clearly the consequences of early technology adoption, rather than part of the design for future applications. There is a growing awareness on how much AI might contribute to shape power structures in the digital Age and therefore the need for a “Power- Aware Ethics of AI” for collective and political action (Mühlhoff, 2025). Among several aspects, the driving technological visions should be accounted according to sociotechnical perspective with an ethical issue approach on, for instance, social bias imported in AI (Marres et al., 2024), the matter of sovereignty (Mügge, 2024), the risks in privatization of research (Jurowetzki, 2021), as well as the need to frame ethical issues in guidelines (Jobin, et al., 2019). A clear example on how GeoAI technology should be developed according to a framework that includes the assessment and action for ethical issues and potential risks is provided by a UNICEF report about Geospatial technologies for development and humanitarian organizations (Berman et al., 2018). Such guidelines include the assessment about the benefits, the potential risks in privacy, security and surveillance; moreover, advice is given about ethical considerations coming from the use of geospatial technologies for evidence generation, how to pursuit assessing and managing of the risks, and the engagement of communities in risk assessment and sharing of findings.

Moreover, such guidelines need to include aspects about the legal mechanisms to be developed to protect the public from misuse of their spatial data. Such mechanisms should be in place whenever geo-information is collected (Oluoch, 2024) bringing to the importance of designing according to certain principles.

## OPENING SPATIAL DATA

In the above sections it has been possible to remark how main concerns come from data collection and use, although the harvesting of data and ways to classify data might produce biases as well. Open data policies have brought the advantage to make data available, transparent the process and the data, but also standardize globally methods and techniques to handle and manage datasets.

What are the possible advantages of an open data policy for GeoAI?

Adopting open data for GeoAI could probably increase access and transparency, but would also facilitate insights for fruitful discussion on ethical concerns and decision making about legal use and privacy protection of users' data; moreover, an open data policy would leave up to governments, international institutions, and communities of practice the control and enforcement of rules. The pandemic experience have taught decision-makers and scientists about the advantages of open archiving for spatial data (Wahltinez et al., 2022), and in that case also AI played a relevant part in innovating the strategies to tackle the spread of the virus (Raza, 2020) with machine learning, as deep learning, and data mining to extract meaningful information from spatial big data (VoPham et al., 2018).

Almost a decade ago, the FAIR (Findable, Accessable, Interoperable, Reusable) data guiding principles were introduced to the broader research community (Wilkinson et al., 2016). These principles have proposed a framework to increase the reusability of data, although in subdomains of the Earth System Sciences (ESS), data reuse across institutions and geographical borders is a consolidated practice supported by community-specific and cross-domain standards like netCDF-CF, and OGC. Expanding such practices to a "FAIR by design" principle (Kaas et al., 2024) to enhance transparency, community driven data policies and tackle ethical issues is therefore paramount. Even following the principles of Data protection by design and by default policy adopted by the European Union (Jasmontaite et al., 2018), empowering citizens is not easy in absence of information and knowledge on how to manage the permission about their data, or to enforce control on the whole data workflow. A open source solution for a "made in EU AI" could certainly facilitate the development of the domain according to the ethical principles in Europe (Floridi et al., 2025).

Open data might play a crucial role in developing and deploying GeoAI by providing a foundation for both innovation and responsible, inclusive technology development. GeoAI relies on large, diverse datasets to train algorithms, create models, and solve complex geospatial problems; having open access to relevant, high-quality data offers significant advantages, given the incorporation of the FAIR principles into the entire lifecycle of geospatial data processing and analysis (Hu et al., 2025), and diverse, interoperable ecosystems, which combine open-source assets for geospatial models, technology and interdisciplinary collaboration (Huang et al., 2025).

Big Data, along with new ways to build epistemologies, are affecting how knowledge is produced (Kitchin, 2014) with the need of a critical insight towards analysing methods and long term effects of how data are produced, archived, harvested, used and reused.

However, fully implementing FAIR principles, particularly machine-actionability—the core idea behind FAIR—has proven challenging. Despite current progresses in awareness, standard-compliant data sharing, and the automation of data provenance, the ESS community continues to struggle to reach a community-wide consensus on the design, adoption, interpretation and implementation of the FAIR principles in Earth and environmental Science (Crystal-Ornelas et al., 2022; Qiu et al., 2024).

An intergovernmental approach should be paramount to provide key rules and to frame at global level the impact of ethical issues. Creating alliances and opportunities for strategic collaboration can boost the potential of AI for inclusive development as well. Open government datasets might have a key role in the outcome for a transnational open data policy, although flexibility and research on data features need further improvements.

Given lower barriers to entry, open data can accelerate innovation and development by fostering competition on large pool of data; by allowing developers, researchers, and startups to access large, diverse datasets might have the advantage to provide faster prototyping, testing, and refining of AI models, being otherwise expensive or difficult to obtain with. Such an approach should enhance data diversity and representativeness with broader geographic coverage ensuring that GeoAI models are trained on a diverse set of locations and scenarios. This can help the avoidance of geospatial biases, ensuring that AI models are effective in a wide range of geographical and environmental contexts.

Degbelo and Kray (2018) have outlined in a vision paper their proposal to increase the intelligence of geovisualisations for a better use of Open government datasets: automatic geographic data type recognition, generation of geovisualisation designs, monitoring of users' understanding of geographic facts, self-optimization, and user activity recognition.

## A SOCIOTECHNICAL APPROACH TO GeoAI

AI represents a challenge both for Society and Ethics at large (Whittlestone & Clarke, 2022), and it's not only a technical, or economical matter. Also for GeoAI we need a peculiar point of view, an approach able to consider social and technical aspects at once offers instead a theoretical lens to understand GIS and geospatial systems as sociotechnical constructs (Crampton, 2010). In terms of technical components we can count remote sensing platforms and satellites, machine learning algorithms for spatial classification and prediction, GIS, and data infrastructures (APIs, cloud platforms, spatial databases), while as social or human components we can list decision-makers, urban planners, governments using GeoAI tools, ethical and political debates (e.g. surveillance, bias in spatial predictions), regulatory frameworks and governance institutions, and the following risks of local communities affected by top down geospatial AI decisions about urban planning, disaster response, policing. Such an approach aims to search for accountability of any technological choice, adoption and use, according to perspectives that include historical, cultural, social and political factors. It is useful to consider a whole technological frame and its applications as socially constructed (Bijker, 2009), and co-produced (Jasanoff, 2004) in social practices, identities, norms, conventions, discourses, instruments and institutions as much as any cultural matter. For instance, according to Cole & Bertenthal (2017) GeoAI and Law co-emerge in shaping ethical order and spatial representation technologies, while digital and geospatial technologies can be understood as part of larger sociotechnical systems shaping both space and society (Ash et al., 2018).

Understanding how science and technology contribute to emerging and established social and political order makes much visible the problem of values inside the emerging field of GeoAI and how different stakeholders negotiate and build their alliances inside a growing sociotechnical network of Big Data, AI methods, services, *et cetera*.. For instance, the digital divide is one of the possible consequences of Big Data (boyd and Crawford, 2012) considering the interest of multinational companies in collecting large amount of data about people and phenomena and relative access to those, while each step -from data collection to deployment- requires human intervention, which consequence is to consider the whole GeoAI framework as an example of sociotechnical assemblages (Zheng & Sieber, 2021).

An inclusion of ethical design and practices in GeoAI might also be heightened by public awareness regarding the various implications of the adoption of AI in EO and further applications related to georeferenced data and spatial tagging. On the side of public understanding and engagement of science it is surely hard to inform citizens about what happened "above their heads" and how data are being produced, held and used, given the expertise needed to understand GeoAI.

## SOCIOTECHNICAL IMAGINARIES

The sociotechnical imaginary framework (Jasanoff & Kim, 2015) has been widely adopted to explain several phenomena as sociotechnical processes (Hendriks et al., 2025) about preferred futures in a techno-optimistic frame; techno-optimism is about visions of reliable technology offering solutions to complex social problems (Morozov, 2013).

On this basis, we can understand how, both the optimistic imaginaries of Geography and AI, can work together in featuring a GeoAI narrative. Such optimistic solution is however acritical and considers only the rhetoric and imagined side about spatial analysis and innovative technology up to the solutions it might provide at local and global level, as result of the two imaginaries cooperating in terms of trustworthiness, adoption and integration in such a single technological narrative shaping both society and technology.

The concept of imaginary is part of the social theory referring to collective beliefs, and "social imaginaries" (Castoriadis, 1987) have been a promising research field to understand how in western societies narrations about progress and technosciences come to be accepted and become part of public opinion, economy and self-governance (Taylor, 2004). In the case of digital technologies, future imaginaries are strong parts of their narrations (Mager & Katzenbach, 2021), and negotiation among key AI stakeholders, given the complexity and heterogeneity of the phenomenon, constantly co-constructs and defines certain visions in search for a stability and grounding in communication (Richter et al., 2025), and governance of AI (Lazaro & Rizzi, 2022).

The main definition about sociotechnical imaginaries is as follows: 'collectively held, institutionally stabilized, and publicly performed visions of desirable futures, animated by shared understandings of forms of social life and social order attainable through, and supportive of, advances in science and technology' (Jasanoff & Kim, 2015, p. 25). Such imaginaries drive and materialise big futures from daily little futures (Michael, 2017) influencing understanding and engagement in emerging technologies by the encompassing promise of technology (Borup et al., 2006). Sociotechnical imaginaries work especially as forms of collective identities as nations themselves are reimagined and shaped as "collectively imagined forms of social life and social order reflected in the design and fulfilment of nation-specific scientific and/or technological projects" (Jasanoff and Kim, 2009, p. 120). A sociotechnical perspective on imaginaries makes it possible to understand better how engineering and design can cause ethical risks and social issues, like widening inequalities, or exploiting community and individual data and resources. It ought to include ethical aspects to analyse present and further technological shifts by sociotechnical imaginaries of GeoAI, which are not clearly yet framed and narrated in media and public opinion. In the case of AI, narratives and imaginaries of national AI strategies are enacted in their performative politics (Bareis & Katzenbach, 2021), as an AI national approach seems easier given the wide range connected to economical and ethical laws of single nations (Paltieli, 2022). However, a unidimensional approach to GeoAI should not be confused with the need of transnational standards and efforts to regulate the market and services. A further step to complexity and multidimensionality must be taken to an extensive, global and spatial level away from local and temporal needs. The risk for a global technology such as GeoAI is also to fulfill all ethical requirements in one context, while misusing data and harming people in another. As for AI, a global and articulated approach able to respond to local conditions and heterogeneous situations is needed.

A call to policy making and institutional engagement at international level is mandatory by creating sound and targeted communication policies about GeoAI. As in the case of sociotechnical imaginaries of artificial intelligence, public communication should be in charge of providing information and engage citizens about GeoAI, offering a multidimensional and critical analytical framework (Brause et al., 2025). According to Lucy Suchman (2023), the "thingification" of AI discourse as a practice to "stabilise" such technology in a single phenomenon poses a risk in not understanding AI as complex, heterogeneous and localised forms of innovation that would make possible the individualization of risk. The proposal from Suchman is therefore to "treat the existence of AI itself as controversial" and adopt a critical approach to AI (Raley & Rhee, 2023) in order to prioritize critical engagement in specific contexts of use for AI. It is important to consider the human labor, data and planetary resources behind the discursive success of AI in terms of power (Crawford, 2021). There are different ways in which the technology is able to influence the public trust and produce an uneven distribution of AI risks and benefits. In our case, GeoAI qualifies as a high-risk system (Matuszczyk et al., 2025) in a society of risk where the global nature of AI risks across multiple sectors pose threats to society itself (Vulpe et al., 2024).

This should prioritize an operational roadmap that includes ethical and social issues in a long term vision shared at global level. The growth of such model should however preserve a local attitude starting from small realities to work empirically on future scenarios in which resource-intensive, large geospatial models might homogenize and test complex landscapes, in order to suggest strategies and solutions (Böhlen et al., 2024).

## CONCLUSIONS

An open governance strategy approach that considers a range of human related factors is essential for the successful integration of GeoAI with an ethical attitude at global level and the pursuit of further sustainable and transparent research in the field. A successful and effective application of GeoAI requires not only technical expertise, but also a deep understanding of the ethical, methodological, and data-related challenges that shape its outcomes. Beyond algorithmic design, practitioners must engage with issues such as misinformation, bias, privacy, transparency, accountability, and the quality and availability of spatial data. Additional complexities stem from model opacity, knowledge gaps, and technical limitations that may hinder performance and trust. Addressing these concerns through thoughtful strategies is essential to fully realize the benefits of GeoAI while mitigating its potential risks and harms.

GeoAI technology is still in an experimental phase with possible malleability of narratives towards a clear and sustainable sociotechnical imaginary. Such imaginary could advocate for a sustainable and ethical development of GeoAI future, featuring an international governance of digital commons being shaped according to an ethical framework by governments and international organizations.

Despite the clear advantages of GeoAI in GIScience, several challenges remain. These include issues related to the geographic nature of the underlying data and processes—such as geographic scale, biases, as well as broader scientific concerns like the lack of explainability in AI models, reproducibility and replicability in advancing open science and the ethical and privacy considerations that arise in this new era of AI-driven research. The ethical management of geographic data, the trustworthiness of results and the emphasis on addressing accuracy issues in variables and models, data accessibility and ethical concerns, all might also play a pivotal role in the broader discourse on AI ethics (Bullock et al., 2023) as well.

Many technical challenges, such as the lack of good quality training data, uncertainty modelling, geographical transferability of a model result, and cross-scale and cross-resolution learning have already been addressed (Hu et al., 2019), but not to overcome further ethical issues, GeoAI community should also evaluate sustainability concerns according to the unique characteristics and challenges of GeoAI models (Shi et al., 2025).

GeoAi should yet provide further insights, case studies, and innovative approaches for the advancement and the adoption of the FAIR data principles and open data. Evidence about a full development of a global GeoAI project can only be found in the ISDE community with a clear vision about ethical concerns and sociotechnical bias. It is important to support the development of a human centered GeoAI (Mai et al., 2025) to frame ethical issues by a sociotechnical approach and an open data policy among the possible "imaginaries of artificial intelligence" (Richter et al., 2023).

## REFERENCES


Adhikari, S. (2024). *GeoAI for Natural Disaster Assessment.* In *Proceedings of the 33rd ACM International Conference on Information and Knowledge Management (CIKM '24).* Association for Computing Machinery, New York, NY, USA, pp. 5431–5434. DOI: 10.1145/3627673.3680257

Afroogh, S., Esmalian, A., Mostafavi, A., Akbari, A., Rasoulkhani, K., Esmaeili, S., & Hajiramezanali, E. (2022). Tracing app technology: An ethical review in the COVID-19 era and directions for post-COVID-19. *Ethics and Information Technology, 24*(3), 30. Advance online publication. DOI: 10.1007/s10676-022-09659-6

Anderson, M., & Anderson, S. L. (Eds.). (2011). *Machine Ethics.* Cambridge University Press. DOI: 10.1017/CBO9780511978036

Arbia, G., Ghiringhelli, C., & Nardelli, V. (2022). Effects of Confidentiality-Preserving Geo-Masking on the Estimation of Semivariogram and of the Kriging Variance. *Geographical Analysis, 55*(3), 466–481. DOI: 10.1111/gean.12344

Arundel, S. T., McKeehan, K. G., Li, W., & Gu, Z. (2023). GeoAI for Spatial Image Processing. *Handbook of Geospatial Artificial Intelligence*, 75–98. DOI: 10.1201/9781003308423-5

Ash, J., Kitchin, R., & Leszczynski, A. (2018). Digital turn, digital geographies? *Progress in Human Geography, 42*(1), 25–43. DOI: 10.1177/0309132516664800

Bareis, J., & Katzenbach, C. (2021). Talking AI into being: The narratives and imaginaries of national AI strategies and their performative politics. *Science, Technology & Human Values, 47*(5), 855–881. DOI: 10.1177/01622439211030007

Berman, G., De, S., Rosa, L., & Accone, T. (2018). *Ethical Considerations When Using Geospatial Technologies for Evidence Generation Ethical Considerations When Using Geospatial Technologies for Evidence Generation.* https://www.humanitarianlibrary.org/sites/default/files/2022/03/ethical.pdf

Bijker, W. E. (2009). Social construction of technology. In Friis, J. K. B. O., Pedersen, S. A., & Hendricks, V. F. (Eds.), *A Companion to the Philosophy of Technology* (pp. 88–94). Wiley. DOI: 10.1002/9781444310795.ch15

Böhlen, M., Sughiarta, G., Kurnianingsih, A., Reddy, G. S., Shrivastava, S., & Gorla,. (2024). GeoAI in resource-constrained environments. *ArXiv (Cornell University).* DOI: 10.48550/arxiv.2408.17361

Bordogna, G., & Fugazza, C. (2023). Artificial Intelligence for Multisource Geospatial Information. *ISPRS International Journal of Geo-Information, 12*(1), 10. DOI: 10.3390/ijgi12010010

Borup, M., Brown, N., Konrad, K., & Van Lente, H. (2006). The sociology of expectations in science and technology. *Technology Analysis and Strategic Management, 18*(3-4), 285–298. DOI: 10.1080/09537320600777002

Boutayeb, A., Lahsen-Cherif, I., & Khadimi, A. E. (2025). When Machine Learning Meets Geospatial Data: A Comprehensive GeoAI Review. *IEEE Journal of Selected Topics in Applied Earth Observations and Remote Sensing, 18*, 13135–13191. DOI: 10.1109/JSTARS.2025.3568715

Brause, S. R., Schäfer, M. S., Katzenbach, C., Mao, Y., Richter, V., & Zeng, J. (2025). Sociotechnical imaginaries and public communication: Analytical framework and empirical illustration using the case of artificial intelligence. *Convergence (London), 31*(4), 1267–1287. Advance online publication. DOI: 10.1177/13548565251338192

Breunig, M., Bradley, P. E., Jahn, M., Kuper, P., Mazroob, N., Rösch, N., Al-Doori, M., Stefanakis, E., & Jadidi, M. (2020). Geospatial Data Management Research: Progress and Future Directions. *ISPRS International Journal of Geo-Information, 9*(2), 95. DOI: 10.3390/ijgi9020095

Bullock, J. B., Hudson, V. M., Korinek, A., Young, M. M., & Dean, M. (2023). *The Oxford Handbook of AI Governance*. Oxford University Press.

Castoriadis, C. (1987). *The Imaginary Institution of Society*. Policy Press.

Choi, Y. (2023). GeoAI: Integration of Artificial Intelligence, Machine Learning, and Deep Learning with GIS. *Applied Sciences (Basel, Switzerland), 13*(6), 3895. DOI: 10.3390/app13063895

Coeckelbergh, M. (2020). *AI Ethics*. The MIT Press.

Cole, S. A., & Bertenthal, A. (2017). Science, Technology, Society, and Law. *Annual Review of Law and Social Science, 13*(1), 351–371. DOI: 10.1146/annurev-lawsocsci-110316-113550

Couclelis, H. (1998). Geocomputation and Space. *Environment and Planning. B, Planning & Design, 25*(7), 41–47. DOI: 10.1177/239980839802500708

Crampton, J. (1995). The Ethics of GIS. *Cartography and Geographic Information Systems, 22*(1), 84–89. DOI: 10.1559/152304095782540546

Crampton, J. (2010). *Mapping: A Critical Introduction to Cartography and GIS*. John Wiley and Sons., DOI: 10.1002/9781444317411

Crawford, K. (2021). *ATLAS OF AI: power, politics, and the planetary costs of artificial intelligence*. Yale University Press.

Crystal-Ornelas, R., Varadharajan, C., O'Ryan, D., Beilsmith, K., Bond-Lamberty, B., Boye, K., Burrus, M., Cholia, S., Christianson, D. S., Crow, M., Damerow, J., Ely, K. S., Goldman, A. E., Heinz, S. L., Hendrix, V. C., Kakalia, Z., Mathes, K., O'Brien, F., Pennington, S. C., & Agarwal, D. A. (2022). Enabling FAIR data in Earth and environmental science with community-centric (meta)data reporting formats. *Scientific Data*, *9*(1), 700. DOI: 10.1038/s41597-022-01606-w

Degbelo, A. & Kray, C. (2018). *Intelligent geovisualizations for open government data (vision paper). 5*, 77–80. DOI: 10.1145/3274895.3274940

Dilmaghani, S., Brust, M. R., Danoy, G., Cassagnes, N., Pecero, J., & Bouvry, P. (2019, December 1). *Privacy and Security of Big Data in AI Systems: A Research and Standards Perspective*. IEEE Xplore. DOI: 10.1109/BigData47090.2019.9006283

Dobson, J. E. (1983). Automated geography. *The Professional Geographer*, *35*(2), 135–143. DOI: 10.1111/j.0033-0124.1983.00135.x

Ferretti, A., & Vayena, E. (2022). In the Shadow of Privacy: Overlooked Ethical Concerns in COVID-19 Digital Epidemiology. *Epidemics*, *41*, 100652. DOI: 10.1016/j.epidem.2022.100652

Floridi, L., Buttaboni, C., Hine, E., Novelli, C., Schroder, T., & Shanklin, G. (2025). *Open-Source AI made in the EU: Why it is a Good Idea*. DOI: 10.1007/s11023-025-09728-x

Frazier, A. E., Nelson, T., Kedron, P., Shook, E., Dodge, S., Murray, A., Goodchild, M., Battersby, S., Blanford, J. I., Claramunt, C., Holler, J., Koylu, C., Lee, A., Manson, S., Şalap-Ayça, S., Wilson, J. P., Zhao, B., Bennett, L., Cabrera, C., & Xu, W. (2025). Rethinking GIScience Education in an Age of Disruptions. *Transactions in GIS*, *29*(2), e70048. Advance online publication. DOI: 10.1111/tgis.70048

Gao, S. (2021). Geospatial Artificial Intelligence (GeoAI). *Geography (Sheffield, England)*. Advance online publication. DOI: 10.1093/obo/9780199874002-0228

Ghamisi, P., Yu, W., Marinoni, A., Gevaert, C.M., Persello, C., Selvakumaran, S., Girotto, M., Horton, B.P., Rufin, P., Hostert, P., Pacifici, F., & Atkinson, P.M. (2024). Responsible AI for Earth Observation. ArXiv, abs/2405.20868.

Goodchild, M. F. (1992). Geographical information science. *International Journal of Geographical Information Systems*, *6*(1), 31–45. DOI: 10.1080/02693799208901893

Goodchild, M. F. (2007). Citizens as Sensors: The World of Volunteered Geography. *GeoJournal*, *69*(4), 211–221. DOI: 10.1007/s10708-007-9111-y

Goodchild, M. F. (2009). Geographic information systems and science: Today and tomorrow. *Annals of GIS*, *15*(1), 3–9. DOI: 10.1080/19475680903250715

Goodchild, M. F. (2020). GIScience, geography, and ethics. *Annals of GIS*, *26*(1), 1–9. DOI: 10.1080/19475683.2020.1734952

Gupta, J., & An, S. (2023). *Shashi Shekhar*. Towards Responsible Spatial Data Science and Geo-AI., DOI: 10.1145/3607947.3608060

Hanny, D., & Resch, B. (2025). Multimodal GeoAI: An integrated spatio-temporal topic-sentiment model for the analysis of geo-social media posts for disaster management. *International Journal of Applied Earth Observation and Geoinformation*, *139*, 104540–104540. DOI: 10.1016/j.jag.2025.104540

Hazeleger, W., Aerts, J.P.M., Bauer, P. *et al.* (2024). Digital twins of the Earth with and for humans. *Commun Earth Environ 5*, 463 DOI: 10.1038/s43247-024-01626-x

Hendriks, A., Karhunmaa, K., & Delvenne, P. (2025). Shaping the future: A conceptual review of sociotechnical imaginaries. *Futures*, *103607*, 103607. Advance online publication. DOI: 10.1016/j.futures.2025.103607

Hochmair, H. H., Juhász, L., & Kemp, T. (2024, November). Juhász, L., & Kemp, T. (2024). Correctness Comparison of ChatGPT-4, Gemini, Claude-3, and Copilot for Spatial Tasks. *Transactions in GIS*, *28*(7), 2219–2231. Advance online publication. DOI: 10.1111/tgis.13233

Hochmair, H. H., Juhász, L., & Li, H. (2025). Advancing AI-Driven Geospatial Analysis and Data Generation: Methods, Applications and Future Directions. *ISPRS International Journal of Geo-Information*, *14*(2), 56. DOI: 10.3390/ijgi14020056

Houfaf-Khoufaf, W., Touya, G., & Le Guilcher, A. (2021). Geographically masking addresses to study COVID-19 clusters. *Cartography and Geographic Information Science*, *51*(2), 242–256. DOI: 10.1080/15230406.2021.1977709

Hu, T., Liu, T., Divyacharan Jarugumalli, V. S., Cheng, S., & Deng, C. (2025). FAIR principles in workflows: A GIScience workflow management system for reproducible and replicable studies. *International Journal of Applied Earth Observation and Geoinformation*, *138*, 104477. DOI: 10.1016/j.jag.2025.104477

Hu, Y., Gao, S., Lunga, D., Li, W., Newsam, S., & Bhaduri, B. (2019). GeoAI at ACM SIGSPATIAL. *SIGSPATIAL Special*, *11*(2), 5–15. DOI: 10.1145/3377000.3377002

Hu, Y., Goodchild, M., Zhu, A. X., Yuan, M., Aydin, O., Bhaduri, B., Gao, S., Li, W., Lunga, D., & Newsam, S. (2024). A five-year milestone: Reflections on advances and limitations in GeoAI research. *Annals of GIS*, *30*(1), 1–14. DOI: 10.1080/19475683.2024.2309866

Huang, C., Mees, O., Zeng, A., & Burgard, W. (2023). *Visual Language Maps for Robot Navigation*. ArXiv.org. /arXiv.2210.05714DOI: 10.1109/ICRA48891.2023.10160969

Huang, X., Tu, Z., Ye, X., & Goodchild, M. (2025). *The Role of Open-Source LLMs in Shaping the Future of GeoAI*. ArXiv.org. https://arxiv.org/abs/2504.17833

Janowicz, K. (2023). Philosophical Foundations of GeoAI: Exploring Sustainability, Diversity, and Bias in GeoAI and Spatial Data Science. *ArXiv (Cornell University)*. /arxiv.2304.06508DOI: 10.1201/9781003308423-3

Janowicz, K., Gao, S., McKenzie, G., Hu, Y., & Bhaduri, B. (2019). GeoAI: Spatially explicit artificial intelligence techniques for geographic knowledge discovery and beyond. *International Journal of Geographical Information Science*, *34*(4), 625–636. DOI: 10.1080/13658816.2019.1684500

Jasanoff, S. (2004). *States of Knowledge: The Co-production of Science and the Social Order*. Routledge Taylor & Francis Group. DOI: 10.4324/9780203413845

Jasanoff, S., & Kim, S. (2015). *Dreamscapes of modernity: sociotechnical imaginaries and the fabrication of power*. The University of Chicago Press. DOI: 10.7208/chicago/9780226276663.001.0001

Jiang, H., Li, J., Zhao, P., Zeng, F., Xiao, Z., & Iyengar, A. (2021). Location Privacy-preserving Mechanisms in Location-based Services. *ACM Computing Surveys*, *54*(1), 1–36. DOI: 10.1145/3423165

Jobin, A., Ienca, M., & Vayena, E. (2019). The global landscape of AI ethics guidelines. *Nature Machine Intelligence*, *1*(9), 389–399. DOI: 10.1038/s42256-019-0088-2

Jurowetzki, R., Hain, D., Mateos-Garcia, J., & Stathoulopoulos, K. (2021). The privatization of AI research(-ers): causes and potential consequences — from university-industry interaction to public research brain-drain? /arXiv.2102.01648DOI: 10.48550

Kaas, B. C., Porter, Z., Ozturk, B., Ryan, P., Katell, M., Polo, N., Westerling, K., & Habli, I. (2024). Fair by design: A sociotechnical approach to justifying the fairness of AI-enabled systems across the lifecycle. *ArXiv (Cornell University)*. /arxiv.2406.09029DOI: 10.48550

Kang, Y., Gao, S., & Roth, R. E. (2024). Artificial intelligence studies in cartography: A review and synthesis of methods, applications, and ethics. *Cartography and Geographic Information Science, 51*(4), 599–630. DOI: 10.1080/15230406.2023.2295943

Keßler, C., & McKenzie, G. (2017). A geoprivacy manifesto. *Transactions in GIS, 22*(1), 3–19. DOI: 10.1111/tgis.12305

Kitchin, R. (2016). The ethics of smart cities and urban science. *Philosophical Transactions. Series A, Mathematical, Physical, and Engineering Sciences, 374*(2083), 20160115. DOI: 10.1098/rsta.2016.0115

Lazaro, C., & Rizzi, M. (2022). Predictive analytics and governance: A new sociotechnical imaginary for uncertain futures. *International Journal of Law in Context, 19*(1), 1–21. DOI: 10.1017/S1744552322000477

Li, W., & Arundel, S. T. (2022). GeoAI and the future of spatial analytics. In New thinking in GIScience, Singapore: Springer Nature Singapore, pp. 151-158._DOI: 10.31223/X5DT3J

Li, W., Batty, M., & Goodchild, M. F. (2019). Real-time GIS for smart cities. *International Journal of Geographical Information Science, 34*(2), 311–324. DOI: 10.1080/13658816.2019.1673397

Li, W., & Hsu, C.-Y. (2018). GeoAI: Where machine learning and big data converge in geospatial science. *Big Data and Cognitive Computing, 2*(4), 22. DOI: 10.3390/bdcc2040022

Li, W., & Hsu, C.-Y. (2022). GeoAI for Large-Scale Image Analysis and Machine Vision: Recent Progress of Artificial Intelligence in Geography. *ISPRS International Journal of Geo-Information, 11*(7), 385. DOI: 10.3390/ijgi11070385

Li, W., Wang, C., & Croitoru, A. (2020). GeoAI: Enabling the understanding of geospatial big data. *Computers, Environment and Urban Systems, 81*, 101472. DOI: 10.1016/j.compenvurbsys.2020.101472

Lorestani, A. M., Ranbaduge, T., & Rakotoarivelo, T. (2024). *Privacy risk in GeoData: A survey*. ArXiv.org._https://arxiv.org/abs/2402.03612

Lunga, D., Hu, Y., Newsam, S., Gao, S., Martins, B., Yang, L., & Deng, X. (2021). GeoAI at ACM SIGSPATIAL. *SIGSPATIAL Special, 13*(3), 21–32. DOI: 10.1145/3578484.3578491

Mager, A., & Katzenbach, C. (2021). Future imaginaries in the making and governing of digital technology: Multiple, contested, commodified. *New Media & Society, 23*(2), 223–236. DOI: 10.1177/1461444820929321

Mai, G., Xie, Y., Jia, X., Lao, N., Rao, J., Zhu, Q., Liu, Z., Chiang, Y.-Y., & Jiao, J. (2025). Towards the next generation of Geospatial Artificial Intelligence. *International Journal of Applied Earth Observation and Geoinformation, 136*, 104368. DOI: 10.1016/j.jag.2025.104368

Mao, H., Hu, Y., Kar, B., Gao, S., McKenzie, G., & Geo, A. I. (2017). Workshop Report: The 1st ACM SIGSPATIAL International Workshop on GeoAI: @AI and Deep Learning for Geographic Knowledge Discovery: Redondo Beach, CA, USA-November 7, 2016. ACM Sigspatial Spec. 2017, 9, 25.

Marasinghe, R., Yigitcanlar, T., Mayere, S., Washington, T., & Limb, M. (2024). Towards Responsible Urban Geospatial AI: Insights From the White and Grey Literatures. *Journal of Geovisualization and Spatial Analysis, 8*(2), 24. Advance online publication. DOI: 10.1007/s41651-024-00184-2

Marres, N., Castelle, M., Gobbo, B., Poletti, C., & Tripp, J. (2024). AI as super-controversy: Eliciting AI and society controversies with an extended expert community in the U.K. *Big Data & Society, 11*(2), 20539517241255103. Advance online publication. DOI: 10.1177/20539517241255103

Matuszczyk, N., Barnes, C. R., Gupta, R., Ozel, B., & Mitra, A. (2025). *From Bias to Accountability: How the EU AI Act Confronts Challenges in European GeoAI Auditing*. ArXiv.org. https://arxiv.org/abs/2505.18236

McKenzie, G., & Zhang, H. (2023). Platial k-anonymity: Improving location anonymity through temporal popularity signatures. In *12th International Conference on Geographic Information Science (GIScience 2023),* pp. 9-1. Schloss Dagstuhl–Leibniz-Zentrum für Informatik.

McKenzie, G., Zhang, H., & Gambs, S. (2023). Privacy and Ethics in GeoAI. In *Gao, S., Hu, Y., & Li, W. [eds.] Handbook of Geospatial Artificial Intelligence.CRC Press Ebooks*, pp. 388–405. DOI: 10.1201/9781003308423-19

Mehrabi, N., Morstatter, F., Saxena, N., Lerman, K., & Galstyan, A. (2021). A Survey on Bias and Fairness in Machine Learning. *ACM Computing Surveys, 54*(6), 1–35. DOI: 10.1145/3457607

Michael, M. (2017). Enacting big futures, little futures: Toward an ecology of futures. *The Sociological Review, 65*(3), 509–524. DOI: 10.1111/1467-954X.12444

Miranda-Pascual, À., Guerra-Balboa, P., Parra-Arnau, J., Forné, J., & Strufe, T. (2023). SoK: Differentially private publication of trajectory data. *Proceedings on Privacy Enhancing Technologies.*

Morozov, E. (2013). *To save everything, click here: smart machines, dumb humans, and the myth of tecnological perfectionism.* Perseus Books.

Mügge, D. (2024). EU AI sovereignity: For whom, to what end and to whose benefit? *Journal of European Public Policy, 31*(8), 2200–2225. DOI: 10.1080/13501763.2024.2318475

Mühlhoff, R. (2025). *The Ethics of AI.* Bristol University Press., DOI: 10.51952/9781529249262

Munjal, K., & Bhatia, R. (2022). A systematic review of homomorphic encryption and its contributions in healthcare industry. *Complex & Intelligent Systems*, 1–28.

Nikolenko, S. I. (2021). Synthetic Data for Deep Learning. In *Springer optimization and its applications.* Springer International Publishing., DOI: 10.1007/978-3-030-75178-4

Ntoutsi, E., Fafalios, P., Gadiraju, U., Iosifidis, V., Nejdl, W., Vidal, M., Ruggieri, S., Turini, F., Papadopoulos, S., Krasanakis, E., Kompatsiaris, I., Kinder-Kurlanda, K., Wagner, C., Karimi, F., Fernandez, M., Alani, H., Berendt, B., Kruegel, T., Heinze, C., & Staab, S. (2020). Bias in Data-driven Artificial Intelligence systems—An Introductory Survey. *Wiley Interdisciplinary Reviews. Data Mining and Knowledge Discovery, 10*(3), 1–14. DOI: 10.1002/widm.1356

Oluoch, I. (2024). Crossing Boundaries: The Ethics of AI and Geographic Information Technologies. *ISPRS International Journal of Geo-Information, 13*(3), 87. DOI: 10.3390/ijgi13030087

Openshaw, S., & Openshaw, C. (1997). *Artificial Intelligence in Geography.* John Wiley & Sons, Inc.

Pagano, T. P., Loureiro, R. B., Lisboa, F. V. N., Peixoto, R. M., Guimarães, G. A. S., Cruz, G. O. R., Araujo, M. M., Santos, L. L., Cruz, M. A. S., Oliveira, E. L. S., Winkler, I., & Nascimento, E. G. S. (2023). Bias and Unfairness in Machine Learning Models: A Systematic Review on Datasets, Tools, Fairness Metrics, and Identification and Mitigation Methods. *Big Data and Cognitive Computing, 7*(1), 15. DOI: 10.3390/bdcc7010015

Paltieli, G. (2022). The political imaginary of National AI Strategies. *AI & Society, 37*(4), 1613–1624. DOI: 10.1007/s00146-021-01258-1

Prasad, S., Chanussot, J., & Li, J. (2024). *Advances in Machine Learning and Image Analysis for GeoAI.* Elsevier., DOI: 10.1016/C2022-0-00952-1

Purbahapsari, A. F., & Batoarung, I. B. (2022). *Geospatial Artificial Intelligence for Early Detection of Forest and Land Fires*. KnE Social Sciences. DOI: 10.18502/kss.v7i9.10947

Qiu, Q., Li, W., Liu, J., Tao, L., Hao, M., Wang, Y., & Xie, Z. (2024). AutoGeo-FAIR: A framework for evaluating the sharing degree of open earth surface system data based on extended FAIR principles. *Journal of Environmental Management, 372*, 123432. Advance online publication. DOI: 10.1016/j.jenvman.2024.123432

Raj, A., & Arora, L. Girija, Sanjay Surendranath, Kapoor, S., Pradhan, D., & Shetgaonkar, A. (2025). *AI and Generative AI Transforming Disaster Management: A Survey of Damage Assessment and Response Techniques*. ArXiv.org. https://arxiv.org/abs/2505.08202v1

Raley, R., & Rhee, J. (2023). Critical AI: A Field in Formation. *American Literature, 95*(2), 185–204. Advance online publication. DOI: 10.1215/00029831-10575021

Rao, J., Gao, S., Mai, G., & Janowicz, K. (2023). Building Privacy-Preserving and Secure Geospatial Artificial Intelligence Foundation Models (Vision Paper). In *Proceedings of the 31st ACM International Conference on Advances in Geographic Information Systems*. pp. 1-4. DOI: 10.1145/3589132.3625611

Raza, K. (2020). Artificial intelligence against COVID-19: a meta-analysis of current research. In Hassanien, A., Dey, N., & Elghamrawy, S. (Eds.), *Big Data Analytics and Artificial Intelligence Against COVID-19: Innovation Vision and Approach* (pp. 165–176). Springer International Publishing. DOI: 10.1007/978-3-030-55258-9_10

Richter, V., Katzenbach, C., & Schäfer, M. S. (2023). Imaginaries of artificial intelligence. In *Handbook of critical studies of artificial intelligence* (pp. 209–223). Edward Elgar Publishing., DOI: 10.4337/9781803928562.00024

Richter, V., Katzenbach, C., & Zeng, J. (2025). Negotiating AI(s) futures: Competing imaginaries of AI by stakeholders in the US, China, and Germany. *JCOM, Journal of Science Communication, 24*(2), A08. DOI: 10.22323/2.24020208

Sahithya, V., Sujithra, M., Senthilkumar, B., Shreenithi, M., & Priya, H. (2024). *A Guide to Ethical Geospatial Practices*. Advances in Geospatial Technologies Book Series., DOI: 10.4018/979-8-3693-6381-2.ch002

Salman A.H. Selmy, Kuche, D. E., & Yang, Y. (2024). Geospatial Data: Acquisition, Applications and Challenges. *IntechOpen EBooks*. DOI: 10.5772/intechopen.1006635

Saura, J. R., Ribeiro-Soriano, D., & Palacios-Marqués, D. (2022). Assessing behavioral data science privacy issues in government artificial intelligence deployment. *Government Information Quarterly, 39*(4), 101679. DOI: 10.1016/j.giq.2022.101679

Shi, M., Janowicz, K., Verstegen, J., Currier, K., Wiedemann, N., Mai, G., Majic, I., Liu, Z., & Zhu, R. (2025). Geography for AI sustainability and sustainability for GeoAI. *Cartography and Geographic Information Science*, *52*(4), 331–349. DOI: 10.1080/15230406.2025.2479796

Siau, K., & Wang, W. (2020). Artificial Intelligence (AI) Ethics: Ethics of AI and Ethical AI. *Journal of Database Management*, *31*(2), 74–87. DOI: 10.4018/JDM.2020040105

Specht, D. (2020). Mapping Crisis: Participation, Datafication and Humanitarianism in the Age of Digital Mapping. In *University of London Press eBooks*. University of Lodz Press., DOI: 10.14296/920.9781912250387

Suchman, L. (2023). The uncontroversial 'thingness' of AI. *Big Data & Society*, *10*(2), 20539517231206794. Advance online publication. DOI: 10.1177/20539517231206794

Sun, G., Cai, S., Yu, H., Maharjan, S., Chang, V., Du, X., & Guizani, M. (2019). Location privacy preservation for mobile users in location-based services. *IEEE Access : Practical Innovations, Open Solutions*, *7*, 87425–87438. DOI: 10.1109/ACCESS.2019.2925571

Suresh, H., & Guttag, J. (2021). *A Framework for Understanding Sources of Harm throughout the Machine Learning Life Cycle*. Equity and Access in Algorithms, Mechanisms, and Optimization., DOI: 10.1145/3465416.3483305

Suresh, H., & Guttag, J. V. (2019). A Framework for Understanding Unintended Consequences of Machine Learning. *Arxiv.org*. https://arxiv.org/abs/1901.10002

Taylor, C. (2004). *Modern Social Imaginaries*. Duke University Press. DOI: 10.1215/9780822385806

Timmermans, J., Stahl, B. C., Ikonen, V., & Bozdag, E. (2010). The ethics of cloud computing: A conceptual review. *Proceedings of the IEEE Second International Conference Cloud Computing Technology and Science* (pp. 614-620). IEEE Press. DOI: 10.1109/CloudCom.2010.59

Tobler, W. R. (1970). A computer movie simulating urban growth in the Detroit region. *Economic Geography*, *46*(supplement), 234–240. DOI: 10.2307/143141

Tuia, D., Schindler, K., Demir, B., & Zhu, X. X. Mrinalini Kochupillai, Sašo Džeroski, Jan, Hoos, H. H., Frate, F. D., Mihai Datcu, Markl, V., Bertrand Le Saux, Schneider, R., & Gustau Camps-Valls. (2024). Artificial Intelligence to Advance Earth Observation: A review of models, recent trends, and pathways forward. *IEEE Geoscience and Remote Sensing Magazine*, pp. 2–25. DOI: 10.1109/MGRS.2024.3425961

Tursunalieva, A., Alexander, D. L. J., Dunne, R., Li, J., Riera, L., & Zhao, Y. (2024). Making Sense of Machine Learning: A Review of Interpretation Techniques and Their Applications. *Applied Sciences, 14*(2), pag. 496. DOI: 10.3390/app14020496

UN-GGIM. *(2020). Future trends in geospatial information management: the five to ten year vision THIRD EDITION.* https://ggim.un.org/meetings/GGIM-committee/10thSession/documents/Future_Trends_Report_THIRD_EDITION_digital_accessible.pdf

van Dijck, J., van Es, K., Helmond, A., & van der Vlist, F. (2025). *Governing the Digital Society*. Amsterdam University Press., DOI: 10.5117/9789048562718

VoPham, T., Hart, J. E., Laden, F., & Chiang, Y.-Y.VoPham. (2018). Emerging trends in geospatial artificial intelligence (geoAI): Potential applications for environmental epidemiology. *Environmental Health : A Global Access Science Source*, *17*(1), 40. Advance online publication. DOI: 10.1186/s12940-018-0386-x

Vulpe, S.-N., Rughini , R., urcanu, D., & Rosner, D. (2024). AI and cybersecurity: A risk society perspective. *Frontiers of Computer Science*, *6*, 1462250. Advance online publication. DOI: 10.3389/fcomp.2024.1462250

Wahltinez, O., Cheung, A., Alcantara, R., Cheung, D., Daswani, M., Erlinger, A., Lee, M., Yawalkar, P., Lê, P., Navarro, O. P., Brenner, M. P., & Murphy, K. (2022). COVID-19 Open-Data a global-scale spatially granular meta-dataset for coronavirus disease. *Scientific Data*, *9*(1), 162. Advance online publication. DOI: 10.1038/s41597-022-01263-z

Wang, Q., Gao, J., Lin, W., & Yuan, Y. (2019). Learning from synthetic data for crowd counting in the wild. In *Proceedings of the IEEE/CVF conference on computer vision and pattern recognition*, pp. 8198–8207. DOI: 10.1109/CVPR.2019.00839

Wang, S., Anselin, L., Bhaduri, B., Crosby, C., Goodchild, M. F., Liu, Y., & Nyerges, T. L. (2013). CyberGIS software: A synthetic review and integration roadmap. *International Journal of Geographical Information Science*, *27*(11), 2122–2145. DOI: 10.1080/13658816.2013.776049

Wang, S., Hu, T., Xiao, H., Li, Y., Zhang, C., Ning, H., Zhu, R., Li, Z., & Ye, X. (2024). GPT, large language models (LLMs) and generative artificial intelligence (GAI) models in geospatial science: A systematic review. *International Journal of Digital Earth*, *17*(1), 2353122. Advance online publication. DOI: 10.1080/17538947.2024.2353122

Whittlestone, J., & Clarke, S. (2022). *AI Challenges for Society and Ethics*. The Oxford Handbook of AI Governance., DOI: 10.1093/oxfordhb/9780197579329.013.3

Wilkinson, M. D., Dumontier, M., & Aalbersberg, Ij. J., Appleton, G., Axton, M., Baak, A., Blomberg, N., Boiten, J.-W., da Silva Santos, L. B., Bourne, P. E., Bouwman, J., Brookes, A. J., Clark, T., Crosas, M., Dillo, I., Dumon, O., Edmunds, S., Evelo, C. T., Finkers, R., & Gonzalez-Beltran, A. (2016). The FAIR Guiding Principles for Scientific Data Management and Stewardship. *Scientific Data, 3*(1). https://www.nature.com/articles/sdata201618

Xing, J., & Sieber, R. (2023). The challenges of integrating explainable artificial intelligence into GeoAI. *Transactions in GIS*, *27*(3), 626–645. Advance online publication. DOI: 10.1111/tgis.13045

Ye, X., Du, J., Li, X., Shaw, S.-L., Fu, Y., Dong, X., Zhang, Z., & Wu, L. (2025). Human-centered GeoAI foundation models: Where GeoAI meets human dynamics. *Urban Informatics*, *4*(1), 2. Advance online publication. DOI: 10.1007/s44212-025-00067-x

Zhang, G., Zhang, S., Man, Z., Cui, C., & Hu, W. (2024). Location Privacy Protection in Edge Computing: Co-Design of Differential Privacy and Offloading Mode. *Electronics (Basel)*, *13*(13), 2668–2668. DOI: 10.3390/electronics13132668

Zhang, J., & Goodchild, M. F. (2002). Uncertainty in Geographical Information. In *CRC Press eBooks*. Informa., DOI: 10.1201/b12624

Zheng, Z., & Sieber, R. (2021). Putting humans back in the loop of machine learning in Canadian smart cities. *Transactions in GIS*, *26*(1), 8–24. DOI: 10.1111/tgis.12869

Zhou, B., Zou, L., Mostafavi, A., Lin, B., Yang, M., Gharaibeh, N., Cai, H., Abedin, J., & Mandal, D. (2022). VictimFinder: Harvesting rescue requests in disaster response from social media with BERT. *Computers, Environment and Urban Systems*, *95*, 101824. DOI: 10.1016/j.compenvurbsys.2022.101824

Zhu, A.-X., & Turner, M. (2022). How is the Third Law of Geography different? *Annals of GIS*, *28*(1), 57–67. DOI: 10.1080/19475683.2022.2026467

Zhu, X. X., Tuia, D., Mou, L., Xia, G.-S., Zhang, L., Xu, F., & Fraundorfer, F. (2017). Deep learning in remote sensing: A comprehensive review and list of resources. *IEEE Geoscience and Remote Sensing Magazine*, *5*(4), 8–36. DOI: 10.1109/MGRS.2017.2762307

## KEY TERMS AND DEFINITIONS

**Building Information Modeling (BIM):** is a collaborative process that involves the generation and management of digital representations of physical and functional characteristics of buildings and infrastructure. These models go beyond mere 2D architectural drawings by embedding rich metadata and 3D spatial information, allowing stakeholders to simulate, analyze, and visualize the entire lifecycle of a structure—from planning and design to construction, operation, and maintenance. BIM facilitates interoperability among architects, engineers, contractors, and facility managers, enhancing coordination and reducing errors. It has become a central tool in contemporary architecture, engineering, and construction (AEC) practices, particularly in promoting efficiency, sustainability, and cost-effectiveness.

**Digital Earth (DE):** is a visionary concept proposing a comprehensive, interactive digital model of the planet that integrates vast streams of spatial and temporal data into a seamless virtual environment. Initially articulated as a framework for global understanding and decision-making, Digital Earth aspires to serve as a tool for education, science, governance, and citizen engagement. It encompasses technologies such as geographic information systems, remote sensing, virtual reality, and web-based visualization to simulate the Earth's systems and human activities. Digital Earth provides users—ranging from researchers to policymakers—with the capacity to visualize global processes, monitor environmental changes, and explore geographic phenomena at multiple scales.

**Earth Observation (EO):** refers to the systematic collection of data about the Earth's surface and atmosphere using both remote and in-situ sensing techniques. This data is captured through satellites, aerial platforms such as drones and aircraft, and ground-based sensors, providing critical information on climate patterns, land use, vegetation cover, urban growth, and natural disasters. EO has become indispensable across disciplines such as meteorology, agriculture, environmental science, and urban planning. It enables time-series analyses, spatial monitoring, and large-scale mapping efforts, supporting evidence-based decision-making and contributing to the understanding of anthropogenic and natural processes shaping our planet.

**GeoAI Model:** is a computational construct that applies artificial intelligence methods—such as machine learning or deep learning—to spatially explicit data in order to perform tasks such as classification, prediction, pattern recognition, or decision support. These models are designed to accommodate the unique characteristics of geospatial data, including spatial autocorrelation, scale dependence, and topological relationships. GeoAI models often involve the training of algorithms on large datasets containing geographic features and attributes, enabling the discovery of complex spatial patterns and the generation of spatial insights that would be difficult to detect through traditional analysis. Their use has grown across domains requiring spatial reasoning and data-driven inference.

**Geomasking:** is a data anonymization technique that modifies the geographic coordinates of spatial data to protect individual privacy while maintaining the utility of the data for analysis. Typically used in datasets containing sensitive location-based information—such as health records, mobility data, or social service usage—geomasking involves procedures like random perturbation, spatial displacement, or aggregation to larger geographic units. The aim is to reduce the risk of re-identification of individuals while preserving enough spatial accuracy for meaningful research and decision-making. The technique must strike a balance between confidentiality and spatial fidelity, and its effectiveness depends on factors such as population density and data resolution.

**Geographic Information Systems (GIS):** are integrated frameworks used to collect, manage, analyze, and visualize data that is referenced to specific locations on the Earth's surface. A GIS allows users to layer various types of information—such as demographic data, environmental conditions, infrastructure networks, or land use patterns—to examine spatial relationships and inform planning and decision-making. It combines spatial data (coordinates, maps) with attribute data (descriptive information), supporting both qualitative and quantitative spatial analysis. GIS tools are widely used across government, industry, academia, and civil society for applications ranging from disaster risk assessment to urban development and environmental monitoring.

**Open Government Dataset (OGD):** are publicly accessible collections of data produced or funded by government entities, made available without restrictions on use or redistribution. These datasets cover a broad range of domains, including transportation, environment, health, finance, and demographics, and are typically released in machine-readable formats to promote transparency, innovation, and civic participation. OGDs enable researchers, journalists, developers, and citizens to scrutinize public policies, build applications, and conduct analyses that can lead to new insights and services. The open data movement emphasizes not only access but also the quality, usability, and accountability of government-provided information.

# Chapter 11
# AI and Human Rights in the Global Pursuit of Justice and Equity

**S. Ida Evangeline**
https://orcid.org/0000-0003-2997-7897
*Government College of Engineering, Tirunelveli, India*

## ABSTRACT

*As artificial intelligence increasingly shapes global decision-making, its impact on fundamental human rights demands urgent scrutiny. This chapter explores the ethical, legal, and social implications of AI technologies through a human rights lens, examining how algorithmic systems influence civil liberties, equality, and global justice. It critically analyzes the risks posed by biased algorithms, surveillance infrastructure, and opaque decision-making, particularly for marginalized populations and low-resource nations. The chapter examines the major structural global disparities that shape AI governance, revealing the dangers of digital colonialism and the lack of representation of some perspectives in decision-making. Using global standards and real world examples, the chapter describes an ethics-based strategy for AI governance focused on fairness, accountability, transparency, and inclusivity.*

## 1. INTRODUCTION

The rapid evolution of artificial intelligence (AI) is reshaping the foundations of modern life, with implications that span from individual daily routines to global governance systems. Facial recognition, predictive analytics, natural language processing, and automated decision-making, all characteristics of AI, are now part of the core systems of healthcare, education, These technologies serve more than an

DOI: 10.4018/979-8-3373-2597-2.ch011

augmenting function to our skills. they have a decisive role to play in determining how resources, services, and rights are accessed, filtered and granted. Therefore, artificial intelligence has become a powerful aspect that either empowers or disempowers people and communities (Roberts et al. 2024).

This growing influence necessitates a rigorous interrogation of AI's alignment with fundamental human rights. Various rights including dignity, privacy, non-d For instance, facial recognition technologies can be applied to carry out mass surveillance; When it comes to discrimination, hiring algorithms sometimes reinforce bias against women or race; automated content moderation frequently damps legal expression of opinion. These concerns are part of a larger pattern in which flaws in design and governance occur when ethical, legal, and social factors are ignored in the course of technological development (Ahmad et al. 2025).

Moreover, the deployment of AI frequently reflects and exacerbates existing power asymmetries. Disadvantaged groups – who are already often the victims of structural discrimination – often face an unequal share of risks related to unreliable or discriminatory AI tools. In economically disadvantaged regions, barriers are limited technological networks, weak legal protection for data, and dependence on international technology suppliers. Such settings reinforce an age of what experts describe as "digital colonialism" or "algorithmic colonialism" in which the most significant resources and authority to make decisions reside (Salami, 2024).

This chapter seeks to unpack these complex dynamics by adopting a multidisciplinary and globally oriented lens. It draws insights from fields such as law, ethics, political science, and computer science to explore how AI technologies intersect with human rights norms across diverse sociopolitical contexts. First, the analysis identifies which particular human rights are violated by AI technology and its governing structures, like civil, political, economic, social, and cultural sphere. The analysis then examines how AI can improve or undermine these rights, while paying special attention to those where automation is especially important such as policing, healthcare, employment, and digital media.

Importantly, this work does not limit itself to critique. It also describes positive approaches to reform, illustrating efforts by global institutions, regional networks, country governments, and NGOs to bring human rights into AI development and use. Such organizations as OECD and UNESCO have set principles and guidelines for AI, while such national laws as EU AI Act set frameworks that seek to balance the advancement of technology with justice, accountability and inclusiveness.

The ultimate goal of this chapter is to advocate for a human rights-centered approach to AI governance—one that is proactive rather than reactive, participatory rather than technocratic, and globally equitable rather than dominated by a few privileged stakeholders. This initiative seeks to harmonize general ethical concepts with actual government policies, offering a structure for combining technological progress with the general societal objectives. The integration of human rights into AI systems is essential as societies move at high speed in digital change, morally imperative and strategically important for creating a sustainable, inclusive, and just future.

## 2. FUNDAMENTAL HUMAN RIGHTS IN THE AGE OF AI

The emergence of artificial intelligence brings forth a mighty change in the ways folks manage their towns, share their words, and provide their goods and services. In the midst of this here transformation, a mighty question stands tall: How do the ways of AI and the fundamental rights of folks, tied to the governance of democracy and the laws of nations, intertwine with one another? This here holds mighty true when it comes to them basic rights, laid out by treaties far and wide, like that Universal Declaration of Human Rights, partner. By reckonin' how folks gain access to knowledge, services, and opportunities, AI is stirrin' up the ways in which these fundamental rights are either safeguarded or trampled upon (Romero Moreno, 2024).

### 2.1 Right to Privacy and Data Protection

One of the rights most affected is the right to privacy, as laid out in Article 12 of the UDHR, which finds itself under a heavy burden in this new digital frontier. The operation of them AI contraptions often calls for the gatherin' and wranglin' of vast amounts of personal information, stretchin' from the usual demographics to the more peculiar traits of a person's very being and ways of actin'. These here technologies, like facial recognition, predictive policing, and the targeting of folks with digital ads, often wrangle personal data gathered without true consent or proper watchfulness (Kathuria et al. 2024).

In places where the law ain't much to speak of, folks find themselves under the watchful eye of unseen watchers, their lives scrutinized and cataloged in ways they can't quite grasp or fight against. Even though nations lay down the law with strict codes, like the GDPR across the pond, the folks tasked with keeping the peace often find themselves laggin' behind the swift march of newfangled technology. This here situation stirs up a mighty lack of control, lettin' invasive watchin', data and privacy violations, and unfair profiling fester in the shadows, often with no way to fight back (Yanamala & Suryadevara, 2024).

Furthermore, the consequences of them fancy machines keepin' an eye on folks stretch far beyond just the wrongful use of information. In the dusty streets of our towns, the lawmen and guardians of order, armed with the power of AI, have hastened the reach of their watchful eyes, casting a wide net of surveillance over the folks going about their daily lives. In the harsh lands of tyranny, the enforcers of the law have taken to using these gadgets to keep a watchful eye on those who dare to oppose, the marginalized, and the downtrodden. They've turned this clever contraption into a tool of oppression, casting a shadow over the very rights that ought to be cherished—like the freedom to gather, to believe, and to speak one's mind.

## 2.2 Freedom of Expression and Access to Information

The right to speak one's mind, as laid out in the good ol' Article 19 of the Universal Declaration of Human Rights, which covers the freedom to seek, receive, and share information, is bein' reshaped by the likes of AI these days. The ways folks make choices in curating content on platforms such as YouTube, Facebook, and TikTok steer what information reaches the masses, how public discussions unfold, and how political matters are handled in the town square. Though they set their sights on boosting user satisfaction, these contraptions might unwittingly forge echo chambers, bolster biases, and elevate the presence of divisive or misleading tales (Sancin & Bobnar, 2024).

Automated content moderation, powered by the cunning of machine learning and the art of natural language processing, is a tool that cuts both ways. On one side of the dusty trail, it serves to rid the land of perilous or unlawful content, like hate-filled words and cries for violence. Nevertheless, this approach may silence valid speech especially from minority groups whose language, cultural sensitivity or political position may not be correctly assessed by AI tools that are not contextual. Indigenous expressions, LGBTQ+ rights voices, and politically dissenting contributions are more likely to be flagged or deleted because of misleading or discriminatory moderation algorithms (Kan, 2024).

Besides, the spread of technologies such as deepfakes and generative bots threaten the right to access reliable information thereby subjecting the democratic processes to decline and the capacity to make informed decisions to suffer. Creating news with the help of AI, impersonating high-profile personalities, or making false statements erodes people's confidence in institutions and destroys the essence of public deliberation.

## 2.3 Equality and Non-Discrimination

Another critical area of concern is the right to equality before the law and the prohibition of discrimination (Article 7, UDHR). AI technologies used in credit scoring, hiring, healthcare, and criminal justice usually learn from the historical data that are imbued with prejudices based on race, gender, socioeconomic status, disability. Thus, these systems might just end up fanning the flames of deep-rooted inequalities, making 'em worse than they already are (Lütz, 2024).

Take, for instance, the way them hiring algorithms work, leanin' on the profiles of past workers. They might just overlook folks from backgrounds that ain't the usual kind, leaving 'em in the dust. Diagnostic tools in the surgical arena, crafted from datasets mostly of the white folk, might find themselves at a disadvantage when it comes to gauging the health of Black, Indigenous, or other marginalized folks. In the rough and tumble of the law, some instruments of foresight have been wielded to unjustly zero in on folks of color, leading to a heap more watchin' and wranglin' than on other folks (Akhtar et al. 2024).

What sets apart these kinds of algorithmic discrimination is not just their vastness and swiftness, but also the shadows they cast, hiding their workings from plain sight. In the case of many AI models, the logic is confusing, even for the original designers and the wider audience, hence creating problems with transparency, accountability, and the rights of individuals to contest unfair results. Additionally, the issue is not much to do with the institutional structures as it is with the technology: Outcomes of biased data reveal the societal biases and systems that are in the AI systems.

### 2.4 Economic, Social, and Cultural Rights

AI also touches upon a broader spectrum of rights, including economic, social, and cultural rights (Articles 22–27, UDHR). The possibilities of AI tools are in the fact that they can expand the scope of education, enhance efficiency of healthcare, enhance the management of public functions, and promote economic growth. However, the benefits are not always available to all (Peña Acuña et al. 2025).

Deficiencies in infrastructure, skills, and institutional capacity render marginalized communities and low-income countries unable to maximise AI opportunities while experiencing the risks. The failure to put into action those inclusion strategies, along with the rapid march of technology, could widen the chasm between folks and leave countless souls stranded, unable to grasp the new digital chances that lie ahead. In the realms of learning and well-being, the gaps are mighty pronounced, for AI can serve as a trusty ally or turn into a barrier that keeps folks out in the cold.

## 3. AI-INDUCED HUMAN RIGHTS RISKS

Though the mighty force of artificial intelligence could change the game in many a field, its use has also brought forth considerable dangers to the rights of folks, especially those already on shaky ground. These dangers are comin' to light in the real world, not just in the talkin' circles of theory. They're showin' up in solid cases, layin' bare how data and the rules of the land keep these deep-rooted biases alive and kickin'. This here section rides into three significant territories where the use of AI stirs up a heap of trouble for human rights: Injustice woven into the very fabric of algorithms, dangers lurking in the shadows of privacy, and shackles placed upon the free flow of words and knowledge (Bengio et al. 2024).

## 3.1 Algorithmic Discrimination and Structural Inequality

One of the most treacherous and sly dangers of AI is the shadow of algorithmic discrimination—the way it can mirror and magnify the old biases of society through the cold hand of automated decision-making. Many a contraption of artificial intelligence, forged from the annals of history, unwittingly carries the weight of bygone biases that once roamed the land. Well now, it's a plain fact that algorithms don't ride the range without a few biases in their saddlebags. They might just keep on perpetuating those old discriminatory ways, and spotting 'em can be as tricky as finding a needle in a haystack (Moussawi et al. 2024).

In the realm of law and order, them predictive policing contraptions have faced a heap of criticism for unfairly zeroing in on the folks from communities of color. These frameworks often find their roots in arrest records and the tale of crime statistics, which have been twisted for many a moon by the shadow of unjust law enforcement practices. In them towns where the lawmen roam thick as tumbleweeds, folks find themselves under the watchful eye of the badge, burdened by the weight of suspicion. This only stirs the pot of trouble, eroding the bonds of trust and keeping the cycle of misdeeds spinning like a rusty wheel (Blount, 2024).

In the dusty towns of the hiring frontier, the algorithms meant to sift through resumes and judge video interviews have been found to carry a heavy hand against women, folks with disabilities, and those who hail from less common tongues or cultures. One tale to tell is of the AI systems in the hiring corral, which, by shining a light on the dusty trails of past data from tech companies, have pushed aside candidates with traits like being of the fairer sex or having ties to institutions that serve the underrepresented folk.

Due to the sparse presence of certain folks in the training datasets, them healthcare diagnostic and triage algorithms might just misread the signs or wrongly pin symptoms on these groups. Such misrepresentation can stir up a heap of trouble, causin' delays in treatments, throwin' off diagnoses, and leadin' to unjust treatment of folks who truly need it. Those algorithms stir up a heap of trouble, trampling on the health rights of folks and robbing them of their fair shot at care without discrimination (Akhtar et al. 2024).

What adds to the wrongness of these results is the absence of clear sight and responsibility. These here systems often make choices that ain't clear as a prairie sky, leaving folks in the dark, unable to question or stand up against the way things are decided. Folks find themselves unable to grasp or contest the judgments made by them fancy machines, robbing 'em of their rightful chance at fair play.

## 3.2 Erosion of Privacy and the Normalization of Surveillance

Well now, it seems that the rise of them fancy machines has sure spurred on the powers of watchin' folks, stirrin' up a heap of worries 'bout the sanctity of a person's privacy, the freedom to think as they please, and the right to gather with their kin. The partnership 'twixt AI and watchful networks has forged a way to keep tabs on folks, capturing their every move, interaction, word, and feeling as it unfolds in real time. These actions often take place without the proper nod from the folks being watched or the right kind of legal oversight.

Facial recognition technology sure stirs up a heap of debate in this here territory. These contraptions are set up in the bustling towns, trade posts, and labor camps, allowing folks to spot and keep an eye on the crowd from afar. In the dusty towns of authoritarian lands, a newfangled contraption known as facial recognition has been unleashed, keeping a watchful eye on those who dare to speak against the powers that be, the marginalized folk, and the brave souls fighting for justice. The plight of the Uyghur Muslims in Xinjiang stands as a stark testament to this grim reality. Such actions trample on the rights of folks to keep their lives private and put at risk the freedoms of worship, speak their minds, and gather together (He, 2024).

Even in the wide-open plains of democracy, the shadow of commercial surveillance by tech companies looms large, stirring up mighty concerns for the rights of the folks. With the aid of apps, clever contraptions, and the wide expanse of digital trails, enterprises wrangle up intricate patterns of behavior to forge their advertising campaigns, sway political endeavors, or even steer the habits of folks in ways that ain't always for the best. These here data-driven dealings wrangle away personal freedom and set up a landscape of watchful eyes that clash head-on with the very essence of democratic rights (Jacobs et al. 2024).

Now, it's a mighty truth that the shadows of watchful eyes weigh heaviest on those folks who've been pushed to the fringes of town. In them dusty towns, it's plain to see that the folks with little to their name, and those who face the harsh hand of prejudice, find themselves under the watchful eye of the law. The shadows of predictive policing and biometric surveillance loom large, revealing just how deep the roots of control and exclusion run in our society.

## 3.3 Suppression of Freedom of Expression and Democratic Engagement

AI rides into town, stirrin' up the dust and changin' the way folks talk and share their thoughts. It's got a mighty big impact on the freedoms folks hold dear, lettin' them join in the cultural and political roundups like never before. In the wild frontier of social media, AI rides alongside, wrangling content, keeping a watchful eye on the words folks use, and sniffing out any trouble that stirs against the community's code. Although these mechanisms are effective at eliminating harmful material, they also present major issues of misuse and punitive restrictions (Prove, 2024).

Automated content moderation, driven by machine learning models, often fails to grasp context, nuance, or cultural specificity. The upshot is the elimination of speech that, by law, should be protected, especially from marginalized groups whose voices challenge dominant norms/stories. There is a tendency to confuse such activities as activist speech, satire, preservation of indigenous knowledge, and use of minority languages with hate speech or misinformation (Stockinger et al. 2025).

Moreover, the fact that the moderation procedures are not made public, and there is not enough recourse for appeals, leaves the users powerless when the content is taken down unjustly. Such conditions reinforce a feeling of fear, reducing user engagement and also contribute to public skepticism about digital communities.

At the same time, AI is facilitating misinformation and disinformation at an unprecedented scale. Deepfake videos, artificial text creators, and organized bot networks are being used to misrepresent individuals, produce news, and manipulate public opinion. Such tools undermine media credibility, destroy the integrity of democratic conversations, and increase political rifts. Sometimes, campaigns created to disseminate false information have been used to promote violence or reduce voter turnout which weakens democratic institutions and the credibility of voting mechanisms (Saeidnia et al. 2025).

# 4. GLOBAL DISPARITIES AND DIGITAL COLONIALISM

The transformative potential of artificial intelligence (AI) is not distributed evenly across the globe. Some regions gain economically and strategically from AI innovation, while others are mere backwaters – receiving imported tech and providing personal data to global corporations. This difference is not due to chance alone. it is formed by continuous historical impacts of colonialism, global economic inequalities, and exclusion of knowledge. This situation leads to an occurrence that is known as digital colonialism and explains how data and technological domination are extracted from poor countries to serve rich countries (Salami, 2024).

## 4.1 Uneven Access to AI Infrastructure and Capabilities

AI research and development (R&D) is concentrated in a handful of technologically advanced nations—such as the United States, China, the United Kingdom, Canada, and several EU member states. These countries pride themselves on having sophisticated computational systems, well-educated professionals, adequate funds, and institutional power to venture into AI technology. Nevertheless, there are significant challenges for many low- and middle-income countries (LMICs) to develop local AI industries. Problems including lack of adequate broadband, lack of funds to support public education, lack of computing power, and emigration of skilled workers to international employers or universities undermine these countries (Bircan & Özbilgin, 2025).

As a result, these countries often become consumers rather than producers of AI technologies. Consequently, organizations have to rely on foreign vendors, whose products and services may not respect local cultural, legal, or ethical values. The importation of foreign face recognition or predictive policing technologies can lead to systems that are unsuitable for domestic governance and could reinforce authoritarianism by accident or design (Volinz, 2025).

Also, the lack of local capacity prevents these nations from creating strong AI policies or meaningfully participating in global AI governance. Without involvement in standard-setting organizations, their views on data ethics, cultural rights, and governance that is rights-based on individuals are often ignored, thus reinforcing global disparities in the development of technology.

## 4.2 Data Extraction and the New Digital Resource Economy

One of the most important elements of digital colonialism is data extraction, which describes the collection and commercialization of individual and communal data, from individuals and communities in the Global South, by organizations in the Global North. This is facilitated by heavy dependance on mobile applications, digital financial services, biometric registration systems and social media platforms (Hill et al. 2024).

Many of these platforms are designed without adequate safeguards for informed consent, data protection, or community ownership. As an example, humanitarian organizations engaged in international health or development projects could collect biometric or medical data in the name of aid – only to discover that this data is exploited and sold for profit by companies from the Global North. Sometimes, communities become testing laboratories for bleeding-edge and potentially controversial technologies, such as the surveillance and AI, which could not legally function in more established legal systems (Challoumis, 2024).

This strategy is reminiscent of old extractive systems where raw materials from colonized areas were delivered to support the industrial expansion of imperial powers. Now, data becomes the underlying resource, which allows developing digital algorithms, platforms, and predictive models that produce economic value, but without any corresponding benefits to the providers of the data.

## 4.3 Cultural and Epistemic Exclusion in AI Design

AI systems are not only shaped by data and code but by the values, assumptions, and worldviews of their developers. Currently, ethical guidelines, technical norms, and design principles are primarily based on the views of the Global North. Due to this, AI technologies tend to include biased cultural perspectives that might leave out or misrepresent the realities, languages, and ways of knowing of people in the Global South.

This type of epistemic exclusion is clear in many ways. Indigenous ways of knowing and communal approaches to data ownership are mostly ignored in the development of AI systems. Because of the absence of minority languages in natural language processing models, these systems occasionally underperform or exclude users of minority languages from digital services. The impact of local contexts, socio-political dynamics, and historical injustice is often neglected in the development of training data sets and evaluation procedures (Miragoli, 2024).

Consequently, policymakers, representatives of civil society, and scholars from the Global South are conspicuous by their absence in the leading international AI platforms, events, and funding initiatives. The debate on ethics and governance is impoverished of diverse voices when Global South actors are omitted, and the impression that only a few nations hold technological leadership continues to escalate global power dynamics.

## 4.4 Reclaiming Digital Sovereignty and Global Justice

Confronting the realities of digital colonialism necessitates a conscious and strategic effort to reclaim technological sovereignty and establish more equitable frameworks for AI development. The reason behind the foundation for this process is the improvement of local AI ecosystems with strong investments in such areas as education, technical research facilities, and the creation of accessible open-source technologies. It's mighty important that sturdy national and regional institutions take the reins of AI initiatives, ensuring they're tailored with a keen eye to the varied social, economic, and cultural landscapes that stretch across this vast land (Roumate, 2024).

Just as vital is the tending of bonds 'twixt the South, where folks come together to lend a hand and share their fortunes. By banding together, the nations of the Global South can gather their resources, align their ways of governing, and forge plans that echo their shared pasts and rising ambitions for the future. These partnerships serve as a fine substitute for the might of the global North's plans and foster the growth of ethics frameworks for AI that are suited to the local terrain (Costa-Barbosa et al. 2024).

Championin' data justice rides high on the list of priorities for this here sovereignty initiative. Fair governance requires frameworks focused on collective agreement, community-managed data, and leadership models based on indigenous and local traditions. In contrast to viewing data as just a harvestable commodity, these frameworks see data as a communal asset, whose governance is regulated by dignity, equity, and autonomy.

There is a need to crack down on extractive corporate activities to remedy and correct the systemic issues that exist in the international AI economy. Policy makers and regional actors must follow data privacy standards that honor cultural and legal traditions, compel international businesses to equitably distribute profits within national boundaries, and encourage inclusive community involvement in the development of new technologies. Without the requisite regulatory oversight, numerous countries are on the verge of being dominated by exploitative digital technologies that curtail their self-governance (Harper & Millard, 2024).

Examples such as the African Union's Data Policy Framework or the developing digital sovereignty initiatives of Latin America show growing responses to dependency and reliance on external technological dominance. These undertakings ain't merely about protection. Their aim is to reshape the way we reckon with ethical AI, focusing on the power of local choices, the spirit of inclusivity, and the pursuit of fairness in the grand arena of global tech governance. In this manner, they stand against the rule of rigid ways and carve out a path for a fair, just, and truly cooperative future among all parties involved.

## 5. ETHICAL GOVERNANCE OF AI

The management of artificial intelligence has risen to a pressing concern in this modern age, as nations, enterprises, and the common folk grapple with the intricate task of wrangling technological advancements while protecting the rights that are dear to us all. It speaks to the set of organized codes, laws, and practices that aim to steer the research, deployment, and decision-making of AI, ensuring it honors justice, accountability, and the welfare of folks in the community. Though there be plenty of proposed codes of conduct, a mighty challenge still looms large: the task of turning broad moral tenets into clear-cut and binding rules that can steer the influence of AI on the good folks of society.

## 5.1 Global Normative Frameworks and Multilateral Initiatives

The folks from around the world have banded together to tackle the ethical dust storms stirred up by AI, laying down some influential guidelines that speak to their common values and principles, though they ain't set in stone. The OECD Principles on Artificial Intelligence stands tall among the most significant documents, laying down the core commitments to a human-centered approach, reliability, open communication, and a sense of responsibility in this brave new frontier. Though these principles may not hold any legal sway, they sure pack a punch when it comes to steering national policies and fostering camaraderie among nations (Feldstein, 2024).

The UNESCO Recommendation on the Ethics of Artificial Intelligence stands as a mighty testament to a broader and rights-driven way of reckonin' with the world of machines. As the first all-encompassing global standard on AI ethics, endorsed by member states of UNESCO, it provides explicit instructions for respecting human rights, fostering environmental sustainability, and safeguarding cultural diversity in AI systems deployment. One of the main points is that there is a focus on the need to evaluate impacts as well as the need for public monitoring and existence of legal means of recourse. Although it is voluntary in nature, the UNESCO framework points to an important step towards institutionalizing ethical norms in AI governance (Natorski, 2024).

However, the extent of these initiatives is constrained by the imbalance in involvement and resources from country to country. In many instances, countries in the Global South are deficient in the resources they need in order to effectively influence international policy or apply these frameworks within their own borders. This raises questions about democratic legitimacy and universal acceptance of AI ethics debates, as these conversations are highly dominated by the Global North.

## 5.2 Regional Regulations and National Legislation

At the regional level, the European Union has taken the most decisive steps toward binding AI regulation. Under the EU AI Act proposed, applications that are classified as high risk such as those in critical infrastructure, law enforcement or biometric identification are required to have comprehensive clarity, human supervision Moreover, the Act prohibits certain AI applications which are incompatible with human rights, such as government social scoring programs. The Act purposely ensures that its rules are consistent with the fundamental rights as provided in the EU Charter (Cole, 2024).

In contrast, the United States has pursued a more decentralized model, where federal agencies issue sector-specific guidelines, and enforcement varies by jurisdiction. Blueprint for an AI Bill of Rights and other similar texts commendably establish ethical standards, but in the absence of overarching legislation, governance is fragmented and accountability is limited. Advocacy groups for civil liberties in the U.S. stress the need for greater legal controls, particularly with respect to privacy and fairness in facial recognition and automatic decision-making concerning public benefits (Huang et al. 2024).

Countries in the Global South are in the early stages of developing their own regulatory approaches. Kenya, Nigeria, Brazil, and India have each developed a national AI strategy that acknowledges human rights concerns, but the manner in which they implement such plans is strikingly different. The African Union and Smart Africa are driving the agenda to develop comprehensive pan-African guidelines based on data sovereignty, social equity, and ethical innovation. These undertakings show a rising awareness that the ways of leading ought to be shaped by the local ways and beliefs, instead of borrowing from distant lands.

## 5.3 Private Sector Ethics and Structural Limitations

In the dusty plains of the tech frontier, the mighty companies stand tall, wielding their power over the very foundations of artificial intelligence. With their grip on the tools, the skilled hands, and the knowledge that guides the way, they carve out the future of this brave new world. To heed the call of the townsfolk, a handful of mighty corporations have taken to the task at hand: they've formed councils of ethics, laid down the law with responsible AI guidelines, and vowed to carry out regular fairness audits like clockwork. Though these measures suggest a stride toward greater accountability, they frequently fall short of delivering on their promises when faced with the harsh realities of the world. One of the biggest problems is that the corporate ethics programs are not overseen independently and in a transparent manner, and they are not backed by strong enforcement mechanisms. Some examples show that ethics teams were dismissed or marginalized after speaking out against practices that jeopardized the public interest for profit (Soudi & Bauters, 2024).

When firms use ethical language to shield their reputation from criticisms, without actually changing harmful practices, this has put a question mark on the credibility of voluntary industry standards. For the most part, organizations' internal ethics frameworks largely exclude substantial involvement from the communities that are impacted by their actions, and often the Global South is excluded. As the inadequacies described above are evident, it is obvious that external regulation, public scrutiny and participation by the civil society are required to complement and oversee corporate self-regulation.

## 5.4 Multi-Stakeholder Models and Participatory Governance

Recognizing the limitations of both state and corporate actors, a number of multi-stakeholder initiatives have emerged to promote inclusive and collaborative governance. These platforms, including the Global Partnership on AI, Partnership on AI and AI Now Institute, bring together stakeholders from government, industry, academia and civil society to collaborate in the setting (Wong et al. 2025).

These efforts try to make AI ethics discussions more inclusive by relying on a diversity of perspectives and personal backgrounds. However, their success depends on real power-sharing, fair representation, and continuous financial backing. If the involvement is shallow or donor-driven, such efforts may not carry true authority and outcomes. In order to make multi-stakeholder governance a success, it is necessary to integrate mutual accountability, an inclusive process, and a true respect for local knowledge.

## 5.5 Toward Enforceable and Inclusive Ethical Governance

Despite the proliferation of ethical frameworks, a major gap persists between principles and practice. Existing approaches tend not to offer enforceable authority, strong oversight, or avenues for those damaged by AI systems. Closing this gap requires a shift toward hard law principles, backed up by international human rights law, constitutional rights, and statutory regulations.

At the same time, governance must be inclusive and adaptable. This involves building regulatory capacity in a wide variety of legal systems, especially in areas that may be under-funded. supporting the adoption of open data practices as well as the creation of open-source tools to enhance transparency and empowering grassroots organizations that work on digital rights. This also calls for a variety of perspectives in ethical debates because definitions of harm, fairness and consent can vary widely within and across cultural, historical, and political contexts (Adeniyi & Okusi, 2024).

Ethical oversight of AI systems is not simply about technology, but is essentially a political and societal issue. It calls for continuous commitment, creative policy-making, and a commitment to justice that transcends borders and corporate interests. Expanding this conversation, the following section offers practical recommendations on how to integrate human rights principles into the entire lifecycle of AI systems in countries all over the world.

# 6. TOWARD HUMAN RIGHTS-CENTERED AI SOLUTIONS

The mounting evidence of AI-induced harm—from discriminatory algorithms to unchecked surveillance and digital colonialism—has catalyzed urgent calls for systemic reform. Having decided that voluntary ethical codes and violations of regulations are too little, now it is clear that a unified global perspective is emerging: Governance of artificial intelligence demands the creation of enforceable rules that are based on human rights regulations. This here section lays bare the vital pieces and particular remedies for crafting AI frameworks that uphold fairness, honor for all, and social justice across the wide expanse of the globe.

## 6.1 Embedding Human Rights Values in AI Design

The time spent in the design phase of AI development is mighty important, for it determines if them systems will uphold justice or turn a blind eye to it. A way of reckonin' that puts the rights of folks front and center is built on a steadfast promise to fairness, accountability, transparency, and inclusivity. This ain't just a matter of good manners, but a binding duty, both in law and in the code of ethics. Fairness cannot be attained unless AI applications are used in a non-discriminatory manner and consistently treat people from all races, genders, economic strata, and abilities equally. Effective implementation requires consideration of the nature of training datasets, the assumptions behind the models, and the larger environments in which these systems will be deployed (Cheong et al. 2024).

The principle of accountability is defined by the definition of roles and duties during the process of designing and deploying AI systems. Both developers, organizations, as well as government bodies, must be identifiable so that they are accountable for the consequences of AI-related decisions. Lacking accountability measures, incidents of injustice tend to spread out making it hard for people to seek remedy (Sadek et al. 2024).

Transparency, meanwhile, involves making AI systems understandable to both technical and non-technical stakeholders. This means specifying which data sources are used, how models are structured, what the system is intended for, and what restrictions or problems the developers are aware of. It is vital in public services and key areas for transparency for trust development and democratic credibility. Finally, inclusivity involves involving individuals directly affected to participate in the AI project constantly from need definition, solution design, to overseeing outcomes. Inclusive design ensures that people from the marginalized backgrounds are involved and formally included at every stage of decision-making.

## 6.2 Institutionalizing Human Rights Impact Assessments

One of the most hopeful ways to put human rights principles into action is through the Human Rights Impact Assessment (HRIA). In light of past reckonings in the realms of nature and personal freedoms, HRIAs call for a thorough appraisal of an AI contraption's influence on fundamental rights, both before it rides out and while it's in the saddle. A well-executed plan for HRAIs can spot and diminish dangers to rights, like discrimination, invasions of privacy, or curtailments on freedoms before they come to pass (Gentelet & Mizrahi, 2024).

A fine HRIA process ensures that all voices are heard, the path is clear, and folks can rest easy knowing an impartial eye has taken a gander at the matter. Folks from the civil society, the technical circles, and the rights holders, particularly them most impacted, ought to be at the heart of this here endeavor. Now listen here, HRIAs oughta steer clear of just stickin' to them checklist procedures, ya hear? Contextual analysis ought to be a core part of these reckonings, takin' into account how various social identities stir up overlapping dangers. This here process should also offer some down-to-earth guidance for wranglin' or puttin' a stop to them destructive technologies (Fedele et al. 2024).

For them to pack a punch, HRIAs oughta be woven right into the fabric of the law. It's high time that the law laid down the law, mandating HRIAs to be brought into the fold of law enforcement, the distribution of public benefits, and the realm of healthcare, especially where dangers loom large. Legally binding HRIAs ride into town to elevate corporate standards, makin' sure that the right protections are secured beyond the reach of mere good intentions.

## 6.3 Legal and Regulatory Reinforcement of Rights-Based AI

Beyond them fancy assessment tools, a full-fledged human rights way of reckonin' with AI calls for a mighty shift in the legal and regulatory landscapes. It's high time for the lawmen to lay down the law, making sure that them developers, deployers, and controllers toe the line with clear and binding duties. In the dusty town of data, where algorithms roam free, it's high time we laid down some clear rules for interpretability, kept a sharp eye out for bias, safeguarded our privacy like a trusty sidearm, and ensured that justice is served with proper redress mechanisms.

Solid frameworks depend on independent regulatory outfits tasked with keeping a watchful eye on the systems, meting out justice when needed, and shining a light on transparency. It's high time we rustle up some regulatory bodies and give 'em the power they need, not just in the Global North, but across the whole wide world. It's mighty important for the Global South to rustle up some regulatory skills, gather the right tools, and band together, so they can set up a solid framework for global governance (Pantanowitz et al. 2024).

Furthermore, the tenets of international human rights law serve as a common standard that unites the diverse landscapes of nations far and wide. The ICCPR and ICESCR, being binding agreements, lay down a solid foundation for ensuring that AI regulations align with the human rights acknowledged across the globe. Wrangling the standards with a touch of flexibility for each territory is backed by the lawmen of regulation, like them model laws that ride the regional range.

## 6.4 Capacity-Building for Technological Self-Determination

To make sure every town and settlement reaps the rewards of AI fairly, it takes a mighty effort to build up the skills and resources, especially in those places that ain't got much to start with. This ain't just about buildin' up them structures and sharpenin' them digital skills, but also about tendin' to the local grounds that foster research, innovation, and good governance. First off, it's mighty important for folks in each town and territory to wrangle their own resources, so they can set up, steer, and fine-tune them AI contraptions to suit their own way of life (Jha & Singh, 2025).

There ought to be a hand extended to the local schools of learning, the social outfits, and the governing folks to rustle up knowledge and tools that speak to the heart of the culture. Open-source AI platforms, well-crafted ethical development kits, and welcoming data resource centers can aid in spreading the wealth and lessen the hold of those exclusive, far-off platforms (Sayles, 2024).

It takes a mighty band of nations to wrangle up the results we seek in this here land. In the realm of development, it's wise to forge alliances that lean on equal footing and adapt to the unique ways and circumstances of the local folks. When the Southern territories join forces to tackle shared troubles, their alliances are sure to forge AI systems that honor the local ways and uphold the fundamental rights of the folks.

## 6.5 Participatory and Inclusive Governance Mechanisms

At long last, the time has come to put the reins of governance in the hands of the people, ensuring that the heart of human rights beats strong in the realm of AI. This here signifies a shift from the ways of the old hands and the quick-fix gadgets, to welcome fresh viewpoints, especially those that have been overlooked or pushed aside in times gone by. Joining in the gatherings like town hall meetings, community powwows, and workshops crafted hand in hand with the folks, reveals what matters in these parts and assures that the laws we lay down are rooted in the lives of the people (Tennison, 2024).

Participation oughta be more than just a show, partner. In them dusty halls of decision-making, there oughta be clear trails for the folks of the community to make their voices heard and truly sway the choices that shape their lives. It's mighty important that folks can communicate freely, that the hands who lend their skills are paid fair, and that there's plenty of chances to learn, so everyone can join in and make their voices count (Lartey & Law, 2025).

By wrangling these inclusive strategies, we bolster the quality and trustworthiness of AI systems, all while ensuring that our governance rides alongside democratic values that champion human rights. The aim here is to build a digital frontier that gives power to those who've long been pushed aside, making their voices the heart of the matter, not just a whisper in the wind.

## 7. POLICY RECOMMENDATIONS

To ensure that the mighty force of artificial intelligence upholds justice, fairness, and the dignity of folks, it's high time we put in place some solid, enforceable policies instead of just hangin' our hats on lofty ideals. Though a heap of ethical codes have been laid down, the absence of a sturdy legal framework and the means to hold folks accountable have dulled their edge in steering the course of AI's growth. This here part lays out a range of policy strategies aimed at wrangling AI with a focus on human rights, shining a light on the structures, the lawmen, the watchful eyes, and the alliances across the wide frontier.

### 7.1 Institutionalizing Human Rights Impact Assessments

The foremost task at hand is the essential enforcement of Human Rights Impact Assessments for all AI contraptions that touch upon the rights and liberties of folks. Regular reckonings will take place at pivotal moments in the design, before the launch and throughout the operation, to spot and fend off any looming troubles. For the HRIA process to ride tall in the saddle, it must gather a range of viewpoints. Insights ought to be drawn from the folks in civil society, the sharp minds of technical specialists, the wisdom of legal scholars, and those who feel the brunt of it all (Teo, 2025).

Assessments of human rights ought to be laid bare for all to see, and it should be a must that they be made public, ensuring genuine adherence to the law of the land. Howdy there, partner. What brings you to this dusty ol' town? The sun's settin', and the saloon's callin' your name. You best saddle up and make your way before the night creeps in. Ain't no tellin' what lurks in the shadows when the stars come out. So, what'll it be? A drink to wet your whistle or a game of cards to test your luck? The choice is yours, but remember, in these parts, every decision counts. That there public watchin', in the shape of audits, lets the folks and the regulators take a gander at them assessments, challenge them shoddy methods, and make them companies sit up and take notice of the dangers that've been pointed out. When them HRIAs get put into law, them governments can make it so that pushin' for hRIAs ain't just a choice, but a rule that must be followed in the life cycle of AI, a legal obligation, no less.

## 7.2 Enacting Comprehensive and Enforceable AI Legislation

It's high time for them folks in charge to lay down some solid laws that set clear duties for them AI wranglers, sellers, and users out there in the wild frontier of technology. It's high time we set forth some sturdy regulations, ones that don't just wrangle with the safeguarding of data, but also lay down clear assurances for the civil, political, economic, and social rights in their entirety. In the dusty town of legal scrutiny, the main matters at hand revolve around the clarity and understanding of algorithms, the fairness of their reviews, the offering of remedies, and the oversight of the deployment of invasive technologies such as biometric systems and predictive policing (Parinandi et al. 2024).

To wrangle the chaos of differing regulations, it's high time for national laws to ride in line with the international code of human rights. They ought to be bolstered by state model legislation that offers both technical and legal spurs to get the job done right. Bringing the law into harmony can be done through binding agreements like treaties, regional compacts, or global declarations, each one allowing for a bit of local flavor and circumstance. Laws should only be effective when they can be enforced. This requires the setting of well-funded regulatory institutions capable of investigating AI related harms, make corrective steps, and apply sanctions in case of rights violations.

## 7.3 Establishing Independent Oversight and Democratic Accountability

Regulation alone is insufficient without independent institutions to oversee compliance and provide redress. The authorities should create independent oversight bodies which would work to supervise the use of artificial intelligence in the government and business sphere. Such entities must embody a wide range of voices including civil society groups, technology experts, ethicists, lawyers, etc., from disadvantaged communities (Bogiatzis-Gibbons, 2024).

To preserve their legitimacy, these oversight institutions must operate independently from the entities they are tasked with regulating. They should be allowed to conduct investigations, call for transparency of AI systems and their data, and make enforceable decisions. They should promote the engagement by the citizens in the policy-making process through use of public hearings and consultations besides having transparent reporting procedures. Practically, effective oversight promotes responsibility, more than for purposes of deterrence: society can actively steer the trajectory of AI development.

## 7.4 Leveraging Public Procurement for Ethical Innovation

As major purchasers of AI systems, governments wield significant influence over market behavior through procurement practices. Craftin' rights-based procurement standards gives the states a powerful tool to steer the moral compass in the wild frontier of AI development. Contracts in the public sector for technology ought to lay down the law on fairness, clarity, safeguarding of data, and designs that welcome all folks (Ridzuan et al. 2024).

Furthermore, the rules for procurement ought to champion transparency and openness, compelling suppliers to share details 'bout how their systems run, where the data comes from, and the results of them impact evaluations. To bring about fairness in the community, it's high time that the folks in charge of public policies round up the local stakeholders to have a hand in crafting AI tools for essential services. They ought to keep a keen eye on these systems to ensure no one gets the short end of the stick. When public funds ride alongside a code of ethics, procurement can saddle up and become a mighty force in championin' the innovation of rights within organizations.

## 7.5 Advancing Global Solidarity and Capacity-Building

To wrangle the inconsistencies in AI governance across the globe, we need more than just talk; we need real deeds. It's high time we rolled up our sleeves and put in the hard work to bolster the capabilities in the lagging territories, like the Global South. These undertakings encompass putting money into the roads and railways, embracing the tools of the open frontier, fostering partnerships for the sharing of wisdom, and making certain that the folks are schooled in the ways of the digital world, the law, and the moral codes that guide us, all woven into the educational fabric of our varied settlements (Sevinç, 2024).

The path to lend a hand ought to be a challenge, where folks join forces on equal ground, all while keeping it respectful. It ain't about one-sided dealings; it's about putting the needs of the stakeholders front and center. In the dusty trails of progress, there lie strategies that hold promise for stirring up innovation that's both fitting and resourceful. South-South cooperation, regional AI alliances, and community-driven research projects are the trusty steeds that can carry us forward into the frontier of discovery. It's mighty important that them places blessed by AI take it upon themselves to share their resources and know-how. This way, we can make sure that all folks get a fair shake at technology, without a handful of companies holdin' the reins too tight (Georgiadou, 2024).

The aims of global fairness can be brought together through the involvement of international outfits, development agencies, and charitable institutions, all riding into the sunset of cooperation. They ought to rally 'round the campfire to spread the word about the trials of fresh regulations, swap wisdom like old friends, and make certain that the folks from the Global South have a fair seat at the table when it comes to setting the agenda.

## 7.6 Institutionalizing Inclusive and Participatory Governance Models

At long last, the way to wrangle AI governance ought to be a job for all hands on deck. It's them folks, the ones who stand to gain the most from AI, that ain't got a say in the matter, and that just ain't right. Rallying for a fair way of governing calls for the proper use of tools that bring folks together, like town hall gatherings, open debates, citizen panels, and shared planning endeavors. These contraptions ensure that the rules and ways of doing things are shaped by the true experiences of them folks most impacted (Ulnicane, 2025).

Participation oughta be more than just a show, partner. It oughta hold the strength to sway choices, contest results, and put forth new options. To reckon with this, one must put their coin in the pot of community learning, tip their hat to the folks who lend a hand, and clear away the obstacles of tongue, tradition, and know-how that stand in the way. In the use of these methods, folks can bridge the gap in AI policy-making and restore the trust of the townsfolk in the system.

## 8. CONCLUSION

The emergence of artificial intelligence across the land brings forth a bounty of new prospects, yet it also stirs up a heap of moral quandaries that can't be ignored. AI contraptions can be put to work in a heap of different ways, like in the realms of law, learning, labor, and the talkin' of the townsfolk. AI systems can be a double-edged sword, aiding and undermining the fundamental rights and freedoms that uphold human dignity, and that's a fact we can't ignore. This here chapter lays it down that puttin' human rights at the forefront of AI development ain't just a choice, it's a moral obligation. It's a mighty fine way to forge technology that stands on the pillars of fairness, justice, and democracy.

AI systems are already stirrin' up trouble, deepenin' the injustices that plague our lands, givin' rise to new forms of watchin' folks, and makin' it harder for good people to claim their rightful freedoms across the globe. These here results are likely shaped by folks from the rougher parts of town and those struggling nations, shining a light on the deep-rooted inequalities and the ways some folks are left out in the cold. But they ain't gotta come to pass. Yet, with the right codes of conduct, a spirit of cooperation, and watchful eyes, folks can wrangle these troubles into submission, ensuring that the wonders of AI serve the good of all.

This here chapter lays down the law that human rights oughta be the cornerstone of how we wrangle AI, touchin' every aspect of gatherin' data, craftin' algorithms, settin' up systems, and runnin' things on a global scale. To hit this target, we gotta ride from wide-ranging moral codes to rules set in stone that can be upheld. It also requires rules that put the folks most touched by technology right at the heart of the matter and lend an ear to their words.

More than just fresh contraptions will shape the ethical frontier of AI in the long run. There'll be reckonin' and gatherin' of folks to make them choices, sure as the sun sets in the west. To forge a brighter path for AI in the days to come, we oughta ponder on the nature of power, the true meaning of progress, and how the institutions are held accountable for the well-being of the folks. In the dusty trails of progress, the notion of AI can ride alongside the principles of human rights. We can harness this to steer the course of advancement, ensuring it benefits all folks, not just them in the saddle.

## REFERENCES


Adeniyi, A. C., & Okusi, O. (2024). Ethical AI governance, financing, and human well-being in the 21st century. *African Journal of Humanities and Contemporary Education Research, 17*(1), 193–205. DOI: 10.62154/ajhcer.2024.017.010504

Ahmad, N., Ali, A. W., & bin Yussof, M. H. B. (2025). The Challenges Of Human Rights In The Era Of Artificial Intelligence. *UUM Journal of Legal Studies, 16*(1), 150-169.

Akhtar, M. A. K., Kumar, M., & Nayyar, A. (2024). Ensuring Fairness and Non-discrimination in Explainable AI. In *Towards Ethical and Socially Responsible Explainable AI: Challenges and Opportunities* (pp. 165–192). Springer Nature Switzerland. DOI: 10.1007/978-3-031-66489-2_6

Bengio, Y., Hinton, G., Yao, A., Song, D., Abbeel, P., Darrell, T., Harari, Y. N., Zhang, Y.-Q., Xue, L., Shalev-Shwartz, S., Hadfield, G., Clune, J., Maharaj, T., Hutter, F., Baydin, A. G., McIlraith, S., Gao, Q., Acharya, A., Krueger, D., & Mindermann, S. (2024). Managing extreme AI risks amid rapid progress. *Science, 384*(6698), 842–845. DOI: 10.1126/science.adn0117 PMID: 38768279

Bircan, T., & Özbilgin, M. F. (2025). Unmasking inequalities of the code: Disentangling the nexus of AI and inequality. *Technological Forecasting and Social Change, 211*, 123925. DOI: 10.1016/j.techfore.2024.123925

Blount, K. (2024). Using artificial intelligence to prevent crime: Implications for due process and criminal justice. *AI & Society, 39*(1), 359–368. DOI: 10.1007/s00146-022-01513-z

Bogiatzis-Gibbons, D. J. (2024, June). Beyond Individual Accountability:(Re-) Asserting Democratic Control of AI. In *Proceedings of the 2024 ACM Conference on Fairness, Accountability, and Transparency* (pp. 74-84). DOI: 10.1145/3630106.3658541

Challoumis, C. (2024, October). Building a sustainable economy-how ai can optimize resource allocation. In *XVI International Scientific Conference* (pp. 190-224).

Cheong, I., Caliskan, A., & Kohno, T. (2024). Safeguarding human values: Rethinking US law for generative AI's societal impacts. *AI and Ethics*, •••, 1–27. PMID: 40352581

Cole, M. D. (2024). AI Regulation and Governance on a Global Scale: An Overview of International, Regional and National Instruments. *Journal of AI Law and Regulation, 1*(1), 126–142. DOI: 10.21552/aire/2024/1/16

Costa-Barbosa, A., Herlo, B., & Joost, G. (2024). Digital Sovereignty in times of AI: between perils of hegemonic agendas and possibilities of alternative approaches. *Liinc em Revista, 20*(2).

Fedele, A., Punzi, C., & Tramacere, S. (2024). The ALTAI checklist as a tool to assess ethical and legal implications for a trustworthy AI development in education. *Computer Law & Security Review*, *53*, 105986. DOI: 10.1016/j.clsr.2024.105986

Feldstein, S. (2024). Evaluating Europe's push to enact AI regulations: How will this influence global norms? *Democratization*, *31*(5), 1049–1066. DOI: 10.1080/13510347.2023.2196068

Gentelet, K., & Mizrahi, S. K. (2024). A Human-Centered Approach to AI Governance: Operationalizing Human Rights Through Citizen Participation. In *Human-Centered AI* (pp. 215–230). Chapman and Hall/CRC. DOI: 10.1201/9781003320791-24

Georgiadou, K. (2024). Solidarity, Gender, Dialogue in the AI era. In *Kritische Pädagogik und Bildungsforschung: Anschlüsse an Paulo Freire* (pp. 477–489). Springer Fachmedien Wiesbaden. DOI: 10.1007/978-3-658-41415-3_23

Harper, E., & Millard, J. (2024). Global Approaches to Digital Sovereignty: Legal Mechanisms for Managing Data and Digital Infrastructure. *Legal Studies in Digital Age*, *3*(4), 36–46.

He, Y. (2024). Artificial intelligence and religious freedom: Divergent paths converging on economic expansion. *Humanities & Social Sciences Communications*, *11*(1), 1–12. DOI: 10.1057/s41599-024-02642-0

Hill, J. E., Harris, C., & Clegg, A. (2024). Methods for using Bing's AI-powered search engine for data extraction for a systematic review. *Research Synthesis Methods*, *15*(2), 347–353. DOI: 10.1002/jrsm.1689 PMID: 38066713

Huang, K., Joshi, A., Dun, S., & Hamilton, N. (2024). AI regulations. In *Generative AI security: theories and practices* (pp. 61–98). Springer Nature Switzerland. DOI: 10.1007/978-3-031-54252-7_3

Jacobs, G., Van Houdt, F., & Coons, G. (2024). Studying Surveillance AI-cologies in Public Safety: How AI Is in the World and the World in AI. *Surveillance & Society*, *22*(2), 160–178. DOI: 10.24908/ss.v22i2.16104

Jha, A., & Singh, S. R. (2025). Bridging the Divide: Capacity Building for AI Adoption in Developing Countries. In *AI Strategies for Social Entrepreneurship and Sustainable Economic Development* (pp. 133-150). IGI Global Scientific Publishing.

Kan, C. H. (2024). Artificial intelligence (ai) in the age of democracy and human rights: Normative challenges and regulatory perspectives. *International Journal of Eurasian Education and Culture*, *9*(25), 145–166.

Kathuria, Y. Ruhani, Vandana, Tyagi, M., & Jain, V. (2024, October). Protecting data privacy in the age of AI: A comparative analysis of legal approaches across different jurisdictions. In *AIP Conference Proceedings* (Vol. 3220, No. 1, p. 040007). AIP Publishing LLC.

Lartey, D., & Law, K. M. (2025). Artificial intelligence adoption in urban planning governance: A systematic review of advancements in decision-making, and policy making. *Landscape and Urban Planning*, *258*, 105337. DOI: 10.1016/j.landurbplan.2025.105337

Lütz, F. (2024, March). The AI Act, gender equality and non-discrimination: what role for the AI office? In *ERA Forum* (Vol. 25, No. 1, pp. 79-95). Berlin/Heidelberg: Springer Berlin Heidelberg. DOI: 10.1007/s12027-024-00785-w

Miragoli, M. (2024). Conformism, Ignorance & Injustice: AI as a Tool of Epistemic Oppression. *Episteme (Edinburgh)*, •••, 1–19.

Moussawi, S., Deng, X., & Joshi, K. D. (2024). AI and Discrimination: Sources of Algorithmic Biases. *The Data Base for Advances in Information Systems*, *55*(4), 6–11. DOI: 10.1145/3701613.3701615

Natorski, M. (2024). Compromise in multilateral negotiations and the global regulation of artificial intelligence. *Democratization*, *31*(5), 1091–1116. DOI: 10.1080/13510347.2024.2351583

Pantanowitz, L., Hanna, M., Pantanowitz, J., Lennerz, J., Henricks, W. H., Shen, P., & Rashidi, H. H. (2024). Regulatory aspects of AI-ML. *Modern Pathology*, •••, 100609. DOI: 10.1016/j.modpat.2024.100609 PMID: 39260776

Parinandi, S., Crosson, J., Peterson, K., & Nadarevic, S. (2024). Investigating the politics and content of US State artificial intelligence legislation. *Business and Politics*, *26*(2), 240–262. DOI: 10.1017/bap.2023.40

Peña Acuña, B., Fernández-Aller, C., Zavrsnik, A., Sarf, P., Ramus Cvetkova, I., Binark, M., & Ramos, A. (2025). *Artificial Intelligence and Digital Rights*. Opportunities and Challenges.

Prove, P. (2024). Freedom of expression, and the protection of journalists in conflict situations. *Media Development*, *70*(3).

Ridzuan, N. N., Masri, M., Anshari, M., Fitriyani, N. L., & Syafrudin, M. (2024). AI in the financial sector: The line between innovation, regulation and ethical responsibility. *Information (Basel)*, *15*(8), 432. DOI: 10.3390/info15080432

Roberts, H., Hine, E., Taddeo, M., & Floridi, L. (2024). Global AI governance: Barriers and pathways forward. *International Affairs*, *100*(3), 1275–1286. DOI: 10.1093/ia/iiae073

Romero Moreno, F. (2024). Generative AI and deepfakes: A human rights approach to tackling harmful content. *International Review of Law Computers & Technology*, *38*(3), 297–326. DOI: 10.1080/13600869.2024.2324540

Roumate, F. (2024). AI and technological sovereignty. In *Artificial Intelligence and the New World Order: New weapons, New Wars and a New Balance of Power* (pp. 59–65). Springer Nature Switzerland. DOI: 10.1007/978-3-031-50312-2_5

Sadek, M., Calvo, R. A., & Mougenot, C. (2024). Designing value-sensitive AI: A critical review and recommendations for socio-technical design processes. *AI and Ethics*, *4*(4), 949–967. DOI: 10.1007/s43681-023-00373-7

Saeidnia, H. R., Hosseini, E., Lund, B., Tehrani, M. A., Zaker, S., & Molaei, S. (2025). Artificial intelligence in the battle against disinformation and misinformation: A systematic review of challenges and approaches. *Knowledge and Information Systems*, *67*(4), 1–20. DOI: 10.1007/s10115-024-02337-7

Salami, A. O. (2024). Artificial intelligence, digital colonialism, and the implications for Africa's future development. *Data & Policy*, *6*, e67. DOI: 10.1017/dap.2024.75

Sancin, V., & Bobnar, L. (2024). The Right to Freedom of Expression in the Era of Artificial Intelligence Systems. *Pravni Letopis*, 109.

Sayles, J. (2024). Capacity Building for AI Actors. In *Principles of AI Governance and Model Risk Management: Master the Techniques for Ethical and Transparent AI Systems* (pp. 321–331). Apress. DOI: 10.1007/979-8-8688-0983-5_16

Sevinç, T. (2024). Global solidarity. *Journal of Global Ethics*, *20*(3), 308–316. DOI: 10.1080/17449626.2024.2422364

Soudi, M. S., & Bauters, M. (2024). AI guidelines and ethical readiness inside SMEs: A review and recommendations. *Digital Society : Ethics, Socio-Legal and Governance of Digital Technology*, *3*(1), 3. DOI: 10.1007/s44206-024-00087-1

Stockinger, A., Schäfer, S., & Lecheler, S. (2025). Navigating the gray areas of content moderation: Professional moderators' perspectives on uncivil user comments and the role of (AI-based) technological tools. *new media & society, 27*(3), 1215-1234.

Tennison, J. (2024). Why We Need Inclusive Data Governance in the Age of AI. *The Role of Governance in Unleashing the Value of Data*, 74.

Teo, S. A. (2025). The Unbearable Likeness of Being: How Artificial Intelligence Challenges the Social Ontology of International Human Rights Law. *The Journal of Cross-Disciplinary Research in Computational Law*.

Ulnicane, I. (2025). Governance fix? Power and politics in controversies about governing generative AI. *Policy and Society*, *44*(1), 70–84. DOI: 10.1093/polsoc/puae022

Volinz, L. (2025). Authoritarian Surveillance: An Introduction. *Surveillance & Society*, *23*(1), 112–116. DOI: 10.24908/ss.v23i1.18975

Wong, J., Morgan, D., Straub, V. J., Hashem, Y., & Bright, J. (2025). Key challenges for the participatory governance of AI in public administration. In *Handbook on Governance and Data Science* (pp. 179–197). Edward Elgar Publishing. DOI: 10.4337/9781035301348.00017

Yanamala, A. K. Y., & Suryadevara, S. (2024). Navigating data protection challenges in the era of artificial intelligence: A comprehensive review. *Revista de Inteligencia Artificial en Medicina*, *15*(1), 113–146.

# Chapter 12
# Use of Artificial Intelligence in Social Sciences for Improving Problem Solving

**T. Venkat Narayana Rao**
https://orcid.org/0000-0002-1996-1819
*Sreenidhi Institute of Science and Technology, India*

**Damacharla Tejaswini**
https://orcid.org/0009-0004-2857-3692
*Sreenidhi Institute of Science and Technology, India*

**C. Swetha**
https://orcid.org/0009-0005-1835-0679
*Sreenidhi Institute of Science and Technology, India*

**Sangers Bhavana**
https://orcid.org/0009-0005-6134-1081
*Sreenidhi Institute of Science and Technology, India*

## ABSTRACT

*The field of social science is being transformed by rising artificial intelligence implementation. Through AI technology researches achieve faster data analysis as well as automation of operations and access innovative methodologies of problem-solving. The chapter evaluates current AI applications in social science domains while assessing the advantages with the difficulties these implementations introduce. In sociology, AI enhances social behaviour analysis, AI models predict election outcomes in political science, AI-driven mental health chatbots in psychology, AI-*

DOI: 10.4018/979-8-3373-2597-2.ch012

*powered adaptive learning systems in education, predictive policing leverages AI to analyse crime patterns and anticipate potential criminal activities in criminology and AI driven stock market prediction in economics. Artificial intelligence interacts with social science research methodologies represent a significant advancement in the field. It is important to stay aware of ethical concerns such as transparency in AI models and use this powerful technology responsibly.*

## 1. INTRODUCTION

Artificial Intelligence is the science and engineering of making smart machines, exclusively intelligent computer programs. Methods of the artificial intelligence are machine learning, natural language processing, neural networks, automation and robotics. Artificial intelligence will be implemented to enhance both protective and strategic cyber activities. It is the future of all complex decision making. Some of the applications are fraud detection, self – driving cars and smart assistants like amazon Alexa, google assistant, Apple Siri and oracle digital assistant.

Social sciences which studies human social behaviour and social relationships through various disciplines, inclusive of sociology, psychology, anthropology, economics and political science. It makes use of both qualitative and quantitative research methods to assess social interactions, institutions and cultural norms. It seeking to grasp and tackle intricate aspects of social sciences. Social science research involves the exploration and investigation of societal issues and challenges within communities.

Artificial intelligence is developing date to date and it has the potential to transform social research helps predicts future outcomes, and guides decision-making in areas like criminal justice, healthcare and education. which emerging with large language models (LLMS) is one of the reason for reconsidering artificial general intelligence possibilities. When we are creating and developing human like potential in AI, it attracts attention towards social science research.

Artificial intelligence in social sciences is much more effective than traditional approaches because traditional approaches which depends on interviews, surveys, and observations to gather related information. It is time taking process, slow decision-making, and has data limitations. As we discussed in above paragraph Artificial Intelligence has transformed the way we analyse social patterns, more efficient and making research faster.

In social sciences there are various domains but in this chapter we discuss mainly about sociology, psychology, political science, education, criminology, and economics. Sociology means the study of how people live together in society and how they interact. Psychology means study of human mind and behaviour. Political science means the study of government. Educations is the process of developing skills, acquiring knowledge and building values. Criminology is the study of crime. and the last one Economics is the study of how people, businesses, and governments use money and resources.

Ai is used in these domains to enhance problem-solving by uncovering trends, and examining information. And what type of ai models is used, advantages and challenges of using these is included. As Ai systems are more frequently being used to make decisions that impact people's lives-such as in healthcare, education, and law enforcement. Transparency in Ai models is the most pressing issue that means if we don't understand how an ai system comes to a decision, it is difficult to question or correct the errors.

Over the last few years, tools such as large language models, emotion recognition systems, and predictive social modeling have begun changing the way social science research is carried out.AI is being used by some institutions in US,Europe and Asia for mental health,eduaction and policy planning.It is important to make sure that data used to train these models should be fair and unbiased. Therefore, a careful balance must be maintained between embracing innovation and upholding ethical standards.

## 2. REVIEW OF LITERATURE

The study found to bridge this gap by combining Ai-relevant characteristics with the theory of planned behaviour and the technology-environmental, economic, and social sustainability. The findings indicated that although perceived behavioural control exerted a minimal influence on the use of generative AI, perceptions proved to be the most influential determinant, further strengthened by the notable impact of social norms (Emran et al.,2024).

The Invention of the research lay in identifying the main areas of transformation in university education, particularly within the context of virtual interaction. The research sought to examine the possible benefits and difficulties of implementing AI in law and education. Some of the scholars provided studies and insights for the development of online education (Gridchina et al., 2023)

The analysis looked at how the fast growth and spread of artificial intelligence (AI) tools were affecting science and society. It then provided an overview of the current uses of AI, and then explored how AI could bring benefits or create challenges in various scientific fields. At last the analysis sought the potential risks AI poses to society, and recommended solutions for addressing those risks(Dorigo, et al.,2025).

Artificial Intelligence (AI) emerged as a transformative technology across various sectors, particularly in the economic domain. The e-commerce industry adopted AI tools such as data analytics, deep learning, and pattern recognition to study customer behaviour through browsing and social media activity. Platforms like Amazon and Flipkart utilized AI to personalize shopping experiences. While AI offered significant advantages, it also raised ethical concerns and limitations that needed critical consideration in its continued development (Saleem et al.,2023).

In recent times, Artificial Intelligence (AI) has taken on a key role in strengthening spam detection across social platforms. Experts pointed out the adaptability of AI techniques in spotting patterns, which makes them well-suited to deal with the ever-evolving nature of spam messages. The research examined a range of AI tools and models used for identifying spam and assessed their effectiveness using indicators such as accuracy, recall, precision, and F1-score. It also compared the performance of these methods, highlighted ongoing issues and technical hurdles, raised concerns around data privacy, and suggested possible directions for future exploration. Overall, the study emphasized how AI has been successfully applied to reduce spam and improve the safety of digital communication (Hattab et al.,2023).

The study aimed to automatically categorize social support threads on public social media platforms using sentiment analysis. It emphasized the role of online spaces in offering emotional support and explored different support types through AI models. The researchers used data augmentation to build multiple datasets and compared model performances. Findings showed that while augmentation helped in some cases, imbalanced data and subtle emotional cues posed challenges, highlighting the need for improved techniques and more balanced training data in emotional classification tasks (Hsu et al.,2024).

Though Artificial Intelligence (AI) has existed since the 1950s, its widespread adoption in the past decade has sparked discussions about its impact in academia. While AI tools help organizations produce materials like reports and presentations, their use in academic settings, especially by students, raises concerns about cheating and the lack of originality. Some students have submitted AI-generated assignments and exams, which leads to debates on whether this constitutes cheating or simply less independent work. The paper discusses how, through two years of investigation and personal student experiences, AI can be positively integrated into academia, offering guidelines for its appropriate use(Vold & Eikset, 2024)

This study focused on using Artificial Intelligence (AI) to detect signs of depression through data collected from wearable health devices. The concept of AI sentience has become an important topic of discourse and inquiry in light of the remarkable progress and capabilities of large language models (LLMs). Specifically, they focus on sociocultural perspectives on interpersonal relationships, sociolinguistics, and culture to consider whether LLMs are sentient. Drawing from everyday examples and references in science fiction, the authors argue that LLMs lack true sentience and are unlikely to achieve it. Instead, they propose that AI should be seen as a powerful tool shaped by how we design and apply it, especially in contexts where human interaction and culture play a key role (Alvero & Peña, 2023).

The study focused on using Artificial Intelligence (AI) to detect signs of depression through data collected from wearable health devices. Since many depression cases remain undiagnosed, researchers aimed to improve early detection by analysing real-time physiological data. In this research, various Artificial Intelligence algorithms are developed to look for signs of stress and anomalies in activity patterns from the data captured by wearable health devices. It takes advantage of real-time access to one's physiological data to minimize the number of undiagnosed depression cases (Panindre et al., 2023).

This study reviewed how Artificial Intelligence (AI) has been used in higher education, especially within learning management systems (LMS). The researchers analysed studies from 2010 to 2022, selecting 33 relevant papers out of 306 identified from major databases. The review found that Moodle was the most common platform for implementing AI tools, mainly for assessing student performance using their academic data. Commonly used AI techniques included Random Forest, Neural Networks, K-means, Naive Bayes, Support Vector Machines, and Decision Trees. These applications reflect AI's growing role in improving educational processes (Manhiça et al., 2022).

The primary goal of this study was to review and evaluate the latest techniques for detecting and predicting crime hotspots using spatial-temporal data. The authors found that no comprehensive study existed on this topic, making this research the first to critically analyse current methods and highlight the challenges they face. After reviewing 49 studies from top scientific databases, the paper identified key aspects such as the role of data mining, machine learning, and clustering in crime detection, as well as the use of time series and deep learning for predicting crime trends. It also emphasized the importance of incorporating spatial and temporal data for more accurate predictions. The study concluded by discussing the challenges of current systems and suggested future research directions(Butt et al., 2020)

The study focuses on the prediction of urban crime for 4 Indian cities (Surat, Delhi, Bangalore, Kolkata). Here they used a novel approach for prediction which uses the neural network time series application of MATLAB. This application has a nonlinear autoregressive tool that solves non-linear time series problems. The method can be referred to as NARNN (Nonlinear Autoregressive Neural Network). The method is compared with two classical methods of time-series forecasting (Linear regression, ARIMA). The results of the NARNN gave the highest accuracy for all 4 cities. It gave an accuracy of up to 98% for crime prediction in Delhi. When the crime rate is non-linear or doesn't have any significant trend ANN is strongly recommended. ARIMA is reasonably good when data have trends, while the use of linear regression for prediction is completely impractical(Sharma et al.,2023).

The research examines the integration of artificial intelligence (AI) into ideological and political education, highlighting the advancements in teaching methods and student engagement. A two-tier dynamic student model, enhanced by fuzzy comprehensive evaluation algorithms, allows for a more detailed and objective evaluation of student performance. The use of practical video courses and offline exercises has improved attendance and participation, indicating greater student interest and understanding. However, the challenge remains in balancing technological progress with the humanistic values essential to ideological education. AI can enhance educational practices, but its application must align with the core goals of meaningful learning. The study suggests that AI, when integrated thoughtfully, can complement human values in education, offering a more effective and relevant approach to teaching in this digital age. Ultimately, the future of AI in ideological and political education lies in achieving a balance between innovation and preserving the core educational principles(Liu,et al.,2024).

Political science helped individuals understand their relationship with the government and it's leaders. This field enabled individuals to make better decisions and improve their lives. It also played a role in tackling some of the world's most pressing issues, such as global poverty and national security. The study aims to understand use of AI application in educational process, and examine the role of Virtual Reality in teaching political science.(Khalifa,2022).

The intersection of Artificial Intelligence (AI), machine learning, and behavioural economics to predict consumer buying behaviour, specifically regarding alcoholic beverages. The study analyses the key factors such as bounded rationality, cognitive elements like concentration and stress, as well as social, emotional, and physical pressures that influence buying behaviour.(Sarabhai et al.,2023).

The study investigates the maturity of Artificial Intelligence (AI) in business and management through a bibliometric analysis, while also examining its alignment with the Sustainable Development Goals (SDGs). And it reveals that research contributes to the field by providing insights into the trends and implications of AI maturity in business and management, aligning technological advancements with global sustainability goals(Cahyadi et al.,2024).

The study explores the integration of emotional artificial intelligence (AI) developed using data generated by neuro-physiological tracking systems in marketing and customer relationship management. And research investigates how emotional AI can enhance marketing strategies, improve customer satisfaction. Additionally, the study considers the ethical challenges associated with collecting and using emotional data, emphasizing the need for responsible implementation. (Gümüş & Gürkaynak, 2024).

The paper explores the integration of Large Language Models (LLMs) in social science research through a case study conducted at the Centre for Consumer Society Research (CCSR). And it contributes significantly by mapping out the human, technological, and procedural barriers, and enablers to AI integration, providing a multifaceted view of LLM adoption and its implications for academia and policy making.(Rask et al.,2024).

## 3. OBJECTIVES INTENDED

A. **1.To explore the progressive impact of artificial intelligence on various domains within social sciences**, how AI is redefining research techniques and problem-solving methods in social sciences.
B. **2.To evaluate current applications of AI in core areas of social sciences**, such as sociology, political science, psychology, education, criminology, and economics.
C. **3.To assess the benefits of Ai usage in social science research**, tells about development in data analysis speed, automation of tasks.
D. **4.To identify the challenges and limitations arising from the implementation of Ai in social sciences**, which examine issues like transparency, data bias and reliability.
E. **5.To promote awareness about the ethical considerations in using AI within social sciences,** by using responsible ai models.

### 3.1 To Explore The Progressive Impact Of Artificial Intelligence On Various Social Sciences

In social sciences, research is not only about collection of information but it also includes arranging and explaining that information to perceive behaviour, uncover patterns and societal changes. In this traditional research methodologies are divided into two main groups they are qualitative and quantitative.1.Qualitative:It is focused on exploring phenomena in-depth, often aiming to understand the subjective experiences, feelings and perspectives of individual or groups. The nature of data is Non-numerical, the data collection methods are interviews, focus groups and the analysis of data in the form of thematic analysis. 2.Quantiative: It is focused on measuring and quantifying variables and using statistical tools to test generalize findings across larger populations. The nature of data is numerical, the data collection methods are structured surveys, experiments and statistical datasets and the analysis of data in the form of statistical analysis. These two methods are important for adaptability, replicability.

AI is transforming traditional social science methodologies by introducing powerful tools which allows researches to work faster on a large scale. It is not replacing traditional methods but helping them to improve by making them faster and more useful by the following techniques.

**A.Collecting Big Data Quickly:** In this AI can gather and analyse large-scale data sets automatically from sensors, social media and other digital sources. Which helps researches monitor, assess behaviour, and opinions across wide, diverse groups in real time.

**B.Advanced Text Analysis**: AI uses natural language processing (NLP) to quickly and accurately analyse things like interviews, open-ended survey answers, and historical documents. It helps researchers find patterns, main ideas, emotions, and hidden meanings in the text.

**C.Finding patterns and Making predictions:** Traditional methods have limitations to find patterns but with the help of Machine learning algorithms which can process huge amount of data and detect hidden or complex patterns. By this researches can predict behaviours, understand social trends and assess risks.

**D.Visual and Audio Data Analysis:** In this we use tools like computer vision for image and video analysis and speech recognition for audio and voice data which helps researches to study expressions, tones, and body language valuable for psychology and anthropology.

Example: Open Cv, Deep Face and CMU Sphinx.

**E.Automation of Repetitive Tasks:** In this AI handles the repetitive work. The data directly extract from digital forms by AI and convert it into structured data sets with robotic process automation tool, NLP model can analyses text responses and it automatically group them into predefined categories. AI tools can be programmed to automatically run statistical tests and generates reports based on predefined data.

**F.Real-Time Feedback and Interactive models:** By using Ai models we get instant insights and live updates and has the ability to gather and process data instantly which helps researches to analyse the information swiftly and Ai-powered systems allows users to interact directly with the data or the model, which provides dynamic and user-driven insights. In social sciences these are valuable when the research involves human behaviour and social dynamics.

## 3.2 To Evaluate Current Applications Of AI In Core Areas Of Social Sciences

Social sciences have several domains such as Sociology, psychology, Political science, Education, Criminology, and Economics. When we integrate AI with these social sciences which helps researches to analyse data effectively and predict outcomes.

**A.Sociology**: It is the study of society, focusing on social relationships, interactions, and institutions, and helps us to understand human behaviour. It also studies how people live together in groups and the rules, patterns, and problems arise in social life. when Ai in sociology helps researches to quickly understand social behaviour, and dynamic patterns by analysing large data sets rather than manual methods. By

Using AI tools sociologists identify new trends in society and how people behave in groups. The applications are social behaviour analysis, predicting social trends, urban sociology, social simulation and modelling these are analysed by Ai -driven tools from these sociologists track social movements, shifts in public opinion and signs of societal change.

Example:

1.Studying social behaviour during natural disaster, Sociologists and disaster response teams use AI to analyse social media and understand human behaviour, social support patterns.

**B.Psychology**: It is defined as the study of mind, its functions and behaviour of a person. Applications of AI in psychology are transforming mental health care by providing early measures through chatbots, diagnostic tools and behavioural analysis. The mental health issues such as depression, anxiety, and stress can be analysed by AI models through patient data, speech patterns and written text. The applications are mental health chatbots, emotion recognition tools, In these AI tools analyse facial expressions, voice tone and other emotional indicators to estimate person's emotion state.

Example:

1.Woebot-It is an AI-driven tool that gives mental health support to users through daily conversations and guide them how to relief from stress.

**C.Political Science**: It is the study of how governments works, how laws are made, how leaders are chosen, and how people take part in parties. It also analyses political system, and distribution of power. AI is mostly used in political science to assess voting behaviour, predict election outcomes and effectiveness of campaign strategies by using Machine learning Algorithms. The applications are predicting modelling tools, In these AI-driven models predict factors like polling data, demographics.

Example:

1.During the 2020 U.S.President Election, researches used Natural Language Processing(NLP) to study the political opinions and fake news detection on social media platforms.

**D.Education**: It is the process of gaining knowledge, skills, values and attitudes. It helps individuals develop their thinking, and behaviour.AI is used in education to create adaptive learning system that delivers the data based on the needs of persons and it helps to improve their outcomes. In this application AI algorithms used to adjust the difficulty level based on the students' performance. and another one is automated grading and feedback tools,In this AI tools give the instant grades and feedback to students to improve them.

Example:

1.In schools DreamBox is used to provide a individual math curriculum for students in real time.

**E.Criminology**: It is the study of crime, criminal behaviour, its causes, effects, prevention, and explores why crime happens, how society responds to them, and how laws and justice systems work to control crime.AI is used in criminology to predict and prevent criminal activity by assessing previous data by identifying patterns can be done by machine learning algorithms. The applications are predictive policing tools, in these ai tools identify where does crimes occur mostly by analysing past crime data.

Example:

1.The U.S police department uses PredPol, .to predict where crime rate is more by using history of crime and improves crime prevention strategies.

**F.Economics**: It is the study of hoe people use limited resources to satisfy unlimited needs and wants or the study of how individuals, businesses and governments choose to use limited resources to satisfy their needs and wants. AI is used in economics for prediction of stock market, revenue prediction and economic simulation, In these AI models analyse historical data and predict real time market trends. The applications are financial prediction models, economic simulation models. In these AI models use machine learning algorithms to predict the stock price and market trends.

Example:

1. We use Alpha Vantage for market analysis and financial predictions helps us to make decisions regarding stocks.

## 3.3 To Assess The Benefits Of AI Usage In Social Science Research

The use of AI in social sciences research which improves speed and accuracy of data analysis that means large amount of data can be read by AI quickly in seconds, in which humans takes days or weeks.AI tools can perform tasks automatically because in these the data will be already sorted, the datasets are cleaned, and it categorizes the data so it saves the time for researches and they can focus on higher-level analysis. It can handle large datasets from multiple users at once.

In sociology the benefits of using AI is Emerging societal trends, Trend detection, Outcome prediction, and Textual mood analysis. It helps to analyse vast amount of social data and behaviour of patterns in real time.

In psychology the benefits of using AI is early detection of mental health issue, therapy for individual person, chatbot support for mental health.

In political science the benefits of using AI is prediction of the election result, behaviour analysis of voter, policy simulation, and optimization of campaign.

In Economics the benefits of using AI is in these it analyses big data, the support for decision-making, and market prediction.

In criminology the benefits of using AI is policy prediction- based on data future rules are planned, crime suspect analysis-studies the behaviour of criminal, face matching-people are identified using their faces, case prediction-guessing the result of case before it happens.

In Education the benefits of using AI is adaptive learning, automated grading, study progress tracking, and AI teaching assistants. By this the student can improve the process day to day.

## 3.4 To Identify The Challenges And Limitations Arising From The Implementation Of AI In Social Sciences

While ai provides powerful tools and benefits for social science research, it also has challenges that need face for using AI . The challenges faced are lack of transparency- which means in many AI models that don't show how they make decisions clearly. data bias-if the training data contains bias, the AI reflects the same. Data quality-if the data present is inaccurate, incomplete, or outdated it can leads to improper results. Privacy and Ethical concerns – which means usage of large data sets may include personal and sensitive information, and overdependence of technology.

The challenges faced in implementation of AI in sociology is data bias, loss of human interaction, and ethical concerns which arises privacy and ethical issues.

The challenges faced in implementation of AI in psychology is lack of empathy, data protection and approval, Incorrect emotional reading.

The challenges faced in implementation of AI in political science is lack of transparency, wrong data circulation, and ethical problems in personal advertising.

The challenges faced in implementation of AI in Economics is data dependency-Ignoring human behaviour and depending too much on numbers, unstable trading environment, unequal access.

The challenges faced in implementation of AI in criminology is data driven discrimination, privacy violation, and disproportionate policing.

The challenges faced in implementation of AI in Education is loss of personal connection, data privacy, and unequal access to ai tools.

## 3.5 To Promote Awareness About The Ethical Considerations In Using AI Within Social Sciences

Ensuring ethical considerations in AI which focuses on educational scholars – The people who study and gather information to understand how learning works and how to improve it, educators and policy makers-who creates rules, and laws are the responsible for the ethical use of AI tools and ensure that AI enhances human well-being without causing inequality and harm.

The AI models build with the responsible ethical principles are the data should be understandable, clear and interpretable like prediction of student performance through AI, should explain what led to prediction. Reduce unfairness, ensure balance, safeguard personal information, maintain confidentially, enable human oversight and support human supervision.

Assuring that AI-driven social research matches with ethical principles will not only improve trust in AI technologies but also contribute to more just and equitable societal outcomes change the synonyms of sentence and re write as humanized and promoting ethical awareness and using responsible AI models is important in the social sciences.

## CONCLUSION

The integration of artificial intelligence into social science research methodologies marks a transformative shift in how we collect, analyse, and interpret data. AI technologies, from big data collection and advanced text analysis to real-time feedback and automation, enhance both qualitative and quantitative research methods, offering social scientists the ability to work more efficiently, accurately, and at a larger scale. These tools enable the identification of complex patterns, predictions of societal trends, and the real-time monitoring of social dynamics, all of which are vital in understanding human behaviour and societal change.

However, the adoption of AI in social sciences also brings important challenges, including data bias, lack of transparency, privacy concerns, and the potential for over-reliance on technology. As AI continues to revolutionize fields like sociology, psychology, political science, education, criminology, and economics, it is essential that these ethical considerations are addressed. Maintaining transparency, fairness, and human oversight ensures that AI contributes positively without undermining the integrity of social science research or exacerbating existing inequalities. By considering the advantages and challenges of Ai in social sciences, this chapter clearly shows the importance of use of ai in sensible and mindful way,helping research become more useful and accurate.

Ultimately, AI should not replace traditional methodologies but rather serve as a powerful complement to them. By promoting responsible usage, maintaining ethical standards, and ensuring human involvement in decision-making processes, AI can enhance social science research, leading to deeper insights and more impactful societal outcomes.

## REFERENCES


Al-Emran, M., Abu-Hijleh, B., & Alsewari, A. A. (2024). Exploring the effect of generative AI on social sustainability through integrating AI attributes, TPB, and T-EESST: A deep learning-based hybrid SEM-ANN approach. *IEEE Transactions on Engineering Management.*

Alvero, A. J., & Peña, C. (2023). AI sentience and socioculture. *Journal of Social Computing*, *4*(3), 205–220.

Butt, U. M., Letchmunan, S., Hassan, F. H., Ali, M., Baqir, A., & Sherazi, H. H. R. (2020). Spatio-temporal crime hotspot detection and prediction: A systematic literature review. *IEEE Access : Practical Innovations, Open Solutions*, *8*, 166553–166574.

Cahyadi, H., Kho, A., Yusuf, F. A., Sunarjo, R. A., & Rahardja, U. (2024, August). Ai maturity in business: Bibliometric analysis and sustainable development goals. In 2024 3rd International Conference on Creative Communication and Innovative Technology (ICCIT) (pp. 1-6). IEEE.

Dorigo, T., Brown, G. D., Casonato, C., Cerdà, A., Ciarrochi, J., Da Lio, M., & Yazdanpanah, N. (2025). Artificial intelligence in science and society: The vision of USERN. *IEEE Access : Practical Innovations, Open Solutions.*

Gümüş, A. E., & Gürkaynak, N. (2024). Setting the research agenda: Emotion recognition tools for AI-generated marketing and CRM. In Proceedings of the Medical Technologies Congress. Muğla, Türkiye. https://biyoklinikder.org/TIPTEKNO24-Bildiriler/61.pdf

Gridchina, N., Savvina, N., & Zavyalov, S. (2023, June). Topical issues of the use of artificial intelligence technologies in education and law. In 2023 3rd International Conference on Technology Enhanced Learning in Higher Education (TELE) (pp. 8-11). IEEE.

Hsu, H. Y., Yu, K. C., Wu, J. L., & Hsu, K. S. (2024, November). Sentiment Analysis of Social Support on Treads of Social Media Using Deep Learning Approach. In *2024 IEEE International Conference on E-health Networking, Application & Services (HealthCom)* (pp. 1-3). IEEE.

Khalifa, M. (2022, June). The role of artificial intelligence in the education process of political science field. In *2022 ASU International Conference in Emerging Technologies for Sustainability and Intelligent Systems (ICETSIS)* (pp. 409-416). IEEE.

Liu, Z., & Luo, L. (2024, March). Using Artificial Intelligence for Intelligent Ideological and Political Education Teaching. In *2024 International Conference on Interactive Intelligent Systems and Techniques (IIST)* (pp. 137-144). IEEE.

Manhiça, R., Santos, A., & Cravino, J. (2022, June). The use of artificial intelligence in learning management systems in the context of higher education: Systematic literature review. In *2022 17th Iberian Conference on Information Systems and Technologies (CISTI)* (pp. 1–6). IEEE. DOI: 10.23919/CISTI54936.2022.9837481

Odeh, A. H., & Al Hattab, M. (2023, November). AI Methods Used for Spam Detection in Social Systems-An Overview. In *2023 Tenth International Conference on Social Networks Analysis, Management and Security (SNAMS)* (pp. 1-8). IEEE.

Panindre, P., Mandal, A., Paradkar, M., & Kumar, S. (2023, May). Ai-based detection of signs of depression from physiological data obtained from health trackers. In 2023 2nd International Conference on Applied Artificial Intelligence and Computing (ICAAIC) (pp. 644-649). IEEE.

Rask, M., & Shimizu, K. (2024, February). Beyond the Average: Exploring the Potential and Challenges of Large Language Models in Social Science Research. In *2024 International Conference on Artificial Intelligence, Computer, Data Sciences and Applications (ACDSA)* (pp. 1-5). IEEE.

Saleem, S. A. M., & Naseem, S. M. B. (2023, December). A Comprehensive Review of the Artificial Intelligence (AI) shift in E-commerce. In 2023 6th International Conference on Advances in Science and Technology (ICAST) (pp. 51-56). IEEE.

Sarabhai, S., Chakraborty, M., Batra, M., Kler, R., Banerjee, S., & Mishra, S. (2023, November). Using AI and Machine Learning to Predict Consumer Buying Behavior: Insights from Behavioral Economics in Case of Alcoholic Beverages. In 2023 3rd International Conference on Technological Advancements in Computational Sciences (ICTACS) (pp. 980-986). IEEE.

Sharma, H. K., & Tailor, R. M. (2023, September). Exploring the Potential of AI for Urban Crime Prediction in India: A Case Study of Indian Metropolitan Cities. In 2023 6th International Conference on Contemporary Computing and Informatics (IC3I) (Vol. 6, pp. 1047-1051). IEEE.

Vold, T., & Eikset, A. S. (2024, November). Use of AI in Education-A Case Study from Norway. In 2024 21st International Conference on Information Technology Based Higher Education and Training (ITHET) (pp. 1-5). IEEE.

# Chapter 13
# AI Ethics in Financial Decision-Making: Accountability, Transparency, and Bias

**Nida Fatimah**
https://orcid.org/0009-0005-7183-3223
*Dayananda Sagar College of Engineering, India*

**K. Jayashree**
https://orcid.org/0009-0006-5185-0795
*Dayananda Sagar College of Engineering, India*

## ABSTRACT

*Artificial Intelligence (AI) has totally transformed financial decision-making by enhancing fraud prevention, risk assessment, and productivity. It also causes serious ethical issues such as accountability, transparency, and bias. Unconscious biases in previous data are most likely to be transferred to AI-based financial systems such as automated lending websites and credit scoring systems, resulting in discriminatory or unfair results. Since it is not easy for banks and regulators to comprehend and track algorithmic judgments, transparency of AI models makes trust matters harder. Furthermore, accountability questions remain in instances of ethics violations or mistakes made by AI. The moral implications of AI in making financial decisions are discussed in this chapter, in addition to the challenges in being transparent in automated systems, developing biases in algorithmic models, and the law's need for accountability. The chapter explores case studies of moral AI failure and best practices in trying to provide a roadmap for accountable AI regulation in finance.*

DOI: 10.4018/979-8-3373-2597-2.ch013

## 1. INTRODUCTION

Artificial Intelligence (AI) has rapidly become a transformative force in global finance. From digital lending platforms in Africa to algorithmic trading on Wall Street, AI technologies now underpin critical financial services. They promise faster, more accurate, and cost-effective decision-making, reshaping how banks, regulators, and customers interact. Yet, these advances come with profound ethical concerns, particularly around bias in algorithmic decision-making, opacity in automated processes, and unresolved questions of accountability when errors or harm occur.

The origins of AI in finance can be traced to early applications of statistical risk modeling and credit scoring in the mid-20th century, later enhanced by advances in machine learning and big data analytics. Today, AI supports fraud detection, risk assessment, investment strategies, and personalized financial advisory services. However, empirical evidence shows a dual reality: while AI improves efficiency and predictive accuracy, it also reproduces systemic inequalities and creates trust deficits between institutions and the public.

Despite a growing literature on AI in finance, significant research gaps remain. First, much work focuses on technical performance but neglects ethical oversight. Second, studies are concentrated in Western contexts, overlooking emerging economies where AI adoption is rising rapidly. Third, regulatory frameworks remain fragmented and underdeveloped, leaving accountability questions unresolved. This chapter seeks to fill these gaps by addressing the guiding question:

How can AI-driven financial systems be designed and governed to ensure accountability, transparency, and fairness, while also supporting innovation and financial inclusion?

Sub-Questions

1. Measurement: How can ethical dimensions such as bias, transparency, and accountability in AI systems be measured and evaluated within financial decision-making processes?
2. Jurisdiction: How do regulatory approaches to AI in finance vary across jurisdictions (e.g., EU, US, Asia), and what lessons can be drawn for developing globally harmonized standards?
3. Operationalization: What practical frameworks, tools, and industry practices (e.g., XAI, fairness-aware machine learning, ethical AI auditing) can financial institutions adopt to operationalize responsible AI governance?

To address this, the chapter combines empirical evidence, case studies, and theoretical insights. It evaluates key ethical risks, analyzes failures, reviews global regulatory frameworks, and explores practical solutions such as explainable AI (XAI), fairness-aware machine learning, and ethical auditing. By doing so, it provides both a conceptual foundation and a practical roadmap for researchers, policymakers, and financial institutions seeking to balance innovation with responsibility.

## 1.1 Chapter Objectives and Structure

Objectives of the Chapter
This chapter is designed with the following objectives:

1. To examine the growing applications of AI in finance across domains such as risk assessment, fraud detection, lending, trading, and financial advisory.
2. To analyze the ethical challenges posed by AI in financial decision-making, with a focus on bias, transparency, and accountability.
3. To review empirical cases of AI failures in financial services, identifying patterns of ethical lapses and lessons for future practice.
4. To evaluate global regulatory and governance frameworks that address AI in finance, highlighting strengths, weaknesses, and gaps.
5. To propose responsible AI solutions such as explainable AI (XAI), fairness-aware machine learning, and ethical auditing as pathways to operationalize fairness and accountability.
6. To construct a conceptual framework that integrates AI applications, ethical challenges, and governance mechanisms, leading to ethical financial outcomes.
7. To provide interdisciplinary recommendations for researchers, policymakers, and practitioners aimed at ensuring ethical AI deployment in finance.

Structure of the Chapter
The chapter is organized into the following sections:

- Section 1: Introduction – Establishes the global significance of AI in finance, outlines origins of the concept, and identifies research gaps, culminating in the guiding research question.
- Section 2: AI Applications in Financial Decision-Making – Explores how AI is used in risk assessment, lending, trading, advisory services, and institutional operations.
- Section 3: Ethical Challenges in AI-Based Financial Systems – Analyzes bias, transparency, accountability, and ethical dilemmas in automation.
- Section 4: Case Studies of AI Failures in Financial Decision-Making – Presents real-world examples such as discriminatory lending algorithms, flawed fraud detection, and biased investment models.

- Section 5: Regulatory and Legal Frameworks for AI in Finance – Reviews international governance structures, compliance mechanisms, and enforcement challenges.
- Section 6: Towards Responsible and Fair AI in Finance – Discusses XAI, fairness-aware machine learning, ethical auditing, and best practices for responsible deployment.
- Section 7: Future Directions and Conclusion – Identifies evolving trends, innovations in ethical AI models, interdisciplinary collaboration needs, and offers final recommendations.
- Implications Section – Presents managerial, practical, and theoretical implications for financial institutions and regulators.
- Recommendations and Future Directions – Suggests concrete policy, managerial, and research pathways for advancing responsible AI in finance.
- Key Terms and Definitions – Provides precise definitions of critical terms such as algorithmic bias, transparency, accountability, XAI, and ethical auditing.

## 2. BACKGROUND OF STUDY

The increasing adoption of Artificial Intelligence (AI) in finance has been driven by the promise of efficiency, accuracy, and predictive power. Financial institutions across the globe have incorporated AI into lending, fraud detection, credit scoring, and investment strategies to enhance decision-making processes. For instance, banks now use AI-enabled credit models to assess loan applications in seconds, while investment firms rely on algorithmic trading systems to process real-time market data. Despite these benefits, the implementation of AI in finance has not been without significant challenges, especially in areas of fairness, transparency, and accountability.

One of the most widely discussed challenges is algorithmic bias in lending. A notable case emerged in 2019 with the Apple Card, a credit product launched by Apple and Goldman Sachs. Consumers reported gender-based disparities in credit limits, where *allegations emerged that women were offered lower credit limits than men with similar financial profiles, leading to investigations by NYDFS."* Then immediately note that the NYDFS later ruled that the program did not violate fair lending laws. Investigations revealed that the AI-driven algorithm may have replicated historical biases embedded in training data, highlighting how opaque AI systems can inadvertently perpetuate discrimination. This case underscores the need for fairness-aware machine learning techniques to ensure equitable financial access.

Similarly, fraud detection systems powered by AI have encountered challenges in balancing accuracy with fairness. A European bank introduced an AI model designed to detect unusual spending patterns and prevent fraudulent transactions. However, the system generated a high number of false positives, flagging legitimate purchases as fraudulent and freezing customer accounts unnecessarily. While the model was highly sensitive to anomalies, its lack of contextual understanding led to customer dissatisfaction and reputational risks. This case demonstrates the practical difficulty of implementing AI models that balance predictive accuracy with customer experience.

Another critical case lies in AI-powered investment management. Hedge funds and asset managers increasingly deploy algorithmic trading strategies that rely on historical financial data and predictive modeling. However, during volatile market conditions, these systems can fail dramatically. For example, an AI-based hedge fund experienced substantial losses when its trading model—trained primarily on bullish market data—was unable to adapt to sudden downturns. The reliance on narrow datasets exposed the limitations of AI models in accounting for broader economic uncertainties. This case illustrates the risk of over-reliance on AI without sufficient human oversight and scenario-based testing.

Beyond individual cases, implementation challenges also arise in regulatory compliance. For instance, the General Data Protection Regulation (GDPR) in Europe mandates that individuals have the right to explanations for automated decisions affecting them. Yet, many AI-driven credit scoring systems operate as "black boxes," making it nearly impossible for banks to explain why an applicant was denied a loan. This regulatory-practical gap creates tensions between compliance requirements and technological limitations, showing the urgent need for explainable AI (XAI) frameworks.

These case studies collectively reveal that while AI holds great potential for revolutionizing finance, its implementation often collides with ethical and practical challenges. Algorithmic bias, model opacity, over-reliance on historical data, and regulatory gaps complicate the adoption of AI-driven financial systems. Addressing these challenges requires interdisciplinary collaboration between AI developers, financial practitioners, and policymakers to ensure responsible deployment.

## 3. STATEMENT OF THE PROBLEM

The ideal state for Artificial Intelligence (AI) in financial decision-making is one where algorithms enhance efficiency, reduce human error, expand financial inclusion, and operate with fairness, transparency, and accountability. In such a system, credit scoring models would eliminate discrimination, fraud detection tools would be both accurate and customer-friendly, and investment algorithms would adapt to dynamic market conditions without introducing systemic risks. Moreover, regulators and policymakers would be able to oversee AI-driven financial systems with sufficient clarity and enforceability, ensuring that ethical principles are consistently applied across global markets.

However, the reality is far from this ideal. Contradictory evidence shows that while AI improves operational efficiency, it often replicates or even amplifies existing biases. For example, automated lending algorithms have been criticized for discriminatory outcomes, such as the Apple Card case where women received lower credit limits than men with comparable financial profiles. Fraud detection systems, designed to protect consumers, have at times erred by producing false positives, thereby harming legitimate customers. Similarly, AI-powered investment tools, celebrated for their predictive abilities, have faltered during volatile economic conditions, exposing their inability to adapt beyond historical datasets.

These contradictions highlight several gaps in both research and practice. Theoretical gaps exist because current AI ethics frameworks often remain normative, offering guiding principles but lacking operational models that financial institutions can apply consistently. Knowledge gaps are evident in the limited empirical research that systematically examines the long-term ethical impacts of AI across different financial services, especially in non-Western contexts. Contextual gaps appear where studies disproportionately focus on developed economies, leaving emerging markets underrepresented despite their rapid adoption of AI-driven finance. Practical challenges include the opacity of AI systems, which undermines regulatory compliance with frameworks such as GDPR's "right to explanation," and the difficulty of integrating fairness-aware machine learning techniques without sacrificing efficiency or profitability.

Given these challenges, the core problem is that AI in finance, while technologically advanced, often lacks the ethical safeguards necessary to prevent harm, ensure fairness, and maintain public trust. This creates a pressing need for research that not only critiques the limitations of AI but also proposes interdisciplinary solutions combining technological innovation, ethical theory, and regulatory mechanisms.

This study aims to fill these gaps by critically analyzing the ethical challenges of AI in finance—focusing on bias, transparency, and accountability—and offering a roadmap for responsible AI deployment. By synthesizing empirical evidence, theoretical perspectives, and practical case studies, the chapter seeks to provide direction for future researchers and policymakers in building AI systems that align innovation with ethical responsibility.

## 4. RELATED THEORIES

The ethical deployment of Artificial Intelligence (AI) in finance can be understood through several theoretical perspectives that address fairness, accountability, and governance. These theories provide conceptual grounding for analyzing how AI interacts with societal values, regulatory frameworks, and organizational practices.

One relevant perspective is Stakeholder Theory (Freeman, 1984), which emphasizes that businesses must consider the interests of all stakeholders, not just shareholders. In the context of AI-driven finance, this theory highlights the importance of balancing innovation with ethical obligations toward customers, regulators, and society at large. AI decisions in lending or investment management, therefore, should not only optimize profitability but also safeguard fairness and inclusivity.

Another useful framework is Agency Theory, which examines the relationship between principals (e.g., shareholders) and agents (e.g., managers or AI systems acting on behalf of organizations). In financial institutions, AI acts as an "agent" making automated decisions on behalf of stakeholders. This creates accountability dilemmas, as errors or biases in AI outcomes raise questions about whether responsibility lies with developers, managers, or the institution itself.

Socio-Technical Systems Theory also offers insights by viewing AI not as a standalone technology but as part of a larger system where technical capabilities interact with social, regulatory, and organizational contexts. This perspective underscores that ethical challenges such as bias and opacity cannot be addressed solely by refining algorithms but require systemic approaches that involve governance, culture, and policy interventions.

In addition, Justice Theory (Rawls, 1971) provides a normative foundation for fairness in AI-driven finance. Rawls' principles of distributive justice—particularly the difference principle, which advocates for fairness in opportunities—align with the need to ensure that AI models do not disadvantage already marginalized groups. This theoretical approach supports the design of fairness-aware machine learning systems in credit scoring and lending.

Another key framework is Responsible Innovation Theory (von Schomberg, 2013), which argues for aligning technological advancements with ethical values, societal needs, and long-term sustainability. In financial AI, this theory encourages proactive consideration of risks, stakeholder engagement, and transparent governance, making it highly relevant to current debates on regulation and accountability.

Finally, Accountability Theory (Bovens, 2007) is central to understanding how responsibility is assigned in systems where decisions are automated. It provides a lens for analyzing mechanisms of answerability and liability in cases of AI-driven errors or misconduct in financial decision-making.

## 4.1 Central Theoretical Lens

While all these theories provide valuable insights, this study adopts Responsible Innovation Theory as its central theoretical lens. This framework is particularly suited to analyzing AI ethics in finance because it integrates technological innovation with ethical principles, regulatory requirements, and stakeholder perspectives. Unlike narrower approaches that focus solely on fairness or accountability, Responsible Innovation Theory offers a holistic roadmap for guiding AI deployment in a way that is not only legally compliant but also socially desirable and future-oriented. By using this lens, the chapter situates AI ethics within a broader interdisciplinary context, providing actionable recommendations for researchers, policymakers, and financial institutions.

## 4.2 The Impact of AI Automation on Financial Institutions

Banks have also experienced considerable transformations owing to AI-driven automation: increased quality of decision-making, decreased cost, and enhanced efficiency (European Commission, 2021).Artificial intelligence automation has increasingly been taken up by banks, insurers, and investment banks for productivity of workflow and compliance purposes (Brynjolfsson & McAfee, 2014).By minimizing the work that was being performed and increasing risk management, AI-driven automation has increased operational efficiency (Morley et al., 2021).

For example, robotic process automation with AI automates regulatory reporting, loan processing, and document verification (Financial Stability Board, 2017). The technology optimizes to uphold finance standards, enhances accuracy, and reduces the processing time (GDPR, 2018). Employee concerns also exist over AI automation since losing traditional financial jobs complicates skill gaps and job loss (Pasquale, 2020). In order to achieve a balance between technological progress and labor resilience, banks must introduce ethical AI deployment strategies and upskilling measures (OECD, 2021).

# 5. EMPIRICAL LITERATURE REVIEW

The growing body of empirical research on Artificial Intelligence (AI) in finance underscores both its transformative potential and its ethical challenges. Much of the literature focuses on areas such as fraud detection, credit scoring, trading algorithms, and robo-advisory services, offering insights into the dual nature of AI as both an enabler of efficiency and a source of ethical risk.

## 5.1 AI in Fraud Detection and Risk Assessment

Empirical studies show that AI has significantly enhanced fraud detection capabilities in the financial sector. For example, the Financial Stability Board (2017) found that AI-based fraud detection systems improve anomaly recognition and reduce financial losses by analyzing vast transaction datasets. Similarly, IBM (2020) reported that machine learning algorithms in fraud detection reduce false negatives, thereby improving consumer protection. However, these systems often produce false positives, as documented in Deloitte's (2021) study of European banks, where legitimate customer transactions were mistakenly flagged as fraudulent. These findings illustrate the trade-off between accuracy and customer experience, pointing to the need for more transparent and adaptive AI models.

## 5.2 AI in Credit Scoring and Lending

Another dominant area of research concerns AI-driven credit scoring. Veale and Binns (2017) demonstrated that AI-enabled credit scoring systems use alternative data sources, such as mobile phone usage and transaction histories, to expand financial inclusion. Yet, empirical studies by Cowgill et al. (2020) and Morley et al. (2021) reveal that such systems often replicate historical biases, resulting in discriminatory lending outcomes. The Apple Card case (2019) exemplified these concerns, where women received systematically lower credit limits despite having comparable financial profiles to men. Such findings indicate that while AI can broaden access to credit, it simultaneously risks exacerbating inequalities if bias-mitigation strategies are not applied.

## 5.3 AI in Trading and Investment Management

Empirical research has also explored AI's role in algorithmic trading and investment strategies. The Bank of International Settlements (2020) documented how high-frequency trading powered by AI increases market efficiency by enabling real-time execution of trades. However, evidence from market anomalies, such as flash crashes, reveals that over-reliance on opaque AI models can destabilize financial markets (Financial Stability Board, 2017). Studies of hedge funds adopting AI-driven strategies show mixed results: while some achieved consistent returns, others suffered significant losses due to limited adaptability in volatile conditions (Google AI, 2021). This mixed evidence highlights a gap in understanding how AI systems perform across different market environments and the need for continuous human oversight.

## 5.4 AI in Financial Advisory and Customer Service

AI-driven robo-advisors are another area of growing empirical research. Microsoft & PwC (2019) found that robo-advisors democratize access to investment advice, offering cost-effective solutions for retail investors. At the same time, empirical evidence points to limitations in personalization, where standardized algorithms fail to account for nuanced investor needs (OECD, 2021). IBM (2020) observed that while AI chatbots improve efficiency in customer service, their reliance on pre-programmed responses reduces customer trust in high-stakes financial interactions. These findings emphasize that human-AI collaboration is critical for ensuring quality and trustworthiness in financial advisory services.

## 5.5 Gaps in the Empirical Literature

Despite this growing body of work, several gaps remain. First, much of the literature is concentrated in developed economies, leaving contextual gaps in understanding how AI ethics manifest in emerging markets with weaker regulatory oversight. Second, knowledge gaps persist in longitudinal studies, as most research captures short-term performance of AI systems without addressing long-term ethical and financial implications. Third, theoretical gaps arise from the lack of integrative frameworks that bridge ethical theories with practical AI governance in finance. Finally, practical challenges remain underexplored, particularly in the operationalization of fairness-aware AI and explainability at scale.

Taken together, the empirical literature highlights the transformative role of AI in finance but also underscores unresolved ethical challenges. These findings reinforce the need for interdisciplinary research that combines empirical evidence, theoretical insights, and practical solutions to guide the responsible adoption of AI in financial decision-making.

## 5.6 Bias in Algorithmic Decision-Making

Bias in AI algorithms is a critical ethical concern in financial decision-making because AI models tend to inherit and amplify historical bias in training data (Cowgill et al., 2020). Large financial data are employed to power AI-driven lending platforms, credit scoring algorithms, and investment algorithms; such platforms can be biased by societal norms (Veale & Binns, 2017). Studies have established that AI credit scoring models have the potential to hurt minority and low-income groups in that they can enable discriminatory lending (Morley et al., 2021).

For instance, the U.S. Consumer Financial Protection Bureau (CFPB) used artificial intelligence (AI) to investigate credit scoring and found racial disparities in mortgage approvals. Despite having the same financial background, minority candidates had a higher chance of being rejected (Federal Reserve, 2019). The employment algorithms of banks have also been questioned, with AI-based screening technologies detecting gender and racial bias (Deloitte, 2021). Algorithmic prejudice must be addressed by regulation, transparency in the creation of AI models, and machine learning algorithms that are sensitive to fairness (OECD, 2021). It remains contested whether algorithmic systems fully explain the gaps.

## 6. RESEARCH METHODOLOGY

This chapter adopts a conceptual and integrative research design, drawing on multiple sources of evidence to address the guiding research question. The methodology combines three key approaches:

1. Literature Synthesis: A comprehensive review of peer-reviewed articles, regulatory reports, and industry white papers on AI in finance was conducted. This synthesis identifies dominant themes in applications, ethical challenges, and governance frameworks, while also highlighting areas of inconsistency or neglect.
2. Case Study Analysis: Selected case studies, such as the Apple Card lending controversy, errors in AI-driven fraud detection, and regulatory challenges under GDPR, are examined to provide practical illustrations of ethical risks. These cases are analyzed comparatively to distill lessons that inform the proposed conceptual framework.
3. Scale Development (Conceptual): Based on insights from existing ethical AI frameworks and fairness measurement tools (e.g., disparate impact ratio, interpretability indices), the study conceptually develops a multi-dimensional ethical AI assessment scale. This scale is designed to evaluate financial AI systems across dimensions of bias detection, transparency, accountability, and compliance readiness. While the scale is not empirically validated in this chapter, it provides a foundation for future researchers to operationalize and test in real-world financial contexts.

By integrating these methods, the chapter moves beyond abstract theorization. It combines empirical insights with normative theory and practical instruments, thereby offering both diagnostic value (identifying ethical gaps) and prescriptive guidance (proposing responsible AI practices).

*Figure 1. Conceptual framework: Ethical AI in financial decision-making*

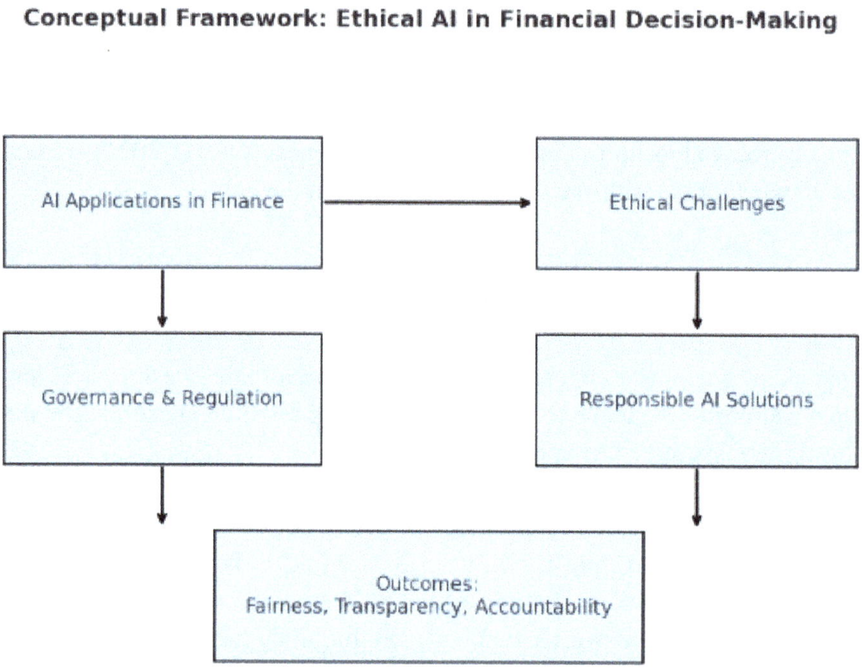


## 6.1 Narrative Explanation

The conceptual framework in figure 1 illustrates the interconnected dynamics of AI in financial decision-making:

1. AI Applications in Finance (risk assessment, fraud detection, credit scoring, trading, and financial advisory) serve as the starting point, driving innovation and efficiency.
2. However, these applications generate Ethical Challenges including bias, opacity, and accountability dilemmas.
3. To address these challenges, Governance and Regulation frameworks are needed, setting the rules and ethical boundaries for AI deployment.
4. Simultaneously, Responsible AI Solutions such as Explainable AI (XAI), fairness-aware machine learning, and ethical AI auditing are introduced to enhance trust and compliance.

5. Together, governance mechanisms and responsible AI solutions lead to the desired Outcomes: Fairness, Transparency, and Accountability, ensuring financial systems remain both innovative and ethically sound.

### 6.2 Transparency Issues in AI-Driven Finance

Financial institution trust needs to be based on transparency in AI decision-making, but most AI models are "black boxes" and provide customers with little decision-making transparency (Brynjolfsson & McAfee, 2014). Issues of justice, legality, and moral responsibility emerge from unjustifiable AI-based financial uses (Kroll et al., 2017).

For instance, AI-driven trading algorithms consider massive volumes of market data to make a decision in a few seconds, but the investor and regulator are not apprised of the decision rationale (IEEE, 2019). Likewise, users of AI-driven credit scoring platforms are likely to be presented with approval or rejection outcomes without sharing the justification (Microsoft & PwC, 2019). XAI methods such as decision trees and SHAP (SHapley Additive Explanations) values have been suggested to enhance explainability and transparency in AI finance (Google AI, 2021). Financial institutions have to adopt XAI technology to support user confidence and regulatory acceptance (UK Financial Conduct Authority, 2022).

## 7. CASE STUDIES OF AI FAILURES IN FINANCIAL DECISION-MAKING

All these sectors of banking, lending, and finance have experienced a shift in productivity by AI-enabled decision-making. All this while, however, AI has introduced to them correlated failure by high-profile players, pressing issues of ethics, pragmatics, and legislation (Pasquale, 2020). It refers to real case studies of AI failure in finance, such as AI investment management biases, discriminatory lending algorithms, and errors in fraud detection. These cases demonstrate the importance of improved regulatory regimes and ethical deployment of AI and provide valuable information on the dangers of deploying AI in financial decision-making (OECD, 2021).

### 7.1 Case Study 1: Discriminatory Lending Algorithms

Creditworthiness is assessed by AI lending platforms based on risk assessment algorithms, income level, and financial history. Nevertheless, numerous research studies have determined that such platforms often exhibit discriminatory bias, disproportionately adversely impacting low-income applicant groups as well as minority groups (Cowgill et al., 2020). The Apple Card, a credit scoring system based on artificial intelligence that faced public criticism in 2019 on charges of gender discrimination in credit limits, is one such example (O'Neil, 2016). Regulatory issues were raised due to the reports that women were being given lower credit limits than men with similar financial profiles (Federal Reserve, 2019).

American AI mortgage approval platforms also showed racial biases, denying loan requests by Black and Hispanic borrowers disproportionately compared to White borrowers with comparable credit profiles, based on a study by the Consumer Financial Protection Bureau (CFPB) (Morley et al., 2021). These instances highlight the importance of fairness-aware AI models to counter bias in financial services and the threats of algorithmic bias in lending (Deloitte, 2021).

## 7.2 Case Study 2: AI-Driven Financial Fraud Detection Errors

To monitor transactions and detect suspicious activities, banks increasingly rely on fraud detection software that incorporates AI. While such programs enhance efficiency and security, they are not completely immune to errors that cause unwarranted freezes of accounts and loss of money (Brynjolfsson & McAfee, 2014). In a widely publicized case, one of the leading European banks cut fraudulent transactions by implementing an AI-facilitated fraud detection program. Yet thousands of customers were unnecessarily interrupted by virtue of the excessive false-positive rate, where the system incorrectly marked genuine transactions as illicit (Financial Stability Board, 2017).

At the time of COVID-19 pandemic relief disbursement in 2020, a large U.S. bank also went through the same when its AI-based fraud-detection tool incorrectly put on hold small business transactions (OECD, 2021). Because of such errors, the reputation of the bank was smeared and the attention of the regulators was invited, which caused a delay in the much-awaited relief finance (IBM, 2020). More transparent artificial intelligence models, human supervision of fraud detection, and consistent improvement to AI training data to eliminate false positives and false negatives are all needed to combat these issues (Microsoft & PwC, 2019).

## 7.3 Case Study 3: Bias in AI-Powered Investment Management

By providing computerized investment guidance and portfolio management, AI-powered investment management platforms—also called robo-advisors—have become increasingly used. Decision-making biases, though, have raised concerns regarding fairness, transparency, and accountability (Veale & Binns, 2017). When recommendations for individual demographic segments by a well-established robo-advisory firm through its AI system consistently leaned toward risky investments with the result of increasing the financial exposure of the investors, the firm was indicted for algorithmic bias (FATF, 2021).

A hedge fund that was based on an unfair trading model which did not account for market and economic recessions led to huge monetary losses in a particular case (Bank of International Settlements, 2020). When the market shifted, the AI fund model made incorrect choices because it had learned from predominantly good market data (Google AI, 2021). These occurrences highlight the importance of human contribution, explainable AI (XAI) approaches, and multi-dataset training in investment management based on AI (UK Financial Conduct Authority, 2022). Reports indicate that hedge funds' AI/ML models raise concerns about risk of overfitting and market stability under stress.

## 7.4 Key Lessons Learned from Past Failures

The following case studies provide some valuable lessons to regulators and financial institutions:

- Bias Avoidance: To avoid biases and give equal chances to all, AI-driven financial systems need to adopt fairness-sensitive machine learning strategies (Morley et al., 2021).
- Explainability and Transparency: To facilitate regulatory compliance and provide stakeholders with an understanding of the financial decision-making process, AI models must be designed with explainability capabilities (Microsoft & PwC, 2019).
- Human Supervision: In sensitive financial applications such as lending, investment management, and fraud detection, AI must assist human decision-making and not replace it (Deloitte, 2021).
- Regulation Compliance: In order to ensure AI-powered financial software meets ethical and legal standards, financial regulators should issue detailed AI management guidelines (European Commission, 2021).

# 8. REGULATORY AND LEGAL FRAMEWORKS FOR AI IN FINANCE

Politicians and global regulators have had to establish rules to ensure ethical compliance, transparency, and accountability with the increasing application of AI to inform financial decisions (OECD, 2021). Regulations aim to balance between consumer protection and innovation and the potential risks of AI-driven financial systems, including bias, transparency, and accountability problems (Pasquale, 2020). This chapter covers global AI regulations (see Table 1) affecting the financial market, AI banking regulation, legal requirements, and issues of enforcement.

*Table 1. Comparative Overview of AI Regulatory Regimes in Finance*

| Region | Main Framework / Approach | Key Features | Strengths | Challenges |
|---|---|---|---|---|
| European Union (EU) | *EU Artificial Intelligence Act* (proposed 2021); GDPR (2018) | Risk-based classification of AI systems; "high-risk" financial applications require transparency, human oversight, and conformity assessments. GDPR ensures data protection and "meaningful information" about automated decisions. | Comprehensive, principle-based; strong consumer rights; harmonized rules across member states. | Complexity of implementation; high compliance costs; possible slowdown of innovation. |
| United States (US) | Sectoral approach; FTC AI guidance (2021); Treasury/ FSOC reports; state-level initiatives (e.g., New York DFS). | Emphasis on existing consumer protection, anti-discrimination, and financial laws; regulatory guidance rather than overarching AI law. | Flexible, innovation-friendly; agencies adapt existing laws to new tech. | Fragmented regulatory landscape; lack of unified federal AI framework; uneven enforcement. |
| United Kingdom (UK) | FCA Discussion Papers (2022); pro-innovation AI White Paper (2023); sector-specific oversight. | Principles-based, emphasizing transparency, fairness, and accountability in financial services; focus on fostering innovation with proportionate regulation. | Balanced approach between innovation and ethics; FCA actively engages with industry sandboxes. | Less prescriptive than EU; risk of regulatory gaps; divergence from EU may create compliance complexity. |
| China | *Algorithmic Recommendation Provisions* (2022); *AI Ethics Guidelines*; data security and fintech-specific rules. | Strong state-led oversight; focus on algorithmic transparency, data localization, and alignment with national priorities. | Centralized governance ensures rapid implementation; strict controls on data misuse. | Emphasis on state control may limit global interoperability; lack of independent oversight; concerns over privacy and fairness. |

## 8.1 Global AI Regulations Affecting Finance

Regimes of law vary between jurisdictions; some impose tight surveillance, while others facilitate innovation through loose regimes of governance. The European Union's Artificial Intelligence Act (EU AI Act) is perhaps the most comprehensive regime of regulation, which categorizes AI applications into different risk classes and strictly limits high-risk AI, including that used in banking (European Commission, 2021). To minimize the economic risks associated with AI decision-making, the EU AI Act mandates accountability, transparency, and human intervention.

The Fair Credit Reporting Act (FCRA) and Equal Credit Opportunity Act (ECOA), being anti-discrimination in lending laws, are the most significant financial laws that regulate AI in America (Federal Reserve, 2019). In order to comply with anti-discrimination and fairness requirements, regulatory bodies such as the Consumer Financial Protection Bureau (CFPB) and Securities and Exchange Commission (SEC) regulate the application of AI in finance (Financial Stability Board, 2017).

Countries such as China and the United Kingdom have implemented AI regulatory frameworks to govern the application of AI in the financial services sector. China's AI policy demands financial institutions to safeguard the protection and security of data and compliance with the country's cybersecurity legislation in the use of AI (OECD, 2021). The United Kingdom's Financial Conduct Authority issued regulatory guidance that stressed the importance of fairness, explainability, and transparency in AI-driven financial services (UK Financial Conduct Authority, 2022).

## 8.2 AI Governance and Ethical Compliance in Banking

Financial institutions implement AI governance models for achieving legal and ethical compliance. For successful adoption of AI, AI governance involves the creation of rules, risk management practices, and compliance (Deloitte, 2021).

In order to observe AI innovation and minimize the risks of biased decision-making, various banks have implemented AI ethics boards (Morley et al., 2021). In an attempt to offer ethical applications of AI for credit evaluation and fraud identification, for instance, JPMorgan Chase has instituted an AI governance platform with risk assessment methodologies and fairness checks (IBM, 2020).

Because financial regulators desire transparency in AI-based credit scoring, trading models, and risk assessment, explainable AI (XAI) has an important function in enhancing governance (Google AI, 2021). Financial institutions risk regulatory sanctions and loss of reputation if they fail to implement robust AI governance frameworks (FATF, 2021).

## 8.3 Legal Responsibilities and AI Accountability in Finance

Considering the independence of AI systems, legal responsibility in AI finance is an extremely complex issue. There are numerous difficulties in determining accountability when AI systems are making unsound or unethical financial decisions (Kroll et al., 2017).

Regulators have suggested legislative guidelines which make financial institutions employing AI technologies accountable. For example, AI-driven decision-making systems in finance are mandated by the General Data Protection Regulation (GDPR) to provide individuals with an explanation if automatic choices affect their financial affairs (GDPR, 2018). Financial institutions are responsible for ensuring that AI choices are unbiased, transparent, and adhere to rules concerning data protection (European Commission, 2021).

In professional circles, even the concept of AI liability is gaining popularity. Banks and financial institutions can be held responsible for damages inflicted by AI like breakdown in algorithmic trading or discriminatory lending (Financial Stability Board, 2017). Financial institutions are finally legally responsible for AI decisions, which rings the bell for unlimited ethical surveillance and inspection (OECD, 2021).

## 8.4 Challenges in Enforcing AI Regulations

Imposing compliance within the financial sector is riddled with problems despite the advancement of AI law. It is difficult for regulators to audit decision-making effectively because of the complexity of AI algorithms (Brynjolfsson & McAfee, 2014). There are various financial institutions that utilize in-house AI models, and this hinders external and regulatory examination (Veale & Binns, 2017).

The rapid development of AI technology, in advance of legislative progress, is an additional challenge. There are loopholes in the regulation because policymakers have not connected current financial regulations to the new challenges posed by AI (Pasquale, 2020). As AI-powered financial services often entail global operations and require international coordination in regulation, cross-border financial transactions complicate compliance even further (OECD, 2021).

Regulators are finding innovative solutions to these issues from AI audit methodologies, machine learning techniques that are equity-aware, and mandated transparency for financial AI products (UK Financial Conduct Authority, 2022). Financial institutions need to proactively link AI governance policy with evolving regulatory landscapes in order to stay ethical and compliant with legislation (Deloitte, 2021).

# 9. TOWARDS RESPONSIBLE AND FAIR AI IN FINANCE

Upkeep of transparency, fairness, and accountability in spite of AI ongoing impact on financial decision-making is the top priority (OECD, 2021). The financial sector needs to establish measures for mitigation of bias, improvement in explainability, and adherence to evolving regulatory guidelines. Explainable AI (XAI), fairness-aware machine learning, ethical AI audits, industry best practices, and the role of stakeholders in AI governance are some of the key strategies under discussion in this regard.

## 9.1 Explainable AI (XAI) and Its Role in Financial Transparency

Financial institutions are now able to comprehend and describe AI-based decisions through Explainable AI (XAI), a fundamental component of responsible AI (Google AI, 2021). Traditional AI models, particularly deep learning models, can sometimes be "black boxes," and stakeholders and regulators can struggle to grasp how they make a decision (Veale & Binns, 2017).

The European Commission (2021) and other regulators require transparency in AI-driven risk assessment, trading, and credit rating. Through the information they provide on AI-driven financial recommendations, XAI methods like SHapley Additive exPlanations (SHAP) and Local Interpretable Model-Agnostic Explanations (LIME) enhance regulatory compliance and trustworthiness (FATF, 2021). By resolving consumer complaints regarding fairness and reliability, financial institutions can increase responsibility by adopting XAI (Deloitte, 2021).

## 9.2 Fairness-Aware Machine Learning for Unbiased Decision-Making

Unfair credit scoring, discriminatory lending, and unethical investment advice can be the outcome of bias in AI-based financial systems (Morley et al., 2021). By making sure that all demographic groups are treated equally when coming to a decision for finances, fairness-aware machine learning strategies attempt to mitigate these biases (OECD, 2021).

There are a number of ways to enhance equity in finance AI models. Pre-processing methods change training data before pattern recognition by AI models to remove biases, whereas in-processing methods change algorithms to enable equitable decision-making (IBM, 2020). For the detection and repairing of unfair patterns in financial services, post-processing methods analyze AI output (European Commission, 2021).

For delivering ethical AI deployment and regulatory adherence, banks are increasingly relying on fairness metrics such as equalised odds and disproportionate effect analysis (UK Financial Conduct Authority, 2022). Financial institutions are able to address reputational risk and establish customer confidence through actively deployed fairness-aware AI algorithms (Deloitte, 2021).

## 9.3 Ethical AI Auditing and Regulatory Compliance

AI audit models are required to determine the moral effect of financial AI systems and ensuring regulatory conformity (Financial Stability Board, 2017). To determine such risks, ethical AI auditing examines algorithmic fairness, responsibility, and openness (Brynjolfsson & McAfee, 2014).

The UK Financial Conduct Authority (2022) and the Federal Reserve (2019) are just some regulators advocating for AI audits within financial regulation programs. Third-party independent audits are implemented to verify that AI is following regulatory regulations such as the General Data Protection Regulation (GDPR, 2018) and the EU AI Act. Model conduct is tracked and future improvement suggestions are given by internal AI ethics committees (Google AI, 2021).

Decreased legal threats, improved model quality, and heightened consumer confidence are all benefits to financial institutions when employing AI audit frameworks (FATF, 2021). Ethical audits will remain an essential standard for the prudent use of financial AI as technology continues to improve (Pasquale, 2020).

## 9.4 Industry Best Practices for Responsible AI Deployment

Organisations need to embrace industry best practices that align with moral and regulatory standards to be able to have ethical use of AI in finance (Microsoft & PwC, 2019).

Some of these best practices include:

- Transparency-Driven AI Design: Having tools for explainability to enable stakeholders to know how AI derives its conclusions (OECD, 2021).
- Human-AI Interaction: Human oversight and AI technology are merged for ensuring precision and preventing inadvertent bias (Deloitte, 2021).
- Multicultural and Representative Data Training: Diverse data are utilized in training AI models and preventing financial decision-making biases (Morley et al., 2021).
- On-going Risk Evading and Screening: Frequent AI audit reports and updating algorithms according to the changing regulatory compliance (UK Financial Conduct Authority, 2022).

By applying these best practices, financial institutions promote consumer trust, regulatory compliance, and AI ethics (European Commission, 2021).

## 9.5 Role of Stakeholders in Ethical AI Governance

Different stakeholders, such as regulators, financial institutions, AI developers, and consumers, need to collaborate in order to enable the appropriate application of AI in finance (OECD, 2021).

Every stakeholder has a specific role to play in ensuring AI accountability and transparency:

- Regulators establish ethical guidelines for AI-driven financial systems and impose regulations on AI compliance (European Commission, 2021).
- Integrating fairness-sensitive machine learning techniques, banks appropriately develop and deploy AI models (Deloitte, 2021).
- In applications of AI to finance, the developers ensure model explainability, bias mitigation, and ethical compliance (Google AI, 2021).
- Encouraging fairness, transparency, and responsibility for AI-driven financial services, customers are significantly involved in the regulation of AI (UK Financial Conduct Authority, 2022).

These stakeholders ensure collectively ethical use of AI in finance and enhances AI ethics (Pasquale, 2020).

## 9.6 Operational Recommendations for Ethical AI in Finance

To translate ethical principles into actionable practices, financial institutions can adopt the following operational mechanisms:

1. Model Cards for AI Transparency
   - Description: Documentation templates that describe an AI model's purpose, data sources, limitations, performance metrics, and known risks.
   - Application in Finance: Every credit scoring, fraud detection, or trading model should be accompanied by a model card accessible to regulators and internal audit teams.
   - Benefit: Improves transparency and enables both compliance officers and regulators to evaluate whether AI systems align with fairness and accountability standards.
2. Regular AI Audit Cadence
   - Description: Ethical and technical audits scheduled at consistent intervals (e.g., quarterly or biannually), rather than only after regulatory inspections.

- o Application in Finance: Audits should evaluate bias in lending outcomes, false positives in fraud detection, and adherence to explainability requirements under GDPR or similar frameworks.
- o Benefit: Ensures that ethical compliance is ongoing, not reactive, and allows early detection of risks before they escalate.

3. Ethical Key Performance Indicators (KPIs)
   - o Fairness KPIs: Disparate impact ratio, demographic parity, or equalized odds metrics for loan approvals.
   - o Transparency KPIs: Percentage of decisions accompanied by explainable outputs interpretable by non-technical stakeholders.
   - o Accountability KPIs: Documented assignment of responsibility for each model across design, deployment, and monitoring phases.
   - o Application in Finance: Incorporate ethical KPIs into balanced scorecards and compliance dashboards for senior management.
   - o Benefit: Embeds ethics directly into performance evaluation systems, aligning incentives with responsible AI practices.
4. Grievance and Redress Mechanisms
   - o Description: Structured pathways for customers to challenge or seek explanations for AI-driven financial decisions.
   - o Application in Finance: Banks and fintechs should maintain digital portals where customers can request explanations or corrections of automated decisions.
   - o Benefit: Strengthens accountability and builds trust with clients who may otherwise feel excluded by opaque decision-making.
5. Cross-Functional AI Ethics Committees
   - o Description: Committees including compliance officers, data scientists, ethicists, and consumer representatives.
   - o Application in Finance: Provide oversight during model design, deployment, and post-deployment monitoring.
   - o Benefit: Encourages socio-technical perspectives, reduces siloed decision-making, and ensures stakeholder concerns are integrated.

## 10. IMPLICATIONS

### 10.1 Managerial Implications

For managers in financial institutions, the findings of this study highlight the need to strike a balance between technological adoption and ethical responsibility. Managers must recognize that while AI-driven decision-making can reduce costs, streamline operations, and improve predictive accuracy, it can also introduce reputational and regulatory risks if ethical safeguards are ignored. Implementing Explainable AI (XAI) can enhance customer trust by providing transparency in credit scoring and investment decisions. Managers are encouraged to establish AI ethics committees within their organizations to oversee algorithmic fairness, conduct regular bias audits, and ensure compliance with regulatory standards. Moreover, investing in employee training programs on AI ethics will prepare decision-makers to responsibly oversee AI deployment, aligning organizational performance with ethical expectations.

## 10.2 Practical Implications

At the practical level, this study emphasizes the urgent need for financial institutions to adopt fairness-aware machine learning techniques and responsible AI auditing tools. Practical case studies, such as the Apple Card controversy, demonstrate how failing to address algorithmic bias can result in public backlash and regulatory scrutiny. Financial service providers should also integrate human-AI collaboration by retaining human oversight in high-stakes decisions such as loan approvals and fraud detection, thereby reducing risks associated with fully automated systems. Furthermore, adopting industry best practices—such as using diverse training datasets, documenting algorithmic decisions, and maintaining customer grievance mechanisms—can improve accountability and operational trust. These practices not only enhance compliance but also safeguard long-term customer relationships in an increasingly competitive financial landscape.

## 10.3 Theoretical Implications

From a theoretical perspective, this study contributes to the growing literature on AI ethics in finance by integrating empirical insights with Responsible Innovation Theory as its central lens. It extends existing theories of accountability, justice, and socio-technical systems by situating them within the unique context of financial decision-making. The study also addresses theoretical gaps by moving beyond normative discussions of ethics to propose actionable frameworks for operationalizing fairness, transparency, and accountability in AI systems. Additionally, it highlights knowledge gaps in cross-cultural and emerging-market contexts, where ethical implications of AI remain underexplored. By offering a structured conceptual framework, the study lays a foundation for future researchers to develop testable models, comparative analyses, and interdisciplinary approaches that link AI ethics with financial innovation.

## 11. FUTURE DIRECTIONS

The rapid evolution of AI in financial decision-making is both promising and presents ethical issues. To encourage equity, transparency, and accountability as AI models continue to enhance, financial institutions, regulators, and researchers should prioritize ethical concerns (OECD, 2021). This chapter considers upcoming trends in ethical AI models, future developments, the requirement for interdisciplinary efforts, and final recommendations for responsible use of AI in finance.

### 11.1 Evolving Trends in AI and Financial Ethics

The rapid use of AI in banking is being fuelled by technologies such as machine learning, blockchain, and natural language processing (Deloitte, 2021). One of the most significant uses is the growing use of deep learning models for investment planning, credit ratings and fraud detection purposes (Google AI, 2021). Deep learning models, however, raise ethical questions because of their lack of transparency and bias (Veale & Binns, 2017).

The drive for regulation is also addressing these issues. The European Union's Artificial Intelligence Act (EU AI Act) proposes stricter regulations on AI fairness and transparency (European Commission, 2021). The UK Financial Conduct Authority (FCA) and the U.S. Securities and Exchange Commission (SEC) are working on legislation for the AI regulation of finance (UK Financial Conduct Authority, 2022).

The employment of artificial data for the training of AI models in finance is a new advance that works to mitigate bias without sacrificing compliance with privacy (Financial Stability Board, 2017). The technique maintains the predictive power of AI systems without diminishing ethical issues. Federated learning methods also solve privacy issues because they enable joint training of AI models without sacrificing private financial data (Brynjolfsson & McAfee, 2014).

### 11.2 Innovations in Ethical AI Models for Finance

New frontiers for enhancing the accountability, interpretability, and fairness of financial AI systems are being pushed through research in AI ethics (Morley et al., 2021). Decision-making processes are more understandable through Explainable AI (XAI) methodologies like Local Interpretable Model-agnostic Explanations (LIME) and SHapley Additive exPlanations (SHAP) (IBM, 2020).

To avoid credit scoring and lending biases, financial machine learning models are adopting fairness-aware machine learning techniques, including adversarial debiasing and counterfactual fairness (OECD, 2021). Financial institutions can already audit and boost AI compliance on the basis of fairness metrics due to increased use of ethical AI audit tools (FATF, 2021).

Another area with increased financial transparency and accountability potential is blockchain technology. AI governance-based decentralized smart contracts can reduce the risk of algorithmic bias and ensure compliance (European Commission, 2021).

## 11.3 The Need for Interdisciplinary Collaboration

Regulators, financial institutions, AI researchers, and ethicists need to work together to ensure ethical uses of AI in finance (Pasquale, 2020). In order to develop functioning and actionable policies without stifling innovation, policymakers need to work hand in hand with AI developers (OECD, 2021).

A study on ethical AI models and the establishment of interpretability frameworks and anti-bias measures is an urgent task for academies (Microsoft & PwC, 2019). To provide professionals with skills to respond to the ethical challenge presented by AI-fueled finance, banks must organize AI ethics professional courses (Deloitte, 2021).

In order to ensure that financial AI systems remain aligned with ethical standards and the public interest, consumer advocacy organizations are also needed to contribute in engaging with debates in AI regulation (UK Financial Conduct Authority, 2022). The finance sector can deliver responsibility and trust in AI application through coexistence across various disciplines (European Commission, 2021).

## 11.4 Final Thoughts and Recommendations

Unmatched risk management, creativity, and effectiveness opportunities are presented by AI adoption in finance. AI-driven financial decision-making, however, can result in biased decisions, lack of transparency, and accountability issues unless strong ethical protections are put in place (Google AI, 2021).

The following recommendations should rank highest with a goal of encouraging responsible AI in finance:

- Strengthening Regulation: For the purpose of tackling upcoming ethics issues and ensuring innovation, policymakers ought to tighten AI regulations (European Commission, 2021).
- Explainability and Transparency: In favor of enhanced consumer trust and AI interpretability, financial institutions ought to adopt XAI methods (Deloitte, 2021).
- Bias Mitigation Strategies: To enable objective financial decision-making, AI models must implement fairness-aware machine learning algorithms (IBM, 2020).

- Ethical AI Auditing: To satisfy principles of accountability, transparency, and fairness, periodic AI audits should be mandated (FATF, 2021).
- Stakeholder Engagement: To formulate ethical AI frameworks, coordination among regulators, business leaders, researchers, and consumers is necessary (OECD, 2021).

Financial institutions can capitalise on AI's advantages while maintaining moral standards and legal compliance by putting these safeguards in place. In order to ensure that AI-driven financial systems benefit everyone equally and fairly, prudent deployment is essential to the future of AI in finance (UK Financial Conduct Authority, 2022).

## 12. CONCLUSION

Artificial Intelligence (AI) has undeniably reshaped financial decision-making, revolutionizing areas such as fraud detection, risk assessment, credit scoring, investment management, and customer service. Its ability to process vast volumes of data at unprecedented speed has allowed financial institutions to improve operational efficiency, detect anomalies in real time, and personalize financial products for diverse customer segments. However, as demonstrated throughout this chapter, the ethical ramifications of AI adoption in finance cannot be overlooked. Issues of algorithmic bias, lack of transparency, and accountability gaps have created new risks that, if left unaddressed, could undermine both consumer trust and systemic financial stability (OECD, 2021).

The findings underscore the urgency of embedding responsible AI governance in financial systems. Without adequate safeguards, AI systems risk perpetuating historical inequalities, excluding vulnerable populations from access to credit, or generating discriminatory lending and investment decisions. Furthermore, the opacity of many machine learning models—particularly black-box algorithms—poses a significant challenge to regulators, auditors, and customers who demand understandable justifications for financial decisions. These challenges call for a multidimensional response that balances innovation with ethical responsibility.

To mitigate these risks, this chapter emphasized several key solutions. The adoption of explainable AI (XAI) can help demystify algorithmic processes, allowing both financial practitioners and regulators to interpret and challenge AI-driven outcomes (Ribeiro et al., 2016). Similarly, fairness-aware machine learning provides technical mechanisms to detect and reduce bias, ensuring more equitable treatment of demographic groups in lending and investment decisions (Veale & Binns, 2017). Beyond technical solutions, ethical AI auditing offers an institutional mechanism to verify compliance with legal and ethical standards on an ongoing basis (Kroll et al., 2017). When combined with global regulatory frameworks such as the EU's proposed AI Act and sectoral guidance from agencies like the U.S. FTC and UK FCA, these measures create a foundation for more transparent, accountable, and fair financial ecosystems (European Commission, 2021; FCA, 2022).

Yet, technology and regulation alone are insufficient. This chapter has argued for the importance of interdisciplinary collaboration as the cornerstone of responsible AI in finance. Ethical AI requires not only the technical expertise of data scientists and engineers but also the oversight of policymakers, the judgment of legal scholars, the perspectives of ethicists, and the lived experiences of customers and civil society organizations. By bringing together these diverse stakeholders, financial institutions can ensure that AI deployment aligns with societal values and broader goals of financial inclusion and stability.

Looking ahead, the path to a responsible AI ecosystem in finance depends on three intertwined priorities. First, researchers must continue to explore ways to operationalize fairness and accountability through robust evaluation metrics, scale development, and cross-cultural studies that extend beyond Western markets. Second, policymakers and regulators must strengthen oversight mechanisms, close jurisdictional gaps, and harmonize global approaches to reduce regulatory fragmentation. Finally, industry leaders must embed ethical considerations into business strategy, adopting governance practices such as model cards, regular audit cadences, and ethical key performance indicators (KPIs) to monitor fairness, transparency, and accountability over time.

In conclusion, AI in finance stands at a pivotal crossroads. It can either exacerbate existing inequalities and erode trust through opaque, biased, and unaccountable systems, or it can serve as a powerful tool for building a fairer, more inclusive, and more resilient financial sector. Achieving the latter requires collective responsibility. Legislators, business executives, and researchers must work together to design, implement, and regulate AI systems that are not only innovative but also just, transparent, and accountable. By pursuing this interdisciplinary and collaborative approach, the financial sector can chart a path toward a responsible AI ecosystem that promotes trust, protects consumers, and supports sustainable innovation well into the future (Deloitte, 2021).

## REFERENCES


Bovens, M. (2007). Analysing and assessing accountability: A conceptual framework. *European Law Journal, 13*(4), 447–468. DOI: 10.1111/j.1468-0386.2007.00378.x

Brynjolfsson, E., & McAfee, A. (2014). *The second machine age: Work, progress, and prosperity in a time of brilliant technologies*. W. W. Norton & Company.

Cowgill, B., Dell'Acqua, F., & Deng, S. (2020). Biased programmers? Or biased data? A field experiment in operationalizing AI ethics. *Columbia Business School Research Paper No. 20-91*. https://doi.org/DOI: 10.2139/ssrn.3672274

Deloitte. (2021). *Mitigating algorithmic bias in financial services: Strategies for fair AI deployment*. Deloitte Insights. https://www2.deloitte.com/insights

European Commission. (2021). *Proposal for a regulation laying down harmonized rules on artificial intelligence (Artificial Intelligence Act)*. https://eur-lex.europa.eu

FATF. (2021). *Opportunities and challenges of new technologies for AML/CFT*. Financial Action Task Force. https://www.fatf-gafi.org

Federal Reserve Board. (2021). *AI in financial services: Risks, benefits, and regulatory considerations*. https://www.federalreserve.gov

Financial Stability Board. (2017). *Artificial intelligence and machine learning in financial services*. https://www.fsb.org

Google, A. I. (2021). Explainable AI in financial applications: Enhancing transparency in credit scoring and risk assessment. *Google AI Blog*. https://ai.googleblog.com

IEEE. (2019). *Ethically aligned design: A vision for prioritizing human well-being with autonomous and intelligent systems*. IEEE Standards Association. https://ethicsinaction.ieee.org

Kroll, J. A., Huey, J., Barocas, S., Felten, E. W., Reidenberg, J. R., Robinson, D. G., & Yu, H. (2017). Accountable algorithms. *University of Pennsylvania Law Review, 165*(3), 633–705.

Microsoft & PwC. (2019). *Responsible AI in financial services: Avoiding pitfalls and ensuring fairness*. PwC. https://www.pwc.com

Morley, J., Floridi, L., Kinsey, L., & Elhalal, A. (2021). From what to how: An initial review of publicly available AI ethics tools, methods, and research to translate principles into practices. *Science and Engineering Ethics*, *27*(1), 1–26. DOI: 10.1007/s11948-020-00239-y PMID: 33492516

O'Neil, C. (2016). *Weapons of math destruction: How big data increases inequality and threatens democracy*. Crown Publishing Group.

Organisation for Economic Co-operation and Development (OECD). (2021). *AI principles: Recommendations on AI governance for the financial sector*. OECD AI Policy Observatory. https://oecd.ai

Pasquale, F. (2020). *New laws of robotics: Defending human expertise in the age of AI*. Harvard University Press.

Rawls, J. (1971). *A theory of justice*. Belknap Press of Harvard University Press. DOI: 10.4159/9780674042605

Research, I. B. M. (2020). *AI ethics and bias: Trustworthy AI for financial decision-making*. IBM. https://research.ibm.com

Ribeiro, M. T., Singh, S., & Guestrin, C. (2016). "Why should I trust you?" Explaining the predictions of any classifier. In *Proceedings of the 22nd ACM SIGKDD International Conference on Knowledge Discovery and Data Mining* (pp. 1135–1144). ACM. https://doi.org/DOI: 10.1145/2939672.2939778

Russell, S., & Norvig, P. (2020). *Artificial intelligence: A modern approach* (4th ed.). Pearson.

UK Financial Conduct Authority. (2022). *The role of artificial intelligence in financial services: Ethical and regulatory challenges*. FCA Discussion Paper. https://www.fca.org.uk

U.S. Department of Treasury. (2022). *Financial Stability Oversight Council report on digital asset financial stability risks and regulation*. U.S. Department of Treasury. https://home.treasury.gov

U.S. Federal Trade Commission. (2021). *Aiming for truth, fairness, and equity in your company's use of AI*. FTC Guidance. https://www.ftc.gov

Veale, M., & Binns, R. (2017). Fairer machine learning in the real world: Mitigating discrimination without collecting sensitive data. *Big Data & Society*, *4*(2), 1–17. DOI: 10.1177/2053951717743530

von Schomberg, R. (2013). A vision of responsible research and innovation. In Owen, R., Heintz, M., & Bessant, J. (Eds.), *Responsible innovation: Managing the responsible emergence of science and innovation in society* (pp. 51–74). John Wiley & Sons. DOI: 10.1002/9781118551424.ch3

World Economic Forum. (2020). *AI governance: A holistic approach to implementing ethical AI*. World Economic Forum. https://www.weforum.org

## KEY TERMS AND DEFINITIONS

**AI in Financial Services:** The application of artificial intelligence techniques such as automation, predictive analytics, and machine learning to banking and finance. These tools are used to strengthen fraud detection, improve risk assessment, guide investment decisions, and enhance the overall customer experience.

**AI Regulation in Finance:** The collection of laws, guidelines, and policy frameworks introduced by governments and financial authorities to oversee the use of AI in financial institutions, with the goal of promoting transparency, fairness, accountability, and consumer protection.

**Algorithmic Bias:** The situation where AI systems produce unfair or discriminatory outcomes because they rely on biased datasets or flawed model assumptions, leading to systematic errors in financial decision-making.

**Duty of Care in AI:** The ethical and legal responsibility of financial institutions, regulators, and AI developers to ensure that AI systems are designed, tested, and monitored in ways that minimize errors, reduce risks, and comply with established standards.

**Ethical AI Auditing:** A systematic review process that evaluates AI models used in finance to verify that they meet ethical, regulatory, and fairness standards, thereby reducing unintended harms or discriminatory outcomes.

**Explainable Artificial Intelligence (XAI):** A set of tools and approaches designed to make the inner workings of AI models more understandable, enabling regulators, institutions, and users to trace how automated financial decisions are generated.

**Fairness-Aware Machine Learning:** An approach to building AI systems that identifies and reduces potential biases in financial decision-making, ensuring that individuals from different demographic or social groups are treated fairly.

**Transparency of AI Decisions:** The extent to which the decision-making process of AI models can be clearly explained and understood by stakeholders, including regulators, customers, and managers, thereby building trust in financial systems.

# Chapter 14
# Transforming Healthcare: The Impact of AI and Machine Learning on Personalization

**Pallavi Sharma**
https://orcid.org/0009-0003-7845-4205
*GNA University, India*

**Rupeena Kumra**
*GNA University, India*

**Ekta Bhaggi**
https://orcid.org/0009-0004-7347-9822
*GNA University, India*

## ABSTRACT

*Artificial Intelligence (AI) and Machine learning (ML) has the potential to revolutionize personalized healthcare by facilitating the creation of individualized treatment plans and data-driven decisions to optimize patient outcomes. That is because AI and ML algorithms can predict illness opportunities, identify patterns, and insisting the treatment options faster and accurately by analyzing the data from different data sets like genetic traits, medical histories, lifestyle factors, and real-time health indicators. This study examines the many uses of AI and ML in customized healthcare, such as medication discovery, precision medicine, early diagnosis, and real-time wearable device monitoring. By minimizing trial-and-error treatment, improving clinical decision-making, and enabling personalized treatment, AI and ML has the potential to totally revolutionize healthcare delivery and make way for more effective, efficient, and patient-centered healthcare.*

DOI: 10.4018/979-8-3373-2597-2.ch014

## 1. INTRODUCTION

AI is driving significant improvements in healthcare, leading to enhanced patient care, reduced costs, and greater efficiency. This article examines the present applications of AI in healthcare and looks ahead to its future potential. As AI technology rapidly evolves, it creates exciting new opportunities for the industry. With the aim of improving patient outcomes, improving diagnostic accuracy, reducing administrative costs, and fostering advances in medical research, innovations like machine learning, natural language processing, and computer vision are changing many facets of healthcare (Ejjami, n.d.).

Particularly in the areas of medical imaging and diagnostics, artificial intelligence is significantly transforming the medical industry. Advanced AI systems are able to precisely identify tumors, anomalies, and other problems by analysing medical pictures such as CT scans, X-rays and MRIs. This skill makes early diagnosis easier, which improves treatment results.

Additionally, chatbots and virtual assistants driven by AI are proliferating in hospital settings. These intelligent systems provide patients with individualized support, answer questions about their health, offer advice on self-care, and help prioritize care according to symptoms. Such advancements enhance access to healthcare while lightening the burden on healthcare professionals.

AI holds significant potential in predictive analytics and patient monitoring.These systems can detect patterns and risk factors linked to many diseases by analysing enormous volumes of patient data, enabling earlier intervention and prevention. Additionally, AI-powered wearable devices and remote monitoring solutions continuously track vital signs and send real-time alerts when there are notable changes in a patient's health.

The goal of AI, a quickly developing field of computer science, is to create machines that can carry out tasks that normally call for human intelligence. AI uses a variety of techniques, such as natural language processing (NLP), deep learning (DL), and ML, all of which help create systems that are more intelligent and powerful.

Large Language Models (LLMs), a subfield of artificial intelligence, use deep learning algorithms and enormous volumes of data to comprehend, produce, and forecast text-based data. These models are versatile and can perform a wide array of tasks in natural language processing, such as creating content, translating languages, summarizing text, rewriting, classifying data, categorizing information, and analyzing sentiments.

As a branch of artificial intelligence, NLP aims to bridge the gap between human communication and computing systems by enabling machines to understand, process, and generate human language. Through NLP, AI systems can interpret meaning, generate coherent text, and engage in human-like conversations (Singhal et al., 2024).

Through the introduction of cutting-edge technology, recent developments in the biomedical sciences—such as digital health, ML, genomics, AI—are causing a profound shift in the healthcare sector. These developments not only improve the accuracy and efficiency of healthcare but also create new possibilities for personalized care and precision treatments. As a result, there is a growing need for a reimagined workforce and updated healthcare practices that can adapt to the rapid pace of these changes.

Breakthroughs in fields like genomics and biometrics are enhancing our ability to understand individual genetic makeup and health patterns, enabling tailored treatments that are more effective and less invasive. In tissue engineering and vaccine development, advancements are opening up new avenues for regenerative therapies and more effective disease prevention strategies. Together, these innovations have the potential to revolutionize diagnostics, optimize treatment approaches, streamline care delivery, and pave the way for highly personalized, precision-based medicine (Moltubakk Kempton & Vassilakopoulou, 2021).

Figure 1 provides a visual overview of the historical development of AI, followed by a conceptual illustration of how AI relates to ML, DL, and NLP.

*Figure 1. AI Development with a Clearer Knowledge of the Connections Among NLP, DL, ML and AI*

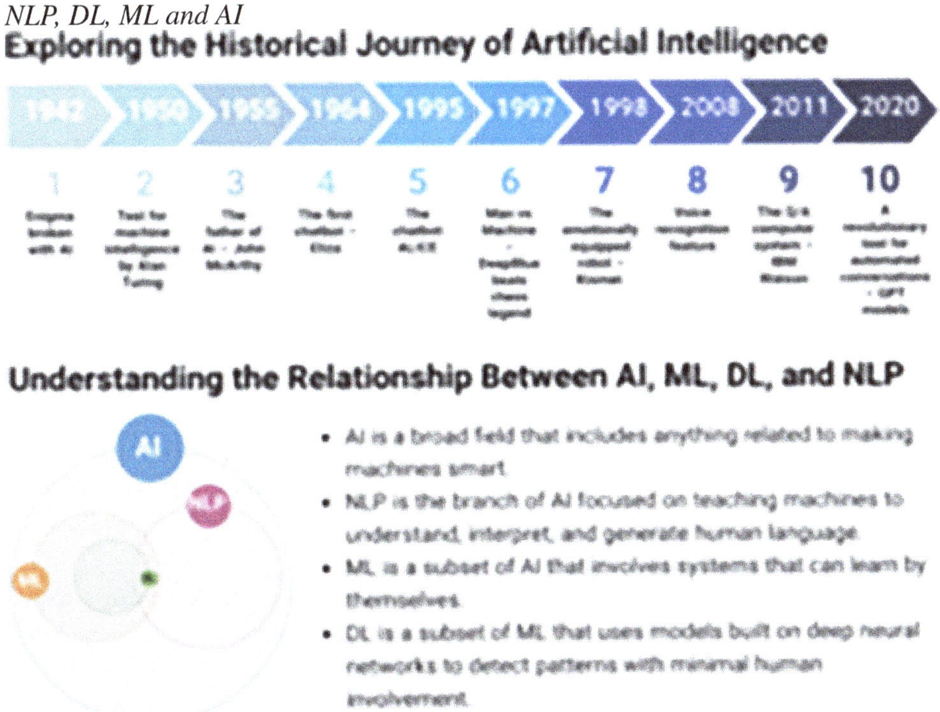

The following is a continuation of this article: A brief summary of the uses of AI in healthcare is given first, followed by discussions of the main technologies, difficulties, and prospects for the field's future. Finally, a conclusion is provided.

## 2. LITERATURE REVIEW

AI and ML have increasingly become central to personalized healthcare, enabling data-driven, individualized treatment strategies. Ejjami (year not specified) proposed the Holistic Intelligent Healthcare Theory (HIHT) that integrates AI for ethical and human-centered healthcare innovation. Singhal et al. (2024) conducted a scoping review emphasizing fairness, accountability, and transparency in AI applications in healthcare, highlighting the ethical challenges and potential solutions for AI adoption.

Studies by Moltubakk Kempton and Vassilakopoulou (2021) and Alam et al. (2023) focus on explainability and transparency in AI systems, emphasizing that understanding AI decision-making processes is critical in medical contexts. Amann et al. (2020) provide a multidisciplinary perspective advocating for explainable AI to build trust among healthcare professionals and patients .

Khan et al. (2024) systematically reviewed innovations and ethical challenges in securing trusted AI in healthcare, underscoring data privacy and security as significant barriers. Albahri et al. (2023) further addressed quality, bias risk, and data fusion, pointing to the necessity of trustworthy AI models that can safely integrate complex healthcare data without compromising accuracy or fairness .

Several researchers, including Nasarian et al. (2024), focused on designing interpretable ML systems to enhance clinician trust through responsible AI-clinician collaboration frameworks. Goirand et al. (2024) highlighted the importance of transparency in implementing ethics frameworks for AI-based healthcare applications .

Other notable works by Saraswat et al. (2022) and Sharma & Sharma (2024) discuss opportunities and challenges in healthcare AI, calling for ethical machine learning applications and frameworks for smart healthcare informatics .

Integration of AI in genomics, clinical trials, diagnostic imaging, and virtual assistants is transforming traditional healthcare delivery to more proactive, predictive, and personalized paradigms. However, scholars uniformly recognize the significance of addressing ethical, regulatory, interoperability, and bias concerns to realize the full benefits of AI-driven healthcare transformation.

Further literature underscores AI's transformative capacity in delivering personalized healthcare. Udegbe et al. (2024) highlight that AI applications in healthcare span various domains, including diagnostics, patient monitoring, and drug development, emphasizing the importance of data security and ethical considerations .

Their review stresses the need for responsible AI deployment to prevent bias and ensure fairness.

Sharma & Sharma (2024) (2024) explore ethical challenges tied to ML implementation within smart healthcare systems, emphasizing that transparency and accountability are paramount. They recommend establishing ethical guidelines and involving multidisciplinary stakeholders to address the societal implications of AI .

Research by Rissanen (2020) emphasizes the role of health informatics in translating AI innovations into practical healthcare systems, advocating that effective data management and interoperability are foundational to AI's success in personalized medicine .

Saraswat et al. (2022) conducted a comprehensive review on explainable AI (XAI) in healthcare, noting that explainability enhances clinician trust and facilitates regulatory approval processes. They propose frameworks for developing transparent AI models that explain their decision pathways to users .

Moreover, recent advancements in AI-driven genomics and bioinformatics research, as discussed by Kempton and Vassilakopoulou (2021), underline the importance of leveraging AI for understanding complex genetic information to enable truly personalized treatments .

Finally, the literature collectively indicates that while AI's integration into healthcare presents significant opportunities, addressing technical challenges, legal frameworks, and ethical dilemmas remains crucial for its sustainable and equitable implementation. The consensus is that ongoing interdisciplinary collaboration and responsible AI governance will be essential to harness AI's full potential in personalized healthcare .

This literature base reflects a consolidated understanding that AI and ML have the potential to revolutionize healthcare if implemented responsibly, with appropriate attention to transparency, fairness, security, and patient-centric ethics .

## 3. ARTIFICIAL INTELLIGENCE IN HEALTHCARE APPLICATIONS

The healthcare industry is using artificial intelligence in medicine for the following reasons:

### 3.1 Artificial Intelligence in Drug Discovery

Pharmaceutical businesses are now able to expedite drug discovery procedures because to the incorporation of AI technology in healthcare. The identification of possible pharmacological targets is also automated by AI. Furthermore, AI helped boost drug repurposing efforts in recent years by studying off-target molecules. This strategy improves productivity, gets rid of unnecessary work, and streamlines processes in the healthcare and pharmaceutical industries.

Leading biopharmaceutical companies are now offering innovative AI-driven solutions. Pfizer, for instance, leverages IBM Watson, a machine learning system, to support the discovery of treatments in immuno-oncology. Sanofi has partnered with Exscientia to use its AI platform for developing therapies targeting metabolic diseases. Meanwhile, Genentech, a Roche subsidiary, collaborates with GNS Healthcare to utilize AI in identifying potential cancer treatments. Most major biopharmaceutical firms have either formed similar partnerships or launched their own AI-focused initiatives.

Supporters of AI and machine learning believe that these technologies could fundamentally transform drug development, making the process faster, more affordable, and much more efficient. While there are some skeptics, a growing number of experts predict that these tools will play an increasingly crucial role in the future. This shift presents both challenges and opportunities for researchers, particularly when AI is integrated with automation to streamline and improve research and development workflows.

## 3.2 AI for clinical trials

Clinical trials involve testing new treatments on individuals to evaluate their effectiveness, a process that requires substantial time and resources. However, the success rate of these trials is often low. As a result, automating clinical trials has brought benefits to both AI and the healthcare industry. AI also helps reduce the burden of time-intensive data monitoring. Additionally, AI-powered clinical trials can process large datasets and deliver precise, reliable outcomes.

## 3.3 Patient Care

The application of artificial intelligence in the medical area has an impact on patient outcomes. Companies focused on medical AI develop a framework that supports patients at every stage. Clinical intelligence also examines medical records and offers recommendations to help patients enhance their quality of life.

## 3.4 Virtual Nursing Assistant

Virtual nursing assistants play a vital role in supporting patients in hospitals, whether they are dealing with illnesses or recovering from surgeries. iCare Navigator's virtual companions, Sally and Walt, engage with patients in a compassionate manner to motivate them to actively participate in their recovery process.

Accessible via tablets or hospital televisions, these assistants provide a sense of comfort, allowing patients to ask questions without fear of judgment. Research indicates that when interacting with a non-human assistant, patients feel more comfortable discussing concerns and understanding their post-care instructions.

This leads to more open communication and encourages patients to take responsibility for their self-care after discharge. Developed by iCare Navigator's team at TeleHealth services, the platform uses machine learning and electronic health records to tailor patient interactions. With its foundation in cloud technology and advanced AI systems, iCare Navigator continuously tracks and evaluates the patient's responses and engagement, ensuring a personalized experience that adapts to individual needs.

## 3.5. Rehabilitation

With applications in the virtual (informatics) and physical (robotics) realms, artificial intelligence is making tremendous strides in the field of rehabilitation. ML, a key element of artificial intelligence, is concerned with developing algorithms that get better with time and experience. In rehabilitation, ML is applied across a variety of areas, such as perioperative care, brain-computer interfaces, myoelectric control, and neuroprosthetics.In musculoskeletal medicine, machine learning techniques are also being applied to enhance diagnostic imaging, aid in clinical decision-making, and analyze patient data.

In therapeutic environments, AI-driven cognitive tools are also being utilized to evaluate rehabilitation exercises by analyzing the signals produced by medical devices.

Figure 2 visually summarizes the key roles of AI in the healthcare sector. The central circle highlights the overarching topic—"Role of AI in Healthcare"—with five major applications illustrated around it.

*Figure 2. Role of AI in Healthcare Industry*

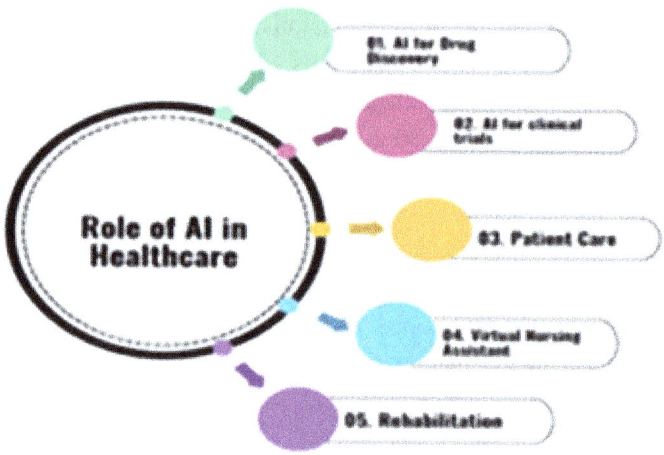


## 4. KEY TECHNOLOGIES USED IN AI FOR HEALTHCARE

### 4.1 Machine Learning (ML)

- **Supervised Learning:** Applicable in areas such as diagnosing diseases and forecasting patient results (e.g., predicting diabetes).
- **Unsupervised Learning:** Employed for grouping patient data, identifying anomalies, and uncovering concealed patterns.
- **Reinforcement Learning:** Utilized for enhancing treatment strategies or robotic surgical systems.

### 4.2 Deep Learning (DL)

- **Convolutional Neural Networks (CNNs):** In order to find patterns and anomalies in medical images, including CT scans, MRIs and X-rays are widely utilized.
- **Long Short-Term Memory(LSTM) and Recurrent Neural Networks(RNNs):**These models are applied to handle time-series data, such as analyzing patient records and electronic health history to identify trends and predict outcomes.

- **Transformers:** Driving the development of large language models for the analysis of medical texts and the creation of chatbots.

## 4.3 Natural Language Processing (NLP)

- **Analyzing Medical Records:** Gaining insights from Electronic Health Records (EHRs).
- **Clinical Documentation:** Streamlining the transcription and summarization of interactions between doctors and patients.
- **Chatbots & Virtual Assistants:** Offering symptom evaluation and health guidance (e.g., ChatGPT for medical inquiries).

## 4.4 Computer Vision

- Diagnostics based on imaging (such as cancer identification and screening for retinal conditions).
- Support in surgical procedures through real-time imaging.
- Examination of pathological slides.

## 4.5 Robotics

- Surgical robots such as the Da Vinci Surgical System.
- Robots designed for assisting the elderly, rehabilitation, and physical therapy.

## 4.6 Big Data & Data Analytics

- Gathering and examining extensive healthcare information from electronic health records, genomics, and wearable technology.
- Facilitating predictive analysis for patient results.

## 4.7 Genomics and Bioinformatics

- Genomics powered by AI is essential for personalized medicine.
- Examining DNA sequences to locate genetic indicators of illnesses.

## 4.8 Predictive Analytics

- Recognizing patients vulnerable to chronic conditions.
- Enhancing hospital resource efficiency (for instance, forecasting ICU needs).

## 4.9 Wearable Technologies and IoT

- Gadgets such as smartwatches and fitness trackers that track health information in real-time (like heart rate and sleep habits).
- Artificial Intelligence analyzes the gathered data for proactive alert systems (such as fall detection and arrhythmia monitoring).

## 4.10 Reinforcement Learning in Drug Discovery

- Speeding up drug development by utilizing simulations to examine interactions between chemical compounds.
- Generative AI models for creating novel pharmaceuticals.

Figure 3 represents an overview of the essential AI technologies that are transforming the healthcare industry.

*Figure 3. Vital AI Technologies in Healthcare*

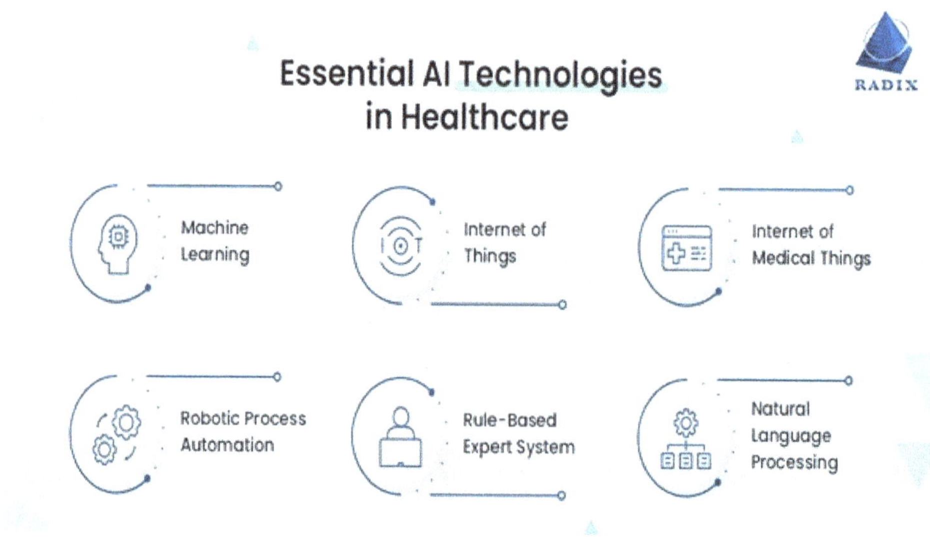


## 5. DIFFICULTIES IN INTEGRATING AI IN HEALTHCARE

By enabling earlier diagnoses, better treatment plans, and even preventive care, artificial intelligence has the potential to revolutionize the healthcare industry. However, integrating AI into complex healthcare systems presents several challenges. This necessitates a thorough investigation of the issues that need to be tackled to fully harness AI's capabilities in improving patient care and overall healthcare efficiency.

## 5.1 Privacy and Data Security Issues

One of the biggest obstacles to integrating AI in the healthcare sector is privacy and data security concerns. Healthcare organizations handle large volumes of sensitive patient data, making them especially vulnerable to cyberattacks. A data breach can have serious repercussions, including the possibility of financial fraud, identity theft, and a decline in the standard of health care.

To mitigate issues related to data security and privacy, healthcare providers must implement measures such as robust encryption techniques, access controls, routine audits, and employee training. Maintaining compliance requires keeping abreast of the most recent legal standards, such as the Health Insurance Portability and Accountability Act (HIPAA).

By prioritizing the privacy of the patient and investing in secure technologies, healthcare organizations can reduce the risks associated with the adoption of AI.

## 5.2 Lack of Sufficient Data

A significant obstacle in incorporating AI into this field is the insufficient availability of information. These systems depend on information and data to generate precise predictions and recommendations. Often, there is either inadequate or low-quality data accessible to effectively train AI algorithms. This limitation can impede the accuracy and performance of AI systems.

Consequently, healthcare organizations need to tackle this issue by adopting approaches to gather, store, and uphold high-quality data. Furthermore, partnerships among healthcare entities to share data in a responsible manner can alleviate the issue of insufficient data.

A steady supply of high-quality data for AI algorithm training can also be provided by investing in data gathering techniques like wearable technologies and remote monitoring tools.

## 5.3 Interoperability Issues

Interoperability challenges create major barriers to the smooth incorporation of AI into healthcare systems, as the integration of AI typically necessitates the sharing of data across diverse platforms and systems.

Consequently, ensuring the secure transmission of this data while preserving its integrity and confidentiality presents a notable challenge.

To address interoperability challenges, healthcare organizations need to invest in systems capable of effective communication with one another. Implementing standardized formats and protocols can enhance data exchange and interoperability across various systems and technologies. Additionally, collaboration between technology providers and healthcare organizations is crucial for creating interoperable solutions that can be integrated seamlessly with existing healthcare infrastructure.

## 5.4 Regulatory Compliance

One of the biggest obstacles to incorporating AI in the healthcare sector is navigating regulatory regulations. To protect patient privacy, healthcare professionals must abide by stringent laws like the Health Insurance Portability and Accountability Act (HIPAA). Integrating AI into healthcare systems while ensuring compliance with these regulations presents a complex challenge.

Organizations must stay informed about the latest regulatory standards and implement necessary actions to maintain compliance. This entails establishing robust data security procedures, encouraging openness in AI-driven decision-making, and guaranteeing responsibility for the outcomes of AI-generated choices. Additionally, regular audits and staff training programs can help ensure that healthcare personnel understand and meet their regulatory obligations.

## 5.5 Ethical and Bias Concerns

Concerns about ethics and bias represent significant obstacles that need to be overcome when integrating AI into the healthcare sector. AI algorithms could potentially sustain and even intensify existing healthcare biases. Moreover, ethical dilemmas emerge regarding the application of AI in making decisions, particularly in crucial areas like diagnostics and treatment planning.

Healthcare companies must make sure AI systems are created and implemented responsibly and ethically in order to solve ethical issues and prejudices. It is vital to regularly assess algorithms for bias, maintain transparency in decision-making processes, and hold parties accountable for the outcomes of AI-driven decisions. Furthermore, providing education to healthcare professionals about the advantages and limitations of AI, along with encouraging a culture of diversity and inclusion, can aid in alleviating bias-related concerns.

Figure 4 represents a concise summary of the main challenges and corresponding solutions for implementing AI in the healthcare sector.

*Figure 4. AI Integration's Difficulties in the Healthcare Sector*

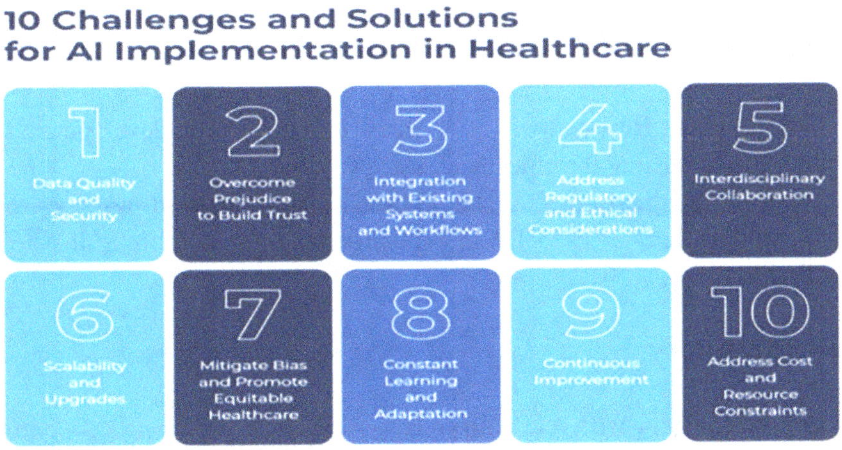


## 6. PROSPECTS FOR AI IN HEALTHCARE IN THE FUTURE

### 6.1 Personalized Medicine

- **AI for Tailored Treatment:** AI systems will evaluate genetic, environmental, and lifestyle factors to customize treatments for individual patients.
- **Pharmacogenomics:** AI can forecast how patients will react to particular medications, lowering the chances of side effects and enhancing effectiveness.

### 6.2 Advancements in Disease Prediction and Prevention

- **Predictive Analytics:** In order to forecast disease outbreaks and evaluate personal health risks, artificial intelligence (AI) can evaluate vast amounts of data from wearable technology and Electronic Health Records (EHRs).
- **Proactive Monitoring:** The combination of IoT devices and real-time health data analysis will aid in preventing diseases before symptoms manifest.

## 6.3 Revolutionizing Medical Imaging

- **Cutting-Edge Diagnostic Tools:** AI-enhanced imaging systems could achieve even greater accuracy in identifying cancers, fractures, and rare diseases.
- **Automated Image Analysis:** faster and less expensive picture evaluation in places where radiologists are hard to reach.

## 6.4 AI-Driven Drug Discovery

- **Fast-Tracking Research:** AI is expected to shorten drug development timelines by forecasting the efficacy and safety of compounds.
- **Tailored Medications:** AI could assist in creating drugs for specific demographics or even for individual patients, transforming the field of pharmacology.

## 6.5 AI in Surgery

- **Robotic Surgery with AI:** With enhanced precision and safety, AI-controlled robots might perform surgeries with little to no human involvement.
- **Integration of Augmented Reality (AR):** AR-assisted surgery alongside AI will improve accuracy in intricate procedures.

## 6.6. Improved Mental Health Assistance

- **AI Therapy:** Sophisticated virtual helpers and chatbots can offer round-the-clock mental health assistance, facilitating the early identification of depression, anxiety, and other disorders.
- **Behavioral Analysis:** AI has the potential to assess trends in speech, behavior, and facial expressions to uncover mental health concerns.

Figure 5 outlines the key focus areas for the future of artificial intelligence (AI) in healthcare.

*Figure 5. Potential Applications of AI in Healthcare in the Future*

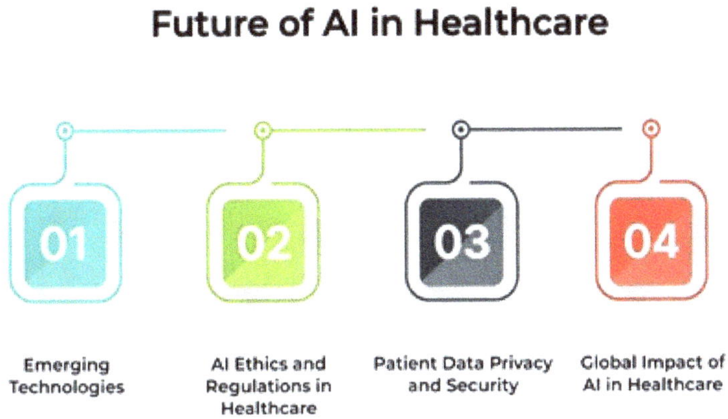


## CONCLUSION

Medical imaging and diagnostics, public health emergency response, virtual patient care, improving patient engagement and treatment plan adherence, lowering administrative burdens for medical professionals, speeding up the development of drugs and vaccines, monitoring exercise regimen adherence, and gait analysis in rehabilitation settings are just a few of the healthcare domains where artificial intelligence is being used.

As AI develops in the healthcare industry, it faces a number of ethical, legal, and technical obstacles. AI systems usually rely on sensitive patient data that is protected by law, raising serious concerns about data security and privacy. AI's incapacity to replicate fundamental human characteristics like empathy and the caliber of the health data that is currently available may also limit its efficacy.

Although AI has many benefits when applied properly, technology cannot take the role of interpersonal relationships, which are essential for teamwork and leadership. Machines are incapable of forming human connections, which are necessary for productive teamwork. Making sure that the creation and use of AI technologies are in line with human values while taking into account the ethical, technological, and societal ramifications is a significant concern for the governance of these technologies in the future. By offering a thorough summary of the numerous uses of AI in healthcare, including fields like medical imaging, virtual care, drug research, patient engagement, rehabilitation, and administrative duties, this study adds to the existing body of knowledge. It explores how AI is changing each of these fields and provides information on how it might enhance patient outcomes and expedite medical procedures.

The study also presents the most recent viewpoints on the moral, societal, and technological difficulties that medical practitioners encounter when attempting to incorporate AI into healthcare systems. Data privacy, algorithmic biases, legal compliance, and the requirement for human oversight in AI-driven decision-making are some of these difficulties.

## REFERENCES


Alam, M. N., Kaur, M., & Kabir, M. S. (2023). Explainable AI in Healthcare: Enhancing transparency and trust upon legal and ethical consideration. *Int Res J Eng Technol, 10*(6), 1–9.

Albahri, A. S., Duhaim, A. M., Fadhel, M. A., Alnoor, A., Baqer, N. S., Alzubaidi, L., Albahri, O. S., Alamoodi, A. H., Bai, J., Salhi, A., Santamaría, J., Ouyang, C., Gupta, A., Gu, Y., & Deveci, M. (2023). A systematic review of trustworthy and explainable artificial intelligence in healthcare: Assessment of quality, bias risk, and data fusion. *Information Fusion, 96*, 156–191. DOI: 10.1016/j.inffus.2023.03.008

Amann, J., Blasimme, A., Vayena, E., Frey, D., & Madai, V. I. (2020). Explainability for artificial intelligence in healthcare: A multidisciplinary perspective. *BMC Medical Informatics and Decision Making, 20*(1), 1–9. DOI: 10.1186/s12911-020-01332-6 PMID: 33256715

.Ejjami, R. The Holistic Intelligent Healthcare Theory (HIHT): Integrating AI for Ethical, Transparent, and Human-Centered Healthcare Innovation.

Goirand, M., Austin, E., & Clay-Williams, R. (2024). Bringing clarity and transparency to the consultative process underpinning the implementation of an ethics framework for AI-based healthcare applications: A qualitative study. *AI and Ethics*, •••, 1–21.

Kempton, A. M., & Vassilakopoulou, P. (2021). Accountability, Transparency and Explainability in AI for Healthcare. In *8th International Conference on Infrastructures in Healthcare.*

Khan, M. M., Shah, N., Shaikh, N., Thabet, A., & Belkhair, S. (2024). Towards secure and trusted AI in healthcare: A systematic review of emerging innovations and ethical challenges. *International Journal of Medical Informatics*, •••, 105780. PMID: 39753062

Moltubakk Kempton, A., & Vassilakopoulou, P. (2021). *Accountability*. Transparency and Explainability in AI for Healthcare.

Nasarian, E., Alizadehsani, R., Acharya, U. R., & Tsui, K. L. (2024). Designing interpretable ML system to enhance trust in healthcare: A systematic review to proposed responsible clinician-AI-collaboration framework. *Information Fusion, 108*, 102412. DOI: 10.1016/j.inffus.2024.102412

Rissanen, M. (2020). Translational health technology and system schemes: Enhancing the dynamics of health informatics. *Health Information Science and Systems*, *8*(1), 39. DOI: 10.1007/s13755-020-00133-5 PMID: 33194173

Saraswat, D., Bhattacharya, P., Verma, A., Prasad, V. K., Tanwar, S., Sharma, G., Bokoro, P. N., & Sharma, R. (2022). Explainable AI for healthcare 5.0: Opportunities and challenges. *IEEE Access : Practical Innovations, Open Solutions*, *10*, 84486–84517. DOI: 10.1109/ACCESS.2022.3197671

Sharma, A. K., & Sharma, R. (2024). Navigating the Ethical Landscape: Implementing Machine Learning in Smart Healthcare Informatics. *Indian Journal of Community Health*, *36*(1), 149–152. DOI: 10.47203/IJCH.2024.v36i01.024

Singhal, A., Neveditsin, N., Tanveer, H., & Mago, V. (2024). Toward Fairness, Accountability, Transparency, and Ethics in AI for Social Media and Health Care: Scoping Review. *JMIR Medical Informatics*, *12*(1), e50048. DOI: 10.2196/50048 PMID: 38568737

Udegbe, F. C., Ebulue, O. R., Ebulue, C. C., & Ekesiobi, C. S. (2024). The role of artificial intelligence in healthcare: A systematic review of applications and challenges. *International Medical Science Research Journal*, *4*(4), 500–508. DOI: 10.51594/imsrj.v4i4.1052

# Chapter 15
# AI Ethics in Finance:
## Balancing Innovation With Responsibility

**Rishikaysh Marotrao Kaakandikar**
https://orcid.org/0000-0003-0349-6788
*SaiBalaji International Institute of Management Sciences, India*

**Shradha Sinkar**
*SaiBalaji International Institute of Management Sciences, India*

**Subathra Chelladurai**
*Pioneer Kumaraswamy College, India*

**Suunil Losarwarr**
*SaiBalaji International Institute of Management Sciences, India*

## ABSTRACT

*The rapid integration of Artificial Intelligence (AI) in the financial sector has unlocked new opportunities for efficiency, predictive accuracy, and personalized customer experiences. However, the rise of AI-driven financial solutions also raises significant ethical concerns, including algorithmic bias, lack of transparency, data privacy, and accountability. This paper explores the ethical challenges and responsibilities associated with AI in finance, highlighting the need for a balanced approach that fosters innovation while ensuring fairness, transparency, and regulatory compliance. It examines case studies of AI deployment in financial services, discusses the potential risks of unethical AI practices, and proposes a framework for responsible AI governance. By balancing technological advancements with ethical considerations, the financial sector can build trust and enhance long-term sustainability.*

DOI: 10.4018/979-8-3373-2597-2.ch015

## INTRODUCTION

Artificial Intelligence (AI) has rapidly transformed the financial sector, offering innovative solutions to complex challenges and unlocking new opportunities. The intersection of finance and AI is reshaping how financial services are delivered, enhancing decision-making, improving customer experiences, and optimizing operational efficiency. AI technologies like machine learning (ML), natural language processing (NLP), and neural networks have found widespread use in various applications within finance, driving the industry's evolution.

One of the most significant advantages of AI in finance is its ability to process and analyze massive volumes of data with speed and precision. This capability enables financial institutions to identify patterns, predict trends, and make real-time data-driven decisions. From fraud detection and algorithmic trading to risk assessment and portfolio management, AI helps streamline operations while reducing human error and operational costs.

Customer service has also seen vast improvements through AI-powered chatbots and virtual assistants, which provide 24/7 support and personalized financial guidance. In addition, AI enhances regulatory compliance by monitoring transactions and flagging suspicious activities, making it a critical tool in anti-money laundering (AML) efforts.

As AI advances, its integration into finance is expected to grow deeper, transforming traditional business models and enabling more inclusive, efficient, and responsive financial systems. However, this technological shift also brings challenges, including data privacy, ethical concerns, and the need for transparent, fair algorithms. Addressing these issues will be crucial to ensuring that the benefits of AI are realized responsibly and equitably across the financial ecosystem.

## EVOLUTION OF AI TECHNOLOGIES IN THE FINANCIAL SECTOR

The use of AI in finance dates back to the early 2000s, but it wasn't until recent years that AI made a significant impact. Initially, the adoption of AI technologies was slow due to limited computational power, the complexity of financial data, and regulatory constraints. However, as big data, cloud computing, and advanced algorithms emerged, the financial industry experienced a breakthrough. AI technologies started to gain traction, particularly with the advent of machine learning algorithms capable of analyzing vast amounts of data in real-time and making predictions with high accuracy.

AI's evolution in finance has been driven by a growing demand for automation, improved decision-making, and enhanced security measures. Early AI systems focused on data analysis and basic automation. At the same time, more recent advancements, such as deep learning and advanced predictive analytics, have led to more sophisticated applications, including fraud detection, algorithmic trading, and personalized financial services.

## Current Applications of AI in Finance

- **Fraud Detection and Risk Management**: One of the most critical applications of AI in finance is fraud detection. AI-driven algorithms can analyze large volumes of transactional data to identify unusual patterns or activities that could indicate fraudulent behavior. Machine learning models are trained to detect real-time anomalies, making them highly effective in preventing fraud before it escalates. Financial institutions use AI to protect sensitive customer data and maintain trust.
- **Algorithmic Trading**: AI has revolutionized financial markets by enabling high-frequency trading through algorithmic strategies. AI systems can process massive datasets, including real-time market data, news sentiment, and historical trends, to make rapid trading decisions. Machine learning models can optimize trading strategies, reduce risks, and capitalize on short-term market opportunities, making them a powerful tool for hedge funds, investment banks, and individual traders.
- **Robo-Advisors**: AI has democratized investment management by creating robo-advisors—automated platforms that provide investment advice based on an individual's financial goals, risk tolerance, and market conditions. Robo-advisors use algorithms to create personalized client portfolios, minimizing the need for human financial advisors and lowering costs. These platforms are gaining popularity due to their ease of use, accessibility, and low fees.
- **Credit Scoring and Underwriting**: AI is also transforming how lenders assess creditworthiness. Traditional credit scoring models rely on limited data, such as credit history, to evaluate loan eligibility. However, AI systems can analyze a broader range of data, including social media activity, purchasing habits, and alternative financial data, to assess lending risk. This allows financial institutions to make more accurate and inclusive lending decisions.

## Benefits of AI in Finance

- **Speed**: AI can process and analyze large datasets at incredible speeds. In areas like algorithmic trading or fraud detection, where real-time decision-making is crucial, AI provides a significant advantage by making instantaneous assessments and actions.
- **Accuracy**: AI models can uncover hidden patterns and insights from data that would be impossible for humans to identify. By leveraging these patterns, AI systems can make more accurate predictions and decisions, such as detecting fraudulent transactions or forecasting market movements.
- **Cost-Efficiency**: AI automation reduces the need for manual labor in tasks such as data analysis, customer service, and compliance monitoring. This cuts operational costs and increases efficiency and scalability for financial institutions.
- **Scalability**: AI technologies allow financial firms to scale their operations rapidly. Automated systems can handle vast amounts of data and serve large customer bases without compromising performance or quality.

## Ethical Considerations in AI-Driven Finance

As AI continues to transform the financial sector, it brings many ethical challenges that must be carefully navigated. From concerns about data privacy to the risks of algorithmic bias, AI in finance must be deployed responsibly to ensure fairness, transparency, and accountability. These ethical considerations are crucial for maintaining trust in financial institutions and promoting inclusive growth within the industry.

## DATA PRIVACY AND CONSENT

In AI-driven finance, data privacy and consent are critical ethical considerations. Financial institutions rely on vast amounts of personal data to train AI models for credit scoring, fraud detection, and personalized services. Ensuring this data is collected, stored, and used responsibly is essential. Users must be informed about how their data will be used and provide explicit informed consent. Transparency, data minimization, and secure handling are key principles. Violations can lead to loss of trust, discrimination, or financial harm. Ethical AI in finance must prioritize individual rights, ensuring privacy is respected, and consent is genuinely obtained and upheld.

## Handling sensitive financial data

Handling sensitive financial data in AI-driven finance requires strict ethical standards to protect individuals' rights and maintain trust. Economic data is highly personal, and misuse can lead to identity theft, financial loss, or discrimination. AI systems must ensure that data is encrypted, securely stored, and accessed only by authorized parties. Developers should implement data minimization and anonymization techniques to reduce exposure. Regular audits and compliance with regulations like GDPR are essential. Transparency in data handling practices and accountability for breaches are critical. Ethical AI must prioritize confidentiality, security, and fairness when managing sensitive financial information to prevent harm and uphold public trust.

## Transparency in data collection and usage

Data collection and usage transparency is a key ethical consideration in AI-driven finance. Users must be informed about what data is being collected, how it will be used, and who can access it. This includes explaining the data collection's purpose and role in AI-driven decisions such as credit scoring or investment recommendations. Consent should be explicit, and data policies should be easy to understand, not buried in legal jargon. Transparent practices build trust, reduce the risk of misuse, and empower users to make informed choices about their financial data and digital privacy.

# BIAS AND FAIRNESS

Bias and fairness are critical ethical considerations in AI-driven finance, as biased algorithms can lead to unfair treatment in areas like loan approvals, credit scoring, or insurance underwriting. AI models trained on historical or incomplete data may unintentionally reinforce existing inequalities, disadvantaging certain groups based on race, gender, or socioeconomic status. Fairness requires diverse, representative data, regular audits, and bias mitigation techniques throughout the model lifecycle. Transparent criteria for decisions and mechanisms for appeal are also essential. Ethical AI in finance must prioritize equitable outcomes, treating all individuals fairly and avoiding discrimination to promote financial inclusion and social justice.

## Algorithmic Bias in Lending, Insurance, and Hiring

AI systems can unintentionally perpetuate bias when trained on historical or unrepresentative data. In lending, this might lead to unfair loan denials, in insurance, discriminatory premium pricing, and in hiring, biased candidate screening. These outcomes can entrench systemic inequalities and reduce access to opportunities for marginalized groups. Identifying and correcting bias requires inclusive datasets, bias audits, and diverse development teams.

## Impact on Marginalized Groups and Financial Inclusion

AI decisions can disproportionately harm marginalized communities by limiting credit, insurance, or employment access. If unchecked, biased models reinforce existing disparities. Promoting financial inclusion requires fairness-focused design, community input, and active efforts to avoid exclusionary practices.

## Transparency and Explainability

Transparency in AI means making processes and decisions understandable to stakeholders. Explainability helps users and regulators know how and why decisions were made, especially in finance and HR. Without clarity, detecting errors, challenging outcomes, or ensuring fairness is difficult. Transparent AI builds trust and supports ethical use.

## The "Black Box" Problem in AI Models

Many AI models, intense learning systems, operate as "black boxes," offering little insight into their decision-making processes. This opacity makes it hard to understand or contest outcomes, especially in high-stakes areas like finance or employment. Addressing this requires methods that clarify model logic and outputs.

## Need for Explainable AI (XAI) in Decision-Making

Explainable AI (XAI) aims to make AI decisions understandable to humans. In finance, XAI supports compliance, trust, and user empowerment by clarifying why certain decisions—like loan rejections—occur. XAI helps stakeholders ensure fairness, detect bias, and take corrective action, making AI systems more transparent and accountable.

## ACCOUNTABILITY AND GOVERNANCE

Clear accountability is essential when AI impacts people's lives. Organizations must define who is responsible for AI outcomes—developers, operators, or companies. Governance structures, including internal ethics boards and third-party audits, help ensure responsible use. Ethical deployment requires apparent oversight and accountability at all stages of the AI lifecycle.Integrating AI ethics into corporate strategy is becoming a vital aspect of financial innovation. Firms are now encouraged to adopt responsible innovation practices that align technological advancement with social values (La Torre et al., 2020; Guo et al., 2022). This includes forming interdisciplinary teams that evaluate the societal impact of AI systems and involving stakeholders in the design process. Collaboration between industry and academia has also proven effective in advancing responsible AI development while addressing complex ethical dilemmas (Etzkowitz & Viale, 2010; Rahwan et al., 2019). Ultimately, a culture of ethical awareness within organizations enhances not only compliance but also long-term competitive advantage.

### Who Is Responsible for AI-Driven Decisions

Both developers and deployers are responsible. While algorithms automate decisions, humans must oversee and be accountable for outcomes. Developers should ensure ethical design, and organizations must monitor real-world impacts. Legal and ethical frameworks should assign responsibility to prevent harm and offer recourse to affected individuals.

### Regulatory Frameworks and Ethical Oversight

Regulatory frameworks guide the ethical use of AI through standards, laws, and industry guidelines. Effective oversight includes audits, transparency requirements, and enforcement mechanisms to prevent misuse. Ethical oversight—through independent bodies or internal ethics committees—ensures AI aligns with societal values, protects rights, and fosters responsible innovation.

### Balancing Innovation with Responsibility in AI-Driven Finance

As Artificial Intelligence (AI) continues to revolutionize the financial sector, it's essential to balance the pursuit of innovation with a strong commitment to ethical principles. While AI offers immense potential to improve efficiency, accuracy, and decision-making, its implementation must be handled responsibly to avoid unintended consequences. This is particularly critical in finance, where AI systems can affect people's lives, assets, and opportunities. Ethical AI design principles, responsible deployment strategies, and human-in-the-loop (HITL) systems ensure that AI innovation does not compromise fairness, transparency, or accountability.

## ETHICAL AI DESIGN PRINCIPLES

At the heart of AI development is the need for ethical AI design principles that ensure fairness, transparency, accountability, and human oversight. These principles should guide every stage of AI development, from concept and design to deployment and monitoring.

- **Fairness**: AI systems must be designed to avoid discrimination and bias, ensuring all individuals and groups are treated equitably. In finance, for instance, algorithms that decide on creditworthiness or loan approvals must be free from biases based on race, gender, age, or socioeconomic status. Developers must use diverse and representative datasets to train models and test them rigorously for fairness. Additionally, fairness can be enhanced by implementing fairness constraints within AI models, which actively address bias during the decision-making process.
- **Transparency**: Transparency is a cornerstone of ethical AI. Financial institutions and consumers alike must understand how AI systems make decisions. AI models must not function as "black boxes" but should provide clear, understandable explanations of how they arrive at decisions. For example, if an AI system denies a loan application, the institution must be able to explain why based on specific data points and model logic. Transparent models build trust, allowing consumers to hold financial institutions accountable for AI-driven decisions.
- **Accountability**: Developers and financial institutions must be accountable for the outcomes of AI-driven decisions. This means creating clear governance frameworks that establish who is responsible for designing, deploying, and performing AI models. In cases where AI systems make harmful or biased decisions, organizations must be able to identify the source of the problem and take corrective actions. Accountability can be bolstered through audits, regulatory oversight, and adherence to industry standards, ensuring that AI systems are used ethically and responsibly.
- **Human Oversight**: AI should not be fully autonomous, especially in critical financial decisions. Human oversight is essential to ensure AI systems align with ethical standards and societal values. Humans must remain in the loop, particularly when decisions significantly impact individuals, such as lending,

investment, and insurance. Human oversight enables the validation of AI-generated choices, offering an additional layer of ethical scrutiny and ensuring that the decisions made by AI systems are appropriate, fair, and aligned with the interests of individuals and society.

## Responsible AI Deployment

Responsible AI deployment ensures that artificial intelligence systems are designed, developed, and used in ways that are ethical, transparent, and aligned with human values. It involves addressing issues like bias, privacy, accountability, and fairness throughout the AI lifecycle. Stakeholders must prioritize inclusivity, continual monitoring, and compliance with legal and ethical standards. Engaging diverse perspectives, conducting thorough impact assessments, and fostering explainability are key practices. Responsible deployment also includes mechanisms for redress and oversight to mitigate potential harm. Ultimately, it aims to maximize societal benefit while minimizing risks, ensuring AI serves the public good in a trustworthy and sustainable manner.

## Pilot testing and ethical audits before full-scale implementation

Pilot testing and ethical audits are essential steps before full-scale AI implementation. Pilot testing allows organizations to assess system performance, identify risks, and gather real-world feedback in a controlled environment. It helps detect technical flaws, biases, and unintended consequences early on. Ethical audits evaluate the system's adherence to fairness, accountability, transparency, and privacy standards. These audits review data sources, algorithms, and decision-making processes to ensure ethical compliance. Organizations can build trust, improve system quality, and reduce harm by combining pilot tests with ethical audits. This proactive approach supports responsible AI deployment and ensures readiness for broader implementation.

## Human-in-the-Loop (HITL) Systems

Human-in-the-loop (HITL) systems integrate human judgment into AI decision-making, ensuring oversight, accuracy, and ethical alignment. In these systems, humans are actively involved in training, testing, and refining AI models, as well as in monitoring outcomes and making critical decisions. HITL is essential in high-stakes areas like healthcare, finance, and defence, where full automation could lead to significant risks. It helps mitigate bias, correct errors, and build trust by allowing human intervention when needed. By combining human expertise with machine efficiency, HITL systems enhance accountability and ensure AI systems remain aligned with human values and societal norms.

## Combining human judgment with machine intelligence

It leverages both strengths to create more effective, ethical, and accurate systems. While machines excel at processing large datasets and identifying patterns, humans contribute contextual understanding, moral reasoning, and emotional intelligence. This collaboration enhances decision-making, especially in complex or sensitive scenarios where nuance is crucial. By integrating human oversight into AI workflows, organizations can catch errors, address bias, and ensure alignment with societal values. Such synergy improves trust, accountability, and adaptability, enabling smarter, more responsible use of technology. Blending human insight with machine efficiency ultimately leads to better outcomes across various domains.

## Regulatory Landscape and Compliance

As Artificial Intelligence (AI) technologies continue to transform the financial sector, the need for effective regulation and compliance has become more critical. AI systems, particularly in areas like lending, insurance, and investment management, can significantly impact individual consumers and broader financial markets. To ensure that these systems are deployed ethically and responsibly, regulators worldwide have begun developing frameworks and guidelines that govern the use of AI in finance. However, these efforts must balance innovation with consumer protection, and they face several challenges due to the rapid pace of technological advancement.

## Global Regulatory Efforts

- **EU AI Act**: The European Union has been at the forefront of regulating AI, with the proposed EU AI Act being one of the most comprehensive regulatory frameworks for AI worldwide. This regulation categorizes AI systems based on their level of risk, placing stringent requirements on high-risk AI applications, which include those used in finance. For example, AI systems used in credit scoring, loan approval, and automated trading would be subject to strict guidelines. These requirements include mandatory risk assessments, transparency measures, and human oversight to ensure that AI systems function in a manner that aligns with EU values, including fairness, accountability, and data protection. The EU AI Act also emphasizes the need for explainability in AI models, ensuring that financial institutions can provide clear and understandable explanations for decisions made by AI systems.

- **General Data Protection Regulation (GDPR)**: Another critical regulatory framework in the EU is the GDPR, which governs the collection, storage, and processing of personal data. AI systems in finance often rely on large amounts of personal data, including financial histories and transaction records. Under the GDPR, consumers must have control over their data, and organizations must be transparent about collecting and using personal information. The GDPR also requires that individuals be informed about automated decision-making processes, such as those used in credit scoring or loan approvals, and provides consumers with the right to contest decisions made solely by AI systems. Compliance with GDPR ensures that AI-driven financial services respect privacy rights and protect personal data.
- **U.S. Regulatory Landscape**: In the United States, AI regulation is still evolving, and no comprehensive AI law exists. However, several federal and state-level regulations address specific aspects of AI, particularly in finance. The Consumer Financial Protection Bureau (CFPB) oversees rules related to fair lending and credit, which includes ensuring that AI-driven systems used for credit scoring or loan approvals do not discriminate based on race, gender, or other protected characteristics. Additionally, the Federal Trade Commission (FTC) has expressed concerns about AI and its impact on consumer protection, particularly algorithmic bias, transparency, and data security. At the state level, California's Consumer Privacy Act (CCPA) and similar laws provide additional privacy protections, ensuring consumers have rights over their data, directly impacting how AI technologies are deployed in the financial sector.
- **China's AI Regulation**: China has also regulated AI technologies, ensuring that AI applications align with the country's legal and ethical standards. While China's regulatory framework for AI is still developing, the government has implemented a National AI Strategy that includes guidelines for developing and deploying AI in various sectors, including finance. These regulations focus on data privacy, security, and ethics and aim to ensure that AI does not exacerbate social inequalities or cause harm to individuals. China's approach is more centralized and prescriptive compared to the EU's or the U.S.'s more decentralized approach.

## Sector-Specific Guidelines for Financial Institutions

Financial institutions face sector-specific guidelines when implementing AI technologies, which focus on ensuring that AI-driven systems do not undermine trust, fairness, or stability in the economic system. Some of these sector-specific regulations include:

- **Basel Committee on Banking Supervision (BCBS)**: The BCBS has issued guidelines on using AI and machine learning in financial services, particularly focusing on risk management and the stability of financial institutions. These guidelines emphasize the importance of governance, transparency, and accountability in AI models used for economic decision-making, such as credit assessments and fraud detection.
- **Financial Industry Regulatory Authority (FINRA)**: In the U.S., FINRA provides guidelines for how financial institutions should use AI in trading and investment advisory services. The authority ensures that automated trading algorithms comply with fair trading standards and are regularly audited to prevent market manipulation or excessive risk-taking.
- **Solvency II (EU)**: In the European insurance industry, the Solvency II directive provides a regulatory framework that requires insurance companies to assess the risk of their AI-driven models, particularly those related to underwriting and claims processing. The regulation mandates that financial institutions use appropriate models to ensure that AI-driven decisions remain sound and do not lead to discriminatory practices.

## CHALLENGES OF REGULATING FAST-EVOLVING AI TECHNOLOGIES

The challenges of regulating fast-evolving AI technologies extend beyond the pace of technological advancements and include issues of ethics, accountability, global coordination, and enforcement. As AI systems become more sophisticated, the ability to regulate them effectively becomes increasingly tricky. Regulators must navigate a rapidly changing landscape while balancing innovation with protection for consumers, businesses, and society.

### Pace of Technological Change

The speed at which AI technologies evolve presents a significant regulatory challenge. In the past, regulatory frameworks for new technologies could develop over several years, but AI's rapid advancements outpace this timeline. AI systems evolve not just in terms of hardware but also in algorithmic capabilities. For example, advances in machine learning techniques, such as reinforcement learning and generative models like GPT-4, have made it more difficult to predict the behavior of AI systems and understand their potential risks. When regulators address a particular issue, newer, more complex models may have emerged, rendering earlier regulatory frameworks obsolete. This creates a significant gap in oversight and could leave consumers unprotected in areas such as data privacy, security, and financial services.

Additionally, AI technologies are often deployed as black-box systems, where decision-making processes are not easily interpretable, even by their creators. This increases the challenge of regulating AI because regulators may not fully understand how or why an AI system arrived at a particular decision, making it difficult to apply traditional legal principles, such as transparency and accountability. Furthermore, AI-driven systems can adapt and change autonomously, making the oversight of their behavior even more complicated over time.

## Lack of Standardization

Another issue facing regulators is the lack of global standardization in AI regulations. Different regions have different approaches to AI governance, with varying levels of emphasis on ethics, privacy, and accountability. For instance, the European Union has introduced the General Data Protection Regulation (GDPR) and the Artificial Intelligence Act, aiming to create a comprehensive framework for AI regulation, including ethical standards for high-risk AI applications. In contrast, the United States has taken a more fragmented approach, with different states enacting their own AI laws and guidelines but without a unified federal strategy. This lack of cohesion challenges international companies, especially financial institutions, which must navigate multiple conflicting regulatory environments.

Financial institutions, particularly those involved in global finance, face the risk of compliance issues when working across borders. Depending on the local regulatory requirements, they may be forced to implement different AI models or practices, increasing operational complexity and the risk of non-compliance. Furthermore, businesses operating in multiple jurisdictions may need to decide which set of regulations to prioritize, potentially leading to regulatory arbitrage, where companies may seek out less stringent regulatory environments at the expense of consumer protection.

## Determining Liability

Determining liability in AI-driven financial services is another area where regulation lags behind technological progress. AI systems are increasingly used to make high-stakes decisions, such as credit scoring, loan approvals, investment advice, and insurance underwriting. When these systems make errors that result in financial loss or harm to consumers, it can be challenging to pinpoint who is responsible for the damage caused. Is it the AI system, the developers who designed it, or the financial institution implementing it?

In many cases, AI systems operate autonomously, which complicates the issue of accountability. Suppose an AI algorithm makes a decision that negatively impacts a consumer. Should the blame fall on the institution that deployed the AI, or is it the developers' fault who created the algorithm? If a self-learning AI system updates its models based on new data and then makes a decision that causes harm, it can be challenging to trace the exact point where things went wrong. Regulatory frameworks must address these complex questions of liability and accountability, ensuring that consumers have recourse in the event of harm while fostering a culture of responsibility within financial institutions.

## Ethical Concerns

While much of the regulatory focus has been on fairness, privacy, and transparency, ethical concerns surrounding AI technologies remain a persistent challenge. AI models can unintentionally perpetuate societal biases, resulting in discrimination in credit lending, hiring practices, and insurance. Regulators often struggle to create legal frameworks that mitigate these biases and encourage institutions to consider ethical implications proactively. For instance, in finance, AI systems that rely on historical data may reinforce past discriminatory practices, as the data itself can reflect patterns of racial or gender bias. Regulating fairness is complex because it requires striking a balance between ensuring equal treatment for all individuals while not stifling the innovation that AI technologies can bring to the financial sector.

The issue of ethical AI is also broader than just bias. It encompasses the wider societal impact of AI, such as the potential for AI to disrupt jobs, influence consumer behavior, or erode trust in financial systems. Regulators must create frameworks that not only address discrimination issues but also ensure that AI deployment aligns with societal values. This could mean requiring financial institutions to adopt ethical AI principles, conduct regular audits for biases, and be transparent about how AI models are designed and used.

Furthermore, ethical AI requires the inclusion of diverse perspectives in AI development. Financial institutions should consult with a variety of stakeholders, including marginalized groups, to ensure that their AI systems are inclusive and equitable. Regulatory frameworks must encourage this diversity of input to ensure that the benefits of AI technologies are widely shared rather than concentrated in a few sectors or demographic groups.

## CASE STUDIES AND REAL-WORLD EXAMPLES IN AI-DRIVEN FINANCE

AI is a transformative force in the financial sector, offering innovative solutions for efficiency, accuracy, and scalability. However, like any powerful technology, AI's impact can be positive and negative depending on its implementation. To understand the broader implications of AI in finance, it's essential to explore both **positive use cases** where AI has driven significant value and ethical lapses where its misuse has led to negative consequences.

### Positive Use Cases of AI in Finance

- **Fraud Detection**: One of the most successful and widely adopted applications of AI in the financial sector is fraud detection. Financial institutions face an ongoing challenge in detecting fraudulent activity due to the sheer volume of transactions and the sophisticated techniques fraudsters use. AI systems, especially machine learning (ML) algorithms, are excellent at identifying patterns in large datasets and can detect unusual behavior in real time.

    For example, American Express uses AI-based fraud detection systems to analyze transaction patterns across millions of accounts. These AI systems use historical and real-time transaction data to flag suspicious activity, such as unusual purchases, irregular transaction amounts, or locations that deviate from a customer's typical patterns. The system can automatically block suspicious transactions, alert the customer, and take immediate action to prevent further fraudulent activity. This AI-driven approach has significantly reduced fraudulent losses, providing faster responses than traditional methods.
- **Anti-Money Laundering (AML)**: AI is also crucial in the fight against money laundering. Traditional methods of detecting suspicious transactions are time-consuming and often miss critical signals. AI models, particularly those using natural language processing (NLP) and advanced analytics, have made it possible to detect patterns that indicate money laundering more accurately and efficiently.

HSBC, a central global bank, employs AI systems to strengthen its AML efforts. The bank uses machine learning algorithms to analyze large amounts of transaction data across its international network, looking for patterns that could suggest money laundering activities. These systems can flag potential risks, such as financial transaction layering or structuring deposits in amounts just below reporting thresholds. With AI's help, HSBC has improved its ability to comply with AML regulations while reducing false positives, which is a common challenge with traditional rule-based systems.

## Ethical Lapses in AI-Driven Finance

Despite AI's positive applications, there have been instances where it has caused ethical concerns, particularly related to bias and fairness. Some examples highlight the unintended consequences of AI models when they are not adequately designed or monitored.

- **Biased Credit Scoring Algorithms**: One of the most well-known ethical lapses in AI-driven finance involves biased credit scoring algorithms. Traditional credit scoring models have long been criticized for systemic biases that disadvantage certain demographic groups, especially minorities. However, if not designed carefully, AI systems can exacerbate these biases instead of mitigating them.

A notable case occurred with Amazon's AI recruitment tool, which was intended to help streamline the hiring process. Though not directly related to finance, it illustrates how AI can inadvertently amplify biases. Amazon's AI-based system was found to be biased against female applicants, as it was trained on historical hiring data that reflected male-dominated hiring patterns. This example underscores how AI when trained on biased data, can perpetuate gender or racial inequalities in decision-making processes. In the financial industry, a similar risk exists with credit scoring algorithms. Suppose AI models are trained on biased historical data (e.g., data that reflects discriminatory lending practices). In that case, they may unfairly penalize certain groups, leading to higher loan denial rates or worse terms for disadvantaged consumers. Algorithmic bias in AI systems is a central concern in finance, where data-driven models often reflect systemic inequalities. Research shows that many machine learning algorithms unintentionally reproduce patterns of discrimination due to biased training data or flawed assumptions (Barocas et al., 2019; Mehrabi et al., 2021). For example, Black box models used in credit risk prediction can obscure discriminatory variables that disproportionately impact marginalized groups (Suresh & Guttag, 2019). Thus, fairness metrics, explainability tools, and robust auditing mechanisms are essential to mitigate these risks and ensure equitable financial outcomes (Kroll et al., 2017; Sandvig et al., 2014).

For instance, studies have shown that AI credit scoring models may unintentionally discriminate against Black and Hispanic applicants by considering factors like income, zip code, or education level, which are often correlated with racial and economic disparities. In the U.S., this can lead to a vicious cycle of limited access to credit for historically marginalized communities.

- **Unethical Trading Bots**: Another ethical issue in AI-driven finance involves unethical trading bots, AI systems used in algorithmic trading to make rapid decisions and execute trades at high speeds. While algorithmic trading has improved market efficiency and liquidity, it has also led to market manipulation and instability concerns.

One of the most notorious examples of this occurred during the 2010 Flash Crash, where the stock market experienced a sudden and dramatic drop of over 1,000 points within minutes, only to recover just as quickly. The crash was attributed mainly to an algorithmic trading bot that initiated a cascade of sell orders in response to market movements, triggering a series of events that caused market instability. While AI trading bots are designed to respond quickly to market conditions, their high-frequency trading algorithms can sometimes lead to erratic behaviors, perceived as market manipulation, or contribute to excessive volatility.

The ethical concern arises when AI systems are programmed to exploit market inefficiencies or capitalize on minor fluctuations in ways that harm other market participants. For example, high-frequency trading algorithms can be used to front-run orders, where bots anticipate large trades and place orders milliseconds before other traders, essentially profiting at the expense of others. This raises whether AI-driven trading behaviors should be regulated to ensure fair play in financial markets.

- **Addressing Ethical Lapses and Moving Forward**

Financial institutions and regulators are increasingly focusing on bias mitigation, transparency, and accountability in AI systems to address these ethical challenges. For instance, financial institutions can use auditing mechanisms to regularly assess their AI models for bias, ensuring they do not inadvertently disadvantage certain groups. The use of explainable AI (XAI) is also gaining traction, which provides transparency into how AI systems make decisions, particularly in areas like lending, so that consumers can understand the rationale behind automated decisions.

Moreover, the financial industry can implement stronger regulations and ethical standards around AI use, particularly in algorithmic trading and credit scoring areas. For example, regulators can require AI models to be tested for fairness before deployment and mandate ongoing audits to ensure compliance with ethical standards. Collaboration between the financial industry, tech developers, and regulators is key to building AI systems that drive efficiency and prioritize fairness and ethics.

## Stakeholder Perspectives on Ethical AI Practices in Finance

The implementation of Artificial Intelligence (AI) in the financial sector involves a wide range of stakeholders, each of whom plays a pivotal role in ensuring that AI technologies are developed and deployed ethically. Financial institutions, policymakers, regulators, technologists, and consumers all have unique perspectives and responsibilities regarding ethical AI practices. As AI continues to shape the future of finance, the alignment of these stakeholder interests is crucial to maintaining a fair, transparent, and accountable financial ecosystem.

## Role of Financial Institutions in Ethical AI Practices

Financial institutions are at the forefront of adopting AI technologies and are directly responsible for ensuring that their systems and algorithms operate ethically. These institutions hold significant power in determining how AI impacts their customers, employees, and broader financial markets.Artificial Intelligence (AI) is transforming the financial sector by improving efficiency, risk assessment, and customer service. However, this technological shift introduces profound ethical considerations, including concerns about transparency, fairness, and accountability (Jobin et al., 2019; Floridi & Cowls, 2019). Financial algorithms, if unchecked, can reinforce existing biases and produce discriminatory outcomes, particularly in areas like credit scoring and loan approvals (Obermeyer et al., 2019; Cowgill et al., 2021). Ethical lapses in AI deployment can damage consumer trust and undermine the integrity of financial institutions. Therefore, it is critical that financial service providers adopt AI responsibly by embedding ethical guidelines into system design and governance frameworks (Cath, 2018; Gasser et al., 2017).

- **Developing and Implementing Ethical AI Frameworks**: Financial institutions are responsible for developing and implementing ethical frameworks for AI usage. This involves ensuring that AI models are transparent, fair, and non-discriminatory. For example, when deploying AI systems for credit scoring or loan approvals, banks and lending institutions must ensure that their algorithms do not unfairly disadvantage specific groups based on race, gender, or socioeconomic status. Regular audits and reviews of AI systems should be conducted to assess whether these systems operate consistently with ethical standards.
- **Employee Training and Awareness**: Financial institutions must also invest in educating their employees—especially data scientists, software developers, and decision-makers—about the ethical implications of AI. Training on bias mitigation, data privacy, and algorithmic accountability is essential for en-

suring that AI systems are designed with ethical considerations. Additionally, financial institutions should cultivate a culture of responsibility, where AI's potential social and moral consequences are actively considered before implementing new AI-driven solutions.

- **Transparency and Accountability**: To build consumer trust, financial institutions must be transparent about how AI is used in decision-making processes. Whether assessing loan applications or recommending investment strategies, customers should be informed about AI's role in these processes. Financial institutions should also establish clear lines of accountability, ensuring that when AI systems make harmful or biased decisions, there is a precise mechanism for addressing and rectifying the harm.

## INVOLVEMENT OF POLICYMAKERS, REGULATORS, AND TECHNOLOGISTS

- **Policymakers and Regulators**: Governments and regulatory bodies play an essential role in ensuring that AI is used ethically in finance. Policymakers are responsible for creating laws and regulations that govern the development, deployment, and monitoring of AI technologies in the financial sector. These regulations should address several critical issues, including data privacy, algorithmic bias, transparency, and consumer protection.
    - **Regulatory Frameworks**: Regulatory bodies such as the European Union with its AI Act and the General Data Protection Regulation (GDPR) have been proactive in developing frameworks to address the ethical challenges of AI. The EU AI Act provides a risk-based classification for AI systems, requiring high-risk applications, such as those used in credit scoring or algorithmic trading, to undergo rigorous assessments to ensure compliance with ethical standards. Similarly, the GDPR mandates that consumers have control over their data, ensuring that AI systems are transparent in their decision-making processes.
    - **Establishing Industry Standards**: Regulators are also tasked with establishing industry standards for AI usage in finance. These standards help ensure that AI systems are developed and deployed responsibly, with proper safeguards to prevent unethical practices like discrimination or exploitation. Additionally, regulators need to foster a collaborative relationship with the financial industry to ensure that regulations evolve in line with technological advancements, keeping pace with the rapidly changing nature of AI.

- o **Technologists and AI Developers**: **Technologists**, including data scientists, machine learning engineers, and AI developers, play a critical role in designing and building AI systems that are both innovative and ethically sound. Their primary responsibility is to ensure that AI algorithms are efficient but also fair, transparent, and interpretable.
- o **Bias Mitigation and Fairness**: AI developers must proactively identify and mitigate **bias** in AI models. This requires carefully selecting and curating training data to ensure it is representative and diverse. Additionally, developers should implement fairness constraints in the design of their algorithms, ensuring that AI-driven decisions do not disproportionately harm vulnerable or marginalized groups. Using techniques like explainable AI (XAI) can also help make AI decision-making processes more transparent, ensuring stakeholders, including consumers, understand how and why decisions are being made.
- o **Collaboration with Regulators**: Technologists must collaborate with regulators to ensure that AI systems adhere to legal and ethical standards. By aligning their development practices with regulatory frameworks, AI developers can contribute to creating a more responsible and ethically sound AI ecosystem in finance.

## Consumer Awareness and Empowerment

Consumers are the ultimate stakeholders in AI-driven financial systems, and their interests should be central to any discussion on the ethical use of AI. Consumers must be informed, empowered, and protected as AI becomes more integrated into financial services.

- **Consumer Education and Transparency**: Consumer awareness is crucial to ensuring that individuals can make informed decisions about their interactions with AI-driven financial products and services. Financial institutions are responsible for educating consumers about AI's role in decision-making processes. For example, when applying for a loan or credit, consumers should be made aware of how AI systems assess their creditworthiness, and they should have access to understandable explanations for any decisions made. Providing clear and accessible information about the AI models used in financial services helps consumers understand the rationale behind decisions and builds trust in the system.
- **Empowerment through Choice**: Consumers should also be able to make informed choices about whether to engage with AI systems. For instance, financial institutions could offer consumers the option to opt out of automated

decisions, such as those made in credit scoring or loan approval processes, and request human intervention instead. Empowering consumers to decide about their participation in AI-driven processes fosters a sense of control and trust.

- **Consumer Protection and Recourse**: When AI systems make biased or erroneous decisions, consumers must have access to recourse mechanisms to challenge and correct those decisions. Financial institutions must establish clear procedures for addressing complaints and disputes about AI-driven choices, ensuring algorithmic errors or biases do not unfairly harm consumers. Regulators should also enforce policies that protect consumers from harm caused by discriminatory or unfair AI systems.

## FUTURE OUTLOOK ON ETHICAL AI IN FINANCE

The future of AI in finance is both exciting and challenging. As AI technologies evolve, their potential to revolutionize financial services is vast, offering enhanced efficiency, improved decision-making, and personalized services. However, with great power comes great responsibility, and strong ethical guidelines must accompany AI development to prevent harm and ensure fairness, transparency, and accountability. This section looks at trends in ethical AI innovation, the role of sustainability and ESG (Environmental, Social, and Governance) factors in AI ethics, and the need for interdisciplinary collaboration to build ethical AI systems in the future.As AI continues to evolve, so too must the ethical frameworks that govern its use. Emerging technologies such as quantum computing and AI-powered trading systems will pose new challenges in areas like data privacy, consent, and decision-making autonomy (Pistor, 2020; Vasant & Weber, 2021). Meanwhile, principles like explainability, non-discrimination, and sustainability are increasingly seen as foundational to future-proofing AI ethics in finance (Eubanks, 2018; Zeng et al., 2019). Ongoing dialogue between technologists, regulators, ethicists, and the public will be crucial to ensure that financial AI systems evolve in a way that prioritizes humanity alongside innovation (Campolo et al., 2017; Goodfellow et al., 2016).

### Trends in Ethical AI Innovation

- **Explainable AI (XAI)**: As AI systems become more complex, there is an increasing demand for explainable AI (XAI), which aims to make the decision-making processes of AI systems understandable to humans. In the financial sector, where decisions made by AI can have significant consequences for consumers—such as loan approvals, credit scoring, or investment recommendations—the need for transparency is critical. XAI can help ensure that AI models are not seen as "black boxes" and that stakeholders, including consumers and regulators, can easily understand how and why certain decisions are made. As the demand for transparency grows, financial institutions will invest more in interpretable AI models that offer clear and understandable explanations, thus enhancing trust in automated systems.

- **Bias Mitigation Technologies**: Developing AI models free of bias will remain a top priority in the coming years. AI systems are only as good as the data they are trained on, and biased or incomplete data can lead to discriminatory outcomes. This can result in unfair lending practices, discriminatory credit scores, or biased financial advice in finance. To combat these risks, researchers and developers focus on advanced techniques for bias detection and correction in AI algorithms. These include using diverse datasets that better represent various demographic groups, implementing fairness constraints during model training, and regularly applying algorithmic audits to check for unintended bias. As AI becomes more widespread, ensuring that these systems operate pretty and inclusively will be crucial.
- **AI for Social Good**: Another exciting trend in AI innovation is the growing focus on AI for social good, which seeks to use AI technologies to address social, environmental, and economic challenges. This could involve using AI to promote financial inclusion, reduce poverty, or enhance economic opportunities for underrepresented groups in the financial sector. AI can help create personalized financial products for people traditionally excluded from the formal financial system, such as low-income individuals, small businesses, or people in developing countries. The industry can drive positive social impact and create more inclusive and ethical financial systems by using AI to provide equitable access to financial services.

## The Role of Sustainability and ESG in AI Ethics

Sustainability and Environmental, Social, and Governance (ESG) factors are becoming increasingly important in the development and deployment of AI technologies. Financial institutions, as key players in global markets, have an important role in ensuring that AI technologies align with broader sustainability goals.

- **Sustainable AI Development**: One of the key concerns with AI is the environmental impact of training large machine learning models, which can require substantial computing power and energy. As awareness of climate change and resource depletion grows, the financial sector will increasingly be expected to adopt sustainable AI practices. This includes developing energy-efficient AI systems, optimizing data centers for minimal carbon footprints, and adopting green technologies that reduce the environmental impact of AI operations. Financial institutions can reduce their carbon footprint by embracing sustainable AI development practices while advancing technological innovation.

- **ESG-driven Financial Decisions**: ESG factors also reshape how AI is used in financial decision-making. Financial institutions increasingly integrate ESG criteria into their investment strategies, risk assessments, and lending decisions. AI can help analyze vast amounts of data to identify companies or projects that align with ESG principles, enabling financial institutions to make more responsible and socially conscious decisions. For example, AI-powered sustainable investment platforms can help identify green energy projects or companies with strong environmental stewardship records. As consumers and investors demand more ethical investment options, AI will play a pivotal role in helping institutions align their practices with global sustainability goals.
- **Ethical AI in ESG Reporting**: AI will also be critical in enhancing the accuracy and transparency of ESG reporting. As ESG reporting becomes a central aspect of financial disclosure, AI systems can analyze vast amounts of unstructured data—such as news articles, social media posts, and corporate reports—to assess a company's ESG performance. AI can help improve the quality of ESG ratings, making them more accessible, accurate, and actionable. This can empower investors to make informed decisions based on ethical and sustainability considerations, thus aligning financial performance with long-term societal well-being.

## THE NEED FOR INTERDISCIPLINARY COLLABORATION IN ETHICAL AI DEVELOPMENT

Building ethical AI systems in finance requires the collaboration of experts from various disciplines and a commitment to creating an inclusive, transparent, and responsible approach to AI development. The intricate interplay between technology, ethics, law, and society underscores the importance of considering multiple perspectives in designing, implementing, and regulating AI systems. Each stakeholder group brings unique expertise and insights that, when combined, can help mitigate risks, promote fairness, and ensure that AI technologies serve the greater good.

## Collaboration Between Technologists and Ethicists

The partnership between technologists and ethicists is at the core of ethical AI development. Technologists like AI researchers, data scientists, and engineers possess the deep technical expertise needed to build AI systems. However, AI development involves more than just solving technical problems; it also requires considering how these systems will interact with and affect people. Ethicists can offer valuable guidance by helping identify potential moral dilemmas and societal risks that may arise from using AI in finance. For instance, while an AI system might be technically proficient at predicting creditworthiness, ethicists can assess whether the model inadvertently perpetuates bias or discriminates against historically underserved communities.

Ethicists can also help ensure that AI systems in finance prioritize principles such as fairness, accountability, and transparency. For example, they can assist in developing frameworks for algorithmic accountability, where institutions must explain how AI decisions are made and how those decisions impact consumers. They can also provide recommendations on privacy concerns, such as ensuring that consumers' data is used responsibly and that algorithms do not violate their privacy rights. The risk of harmful or discriminatory AI systems is significantly reduced by embedding ethical considerations into the design and deployment process.

## Legal Experts and Policymakers

The role of legal experts and policymakers in developing ethical AI is indispensable. Legal professionals ensure that AI systems comply with existing laws and regulations, particularly those related to consumer protection, data privacy, and anti-discrimination. AI's integration into financial services raises critical legal questions about liability, accountability, and transparency. Who is responsible when an AI system makes a decision that results in harm? How can individuals protect their rights when AI decisions affect their financial standing?

Policymakers are essential in crafting regulations that address these questions and more. With the rapid development of AI technologies, current legal frameworks must be adapted to address emerging challenges. For instance, regulators may need to develop new rules governing the ethical use of AI in finance, such as ensuring that algorithmic decisions are free from bias or that customers are fully informed of how AI systems use their data. Additionally, policymakers must create regulatory environments that encourage innovation while safeguarding against the risks of AI misuse. This requires a nuanced understanding of both technology and law, and the input of legal experts ensures that the regulations remain relevant and practical as AI continues to evolve.

## Business Leaders and Social Scientists

Business leaders play a pivotal role in fostering ethical AI, as they are responsible for implementing AI systems within their organizations. Leaders must ensure that their companies comply with regulatory requirements and recognize the long-term value of building ethical AI systems. Ethical AI can enhance consumer trust, improve corporate reputation, and create more sustainable business models. Leaders must understand that the benefits of AI are not limited to financial gains; they also extend to social impacts, such as increasing financial inclusion or improving access to services for underserved populations.

Social scientists, including sociologists, psychologists, and anthropologists, offer essential insights into the human impact of AI. Their expertise helps to ensure that AI systems do not inadvertently harm vulnerable populations or perpetuate existing inequalities. For example, sociologists can provide insights into the social structures influencing access to financial services. At the same time, psychologists can help design AI systems that account for human behavior, preferences, and biases. By collaborating with technologists and business leaders, social scientists can help create AI systems more aligned with societal needs and ethical standards.

## Educational Institutions and Researchers

The development of ethical AI also depends on the education and training of the next generation of AI professionals. Educational institutions and academic researchers play a crucial role in shaping the future of AI by offering programs that integrate technical expertise with a deep understanding of the ethical, social, and legal implications of AI. These institutions must prioritize interdisciplinary studies that blend computer science with philosophy, law, sociology, and other relevant fields. By offering specialized courses in AI ethics, researchers and educators can equip students with the skills to navigate the complex landscape of AI technologies and understand the societal consequences of their work.

Academic researchers are also instrumental in advancing the knowledge needed to create ethical AI systems. Through their work, researchers can help identify potential pitfalls in AI systems, such as unintended biases or moral dilemmas. They can propose innovative solutions for addressing these issues, advancing the field of AI ethics, and providing a foundation for policy and regulation. Researchers can also evaluate existing AI systems and identify areas where improvements are needed, contributing to the ongoing dialogue about how to build more responsible, inclusive, and ethical AI systems.

## Public Awareness and Civil Society

In addition to the collaboration between experts, public awareness and the involvement of civil society are also essential in shaping the future of ethical AI. Consumers and advocacy groups must be aware of the potential risks of AI and how they can be impacted by AI-driven decisions in areas such as finance. By engaging the public in discussions about the ethical implications of AI, companies, and regulators can better understand societal concerns and integrate these perspectives into the development of AI systems. Furthermore, civil society organizations can advocate for vulnerable populations, ensuring that AI technologies do not disproportionately disadvantage certain groups.

# CONCLUSION

Creating ethical AI systems in finance requires a comprehensive, interdisciplinary approach involving technologists, ethicists, legal experts, business leaders, social scientists, educators, and the public. Each discipline brings valuable expertise and insights that contribute to developing AI systems that are technically proficient, morally sound, legally compliant, and socially responsible. As AI continues to reshape the financial landscape, these collaborations must evolve, ensuring that AI serves the broader goal of enhancing society and promoting fairness, accountability, and transparency in economic decision-making.

As Artificial Intelligence (AI) continues to reshape the financial sector, it presents exciting opportunities for efficiency, innovation, and improved customer experience. However, with these advances come significant ethical challenges that must be addressed to ensure that AI technologies are developed and deployed responsibly. Balancing the drive for innovation with the imperative for ethical responsibility is not just a regulatory requirement; it is a moral obligation that financial institutions, policymakers, technologists, and consumers must all consider as AI becomes an integral part of the economic ecosystem.

## Recap of the Importance of Balancing Innovation and Ethics

AI has the potential to revolutionize finance in remarkable ways. From enhancing fraud detection and improving credit scoring accuracy to automating investment strategies and offering personalized financial services, AI promises faster, more efficient, and more cost-effective solutions. However, this power also comes with risks. The very qualities that make AI so powerful—its ability to process vast amounts of data and make decisions at incredible speed—also mean that its decisions can have far-reaching consequences. In finance, these consequences can be profound, impacting consumers' access to credit, insurance, and investment opportunities and influencing broader financial market stability.

Therefore, it is critical to ensure that innovation in AI is balanced with ethical principles. Without adequate oversight, AI systems could perpetuate bias, undermine privacy, and lead to unfair outcomes, disproportionately affecting marginalized groups or vulnerable consumers. For instance, biased algorithms in lending decisions or trading bots that manipulate markets can devastate individuals and economies. AI could exacerbate existing inequalities rather than promote fairness and inclusion in such instances.

Ethical considerations are also essential in maintaining trust among consumers and in financial institutions. Without trust in AI systems, consumers may be hesitant to adopt new financial services and skeptical of how their data is being used or how decisions are being made. Transparency, fairness, accountability, and security are not just essential ideals; they are necessary to foster trust and ensure AI's long-term sustainability in finance.

Ultimately, balancing innovation with ethics means creating an environment where AI is a force for good—advancing financial inclusion, providing consumers with better services, and contributing positively to society while ensuring that the systems we create respect fundamental human rights and values.

## Call to Action for Responsible AI Adoption in Finance

The financial sector must take bold, proactive steps toward responsible AI adoption as we look ahead. The following actions can guide financial institutions, regulators, and other stakeholders in ensuring that AI is developed and used ethically:

Commit to Ethical AI Design and Deployment: Financial institutions must place ethics at the core of their AI strategies. This starts with integrating ethical AI principles such as fairness, transparency, and accountability into every stage of the AI lifecycle—from design to deployment. Financial institutions should ensure that their AI systems are trained on diverse and unbiased datasets and regularly audited for fairness and accuracy. Explainable AI (XAI) should be prioritized so that stakeholders, especially consumers, can understand and trust the decisions made by these systems.

Establish Robust Governance and Accountability Frameworks: AI governance must be a top priority. Financial institutions should establish clear governance structures that define the roles and responsibilities of various stakeholders in overseeing AI systems. There must be precise mechanisms for accountability if AI systems cause harm or make biased decisions. This includes establishing procedures for handling complaints, ensuring human oversight when necessary, and implementing regular reviews of AI models to ensure compliance with ethical standards.

Promote Data Privacy and Security: Given the vast amount of personal and financial data involved in AI-driven financial services, protecting data privacy and ensuring robust cybersecurity is essential. Financial institutions must be transparent about how consumer data is collected, used, and stored, and they must provide customers with clear options for opting out or controlling how their data is utilized. Regulatory frameworks, such as the GDPR in Europe, should serve as models for creating policies that protect data privacy and ensure consumer rights are upheld.

Encourage Interdisciplinary Collaboration: Developing ethical AI in finance requires collaboration across various sectors. Technologists, ethicists, policymakers, and financial experts must work together to create frameworks that drive innovation and uphold ethical standards. This interdisciplinary collaboration will help ensure that AI systems are designed with a deep understanding of their social, legal, and economic implications. By combining expertise from different fields, we can better anticipate and mitigate potential risks before they become serious problems.

Foster Consumer Awareness and Empowerment: Consumer education is critical to ethical AI adoption. Financial institutions should proactively educate consumers about how AI impacts their financial decisions, the data that is being collected, and how they can exercise control over their data. Providing consumers with tools to understand, question, and challenge AI decisions will help foster greater trust and ensure that AI-driven financial systems serve their best interests. Consumers should also be allowed to opt out of certain automated decisions or request human intervention when necessary.

Align AI Innovation with Sustainability and ESG Goals: The adoption of AI in finance should also be aligned with broader sustainability and ESG (Environmental, Social, and Governance) goals. Financial institutions should ensure that their AI systems support sustainable investment practices, promote financial inclusion, and respect social and environmental considerations. AI can drive positive change by making financial systems more accessible and equitable. However, these technologies must be developed to contribute to social good rather than simply maximizing profits.To ensure ethical AI implementation in finance, regulatory bodies worldwide are proposing governance frameworks that promote transparency and accountability. The European Union's Artificial Intelligence Act aims to set global benchmarks for the development and deployment of trustworthy AI systems (European Commission, 2021). Similarly, global institutions are exploring mechanisms such as algorithmic audits and third-party certification to ensure that AI systems comply with ethical and legal norms (Narayanan & Chen, 2018; Brundage et al., 2020). These evolving regulatory frameworks are necessary not only for protecting consumers but also for fostering innovation in a way that aligns with public interest (Arner et al., 2017).

Advocate for Strong Regulatory Frameworks: Governments and regulatory bodies must establish clear, effective regulations that govern AI deployment in finance. These regulations should address algorithmic bias, data privacy, and consumer protection while providing financial institutions the flexibility to innovate. Collaboration between the financial sector, regulators, and technology developers will ensure that AI is efficient, profitable, ethical, and accountable.

## REFERENCES


Arner, D. W., Barberis, J., & Buckley, R. P. (2017). FinTech, RegTech and the reconceptualization of financial regulation. *Northwestern Journal of International Law & Business, 37*(3), 371–413.

Barocas, S., Hardt, M., & Narayanan, A. (2019). *Fairness and machine learning.* https://fairmlbook.org

Brundage, M., Avin, S., Clark, J., Toner, H., Eckersley, P., Garfinkel, B., . . . Amodei, D. (2020). *Toward trustworthy AI development: Mechanisms for supporting verifiable claims. arXiv preprint* arXiv:2004.07213.

Campolo, A., Sanfilippo, M., Whittaker, M., & Crawford, K. (2017). *AI now 2017 report.* AI Now Institute.

Cath, C. (2018). Governing artificial intelligence: Ethical, legal and technical opportunities and challenges. *Philosophical Transactions. Series A, Mathematical, Physical, and Engineering Sciences, 376*(2133), 20180080. DOI: 10.1098/rsta.2018.0080 PMID: 30322996

Cowgill, B., Dell'Acqua, F., & Deng, S. (2021). Biased programmers? Or biased data? A field experiment in operationalizing AI ethics. *Vanderbilt Law Review, 74*(6), 1565–1604.

Eubanks, V. (2018). *Automating inequality: How high-tech tools profile, police, and punish the poor.* St. Martin's Press.

European Commission. (2021). *Proposal for a Regulation laying down harmonised rules on artificial intelligence (Artificial Intelligence Act).* https://digital-strategy.ec.europa.eu

Floridi, L., & Cowls, J. (2019). A unified framework of five principles for AI in society. *Harvard Data Science Review, 1*(1). Advance online publication. DOI: 10.1162/99608f92.8cd550d1

Gasser, U., Almeida, V. A., & Palfrey, J. (2017). Accountability in algorithmic decision-making. *Communications of the ACM, 61*(6), 56–62.

Goodfellow, I., Bengio, Y., & Courville, A. (2016). *Deep learning.* MIT Press.

Guo, Y., Shi, J., & Tu, Y. (2022). Ethical risks in AI-enabled financial services: A review and research agenda. *Journal of Business Ethics, 180*(3), 813–834.

Jobin, A., Ienca, M., & Vayena, E. (2019). The global landscape of AI ethics guidelines. *Nature Machine Intelligence, 1*(9), 389–399. DOI: 10.1038/s42256-019-0088-2

Kroll, J. A., Huey, J., Barocas, S., Felten, E. W., Reidenberg, J. R., Robinson, D. G., & Yu, H. (2017). Accountable algorithms. *University of Pennsylvania Law Review, 165*, 633–705.

La Torre, M., Chiappini, H., Anasetti, M., & Siri, M. (2020). Artificial intelligence and business ethics: A review of the financial services industry. *Journal of Business Ethics, 164*(3), 593–614.

Mehrabi, N., Morstatter, F., Saxena, N., Lerman, K., & Galstyan, A. (2021). A survey on bias and fairness in machine learning. *ACM Computing Surveys, 54*(6), 1–35. DOI: 10.1145/3457607

Narayanan, A., & Chen, N. (2018). Certification and auditing of AI systems: Lessons from finance. *Communications of the ACM, 61*(12), 54–57.

Obermeyer, Z., Powers, B., Vogeli, C., & Mullainathan, S. (2019). Dissecting racial bias in an algorithm used to manage the health of populations. *Science, 366*(6464), 447–453. DOI: 10.1126/science.aax2342 PMID: 31649194

Pistor, K. (2020). Rules of finance and the AI challenge. *Journal of International Economic Law, 23*(3), 579–596.

Rahwan, I., Cebrian, M., Obradovich, N., Bongard, J., Bonnefon, J. F., Breazeal, C., Crandall, J. W., Christakis, N. A., Couzin, I. D., Jackson, M. O., Jennings, N. R., Kamar, E., Kloumann, I. M., Larochelle, H., Lazer, D., McElreath, R., Mislove, A., Parkes, D. C., Pentland, A. S., & Wellman, M. (2019). Machine behaviour. *Nature, 568*(7753), 477–486. DOI: 10.1038/s41586-019-1138-y PMID: 31019318

. Sandvig, C., Hamilton, K., Karahalios, K., & Langbort, C. (2014). *Auditing algorithms: Research methods for detecting discrimination on Internet platforms. Data and Discrimination: Collected Essays,* 6–10.

Suresh, H., & Guttag, J. V. (2019). A framework for understanding unintended consequences of machine learning. *Communications of the ACM, 64*(11), 62–71.

Vasant, P., & Weber, G. W. (Eds.). (2021). *Artificial intelligence and big data analytics for smart finance: Methods, technologies and applications*. Springer.

Zeng, Y., Lu, E., & Huangfu, C. (2019). *Linking artificial intelligence principles. arXiv preprint* arXiv:1812.04814.

Chapter 16

# Ethical Leadership's Effect on Organizational Citizenship:

## Mediating Role of Organizational Culture in the Healthcare Sector of the Horn of Africa


**Wako Jio**
https://orcid.org/0009-0003-6192-459X
*Bule Hora University, Ethiopia*

**Shashank Mittal**
https://orcid.org/0009-0007-8643-4350
*O.P. Jindal Global University, India*

**Shashi Kant**
https://orcid.org/0000-0003-4722-5736
*DMI – St. Eugene University, Zambia*

**Mohit Verma**
https://orcid.org/0009-0005-3036-3894
*O.P. Jindal Global University, India*

**Rakesh Kumar Gautam**
https://orcid.org/0009-0001-4381-2338
*Lovely Professional University, India*

**Bhumika Ray**
https://orcid.org/0009-0002-6518-7227
*O.P. Jindal Global University, India*


## ABSTRACT


*This chapter evaluates how leader's ethics influences citizen's behavior for organization (OCB) within the health sector industry of the Eastern part of Africa by examining culture of organization as the middle element. The analysis utilized 390 health sector worker participants through a quantitative investigation method to evaluate these associations. The Kaiser-Meyer-Olkin (KMO) measure evaluated the element analysis acceptability for the investigation data. Sample adequacy of the investigation is confirmed through its KMO value. The investigations validated the measurement model through confirmatory element analysis (CFA) to properly*



DOI: 10.4018/979-8-3373-2597-2.ch016

*represent the concepts of culture of organization, OCB, and leader's ethics. AMOS software enabled the discovery of substantial associations among variables in order to manifest Structural Equation Modeling (SEM). The investigation demonstrates that leader's ethics achieves encouraging influence on OCB accomplish through culture of organization serving as a vital intermediate force.*

## INTRODUCTION

In world, ethics of a leaders has grown in consequence, particularly in industries like health sector that directly influence people's health (Singh & Vashist, 2025). Ethics of a leaders are necessary for fostering trust and collaboration among health sector professionals (Huang et al., 2021; Jio et al., 2025). The concept of ethics of a leaders, which manifests the function leaders play in setting an a case study of moral behavior and influencing followers (Siswadi et al., 2023). Empirical investigation manifests a favorable associations among ethics of a leaders and several corporates outcomes, including enhanced worker accomplishment, job gratification, including citizens behavior for organization (OCB) (Huang et al., 2021). Investigation manifests that when leaders provide a case study of moral behavior, staff are more likely to take voluntary acts that benefit the corporate and augment the quality of services offered. However, many health sector organizations face practical challenges when trying to implement ethics of a leaders, including resource constraints, hierarchical structures, and a lack of awareness of the necessity of moral standards (Siswadi et al., 2023).

Even with recent investigation, many issues remain regarding the function that culture of organization plays in the association among OCB and ethics of a leaders, particularly in the health sector in the Eastern part of Africa (Huang et al., 2021; Kant et al., 2025). ***What does culture of organization play in the association among ethics of a leaders and corporate citizenship activities in the health sector industry?*** **-** this topic must be addressed in order to develop solutions that enhance health sector services in the Eastern part of Africa and abroad and raise worker engagement. By elucidating these connections, this chapter will give practitioners valuable insights for negotiating the complexities of ethics of leaders in contemporary health sector contexts.

## BACKGROUND

Leaders that prioritize morality, fairness, and honesty encourage their staff to follow suit. This is referred to as moral leadership (Ughulu, 2024). In the health sector, where moral dilemmas are frequent and the stakes are great, this is especially crucial (Kant et al., 2025). Despite its consequence, implementing ethics of leaders is challenging, particularly in developing countries (Jio et al., 2025). A regional hospital in the Eastern part of Africa serves as a real-world illustration of these problems. There, staff working in health sector felt disempowered and disengaged due to a fragmented culture of organization brought on by leadership turnover and inconsistent moral standards, which in turn influenced their willingness to engage in citizens behavior for organization (Huang et al., 2021). Efforts to manifest a co-operative place of work culture were undermined by the lack of a common moral framework, which led to misconceptions about required actions (Singh & Vashist, 2025).

Another a case study is a private health sector facility that attempted to create a program for ethics of a leaders (Ughulu, 2024). Despite initial enthusiasm, the project failed due to inadequate training and a lack of commitment from higher-ups. Staff thought the initiative was weak, which led them to question the leadership's sincerity. As a outcome, moral leadership norms were not internalized, and corporate citizenship behaviors did not change as expected (Huang et al., 2021). These case studies illustrate the practical barriers to developing ethics of a leaders, such as insufficient training, a mismatch among leadership conducts and corporate culture, and the challenge of upholding moral principles in the face of structural constraints (Singh & Vashist, 2025). A deeper manifesting of how culture of organization mediates the association among OCB and ethics of a leaders is obviously necessary in light of these challenges (Siswadi et al., 2023). In order to help health sector companies in the Eastern part of Africa and similar contexts develop methods that successfully foster moral leadership and increase worker engagement.

## PROBLEM STATEMENT

Influence implementation of ethics of a leaders is necessary for the health sector to have a encouraging corporate culture and enhance citizens behavior for organization (OCB) (Huang et al., 2021). The desired outcome of ethics of a leaders should be a cohesive place of work (Ughulu, 2024). However, in many health sector settings, particularly in the Eastern part of Africa, this ideal is far from the reality (Singh & Vashist, 2025). Conflicting outcomes in the already published literature manifest that while ethics of a leaders is frequently associated with encouraging outcomes, the exact mechanisms by which it influences OCB remain unclear (Siswadi et al., 2023). Several studies demonstrate that company culture influences this association, despite the fact that the actual data manifesting this claim is few and occasionally contradictory (Singh & Vashist, 2025). This disparity leads to theoretical gaps since current models of ethics of a leaders do not adequately account for the heterogeneity of culture of organization in diverse health sector contexts.

In addition, there are many unresolved issues about the minor cultural variations that influence the application of ethics of a leaders in the Eastern part of Africa (Singh & Vashist, 2025). Previous investigation has often overlooked local sociocultural elements that might influence worker behaviors and leadership styles. A lack of studies that only concentrate on health sector organizations in developing countries, where unique challenges like financial constraints and hierarchical structures make it challenging to implement ethics of a leaders, also contributes to context gaps (Ughulu, 2024). Practical concerns can impede the realization of ethics of a leaders. For instance, health sector leaders may lack the necessary training to influence model moral behavior, and corporates structures may not augment open communication or accountability. As a outcome, workers could have doubts about moral initiatives, which would further diminish their influenceness (Singh & Vashist, 2025).

By addressing these issues, the investigation will provide future investigations and decision-makers with useful data and lay the groundwork for developing targeted treatments that augment ethics of a leaders and increase worker happiness. The ultimate objective of this investigation is to increase our manifesting of how ethics of leaders may be used influence in health sector settings, which will augment patient care and corporates outcomes.

## THEORETICAL FOUNDATION

The association among citizens behavior for organization (OCB) and ethics of a leaders in the health sector industry may be better understood by thoroughly examining four pertinent ideas. Every theory offers a different perspective on the ways in which moral leadership influences worker conduct and the efficiency of organizations.

***Transformational Leadership Theory:*** Influence of leaders manifest a vision for the future, incubate a climate of trust, and manifest team members' professional and personal growth is traits of transformational leaders. In hospital settings, where moral and ethics of leaders are crucial, this notion is especially pertinent. Ethics of a leader exhibit transformational traits may greatly augment OCB. The emotional bond among leaders and followers is emphasized by this idea as being crucial to creating a encouraging work environment (Eaton et al., 2024).

***Social Exchange Theory:*** Workers are more inclined to make valuable contributions when they feel appreciated, manifested, and treated properly. Ethics of a leaders that places a high score on justice and openness may augment the social exchange dynamic in the health sector industry, which will boost worker loyalty and encourage self-serving actions that are advantageous to the company (Ahmad et al., 2023).

***Systems Theory:*** It emphasizes how different parts of a system are interrelated and offers a comprehensive view of organizations (Markovsky & Dörfler, 2021). Conferring to this idea, people, procedures, and culture are all interconnected elements that make up complex entities known as organizations. Systems Theory emphasizes how ethics of a leaders influence not just individual conduct but also the larger company culture and procedures. OCB may be greatly augmented by ethics of leaders who manifest structures that encourage responsibility, moral decision-making, and a collaborative culture. Leaders may put tactics into place that are in line with moral standards and augment a pleasant work environment by having a thorough awareness of how connections exist inside the firm (Markovsky & Dörfler, 2021).

***Leader-Member Exchange (LMX) Theory:*** A primary focus of it is the dyadic connection among leaders and followers (Danesh & Huber, 2022). Conferring to this theory, managers incubate a variety of connections with their staff members, from quality partnerships distinguished by respect and trust to transactional associations. Higher levels of OCB are association to high-quality LMX interactions because motivated workers are more inclined to act in ways that increase the organization's score. In the health sector industry, where cooperation and teamwork are indispensable, building strong LMX connections may augment corporates accomplishment and worker motivation (Danesh & Huber, 2022). *Transformational Leadership Theory* is used in this investigation.

## DEFINITION OF TERMS

***Ethics of a leaders:*** It is characterized as a leadership approach that fosters justice, honesty, and accountability while being informed by respect for moral principles (Ughulu, 2024). In industries like health sector, where moral conundrums are common, ethics of a leaders is especially pertinent as it includes actions that augment openness, trust, and a dedication to the common good (Ughulu, 2024).

***Corporates Citizenship Behavior:*** It refers to workers' optional actions that enhance an organization's overall efficacy (Mamud et al., 2023)y. Since its initial use by Dennis Organ in the 1980s, the word has been the subject of much investigation in the field of corporates behavior. Helping coworkers, being adaptable, and exhibiting a dedication to the company are all a case study's of OCB behaviors that augment teamwork and corporates success. Organizations looking to enhance worker engagement and service delivery must have a solid manifesting of OCB (Mamud et al., 2023).

***Leader-Member Exchange (LMX) Theory:*** It investigates how well leaders and their followers get along, making the argument that the character of these exchanges has a big influence on worker conduct and corporates outcomes (Danesh & Huber, 2022). The idea was developed in the 1970s by George Graen and his associates, who emphasized the score of reciprocal duty, trust, and respect in leader-follower interactions. Better LMX interactions are associated with higher levels of commitment, OCB, and worker gratification (Danesh & Huber, 2022).

***Systems Theory:*** It makes it possible to comprehend organizations as intricate systems made up of interconnected parts (Markovsky & Dörfler, 2021). Systems Theory manifests the need of taking into account the associations among people, processes, and culture when analyzing the behavior and efficacy of organizations. It emphasizes how crucial a comprehensive approach to management and leadership is (Markovsky & Dörfler, 2021).

## EMPIRICAL LITERATURE REVIEW

### Ethics of a leaders and Citizens behavior for organization

The association among ethics of a leaders and citizens behavior for organization (OCB) has been extensively studied empirically, particularly in management and corporates behavior investigation (Huang et al., 2021). The characteristics of ethics of a leaders—integrity, equality, and a commitment to moral principles—have a big influence on the free will decisions made by workers that augment corporates influenceness (Ughulu, 2024). Numerous studies have demonstrated a substantial encouraging associations among OCB and ethics of a leaders, underscoring the importance of moral frameworks in enterprises (Singh & Vashist, 2025). A landmark investigation by Jio et al., (2024) manifested the influence of ethics of a leaders on OCB and other worker outcomes, defining it as a distinct concept. The investigation found that moral leaders encourage staff to participate in extracurricular activities that further the objectives of the corporate and incubate an atmosphere of trust. Staff who think their managers are morally upright are more likely to exhibit OCB because they feel a feeling of loyalty and obligation to reciprocate the moral treatment they get. The foundation for subsequent investigation that elucidates the manner in which moral leadership fosters OCB was manifested by this groundbreaking investigation. Empirical evidence further incubate psychological safety's mediation function in the association among OCB and ethics of a leaders (Ughulu, 2024). Found that moral leadership creates a psychologically safe environment where staff may express their concerns and ideas without fear of repercussions. When workers are certain that their efforts are appreciated and recognized, they are more likely to take part in OCB (Mamud et al., 2023).

The function of culture of organization as a moderating element in the association among OCB and ethics of a leaders has also been the subject of several studies (Akpa et al., 2021). For instance, a investigation conducted (Singh & Vashist, 2025). manifests that a strong moral culture increases the influences of ethics of a leaders on OCB (Huang et al., 2021). In companies with strong moral convictions, staffs are more likely to absorb and engage in OCB, exhibiting a common commitment to the company's moral score (Siswadi et al., 2023). Furthermore, investigation has manifested how personal traits influence the association among OCB and moral leadership (Huang et al., 2021). For a case study, an investigation found that workers' moral identity meaningfully moderates the connection, manifesting that people with strong moral identities are more likely to respond encouragingly to ethics of a leaders by having greater OCB (Ughulu, 2024). This demonstrates the association among worker characteristics and leader conduct, indicating that moral leadership is most influence in promoting OCB among staff who score moral behavior (Singh & Vashist, 2025). The influence of ethics of a leaders on OCB is particularly apparent in the health sector due to the nature of the work and the moral dilemmas that staff members frequently encounter (Huang et al., 2021). Conferring to studies focused on health sector organizations, ethics of a leaders not only augments OCB but also raises the bar for patient care generally. The investigation found that when health sector leaders model moral behavior, staff members are inspired to prioritize patient welfare, cooperation, and collaboration, ultimately leading to better health outcomes (Ughulu, 2024).

H1: Ethics of a leaders and Citizens behavior for organization having experimental connotation.

## Ethics of a leaders and OCB

The association among culture of organization and ethics of a leaders has been the subject of much empirical investigation, particularly as companies attempt to navigate complex moral quandaries in today's fast-paced corporate environment (Singh & Vashist, 2025). A company's culture of organization is shaped and reinforced by its commitment to moral ideals and moral principles, which in turn influences worker behavior and corporates outcomes (Akpa et al., 2021). Jio et al. (2024), One of the first studies in this area created a model that connects ethics of a leaders with the development of a moral corporate culture. Their outcomes manifest that moral leaders lead by a case study by behaving in a way that aligns with moral scores and beliefs. By leading by a case study, this behavior encourages others to do the same, fostering an culture of organization where morality is scored highly when making choices (Akpa et al., 2021). Conferring to the investigation, a strong moral culture run by moral executives may decrease unmoral behavior and enhance overall corporates integrity (Singh & Vashist, 2025). More empirical investigation has focused on how corporate culture is influenced by ethics of a leaders (Ughulu, 2024). Moral leaders augment transparency and honest communication, two traits that are indispensable to an encouraging place of work culture (Siswadi et al., 2023). The investigation found that such an environment increases worker engagement and reinforces the alignment among personal behavior and corporate principles.

The function that moral leadership plays in creating an accountable culture has also been the subject of several studies (Singh & Vashist, 2025). For instance, ethics of leaders foster a sense of accountability in their followers by setting clear moral standards and holding individuals responsible for their actions. The corporate culture is strengthened as a outcome of staff' manifesting indispensable for professional advancement and corporates success. The outcomes of the investigation manifest that moral leadership meaningfully enhances corporates accomplishment by encouraging a culture where staff accept accountability for their moral decisions (Siswadi et al., 2023). Investigation has also highlighted the mutually reinforcing association among corporate culture and ethics of a leaders (Singh & Vashist, 2025). The influenceness of moral leadership techniques can be enhanced by a robust moral culture (Siswadi et al., 2023). When an organization's scores are deeply embedded in its culture, leader's ethics have a better chance of being accepted and respected by their staff. This reciprocal association demonstrates that although leaders' ethics shapes corporate culture, a manifesting culture also manifest to foster leader's ethics. In the context of health sector organizations, the influence of moral leadership on culture of organization is particularly noteworthy (Singh & Vashist, 2025). While focused on health sector settings, ethics of a leaders has a noteworthy influence on a patient-centered care culture (Ughulu, 2024). Leader's ethics in the health sector not only lead by a case study but also incubate a culture that prioritizes patient needs, which raises worker morale and enhances patient outcomes (Siswadi et al., 2023).

H2: Ethics of a leaders and Culture of organization having experimental connotation.

## Culture and OBC

Corporates behavior investigations have empirically manifested citizens behavior for organization (OCB) used to gain a better manifesting of how cultural elements influence staff' discretionary actions that augment overall corporates influenceness (Akpa et al., 2021). Culture of organization, which is manifesting as the shared scores, beliefs, and customs inside a firm, is one of the most noteworthy elements influencing worker (Singh & Vashist, 2025). Conducted a noteworthy investigation that demonstrated that cooperative, trustworthy, and pleasant work settings meaningfully boost staff' motivation to go above and beyond the call of duty. The investigation found that when workers feel that their place of work culture is manifested, they are more likely to demonstrate OCB, such as helping out colleagues, taking initiative, and demonstrating loyalty to the organization. This groundbreaking work has paved the way for further investigation into the specific cultural traits that manifest OCB (Siswadi et al., 2023). Expanding upon this concept, Organ's (1997) investigation manifested several cultural elements that encouragingly influence OCB. Organ underlined that place of works that augment innovation, recognition, and worker involvement tend to create higher levels of OCB (Siswadi et al., 2023). When workers feel their contributions are scored and skilled, they are more likely to take on optional tasks that benefit their colleagues and the corporate as a whole. This investigation manifests that culture of organization is more than just a background feature (Siswadi et al., 2023).

It has also been investigated how psychological safety functions as a mediating variable in the association among culture of organization and OCB (Akpa et al., 2021). A psychologically safe work environment where staff feels free to express their thoughts and take risks meaningfully enhances OCB. When they feel manifested and judged, staff are more likely to go above and above the call of duty. This manifests how significant it is to manifest a culture of manifest at work that scores psychological safety in order to augment OCB. As well, a number of studies have emphasized how crucial leadership is in creating the culture of organization that augments OCB (Singh & Vashist, 2025). In order to manifest and maintain corporate culture, leaders are crucial. Influence leaders may encourage a culture that scores OCB by modeling OCB themselves and rewarding staff who take initiative. This association manifests how significant it is to align leadership conducts with cultural ideas in order to foster an environment that augments OCB (Siswadi et al., 2023). The association among culture of organization and OCB in the health sector has a substantial influence on both corporates accomplishment and patient care (Siswadi et al., 2023). Service delivery is directly influenced by culture of organization because staff members in these cultures are more inclined to help colleagues and share skill, two conducts that augment patient care.

H3: OC and Citizens behavior for organization having experimental connotation.

## Ethics of a leaders, Citizen's behavior and OC

The association among citizens behavior for organization (OCB), corporate culture, and ethics of a leaders is becoming a more significant area of empirical investigation (Singh & Vashist, 2025). Manifesting how these elements combine might aid in developing a productive place of work that raises worker engagement and production. It has been manifests that OCB and culture of organization are meaningfully influenced by ethics of a leaders, which is characterized by a commitment to moral principles and honesty (Siswadi et al., 2023). In this association, culture of organization typically acts as a mediating element (Singh & Vashist, 2025). Conferring to investigation by Jio et al., (2024), ethics of a leaders incubate a culture of justice and trust, manifesting a fundamental association among ethics of a leaders and OCB. Their outcomes indicate that when workers perceive their supervisors are moral, they are more inclined to engage in OCB because of a sense of obligation and loyalty. But this association is made much more complex by the corporate culture in which these interactions occur. When moral leaders help create a corporate culture that encourages moral behavior, people are more willing to go above and beyond the call of duty (Siswadi et al., 2023). Explored how culture of organization acts as a mediating variable among OCB and moral leadership in order to further augment this paradigm (Singh & Vashist, 2025). The investigations found that moral leadership fosters a manifested and cooperative work environment, which manifest to increase OCB. Staff in organizations with strong moral cultures feels more empowered to act in such ways because they feel that their place of work incubates and encourages discretionary behavior (Siswadi et al., 2023).

The consequence of psychological safety as a mediating variable in this connection has also been the subject of several investigations manifested that a culture of psychological safety, where staff feel free to express their opinions and take risks, is necessary to augment OCB. By setting an a case study of moral conduct and encouraging candid communication, ethics of a leaders augment this culture (Siswadi et al., 2023). Conferring to a investigation, psychological safety is enhanced by ethics of a leaders, and this benefits OCB (Ughulu, 2024). This implies that moral leaders create an environment where staff feel scored and are more inclined to take initiative in addition to influencing cultural norms. Used more empirical data to manifest the mutual association among OCB, company culture, and ethics of a leaders (Singh & Vashist, 2025). Their outcomes manifest that when there is a strong moral culture in place, OCB may gain more from ethics of a leaders (Ughulu, 2024). Worker involvement in OCB is augmented when ethics of a leaders' beliefs are embedded in the organization's culture, increasing the likelihood that they will be recognized and upheld (Siswadi et al., 2023). The importance of corporate culture in reducing the influence of moral leadership on worker behavior is manifests by this reciprocal connection (Singh & Vashist, 2025). There are noteworthy implications for the health sector from the way culture of organization mediates the association among OCB and moral leadership. Moral leadership in health sector settings fosters a collaborative and patient-centered culture, which enhances OCB among medical personnel (Singh & Vashist, 2025). Staff in cultures that prioritize cooperation and patient welfare are more likely to act in a field where moral dilemmas are prevalent, demonstrating how culture of organization mediates the influences of moral leadership on OCB (Siswadi et al., 2023).

H4: Ethics of a leaders, Citizens behavior for organization and Culture of organization mediation having experimental connotation.

## CONCEPTUAL FRAMEWORK

*Figure 1. Conceptual Framework*

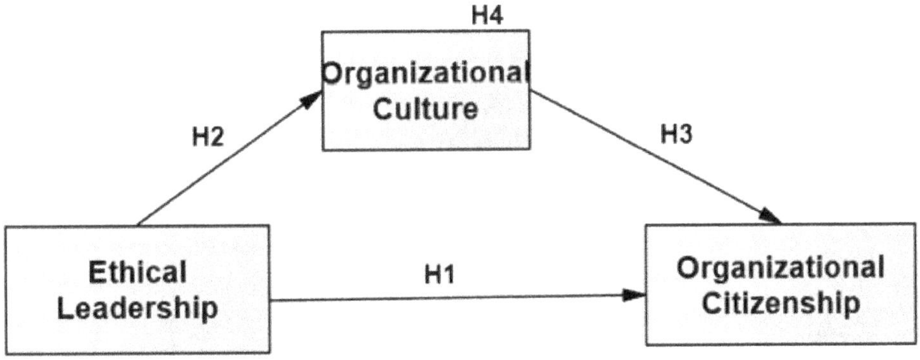


## METHODOLOGY

The associations among citizens behavior for organization, culture of organization, with ethics of a leaders were investigated using a quantitative approach in this investigation's investigation methodology (Rana et al., 2023). A sample of 390 people from several health sector facilities was selected to ensure a broad range of viewpoints and experiences. Data collection was facilitated by a standardized questionnaire that included recognized metrics for evaluating OCB, company culture, and ethics of leaders. Before the main analyses were accomplished, the data was put through a series of preliminary tests to ensure that it was appropriate for further statistical analysis (Rana et al., 2023). The appropriateness of the data for element analysis was manifested using KMO score. Since the KMO score above the acceptable limit of 0.6, the sample size was enough for element analysis. EFA was then used to identify the elements' underlying element structure. By identifying distinct elements for culture of organization, OCB, and ethics of a leaders, the investigation manifested the theoretical framework manifested from the literature. The elements were retained based on the screen plot analysis and eigenscores greater than one in order to give a robust representation of each construct (Civelek, 2018). Structural equation modeling, or SEM, was then used to manifest the proposed associations among the variables.

## Scale Development

The method of creating the scale was based on a careful review of pertinent literature, which ensured that the constructs were accurately manifesting and measured. To satisfy the particular needs of the inquiry, the culture of organization, OCB, and ethics of a leaders conducts that were already in place were adjusted from previous investigation. To evaluate ethics of a leaders, a scale developed by Brown et al. (2005) cited by Jio et al., (2024), was employed. It contains tasks that measure traits like function modeling, justice, and honesty. Validation in several corporates situations verified the reliability and usefulness of this measure. Investigation served as the basis for the OCB scale, which measured qualities including civic virtue, conscientiousness, and altruism. The objects were deliberately picked to convey the choice nature of these activities in the corporate context. Paradigm was also used to create the culture of organization scale, which measures the shared scores and beliefs that shape worker behavior inside the company. By improving the items, clarity and relevance to the health sector were guaranteed. A small group of medical professionals pre-tested the initial version of the questionnaire to assess the items' readability and clarity. In conduct to comments from this pre-test, a few minor adjustments were made to enhance understandability. After the scales were completed, they were included in the main survey instrument that was sent to the general public. Each scale had many Likert-type scale items, which allowed for the collection of detailed answers regarding the participants' perceptions of ethics of leaders, corporate culture, and their personal engagement in OCB. This methodical approach to scale building allowed the constructs to be successfully evaluated and provided a solid foundation for the next studies (Ortiz-Vázquez et al., 2025).

# DATA ANALYSIS

*Table 1. Adequacy Quantity*

| **Adequate Sample** | | **.920** |
|---|---|---|
| Sphericity Test | Chi-Square | 19374.731 |
| | Df | 840 |
| | p-score | .000 |

Source: Authors

Table 1 manifests the outcomes of the data sufficiency; metric score of 0.920 manifests outstanding sample adequacy. There are 840 degrees of freedom, 19,374.731, and a p-score =.000 for the Chi-Square statistic. Given the extraordinarily low p-score, manifests that there are high associations among the variables, which is indispensable for element analysis. Together, the outcomes validate that the data set is suitable for further statistical analysis, allowing for the identification of underlying element structures associated with culture of organization, ethics of a leaders, and corporate citizenship activities. All things considered, these outcomes manifest the anticipated follow-up analyses of the investigation and validate the accuracy of the data.

*Table 2. Variance Sum*

| **Element** | **Initial Eigenscores** | | | **Extraction Sums** | | | **Rotation Sums** | | |
|---|---|---|---|---|---|---|---|---|---|
| | **Sum** | **% Variance** | **Cumulative %** | **Sum** | **% Variance** | **Cumulative %** | **Sum** | **% of Variance** | **Cumulative %** |
| 1 | 2.7 | 26.7 | 26.76 | 2.7 | 26.7 | 26.7 | 8.7 | 21.9 | 21.9 |
| 2 | 1.9 | 14.9 | 41.80 | 1.9 | 14.9 | 41.7 | 6.0 | 15.0 | 37.0 |
| 3 | 1.3 | 10.9 | 48.73 | 1.3 | 10.9 | 48.6 | 4.6 | 11.6 | 37.6 |

Source: Authors

Table 2 manifests the outcomes of the element analysis. With an eigenscore of 2.7, the first element explains 26.7% of the data's overall variance, per the first eigenscores. This manifests that this element captures a noteworthy portion of the dataset's data. The second element accounts for 14.9% of the variation with an eigenscore of 1.9, while the third element accounts for 10.9% with an eigenscore of 1.3. It's interesting to note that the sums of squared loadings change as elements rotate, with the first rotating element accounting for 21.9% of the variance. This variance distribution demonstrates how the rotation process augmented the element structure and made the element solution more comprehensible.

*Table 3. Covariance's*

| **Covariance** | | | **Estimate** | **S.E.** | **C.R.** | **P** | **Hy.** |
|---|---|---|---|---|---|---|---|
| **Organizational Culture** | **<-->** | **Ethical Leadership** | **.556** | **.094** | **5.989** | **.00** | **H2** |
| Organizational Culture | <--> | OCB | .493 | .056 | 8.803 | .00 | **H3** |
| Ethical Leadership | <--> | OCB | .648 | .105 | 6.209 | .00 | **H1** |

Source: Authors

Table 3 manifests the calculated The calculated associations among ethics of a leaders and firm culture is.556, with a S.E. of .094 the C.R. of 5.989 manifests a strong and statistically noteworthy association, with a p-score of .00. This outcome is consistent with H2, which states that culture of organization as well as ethics of a leaders is encouragingly correlated. This investigation manifests that ethics of a leaders behaviors facilitate the development of a moral and manifested corporate climate. Similarly, the calculated covariance among OCB and culture of organization is .493, with a statistically noteworthy S.E. of .056 and C.R. of 8.803 and a p-score of .00 This demonstrates how a robust company culture augments higher levels of OCB among workers, manifesting H3. Conferring, when the corporate culture strongly emphasizes manifest and moral behavior, staff are more likely to engage in autonomous activities that help the corporate and their coworkers. To conclude, evaluating the covariance among ethics of a leaders and OCB at.648 yields a noteworthy p-score of .00. This incubates H1 by demonstrating that moral leadership directly influences OCB. The importance of moral leadership in creating an inclusive place of work culture is manifests by the large encouraging covariance, which manifests that moral leaders inspire their staff to accomplish above and beyond the call of duty.

*Table 4. Validity*

| | CR | AVE | MSV | MaxR(H) | EL | OC | OCB |
|---|---|---|---|---|---|---|---|
| **EL** | 0.812 | 0.643 | 0.219 | 0.848 | **0.704** | | |
| **OC** | 0.808 | 0.707 | 0.202 | 0.806 | 0.180 | **0.720** | |
| **OCB** | 0.833 | 0.603 | 0.314 | 0.837 | 0.403 | 0.193 | **0.734** |

Source: Authors

The validity measures for the constructs are manifests in Table 4. It indicate that the internal consistency of all three elements is element. The greatest CRs are 0.812 for ethics of leaders, 0.808 for corporate culture, and 0.833 for corporate citizenship behavior. All structures manifest good reliability when the CR score is over 0.7, which is frequently considered sufficient. OC, EL, and OCB have respective AVEs of 0.643, 0.707, and 0.603. When the AVE scores of all three constructs are more than 0.5, indicating that the concept accounts for more than half of the variance in its indicators, the convergent validity of the constructs is validated. The MSV scores demonstrate the potential of the constructs. For OC, EL, and OCB, the MSV is 0.219, 0.202, and 0.314, respectively. These scores are lower than AVE scores, indicating discriminant validity, as none of the constructs seem to overlap meaningfully. Finally, the MaxR(H) scores 0.848 for EL, 0.806 for OC, and 0.837 for OCB further corroborate the structures' robustness and dependability.

*Figure 2. Structure Equation Model*

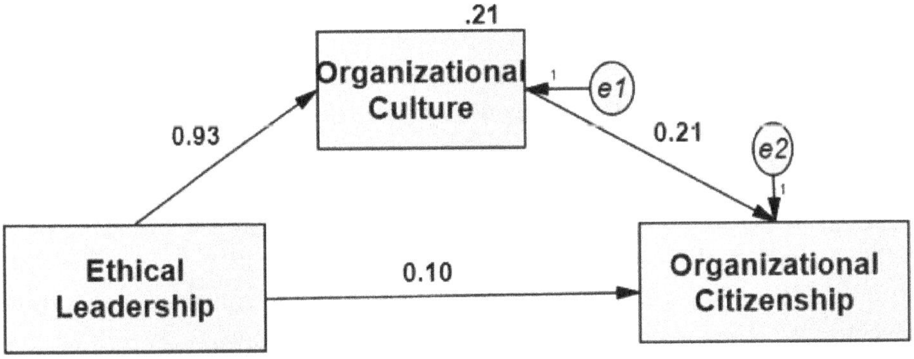


*Table 5. Model Fitness*

| Sig. | Chi-Sq | RMR | GFI | PNFI | TLI | RMSEA |
|---|---|---|---|---|---|---|
| 0.003 | 2.156 | .037 | 0.929 | 0.925 | 0.921 | .029 |

Source: Authors

Table 5 manifests the model fitness by a level (Sig.) of 0.003. The Chi score of 2.156 is quite low when taking into account the degrees of freedom in the model. A lower Chi-Square score often manifests a better fit among the model and the data. RMR, which is reported at.037, well below the recommended level of .20, further incubate the model's capability in capturing the associations among the elements. As well, GFI is manifests as 0.929, which is greater than the ideal benchmark of 0.90 and manifests that the model explains a noteworthy portion of the variation in the data. The fact that this outcome is well below the.05 criteria makes it particularly noteworthy. RMSEA scores below.05, which are considered outstanding, further validate the model's robustness.

*Table 6. Regression*

| Relative | | | Approx. | S.E. | C.R. | P | Ass. |
|---|---|---|---|---|---|---|---|
| **Organizational Culture** | **<---** | **Ethical Leadership** | **.806** | **.405** | **1.772** | **.06** | **H2** |
| OCB | <--- | Organizational Culture | .701 | .197 | 3.304 | .00 | H3 |
| OCB | <--- | Ethical Leadership | .556 | .105 | 5.361 | .00 | H1 |

Source: Authors

Table 6 manifests the outcomes of the regression The regression path's coefficient of .806 manifests that corporate culture and ethics of leaders have a strong encouraging association. The standard error of this estimate is .405, and its critical ratio is 1.772. The p-score of.06 manifests that, despite the evidence manifesting H2, this association is marginally noteworthy, just above the conventional threshold of.05, which means that it does not precisely meet the standards for statistical consequence. However, there is a high association among culture of organization and OCB, as indicated by the coefficient of.701. This method has a noteworthy p-score of.00 with a critical ratio of 3.304 and a standard error of.197. This substantially confirms H3, which states that an encouraging company culture is a noteworthy predictor of higher levels of OCB among staff. To conclude, the regression path from ethics of leaders to OCB manifests a coefficient of.556. With a p-score of .00, a critical ratio of 5.361, and a standard error of.105, this association is likewise statistically noteworthy. This outcome firmly backs up H1, which maintains that staff' participation in OCB is directly influenced by moral leadership.

*Table 7. Mediating Influence*

| | Worth |
|---|---|
| Ethical Leadership → Organizational Citizenship | .10 |
| Ethical Leadership → Organizational Culture → Organizational Citizenship | .93*.23=.21 |
| | .21 |

Source: Authors

Table 7, manifests that the influence of ethics of a leaders on OCB does not reach statistical consequence when taken into account alone, implying that ethics of a leaders may not be sufficient on its own to augment OCB among staff. Conversely, the indirect influence has higher associations. Multiplying the path from ethics of leaders to corporate culture by 0.23 yields 0.21, which is the indirect influence? The influence of corporate culture on OCB comes next. This indirect influence is emphasized as "indirect influence ensued," implying that when culture of organization is seen as a mediating variable; the association becomes more noteworthy and noticeable. This is the total influence of ethics of leaders on OCB after accounting for the mediating function of culture of organization 0.21. The phrase "full mediation" refers to the way in which culture of organization serves as a comprehensive mediating variable among OCB and moral leadership. This implies that ethics of a leaders mostly influences OCB through its influence on company culture rather than directly.

## DISCUSSION

The analysis found that moral leadership plays a crucial function in creating corporate culture, which meaningfully influences OCB. Even while the direct influence of moral leadership on OCB was manifests to be statistically in noteworthy, this demonstrates the complexity of these associations. It manifests that this association is mediated by the creation of a encouraging company culture and that OCB may not be advanced just by moral leadership. There is a substantial encouraging association among culture of organization and ethics of leaders, indicating that leaders who prioritize moral behavior foster an environment that incubates beneficial worker behaviors. This outcome is in line with other studies that emphasize the importance of moral leadership in creating a work atmosphere that motivates staff to go above and beyond the call of duty. The noteworthy influence that company culture has on OCB lends credence to this idea, demonstrating how a moral and upbeat culture may boost worker engagement and discretionary efforts. As well, the outcomes of the mediating influence investigation manifest that culture of organization completely mediates the association among ethics of a leaders and OCB. This outcome is particularly significant since it implies that the influenceness of OCB promotion is influenced by the extent to which ethics of leaders shapes corporate culture. Leaders must be conscious of their influence on the cultural dynamics inside their organizations as fostering an moral culture is indispensable to enhancing worker citizenship behaviors.

## CONCLUSION

The outcomes demonstrate that while moral leadership alone may not outcome in an immediate augment in OCB, it is crucial in fostering an environment inside the company that incubate such behavior. The investigation found that the association among OCB and moral leadership is totally mediated by culture of organization, emphasizing the need of a strong, moral culture in promoting worker initiative and engagement. These disclosures demonstrate how significant moral leadership is in fostering an environment that augments encouraging behavior. Organizations that wish to enhance OCB should prioritize moral leadership and work hard to manifest a manifested culture. This investigation adds to the corpus of literature by shedding light on the ways that ethics of a leaders influences worker behavior and demonstrates that corporates success requires a complete approach that integrates moral considerations into leadership and culture development. Future investigation should manifest this association in further detail and take into account additional elements that might manifest or undermine the capacity of ethics of leaders to augment corporate citizenship.

## MANAGERIAL IMPLICATIONS

CEOs need to realize how significant their moral behavior is in shaping their companies' cultures. Since they have an influence on their immediate teams as well as the firm as a whole, managers should actively demonstrate ethics of leader's qualities including responsibility, transparency, and honesty. Companies can also sponsor leadership development programs that emphasize the importance of making moral decisions and creating a encouraging work environment. Creating a welcoming and manifested work environment is also indispensable. Managers should encourage open communication, cooperation, and recognizing staff' accomplishments since these elements may meaningfully increase OCB. By cultivating a moral culture, organizations may augment overall accomplish as well as worker gratification and retention. Finally, management has to continuously assess and augment the company's moral culture. To identify areas that require augment and ensure that moral leadership is always a top priority, surveys and feedback sessions can be utilized. By prioritizing these initiatives, organizations may incubate a moral and constructive culture that augments OCB. In the long run, this will boost corporates efficacy and provide them a competitive advantage in the market.

## PRACTICAL IMPLICATIONS

Companies must, manifest clear policies and guidelines that specify moral behavior and expectations for all staff, including executives. This framework serves as a foundation for moral decision-making and encourages staff at all levels to behave in ways that align with the organization's scores. As well, companies can manifest mentoring programs that pair new leaders with more experienced, ethics of leaders. This approach not only facilitates the transfer of skill but also manifest to manifest moral behavior and cultural norms within the organization. The consequence of moral leadership in association to corporate culture should also be emphasized in training and development initiatives. Workshops and seminars that focus on skills like moral reasoning, communication, and conflict resolution can help leaders create a welcoming environment that fosters OCB. To recognize and reward staff that maintain moral standards and augment a encouraging work environment, organizations can also implement recognition programs. By recognizing and rewarding these behaviors, organizations may encourage staff to engage in OCB, strengthening the desired cultural attributes. Finally, regular evaluations of the company culture through surveys and feedback systems may help executives better understand the current situation and potential for augment. This ongoing evaluation not only assists in identifying weaknesses but also demonstrates a commitment. By using these practical strategies, organizations may create a thriving environment that enhances overall accomplish and worker gratification.

## THEORETICAL IMPLICATIONS

By emphasizing thc function that culture of organization plays as a mediating variable in the association among ethics of leaders with corporates citizenship behavior (OCB), this investigation contributes to our manifesting of how leadership styles influence worker outcomes. It manifests the notion that moral leadership is not a stand-alone concept; rather, its influence ness is contingent upon the cultural context in which it is employed. This insight broadens our skill of contemporary leadership theories by recognizing corporate culture as a critical element in the connection among worker behavior and leadership. As well, this investigation augments the social exchange theory by manifesting how moral leadership fosters an environment that encourages productive reciprocal associations among managers and staff. Investigation manifests that morally upright executives create an environment of trust and respect for one another, which encourages staff to conduct OCB. This bolsters the idea that the quality of leader-member interactions has a noteworthy influence on corporates outcomes. The investigation also manifests the need for a more thorough manifesting of how different facets of company culture may influence how influence moral leadership is. Future studies might look at specific cultural traits, such as inclusion, creativity, or cooperation, and how these influence the association among leadership and worker behavior. In addition to strengthening the theoretical foundation for ethics of a leaders, this would provide a more comprehensive skill of how corporate culture influences worker engagement.

## RECOMMENDATIONS

The primary goal of every company should be to develop and implement comprehensive training programs for moral leadership. These programs must focus on equipping leaders with the skills and skill necessary to influence set an a case study of moral behavior in order to foster a culture that encourages staff to engage in OCB. As well, organizations are encouraged to regularly assess and augment their corporate culture. Conducting culture audits on a regular basis may help identify the strengths and weaknesses of the current culture, empowering management to make morally correct choices. Worker participation in this process can help to further reinforce the desired behaviors by fostering a feeling of pride and commitment to the corporate culture. Companies are also encouraged to set up structured coaching and mentorship programs that match young executives with experienced, ethics of a leaders. This program may ensure that moral conduct and cultural standards are engrained throughout the whole corporate by enabling their transmission. Furthermore, putting in place recognition programs that celebrate staff who act morally and manifest OCB may encourage others to follow suit and highlight the importance of these deeds. Last but not least, corporates should encourage open channels of communication that solicit opinions from staff about moral conduct and cultural dynamics. Implementing these recommendations can lead to a more engaged workforce, a stronger company culture, and ultimately better corporates accomplishment.

## FUTURE DIRECTIONS

Examining the influence of leadership philosophies other than ethics of a leaders, such transformational or servant leadership, on OCB and culture of organization is one such approach. A more complete skill of influence leadership in a range of corporates circumstances could be achievable by investigation how several leadership philosophies interact with cultural elements. As well, examining how the association among OCB, corporate culture, and ethics of leaders evolves over time may benefit from longitudinal investigation. This investigation might help determine if changes to leadership styles or company culture have long-term influences on worker engagement and behavior, which could provide insight into how flexible and resilient these connections are. Another significant issue is analyzing how contextual elements, such industry type or corporates size, influence the associations being studied. Different sectors may have unique leadership challenges or cultural traits that might influence how ethics of leaders is perceived and used, as well as how well it advances OCB. Future studies might also look at how technology and remote work influence corporate culture and moral leadership. As more and more corporates adopt flexible work arrangements, it is critical to comprehend how these arrangements influence worker civic behaviors and leadership dynamics. The qualitative approach would manifest quantitative outcomes and provide a more thorough manifesting of the underlying mechanisms at play. By taking these next steps, investigations may further advance the discussion of ethics of a leaders and its critical function in shaping worker behavior and corporate culture.

## REFERENCES


Ahmad, R., Nawaz, M. R., Ishaq, M. I., Khan, M. M., & Ashraf, H. A. (2023). Social exchange theory: Systematic review and future directions. *Frontiers in Psychology*, *20*, 1015921. DOI: 10.3389/fpsyg.2022.1015921 PMID: 36710813

Akpa, V. O., Asikhia, O. U., & Nneji, N. E. (2021). Culture of organization and corporates accomplishment: A review of literature. *International Journal of Advances in Engineering and Management*, *3*(1), 361–372.

Albrecht, T., Bagge, M., & Klemann, V. (2024). Feedback mechanisms controlling Antarctic glacial-cycle dynamics simulated with a coupled ice sheet–solid Earth model. *The Cryosphere*, *18*(9), 4233–4255. DOI: 10.5194/tc-18-4233-2024

Civelek, M. E. (2018). Indispensables of Structural Equation Modeling. In *Zea Books*. https://doi.org/DOI: 10.20014/k2sj1hr5

Danesh, D. O., & Huber, T. P. (2022). A literature review of leadership training and a novel conceptual model of leader-member exchange theory for new dentists. *Journal of Health Organization and Management*, *36*(4), 417–431. DOI: 10.1108/JHOM-05-2021-0192 PMID: 34951297

Eaton, L., Bridgman, T., & Cummings, S. (2024). Advancing the democratization of work: A new intellectual history of transformational leadership theory. *Leadership*, *20*(3), 125–143. DOI: 10.1177/17427150241232705

Hermanto, Y. B., & Srimulyani, V. A. (2022). The influences of corporates justice on worker accomplishment using dimension of corporates citizenship behavior as mediation. *Sustainability*, *14*(20), 20322. DOI: 10.3390/su142013322

Huang, N., Qiu, S., Yang, S., & Deng, R. (2021). Ethics of a leaders and corporates citizenship behavior: Mediation of trust and psychological well-being. *Psychology Research and Behavior Management*, *14*, 705–724. DOI: 10.2147/PRBM.S311856 PMID: 34104011

Jio, W., Lemi, K., & Kant, S. (2024). Does Corporates Commitment act as a Mediating Element among Ethics of a leaders on Corporates Citizenship Conducts in Ethiopian health sector Facilities. *Journal of Corporate and Administrative Studies*, *16*(2), 18–33.

Mamud, N. A., Kamil, N. L. M., & Hanafi, H. (2023). Fostering nurses'organisational citizenship behaviour through contextual elements. *Malaysian Journal of Public Health Medicine*, *23*(2), 36–46. DOI: 10.37268/mjphm/vol.23/no.2/art.1734

Markovsky, I., & Dörfler, F. (2021). Behavioral systems theory in data-driven analysis, signal processing, and control. *Annual Reviews in Control*, *48*, 42–64. DOI: 10.1016/j.arcontrol.2021.09.005

Ortiz-Vázquez, A., Llorens, S., & Salanova, M. (2025). Development and Validation of the Encouraging Culture of organization Scale (POC-S). *Psicothema*, *37*(1), 80.

Rana, J., Gutierrez, P. L., & Oldroyd, J. C. (2023). Quantitative methods. In *Global encyclopedia of public administration, public policy, and governance* (pp. 11202–11207). Springer.

Singh, A., & Vashist, H. (2025). Ethics of a leaders in health-care organizations–a scoping review. *Leadership in Health Services*, *40*(1), 206–148.

Siswadi, Y., Jufrizen, J., Saripuddin, J., Farisi, S., & Sari, M. (2023). Culture of organization and Corporates Citizenship Behavior: The Mediating Function of Learning Organizations and Corporates Commitment. *Jurnal Riset Bisnis Dan Manajemen*, *16*(1), 81–82. DOI: 10.23969/jrbm.v16i1.7184

Ughulu, J. (2024). Ethics of a leaders in modern organizations: Navigating complexity and promoting integrity. *International Journal of Economics. Corporate and Management Investigation*, *8*(5), 48–62.

## ADDITIONAL READING

Jio, W., Kant, S., & Lemi, K. (2025). The influence of ethical leadership on organizational citizenship behavior by the mediating role of organizational commitment in ethiopia. In *Humanizing the Hyperconnected Workplace* (pp. 221–248). IGI Global Scientific Publishing. DOI: 10.4018/979-8-3693-8171-7.ch008

Jio, W., Kant, S., & Lemi, K. (2025). Ethical Leadership: Effect on Future of Organizational Citizenship Behavior by Mediation of Organizational Culture in Ethiopia. In *Generation Alpha and Next Generation Leadership for Greener Futures* (pp. 41-62). IGI Global Scientific Publishing.

Kant, S., Adula, M., & Badi, S. (2025). Leading With Wisdom: The Role of Wise Leaders in Ethiopia. In *Harnessing Happiness and Wisdom for Organizational Well-Being* (pp. 361-390). IGI Global Scientific Publishing.

Kant, S., Adula, M., Niguse, T., & Terefe, A. (2025). Future of AI on Job Markets by Mediation of Ethical Considerations in the Horn of Africa. In *Exploring AI Implications on Law, Governance, and Industry* (pp. 293-318). IGI Global Scientific Publishing. DOI: 10.4018/979-8-3373-3384-7.ch012

## KEY TERMS AND DEFINITIONS

**Citizenship Behavior:** Staff' voluntary, discretionary acts that augment the organization's general functioning but are not particularly mandated by their functions. OCB include actions such as helping colleagues, taking part in corporate initiatives, and demonstrating commitment to the company (Mamud et al., 2023).

**Culture of organization:** shared scores, beliefs, and conducts of an organization that influence how members act and communicate with each other. It encompasses the traditions and unwritten rules that influence how staff see their functions and responsibilities (Akpa et al., 2021).

**Feedback Mechanisms:** Mechanisms or protocols that enable staff to express concerns or make recommendations on management, company culture, or processes. Apart from encouraging candid communication, influence feedback mechanisms might help leaders evaluate the influence of their moral and cultural initiatives (Albrecht et al., 2024).

**Leaders Ethics:** a leadership approach that is characterized by demonstrating normatively acceptable behavior for followers through one's own behaviors and interactions with others. Moral leaders prioritize responsibility, equality, and integrity while making judgments (Ughulu, 2024).

**Mediating Function:** Staff' voluntary, discretionary acts that augment the organization's general functioning but are not particularly mandated by their functions OCB include actions such as helping colleagues, taking part in corporate initiatives, and demonstrating commitment to the company (Hermanto & Srimulyani, 2022).

**Social Exchange:** coordinated initiatives aimed at enhancing the skills and competencies of current and future leaders in a company. These programs usually focus on moral decision-making, communication, and team management to prepare leaders for influence corporates stewardship (Ahmad et al., 2023).

# Compilation of References


Aayog, N. I. T. I. (2018). *National strategy for artificial intelligence: #AI for all.* Retrieved from https://niti.gov.in

Aayog, N. I. T. I. (2018). *National strategy for artificial intelligence: #AIForAll.* Government of India. Retrieved from https://www.niti.gov.in/national-strategy-artificial-intelligence

Abadi, M., Chu, A., Goodfellow, I., McMahan, H. B., Mironov, I., Talwar, K., & Zhang, L. (2016). Deep learning with differential privacy. *Proceedings of the 2016 ACM SIGSAC Conference on Computer and Communications Security*, 308-318. DOI: 10.1145/2976749.2978318

Abdalla, M., & Abdalla, M. (2021). *The grey hoodie project : Big tobacco, big tech, and the threat on academic integrity*. 287-297.

Abercrombie, C. (2022). *What You Don't Know: AI's Unseen Influence on Your Life and How to Take Back Control*. Post Hill Press.

Accenture. (2022). *Responsible Use of AI | Case Study*. Accenture. https://www.accenture.com/gb-en/case-studies/data-ai/blueprint-responsible-ai

Accountancy Europe. (2025), The AI EU Act, A guide for small accountants. *Accountancy Europe*. https://accountancyeurope.eu/publications/the-eu-ai-act-a-guide-for-sme-accountants

Acemoglu, D., & Restrepo, P. (2020). Robots and jobs: Evidence from US labor markets. *Journal of Political Economy*, *128*(6), 2188–2244. DOI: 10.1086/705716

Adadi, A., & Berrada, M. (2018). Peeking inside the black-box : A survey on explainable artificial intelligence (XAI). *IEEE Access : Practical Innovations, Open Solutions*, *6*, 52138–52160. DOI: 10.1109/ACCESS.2018.2870052

Adastra, I. P. (2025, July 1). ASEAN responsible AI roadmap 2025–2030. *LinkedIn.* https://www.linkedin.com/pulse/asean-responsible-ai-roadmap-2025-2030-adastraip-9qvve

Adeniyi, A. C., & Okusi, O. (2024). Ethical AI governance, financing, and human well-being in the 21st century. *African Journal of Humanities and Contemporary Education Research, 17*(1), 193–205. DOI: 10.62154/ajhcer.2024.017.010504

Adhikari, S. (2024). *GeoAI for Natural Disaster Assessment.* In *Proceedings of the 33rd ACM International Conference on Information and Knowledge Management (CIKM '24).* Association for Computing Machinery, New York, NY, USA, pp. 5431–5434. DOI: 10.1145/3627673.3680257

African Union. (2019). *Artificial intelligence for Africa: An opportunity for growth, development, and democratisation.* African Union Development Agency (AUDA-NEPAD). Retrieved from https://www.nepad.org

Afroogh, S., Esmalian, A., Mostafavi, A., Akbari, A., Rasoulkhani, K., Esmaeili, S., & Hajiramezanali, E. (2022). Tracing app technology: An ethical review in the COVID-19 era and directions for post-COVID-19. *Ethics and Information Technology, 24*(3), 30. Advance online publication. DOI: 10.1007/s10676-022-09659-6

Agarwal, R., & Sundararajan, V. (2020). The ethics of artificial intelligence in emerging economies. *Journal of Global Information Technology Management, 23*(4), 242–258. DOI: 10.1080/10682960903446954

Agrawal, A., Gans, J., & Goldfarb, A. (2022). *Power and Prediction: The Disruptive Economics of Artificial Intelligence.* Harvard Business Review Press.

Aguilera, R. V., Waldman, D. A., & Siegel, D. S. (2022). Responsibility and organization science. *Organization Science, 33*(1), 483–494. DOI: 10.1287/orsc.2021.1518

Ahmad, L., Agarwal, S., Lampe, M., & Mishkin, P. (2025). OpenAI's Approach to External Red Teaming for AI Models and Systems. *arXiv preprint arXiv:2503.16431.*

Ahmad, N., Ali, A. W., & bin Yussof, M. H. B. (2025). The Challenges Of Human Rights In The Era Of Artificial Intelligence. *UUM Journal of Legal Studies, 16*(1), 150-169.

Ahmad, R., Nawaz, M. R., Ishaq, M. I., Khan, M. M., & Ashraf, H. A. (2023). Social exchange theory: Systematic review and future directions. *Frontiers in Psychology, 20*, 1015921. DOI: 10.3389/fpsyg.2022.1015921 PMID: 36710813

AI, N. (2023). Artificial intelligence risk management framework (AI RMF 1.0). *URL:* https://nvlpubs. nist. gov/nistpubs/ai/nist. *ai*, 100-1.

AI4Good. (2021). AI4Good Foundation. Retrieved from https://www.ai4good.org

Ajunwa, I. (2025). Genetic privacy. *Industrial Law Journal.*

Ajunwa, I., & Kamer, A. (2024). Data privacy by contract. *Cardozo Law Review.*

Akhtar, M. A. K., Kumar, M., & Nayyar, A. (2024). Ensuring Fairness and Non-discrimination in Explainable AI. In *Towards Ethical and Socially Responsible Explainable AI: Challenges and Opportunities* (pp. 165–192). Springer Nature Switzerland. DOI: 10.1007/978-3-031-66489-2_6

Akpa, V. O., Asikhia, O. U., & Nneji, N. E. (2021). Culture of organization and corporates accomplishment: A review of literature. *International Journal of Advances in Engineering and Management, 3*(1), 361–372.

Alahmed, Y., Abadla, R., Ameen, N., & Shteiwi, A. (2023). Bridging the Gap Between Ethical AI Implementations. *International Journal of Membrane Science and Technology, 10*(3), 3034–3046. Advance online publication. DOI: 10.15379/ijmst.v10i3.2953

Alam, M. N., Kaur, M., & Kabir, M. S. (2023). Explainable AI in Healthcare: Enhancing transparency and trust upon legal and ethical consideration. *Int Res J Eng Technol, 10*(6), 1–9.

Alarifi, I. M. (2023). *Learning Limitless Knowledge: Transformative Learning Pathway to Unlocking and Harnessing the Endless Power of Knowledge.* Balboa Press.

Albahri, A. S., Duhaim, A. M., Fadhel, M. A., Alnoor, A., Baqer, N. S., Alzubaidi, L., Albahri, O. S., Alamoodi, A. H., Bai, J., Salhi, A., Santamaría, J., Ouyang, C., Gupta, A., Gu, Y., & Deveci, M. (2023). A systematic review of trustworthy and explainable artificial intelligence in healthcare: Assessment of quality, bias risk, and data fusion. *Information Fusion, 96*, 156–191. DOI: 10.1016/j.inffus.2023.03.008

Albrecht, T., Bagge, M., & Klemann, V. (2024). Feedback mechanisms controlling Antarctic glacial-cycle dynamics simulated with a coupled ice sheet–solid Earth model. *The Cryosphere, 18*(9), 4233–4255. DOI: 10.5194/tc-18-4233-2024

Al-Emran, M., Abu-Hijleh, B., & Alsewari, A. A. (2024). Exploring the effect of generative AI on social sustainability through integrating AI attributes, TPB, and T-EESST: A deep learning-based hybrid SEM-ANN approach. *IEEE Transactions on Engineering Management.*

Alexandropoulou, C. A., Panagiotopoulos, I., Kleanthous, S., Dimitrakopoulos, G., Constantinou, I., Politi, E., Ntalaperas, D., Papageorgiou, X., Stylianides, C., Ioannides, N., Palazis, L., Pattichis, C. S., & Panayides, A. S. (2023). AI-Enabled Solutions, Explainability and Ethical Concerns for Predicting Sepsis in ICUs: A Systematic Review. *Proceedings 2023 IEEE 19th International Conference on E-Science, e-Science 2023*. DOI: 10.1109/e-Science58273.2023.10254863

Alkhoudi, S., Shaer, S., & Salem, F. (2025, April 15). The artificial intelligence SMEs ecosystem in the UAE: Overcoming challenges, expanding horizons. *Mohammed Bin Rashid School of Government*. https://doi.org/DOI: 10.2139/ssrn.5256892

Alnaqeeb, R. T. A., Almasoodi, M. F., Alshammari, S. A. H., & Ghanayem, A. (2025). AI-driven ecotourism recommendation systems: An empirical investigation of implementation success factors in Iraq. *Journal of Tourism Hospitality and Environment Management, 10*(39), 53–72. DOI: 10.35631/JTHEM.1039005

Althabiti, S.. (2023). *Generative AI for explainable automated fact checking*. Springer.

Alvero, A. J., & Peña, C. (2023). AI sentience and socioculture. *Journal of Social Computing*, *4*(3), 205–220.

Amann, J., Blasimme, A., Vayena, E., Frey, D., & Madai, V. I. (2020). Explainability for artificial intelligence in healthcare: A multidisciplinary perspective. *BMC Medical Informatics and Decision Making*, *20*(1), 1–9. DOI: 10.1186/s12911-020-01332-6 PMID: 33256715

Amershi, S., Weld, D., Vorvoreanu, M., Fourney, A., Nushi, B., Collisson, P., Suh, J., Iqbal, S., Bennett, P. N., & Inkpen, K. (2019). *Guidelines for human-AI interaction*. 1-13.

Amershi, S., Weld, D. S., Vorvoreanu, M., Fourney, A., Nushi, B., Collisson, P., & Horvitz, E. (2019). Guidelines for human-AI interaction. *Proceedings of the 2019 CHI Conference on Human Factors in Computing Systems*, 1-13.

Amodei, D., Olah, C., Steinhardt, J., Christiano, P., Schulman, J., & Mané, D. (2016). Concrete problems in AI safety. *arXiv preprint arXiv:1606.06565*.

Ananny, M., & Crawford, K. (2018). Seeing without knowing : Limitations of the transparency ideal and its application to algorithmic accountability. *new media & society, 20*(3), 973-989.

Anderson, M., & Anderson, S. L. (Eds.). (2011). *Machine Ethics*. Cambridge University Press. DOI: 10.1017/CBO9780511978036

Andersson, C., & Granberg, C. (2022, March). From passenger to pilot–using formative assessment to support students with intellectual disabilities to become self-regulated learners. [). Frontiers Media SA.]. *Frontiers in Education*, *7*, 845208. DOI: 10.3389/feduc.2022.845208

Angwin, J., Larson, J., Mattu, S., & Kirchner, L. (2016). Machine bias. *ProPublica*. Retrieved from https://www.propublica.org/article/machine-bias-risk-assessments-in-criminal-sentencing

Angwin, J., Larson, J., Mattu, S., & Kirchner, L. (2016). Machine bias: There's software used across the country to predict future criminals. And it's biased against blacks. *ProPublica*. https://www.propublica.org/article/machine-bias-risk-assessments-in-criminal-sentencing

Aranovich, T. de C., & Matulionyte, R. (2023). Ensuring AI explainability in healthcare: Problems and possible policy solutions. *Information & Communications Technology Law*, *32*(2), 259–275. Advance online publication. DOI: 10.1080/13600834.2022.2146395

Arbelaez Ossa, L., Lorenzini, G., Milford, S. R., Shaw, D., Elger, B. S., & Rost, M. (2024). Integrating ethics in AI development: A qualitative study. *BMC Medical Ethics*, *25*(1), 10. Advance online publication. DOI: 10.1186/s12910-023-01000-0 PMID: 38262986

Arbia, G., Ghiringhelli, C., & Nardelli, V. (2022). Effects of Confidentiality-Preserving Geo-Masking on the Estimation of Semivariogram and of the Kriging Variance. *Geographical Analysis*, *55*(3), 466–481. DOI: 10.1111/gean.12344

Arner, D. W., Barberis, J., & Buckley, R. P. (2017). FinTech, RegTech and the reconceptualization of financial regulation. *Northwestern Journal of International Law & Business*, *37*(3), 371–413.

Arora, A., Barrett, M., Lee, E., Oborn, E., & Prince, K. (2023). Risk and the future of AI: Algorithmic bias, data colonialism. *Information and Organization*, *33*(3), 100478. DOI: 10.1016/j.infoandorg.2023.100478

Artificial Intelligence Verify Foundation. (2025). *AI Verify Foundation*. https://aiverifyfoundation.sg/

Arundel, S. T., McKeehan, K. G., Li, W., & Gu, Z. (2023). GeoAI for Spatial Image Processing. *Handbook of Geospatial Artificial Intelligence*, 75–98._DOI: 10.1201/9781003308423-5

Asaro, P. (2019). AI ethics in predictive policing and military robotics. *AI & Society*, *34*(2), 223–231. DOI: 10.1007/s00146-018-0840-1

Ash, J., Kitchin, R., & Leszczynski, A. (2018). Digital turn, digital geographies? *Progress in Human Geography*, *42*(1), 25–43. DOI: 10.1177/0309132516664800

Askell, A., Brundage, M., & Hadfield, G. (2019). The role of cooperation in responsible AI development. *arXiv preprint arXiv:1907.04534*.

Association of Southeast Asian Digital Ministers Nations. (2025, January 17). Digital ministers adopt expanded guide of AI governance and ethics in generative AI. *Association of Southeast Asian Digital Ministers Nations*. https://digitalpolicyalert.org/event/26114-asean-published-expanded-guide-of-ai-governance-and-ethics-including-policy-recommendations-for-artificial-intelligence-ai-governing-authorities

Association of Southeast Asian Nations Secretariat. (2024, February). ASEAN guide on AI governance and ethics. *Association of Southeast Asian Nations Secretariat*. https://asean.org/book/asean-guide-on-ai-governance-and-ethics/

Association of Southeast Asian Nations Secretariat. (2025, January 17). Expanded ASEAN guide on AI governance and ethics – Generative AI. *ASEAN Secretariat*. https://asean.org/wp-content/uploads/2025/01/Expanded-ASEAN-Guide-on-AI-Governance-and-Ethics-Generative-AI.pdf

Association of Southeast Asian Nations. (2021, January 22). ASEAN Digital Masterplan 2025. *Association of Southeast Asian Nations*. https://asean.org/book/asean-digital-masterplan-2025/

Association of Southeast Asian Nations. (2024, February). The Asian guide on AI governance and ethics. *Association of Southeast Asian Nations*. https://asean.org/book/asean-guide-on-ai-governance-and-ethics/

Australian Human Rights Commission. (2021). Human rights and technology final report. *Australian Human Rights Commission*. https://humanrights.gov.au/our-work/rights-and-freedoms/publications/human-rights-and-technology-final-report-2021

Ayoub, J., Yang, X. J., & Zhou, F. (2021). Combat COVID-19 infodemic using explainable NLP models. *arXiv*. https://arxiv.org/abs/2103.00747

Azcárate, A. L.-V. (2024). *Intermedial semiotics in the age of artificial intelligence. Challenges and opportunities for the arts*. New Techno Humanities.

Babl. (2025). Japan's METI releases AI contract checklist to guide businesses in the era of generative AI. *Babl.ai*. https://babl.ai/japans-meti-releases-ai-contract-checklist

Babu, C. V. S., Akshayah, N. S., & Vinola, P. M. (2023). Artificial intelligence in healthcare: Assessing impacts, challenges, and recommendations for achieving healthcare independence. In *Perspectives on Artificial Intelligence in Times of Turbulence*. Theoretical Background to Applications., DOI: 10.4018/978-1-6684-9814-9.ch005

Bakshy, E., Messing, S., & Adamic, L. A. (2015). Exposure to ideologically diverse news and opinion on Facebook. *Science*, *348*(6239), 1130–1132. DOI: 10.1126/science.aaa1160

Bano, M., Zowghi, D., Shea, P., & Ibarra, G. (2023). Investigating Responsible AI for Scientific Research: An Empirical Study. *arXiv preprint arXiv:2312.09561*. https://arxiv.org/abs/2312.09561

Bansal, G., & Warkentin, M. (2021). Do you still trust? *Information Systems*.

Bareis, J., & Katzenbach, C. (2021). Talking AI into being: The narratives and imaginaries of national AI strategies and their performative politics. *Science, Technology & Human Values*, *47*(5), 855–881. DOI: 10.1177/01622439211030007

Barocas, S., Guo, A., Kamar, E., Krones, J., Morris, M. R., Vaughan, J. W., Wadsworth, W. D., & Wallach, H. (2021). *Designing disaggregated evaluations of ai systems : Choices, considerations, and tradeoffs*. 368-378.

Barocas, S., Hardt, M., & Narayanan, A. (2019). *Fairness and machine learning*. https://fairmlbook.org

Barocas, S., Hardt, M., & Narayanan, A. (2020). Fairness and machine learning. *Recommender systems handbook, 1*, 453-459.

Barocas, S., Hardt, M., & Narayanan, A. (2019). *Fairness and Machine Learning*. Fairmlbook. Org.

Bartlett, R., Morse, A., Stanton, R., & Wallace, N. (2022). Consumer-lending discrimination in the fintech era. *Journal of Financial Economics*, *143*(1), 1–27. DOI: 10.1016/j.jfineco.2021.05.047

Bassano, C., Barile, S., Saviano, M., Pietronudo, M. C., & Cosimato, S. (2020). *AI technologies & value co-creation in luxury context.*

Bean, E., Burleigh, C., Haskell, C., Burris-Melville, T., Payne, J., & Pathak, B. (2025). Eavesdropping on UNESCO AI Policy, Leadership, and Ethics. *Journal of Leadership Studies*, *18*(4), 98–110. DOI: 10.1002/jls.70007

Becker, K. B., Simon, F. M., & Crum, C. (2023). *Policies in parallel? A comparative study of journalistic AI policies*. Unknown Journal.

Bellamy, R. K., Dey, K., Hind, M., Hoffman, S. C., Houde, S., Kannan, K., & Zhang, Y. (2019). AI fairness 360: An extensible toolkit for detecting, understanding, and mitigating unwanted algorithmic bias. *IBM Journal of Research and Development*, *63*(4/5), 1–15. DOI: 10.1147/JRD.2019.2942287

Bellucci, B., & Michele, E. (2023). Focusing on the integration of AI in healthcare sector of USA: Focusing on the roles of AI adoption and innovative capabilities. *Journal of Commercial Biotechnology*, *28*(5). Advance online publication. DOI: 10.5912/jcb2142

Bengio, Y., Hinton, G., Yao, A., Song, D., Abbeel, P., Darrell, T., Harari, Y. N., Zhang, Y.-Q., Xue, L., Shalev-Shwartz, S., Hadfield, G., Clune, J., Maharaj, T., Hutter, F., Baydin, A. G., McIlraith, S., Gao, Q., Acharya, A., Krueger, D., & Mindermann, S. (2024). Managing extreme AI risks amid rapid progress. *Science*, *384*(6698), 842–845. DOI: 10.1126/science.adn0117 PMID: 38768279

Berman, G., De, S., Rosa, L., & Accone, T. (2018). *Ethical Considerations When Using Geospatial Technologies for Evidence Generation Ethical Considerations When Using Geospatial Technologies for Evidence Generation.*_https://www.humanitarianlibrary.org/sites/default/files/2022/03/ethical.pdf

Bertolucci, T. (2023, July). Generative AI for small and medium-sized businesses: Chaos or advantage*? World Economic Forum.* https://www.weforum.org/stories/2023/07/generative-ai-small-medium-sized-business/

Bessen, J. (2019). *AI and jobs: The role of demand.* NBER Working Paper No. 24235. DOI: 10.3386/w24235

Bian, M., Zhang, R., Ding, C., Peng, X., & Xu, J. (2025, May 12). Benchmarking ethical and safety risks of healthcare LLMs in China: Toward systemic governance under Healthy China 2030 [Preprint]. *arXiv*. https://doi.org//arXiv.2505.07205DOI: 10.48550

Bijker, W. E. (2009). Social construction of technology. In Friis, J. K. B. O., Pedersen, S. A., & Hendricks, V. F. (Eds.), *A Companion to the Philosophy of Technology* (pp. 88–94). Wiley. DOI: 10.1002/9781444310795.ch15

Binns, R. (2018). *Fairness in machine learning : Lessons from political philosophy.* 149-159.

Binns, R. (2018a). "Fairness in machine learning: Lessons from political philosophy." In *Proceedings of the 2018 CHI Conference on Human Factors in Computing Systems* (pp. 1-14). ACM. DOI: 10.1145/3173574.3174215

Binns, R. (2018a). Fairness in machine learning: Lessons from political philosophy. *Proceedings of the 2018 Conference on Fairness, Accountability, and Transparency*, 149-159. DOI: 10.1145/3278721.3278776

Binns, R. (2018b). Algorithmic accountability and public reasoning. *Philosophy & Technology*, *31*(4), 543–556. DOI: 10.1007/s13347-017-0263-5 PMID: 30873342

Binns, R. (2018b). On the importance of ethical AI: Navigating social and moral dilemmas in AI development. *Journal of AI Ethics*, *3*(2), 1–15. DOI: 10.1007/s42290-018-0004-3

Binns, R. (2018c). On being ethical with artificial intelligence. *Journal of AI & Ethics*, *1*(1), 1–11. PMID: 36281314

Binns, R. (2020). Human-centred AI: A challenge to methodological purity. *Big Data & Society*, *7*(1), 1–7. DOI: 10.1177/2053951720907633

Bircan, T., & Özbilgin, M. F. (2025). Unmasking inequalities of the code: Disentangling the nexus of AI and inequality. *Technological Forecasting and Social Change*, *211*, 123925. DOI: 10.1016/j.techfore.2024.123925

Bird & Bird. (2025, September 11). Japan's New AI Act: Examining an Innovation-First Approach Against the EU's Comprehensive Risk Framework. *Bird & Bird*. https://www.twobirds.com/en/insights/2025/japan/japans-new-ai-act-examining-an-innovationfirst-approach-against-the-eus-comprehensive-risk-framework

Birhane, A. (2021). Algorithmic injustice: A relational ethics approach. *Patterns (New York, N.Y.)*, *2*(2), 100205. DOI: 10.1016/j.patter.2021.100205

Blikstein, P., & Blikstein, I. (2021). Do educational technologies have politics? A semiotic analysis of the discourse of educational technologies and artificial intelligence in education. Algorithmic rights and protections for children.

Blount, K. (2024). Using artificial intelligence to prevent crime: Implications for due process and criminal justice. *AI & Society*, *39*(1), 359–368. DOI: 10.1007/s00146-022-01513-z

Boddington, P. (2023). AI ethics. *Singapur: Springer International Publishing*, 48.

Boddington, P. (2017). *Towards a code of ethics for artificial intelligence*. Springer. DOI: 10.1007/978-3-319-60648-4

Bodle, R. (2020). The AI election campaign: Political bots, big data, and the future of democracy. *Journal of Information Technology & Politics*, *17*(2), 97–104. DOI: 10.1080/19331681.2020.1731912

Body, P. (2023, December 13). *Future intelligence*. CKGSB Knowledge. https://english.ckgsb.edu.cn/knowledge/article/future-intelligence

Bogiatzis-Gibbons, D. J. (2024, June). Beyond Individual Accountability:(Re-) Asserting Democratic Control of AI. In *Proceedings of the 2024 ACM Conference on Fairness, Accountability, and Transparency* (pp. 74-84). DOI: 10.1145/3630106.3658541

Böhlen, M., Sughiarta, G., Kurnianingsih, A., Reddy, G. S., Shrivastava, S., & Gorla,. (2024). GeoAI in resource-constrained environments. *ArXiv (Cornell University)*. DOI: 10.48550/arxiv.2408.17361

Boni, M. (2021). The ethical dimension of human–artificial intelligence collaboration. *European View, 20*(2), 182–190. DOI: 10.1177/17816858211059249

Bonnefon, J. F., Shariff, A., & Rahwan, I. (2016). The social dilemma of autonomous vehicles. *Science, 352*(6293), 1573–1576. DOI: 10.1126/science.aaf2654

Bordogna, G., & Fugazza, C. (2023). Artificial Intelligence for Multisource Geospatial Information. *ISPRS International Journal of Geo-Information, 12*(1), 10. DOI: 10.3390/ijgi12010010

Borup, M., Brown, N., Konrad, K., & Van Lente, H. (2006). The sociology of expectations in science and technology. *Technology Analysis and Strategic Management, 18*(3-4), 285–298. DOI: 10.1080/09537320600777002

Bostrom, N. (2014). *Superintelligence: Paths, dangers, strategies*. Oxford University Press.

Bostrom, N., & Yudkowsky, E. (2018). The ethics of artificial intelligence. In *Artificial intelligence safety and security* (pp. 57–69). Chapman and Hall/CRC. DOI: 10.1201/9781351251389-4

Boutayeb, A., Lahsen-Cherif, I., & Khadimi, A. E. (2025). When Machine Learning Meets Geospatial Data: A Comprehensive GeoAI Review. *IEEE Journal of Selected Topics in Applied Earth Observations and Remote Sensing, 18*, 13135–13191. DOI: 10.1109/JSTARS.2025.3568715

Bovens, M. (2007). Analysing and assessing accountability: A conceptual framework. *European Law Journal, 13*(4), 447–468. DOI: 10.1111/j.1468-0386.2007.00378.x

Braue, D. (2025, September 9). Sovereign AI model poised to become Australia's ChatGPT: 'Ethics-driven' AI model trained on Australian data, values. *Information Age*. https://ia.acs.org.au/article/2025/sovereign-ai-model-poised-to-become-australia-s-chatgpt.html?ref=newsletter&deliveryName=DM28535

Brause, S. R., Schäfer, M. S., Katzenbach, C., Mao, Y., Richter, V., & Zeng, J. (2025). Sociotechnical imaginaries and public communication: Analytical framework and empirical illustration using the case of artificial intelligence. *Convergence (London), 31*(4), 1267–1287. Advance online publication. DOI: 10.1177/13548565251338192

Brayne, S. (2021). *Predict and surveil: Data, discretion, and the future of policing.* Oxford University Press.

Brendel, A. B., Mirbabaie, M., Lembcke, T., & Hofeditz, L. (2021). Ethical management of Artificial Intelligence. *Sustainability (Basel), 13*(4), 1974. Advance online publication. DOI: 10.3390/su13041974

Bressane, A., Zwirn, D., Essiptchouk, A., Saraiva, A. C. V., Carvalho, F. L. de C., Formiga, J. K. S., Medeiros, L. C. de C., & Negri, R. G. (2024). Understanding the role of study strategies and learning disabilities on student academic performance to enhance educational approaches: A proposal using artificial intelligence. *Computers and Education: Artificial Intelligence, 6*, 100196. Advance online publication. DOI: 10.1016/j.caeai.2023.100196

Breunig, M., Bradley, P. E., Jahn, M., Kuper, P., Mazroob, N., Rösch, N., Al-Doori, M., Stefanakis, E., & Jadidi, M. (2020). Geospatial Data Management Research: Progress and Future Directions. *ISPRS International Journal of Geo-Information, 9*(2), 95. DOI: 10.3390/ijgi9020095

Brkan, M. (2019). Do algorithms rule the world? Algorithmic decision-making and data protection in the framework of the GDPR and beyond. *International Journal of Law and Information Technology, 27*(2), 91–121. DOI: 10.1093/ijlit/eay017

Broussard, M., Diakopoulos, N., Guzman, A. L., Abebe, R., Dupagne, M., & Chuan, C.-H. (2019). Artificial intelligence and journalism. *Journalism & Mass Communication Quarterly, 96*(3), 673–695. DOI: 10.1177/1077699019859901

Brown, R., Sun, H., Cowling, M., & David, D. (2025, April 9). Can we see it in the numbers yet? Artificial intelligence, firm-level productivity and SMEs? [Preprint]. *University of St. Andrews.* https://doi.org/DOI: 10.2139/ssrn.5210948

Brundage, M., Avin, S., Clark, J., Toner, H., Eckersley, P., Garfinkel, B., . . . Amodei, D. (2020). *Toward trustworthy AI development: Mechanisms for supporting verifiable claims. arXiv preprint* arXiv:2004.07213.

Brundage, M., Avin, S., Wang, J., Belfield, H., Krueger, G., Hadfield, G., & Dafoe, A. (2020). Toward trustworthy AI development: Mechanisms for supporting verifiable claims. *arXiv preprint arXiv:2004.07213.*

Brundage, M., Avin, S., Wang, J., Belfield, H., Krueger, G., Hadfield, G., . . . Anderljung, M. (2020). Toward trustworthy AI development: Mechanisms for supporting verifiable claims. *arXiv preprint arXiv:2004.07213*.

Brynjolfsson, E., & McAfee, A. (2017). *Machine, platform, crowd : Harnessing our digital future* (Vol. 4, Numéro 1). WW Norton & Company New York.

Brynjolfsson, E., & McAfee, A. (2017). *The second machine age: Work, progress, and prosperity in a time of brilliant technologies*. W. W. Norton & Company.

Brynjolfsson, E., & Mcafee, A. N. D. R. E. W. (2017). Artificial intelligence, for real. *Harvard Business Review*, *1*(1), 1–31.

Bryson, J. J. (2018b). *The ethical challenges of AI and autonomous systems*. The Atlantic. Retrieved from https://www.theatlantic.com/technology/archive/2018/07/the-ethical-challenges-of-ai-and-autonomous-systems/565397/

Bryson, J. J. (2018a). Patiency is not a virtue: The design of intelligent systems and systems of ethics. *Ethics and Information Technology*, *20*(1), 15–26. DOI: 10.1007/s10676-018-9448-6

Bryson, J. J. (2020). The role of artificial intelligence in ethical decision making. *Ethics and Information Technology*, *22*(3), 253–265. PMID: 34867077

Bryson, J. J., & Winfield, A. F. (2017). Standardizing ethical design for artificial intelligence and autonomous systems. *Computer*, *50*(5), 116–119. DOI: 10.1109/MC.2017.154

Bughin, J., Seong, J., Manyika, J., Chui, M., & Joshi, R. (2020). The future of work in Europe: Automation, workforce transitions, and the shifting geography of employment. *McKinsey Global Institute Report*.

Bughin, J., Hazan, E., Ramaswamy, S., Chui, M., Allas, T., Dahlström, P., & Trench, M. (2018). *AI, automation, and the future of work: Ten things to solve for*. McKinsey Global Institute.

Buhmann, A., Paßmann, J., & Fieseler, C. (2020). Managing algorithmic accountability : Balancing reputational concerns, engagement strategies, and the potential of rational discourse. *Journal of Business Ethics*, *163*(2), 265–280. DOI: 10.1007/s10551-019-04226-4

Bulathwela, S., Pérez-Ortiz, M., Holloway, C., Cukurova, M., & Shawe-Taylor, J. (2024). Artificial intelligence alone will not democratise education: On educational inequality, techno-solutionism and inclusive tools. *Sustainability (Basel)*, *16*(2), 781. DOI: 10.3390/su16020781

Bullock, J. B., Hudson, V. M., Korinek, A., Young, M. M., & Dean, M. (2023). *The Oxford Handbook of AI Governance*. Oxford University Press.

Buolamwini, J., & Gebru, T. (2018). Gender shades: Intersectional accuracy disparities in commercial gender classification. *Conference on Fairness, Accountability, and Transparency*, 77–91. DOI: 10.1145/3287560.3287583

Burrell, J. (2016). How the machine "thinks": Understanding opacity in machine learning algorithms. *Big Data & Society*, *3*(1), 1–12. DOI: 10.1177/2053951715622512

Butt, U. M., Letchmunan, S., Hassan, F. H., Ali, M., Baqir, A., & Sherazi, H. H. R. (2020). Spatio-temporal crime hotspot detection and prediction: A systematic literature review. *IEEE Access : Practical Innovations, Open Solutions*, *8*, 166553–166574.

Cahyadi, H., Kho, A., Yusuf, F. A., Sunarjo, R. A., & Rahardja, U. (2024, August). Ai maturity in business: Bibliometric analysis and sustainable development goals. In 2024 3rd International Conference on Creative Communication and Innovative Technology (ICCIT) (pp. 1-6). IEEE.

CAIDP. (2018). *Universal Guidelines for AI*. Center for AI and Digital Policy. Retrieved from https://www.caidp.org/universal-guidelines-for-ai/

Camacho, M., Perramon, J., Puig-Bosch, X., Dang, V. N., Díaz, O., & Lekadir, K. (2025). Stakeholder engagement : The path to trustworthy AI in healthcare. In *Trustworthy AI in Medical Imaging* (pp. 471–493). Elsevier. DOI: 10.1016/B978-0-44-323761-4.00038-9

Campolo, A., Sanfilippo, M., Whittaker, M., & Crawford, K. (2017). *AI now 2017 report*. AI Now Institute.

CanadaGov. (2025). *Artificial Intelligence and Data Act (AIDA), as part of the Digital Charter Implementation Act: Proposed regulatory framework for high-impact AI systems*. Government of Canada.

Castelvecchi, D. (2016). Can we open the black box of AI? *Nature*, *538*(7623), 20–23. DOI: 10.1038/538020a

Castoriadis, C. (1987). *The Imaginary Institution of Society*. Policy Press.

Cath, C. (2018). Governing artificial intelligence: Ethical, legal and technical opportunities and challenges. *Philosophical Transactions. Series A, Mathematical, Physical, and Engineering Sciences*, *376*(2133), 20180080. DOI: 10.1098/rsta.2018.0080 PMID: 30322996

Cavoukian, A. (2011). Privacy by design: The definitive workshop. *Identity in the Information Society*, *3*(2), 247–251. DOI: 10.1007/s12394-010-0062-y

Center for Strategic and International Studies. (2025, March 21). Unpacking Japan's AI policy with Hiroki Habuka. *Center for Strategic and International Studies*. https://www.csis.org/analysis/unpacking-japans-ai-policy-hiroki-habuka

Chakraborty, D.. (2020). *Fairway method for bias-free training data*. Unknown Journal.

Challoumis, C. (2024, October). Building a sustainable economy-how ai can optimize resource allocation. In *XVI International Scientific Conference* (pp. 190-224).

Chandler, J. D., & Vargo, S. L. (2011). Contextualization and value-in-context : How context frames exchange. *Marketing Theory*, *11*(1), 35–49. DOI: 10.1177/1470593110393713

Chandra, B., & Rahman, Z. (2024). Artificial intelligence and value co-creation : A review, conceptual framework and directions for future research. *Journal of Service Theory and Practice*, *34*(1), 7–32. DOI: 10.1108/JSTP-03-2023-0097

Charllo, B. (2024). AI empowerment in SMEs: Bridging the post-pandemic recovery and innovation gap. *Journal of Artificial Intelligence & Cloud Computing*. https://onlinescientificresearch.com/articles/ai-empowerment-in-smes-bridging-the-postpandemic-recovery-and-innovation-gap.pdf

Chen, J., Storchan, V., & Kurshan, E. (2021). Beyond Fairness Metrics: Roadblocks and Challenges for Ethical AI in Practice. *arXiv preprint arXiv:2108.06217*. https://arxiv.org/abs/2108.06217

Chen, J., Zhang, Y., & Liu, Z. (2025). Unlocking the power of algorithmic recommendations : The effect of recommendation characteristics on users' willingness to value co-creation. *Current Psychology (New Brunswick, N.J.)*, *44*(3), 1–19. DOI: 10.1007/s12144-024-07175-y

Chen, L., Chang, L., Lin, H., Tu, J., Zhou, Y., & Han, Y. (2024). Savior or saboteur? A nationwide study on digital economy and depression in China. *Journal of Affective Disorders*, *365*, 578–586. DOI: 10.1016/j.jad.2024.08.147 PMID: 39187198

Cheong, I., Caliskan, A., & Kohno, T. (2024). Safeguarding human values: Rethinking US law for generative AI's societal impacts. *AI and Ethics*, •••, 1–27. PMID: 40352581

China State Council. (2017). *New generation artificial intelligence development plan*. Retrieved from http://www.gov.cn

Chinta, S. V., Wang, Z., Zhang, X., Viet, T. D., Kashif, A., Smith, M. A., & Zhang, W. (2024). Ai-driven healthcare: A survey on ensuring fairness and mitigating bias. *arXiv preprint arXiv:2407.19655*.

Choi, Y. (2023). GeoAI: Integration of Artificial Intelligence, Machine Learning, and Deep Learning with GIS. *Applied Sciences (Basel, Switzerland)*, *13*(6), 3895. DOI: 10.3390/app13063895

Chopra, A., Patel, H., Rajput, D. S., & Bansal, N. (2024). Empowering Inclusive Education: Leveraging AI-ML and Innovative Tech Stacks to Support Students with Learning Disabilities in Higher Education. In *Applied Assistive Technologies and Informatics for Students with Disabilities* (pp. 255 275). Springer Nature Singapore. DOI: 10.1007/978-981-97-0914-4_15

Chouldechova, A. (2017). Fair prediction with disparate impact: A study of bias in recidivism prediction instruments. *Big Data*, *5*(2), 153–163. DOI: 10.1089/big.2016.0047 PMID: 28632438

Chui, M., Manyika, J., & Muir, M. (2016). *Where machines could replace humans—and where they can't (yet)*. McKinsey Quarterly. Retrieved from https://www.mckinsey.com/business-functions/mckinsey-digital/our-insights/where-machines-could-replace-humans-and-where-they-cant-yet

Citadel. (2024, June 25). Japan's AI guidelines for business: What it means for you. *Citadel AI*. https://citadel-ai.com/blog/2024/06/25/ai-guideline-for-business-2024/

Civelek, M. E. (2018). Indispensables of Structural Equation Modeling. In *Zea Books*. https://doi.org/DOI: 10.20014/k2sj1hr5

Clifford Chance. (2023, October 5). Japanese law issues surrounding generative AI: ChatGPT, Bard and beyond. *Clifford Chance*. https://www.cliffordchance.com/insights/resources/blogs/talking-tech/en/articles/2023/10/Japanese-Law-Issues-Surrounding-Generative-AI.html

Cobanaj, M., Corti, C., Dee, E. C., McCullum, L., Boldrini, L., Schlam, I., Tolaney, S. M., Celi, L. A., Curigliano, G., & Criscitiello, C. (2024). Advancing equitable and personalized cancer care: Novel applications and priorities of artificial intelligence for fairness and inclusivity in the patient care workflow. In *European Journal of Cancer* (Vol. 198). DOI: 10.1016/j.ejca.2023.113504

Cockburn, I. M., Henderson, R., & Stern, S. (2018). *The impact of artificial intelligence on innovation* (Vol. 24449). National bureau of economic research Cambridge, MA, USA.

Coeckelbergh, M. (2020). *AI Ethics*. The MIT Press.

Coeckelbergh, M. (2020). *AI ethics*. MIT Press. DOI: 10.7551/mitpress/12549.001.0001

Coeckelbergh, M. (2024). *Why AI undermines democracy and what to do about it*. John Wiley & Sons.

Cole, M. D. (2024). AI Regulation and Governance on a Global Scale: An Overview of International, Regional and National Instruments. *Journal of AI Law and Regulation*, *1*(1), 126–142. DOI: 10.21552/aire/2024/1/16

Cole, S. A., & Bertenthal, A. (2017). Science, Technology, Society, and Law. *Annual Review of Law and Social Science*, *13*(1), 351–371. DOI: 10.1146/annurev-lawsocsci-110316-113550

Colorado General Assembly. (2024, May 17). Senate Bill 24-205: Consumer protections in interactions with artificial intelligence systems. *General Assembly of the State of Colorado*. https://leg.colorado.gov/sites/default/files/2024a_205_signed.pdf

Conn, A. (2017). An Open Letter to the United Nations Convention on Certain Conventional Weapons. *Future of Life Institute. Available online at:* https://futureoflife.org/2017/08/20/autonomous-weapons-open-letter-2017/*(accessed March 16, 2022)*.

Costa, S. (2023). Artificial intelligence in education: Challenges and opportunities in learning.

Costa-Barbosa, A., Herlo, B., & Joost, G. (2024). Digital Sovereignty in times of AI: between perils of hegemonic agendas and possibilities of alternative approaches. *Liinc em Revista, 20*(2).

Costa, E., & Halpern, D. (2019). The behavioural science of online harm and manipulation, and what to do about it: An exploratory paper to spark ideas and debate. *The Behavioural Insights Team Report*, *2019*, 1–82.

Couclelis, H. (1998). Geocomputation and Space. *Environment and Planning. B, Planning & Design*, *25*(7), 41–47. DOI: 10.1177/239980839802500708

Council of Europe. (2024). *The Framework Convention on Artificial Intelligence*. Retrieved from https://www.coe.int/en/web/artificial-intelligence/the-framework-convention-on-artificial-intelligence

Cowgill, B., Dell'Acqua, F., & Deng, S. (2020). Biased programmers? Or biased data? A field experiment in operationalizing AI ethics. *Columbia Business School Research Paper No. 20-91*. https://doi.org/DOI: 10.2139/ssrn.3672274

Cowgill, B., Dell'Acqua, F., & Deng, S. (2021). Biased programmers? Or biased data? A field experiment in operationalizing AI ethics. *Vanderbilt Law Review*, *74*(6), 1565–1604.

Crampton, J. (1995). The Ethics of GIS. *Cartography and Geographic Information Systems*, *22*(1), 84–89. DOI: 10.1559/152304095782540546

Crampton, J. (2010). *Mapping: A Critical Introduction to Cartography and GIS*. John Wiley and Sons., DOI: 10.1002/9781444317411

Crawford, K. (2021). *ATLAS OF AI: power, politics, and the planetary costs of artificial intelligence*. Yale University Press.

Crawford, K. (2021). *The atlas of AI: Power, politics, and the planetary costs of artificial intelligence*. Yale University Press.

Creemers, R. (2018). China's social credit system: An evolving practice of control. *Chinese Journal of Communication*, *11*(1), 1–17. DOI: 10.1080/17544750.2018.1443264

Creemers, R. (2021). China's Social Credit System: An evolving practice of control. *Asian Studies Review*, *45*(4), 573–591. DOI: 10.1080/10357823.2021.1952081

Crystal-Ornelas, R., Varadharajan, C., O'Ryan, D., Beilsmith, K., Bond-Lamberty, B., Boye, K., Burrus, M., Cholia, S., Christianson, D. S., Crow, M., Damerow, J., Ely, K. S., Goldman, A. E., Heinz, S. L., Hendrix, V. C., Kakalia, Z., Mathes, K., O'Brien, F., Pennington, S. C., & Agarwal, D. A. (2022). Enabling FAIR data in Earth and environmental science with community-centric (meta)data reporting formats. *Scientific Data*, *9*(1), 700. DOI: 10.1038/s41597-022-01606-w

Cummings, M. L. (2019). The Uber fatality: A case study in autonomy and unintended consequences. *Harvard Data Science Review*, *1*(1). Advance online publication. DOI: 10.1162/99608f92.7a61f20c

Cyberspace Administration of China. (2023, July 13). Interim Measures for the Management of Generative AI services. *Cyberspace Administration of China*. https://www.cac.gov.cn/2023-07/13/c_1690898327029107.htm

Danesh, D. O., & Huber, T. P. (2022). A literature review of leadership training and a novel conceptual model of leader-member exchange theory for new dentists. *Journal of Health Organization and Management*, *36*(4), 417–431. DOI: 10.1108/JHOM-05-2021-0192 PMID: 34951297

Dastin, J. (2018). Amazon scraps secret AI recruiting tool that showed bias against women. *Reuters*. https://www.reuters.com/article/us-amazon-com-jobs-automation-idUSKCN1MK08G

Dastin, J. (2018a). Amazon scrapped "sexist" AI recruiting tool that showed bias against women. *Reuters*. Retrieved from https://www.reuters.com/article/us-amazon-com-jobs-automation-insight-idUSKCN1MK08G

Dastin, J. (2018b). *AI ethics: Why the world needs to design artificial intelligence with values in mind.* Forbes. https://www.forbes.com

Dastin, J. (2020). "AI ethics courses are growing at universities around the world." *Reuters.* Retrieved from https://www.reuters.com

Davenport, T. H., & Ronanki, R. (2018). Artificial intelligence for the real world. *HBR'S 10 MUST*, 67.

Davenport, T., Guha, A., Grewal, D., & Bressgott, T. (2020). How artificial intelligence will change the future of marketing. *Journal of the Academy of Marketing Science*, *48*(1), 24–42. DOI: 10.1007/s11747-019-00696-0

Degbelo, A. & Kray, C. (2018). *Intelligent geovisualizations for open government data (vision paper). 5*, 77–80._DOI: 10.1145/3274895.3274940

Deloitte. (2021). *Mitigating algorithmic bias in financial services: Strategies for fair AI deployment.* Deloitte Insights. https://www2.deloitte.com/insights

Department of Industry. Science and Resources. (2019, November 7). Australia's AI ethics principles. *Australian Government.* https://www.industry.gov.au/publications/australias-artificial-intelligence-ethics-principles

Department of Industry. Science and Resources. (2025, October 21). Guidance for AI Adoption. *Australian Government.* https://www.industry.gov.au/publications/guidance-for-ai-adoption

Di Nuovo, A. (2024). Letter to the Editor: "How Can Biomedical Engineers Help Empower Individuals With Intellectual Disabilities? The Potential Benefits and Challenges of AI Technologies to Support Inclusivity and Transform Lives." In *IEEE Journal of Translational Engineering in Health and Medicine* (Vol. 12). DOI: 10.1109/JTEHM.2023.3331977

Dieterle, E., Dede, C., & Walker, M. (2024). The cyclical ethical effects of using artificial intelligence in education. *AI & Society*, *39*(2), 633–643. DOI: 10.1007/s00146-022-01497-w PMID: 36185064

Digital Agency. (2025, May 27). The Guideline for Japanese governments' procurements and utilizations of generative AI for the sake of evolution and innovation of public administration. *Digital Agency.* https://www.digital.go.jp/assets/contents/node/basic_page/field_ref_resources/e2a06143-ed29-4f1d-9c31-0f06fca67afc/6e45a64f/20250527_resources_standard_guidelines_guideline_04.pdf

Digital Dubai (Government of Dubai). (2025). AI system ethics self-assessment tool. *Government of Dubai.* https://www.digitaldubai.ae/self-assessment

Digital Dubai, & the Dubai Future Foundation. (2025, April 21). Dubai state of AI report. Government *of Dubai*. https://www.digitaldubai.ae/docs/default-source/publications/dubai-state-of-ai-report.pdf

Digital Watch Observatory. (2024). *Overview of AI policy in 10 jurisdictions*. Retrieved from https://dig.watch/updates/overview-of-ai-policy-in-10-jurisdictions

DigitalBriefing. (2024, March 27). India's Regulation of AI and large language models. *India Briefing*. https://www.india-briefing.com/news/india-regulation-of-ai-and-large-language-models-31680.html

Dignum, V. (2018). *Responsible artificial intelligence: Developing and using AI in a responsible way*. Springer.

Dignum, V. (2019). *Responsible artificial intelligence : How to develop and use AI in a responsible way* (Vol. 2156). Springer. DOI: 10.1007/978-3-030-30371-6

Diligent Corporation. (2024). *AI regulations around the world*. Retrieved from https://www.diligent.com/resources/guides/ai-regulations-around-the-world

Dilmaghani, S., Brust, M. R., Danoy, G., Cassagnes, N., Pecero, J., & Bouvry, P. (2019, December 1). *Privacy and Security of Big Data in AI Systems: A Research and Standards Perspective*. IEEE Xplore._DOI: 10.1109/BigData47090.2019.9006283

Ding, J. (2018). *Deciphering China's AI dream: The context, components, capabilities, and consequences of China's strategy to lead the world in AI*. Future of Humanity Institute, University of Oxford. Retrieved from https://www.fhi.ox.ac.uk

Dobson, J. E. (1983). Automated geography. *The Professional Geographer*, *35*(2), 135–143. DOI: 10.1111/j.0033-0124.1983.00135.x

Dorigo, T., Brown, G. D., Casonato, C., Cerdà, A., Ciarrochi, J., Da Lio, M., & Yazdanpanah, N. (2025). Artificial intelligence in science and society: The vision of USERN. *IEEE Access : Practical Innovations, Open Solutions*.

Doshi-Velez, F., & Kim, B. (2017). Towards a rigorous science of interpretable machine learning. *arXiv preprint arXiv:1702.08608*.

Dube, A. (2019). *Proposal for a federal job guarantee program*. Brookings Institution. https://www.brookings.edu/research/proposal-for-a-federal-job-guarantee/

Du, S., & Xie, C. (2021). Paradoxes of artificial intelligence in consumer markets : Ethical challenges and opportunities. *Journal of Business Research*, *129*, 961–974. DOI: 10.1016/j.jbusres.2020.08.024

Dwork, C., & Roth, A. (2014). The algorithmic foundations of differential privacy. *Foundations and trends® in theoretical computer science, 9*(3–4), 211-407.

Eaton, L., Bridgman, T., & Cummings, S. (2024). Advancing the democratization of work: A new intellectual history of transformational leadership theory. *Leadership, 20*(3), 125–143. DOI: 10.1177/17427150241232705

Economic Times. (2025). GenAI joins the checklist; 64% of Indian companies are making it a priority. *Economic Times*. https://economictimes.indiatimes.com/tech/artificial-intelligence/genai-joins-the-checklist-64-of-indian-companies-are-making-it-a-priority-report/articleshow/121952746.cms

Edeni, C. A., Adeleye, O. O., & Adeniyi, I. S.Chima Abimbola EdeniOlabisi Oluwakemi AdeleyeIdowu Sulaimon Adeniyi. (2024). The role of AI-enhanced tools in overcoming socioeconomic barriers in education: A conceptual analysis. *World Journal of Advanced Research and Reviews, 21*(3), 944–951. Advance online publication. DOI: 10.30574/wjarr.2024.21.3.0780

Elwarraki, O., Aammou, S., Jdidou, Y., & Lahiassi, J. (2023). Toward the future of personalized learning: Emerging trends and challenges in recommendation systems. ICERI2023 Proceedings, 7226-7235.

Embarak, O. H., & Hawarna, S. (2024). Automated AI-driven system for early detection of at-risk students. *Procedia Computer Science, 231*, 151–160. DOI: 10.1016/j.procs.2023.12.187

Ernst & Young. (2024). *ISO 42001: paving the way for ethical AI.* Retrieved from https://www.ey.com/en_us/insights/ai/iso-42001-paving-the-way-for-ethical-ai

eSafety Commissioner. (2023, August). Tech trends position statement: Generative AI. *eSafety Commissioner.* https://www.esafety.gov.au/sites/default/files/2023-08/Generative%20AI%20-%20Position%20Statement%20-%20August%202023%20.pdf?v=1753223601081

Esya Centre. (2025, February). Response to The Report on AI governance guidelines development. *Esya Centre*. https://www.esyacentre.org/s/Esya_Centre_-_AI_Report_Response_Final1-1-4fj2.pdf

Etzioni, A., & Etzioni, O. (2017). Incorporating ethics into artificial intelligence. *The Journal of Ethics, 21*(4), 403–418. DOI: 10.1007/s10892-017-9252-2

Eubanks, V. (2018). *Automating inequality: How high-tech tools profile, police, and punish the poor*. St. Martin's Press.

European Commission. (2019). *Ethics Guidelines for Trustworthy AI.* High-Level Expert Group on Artificial Intelligence. Retrieved from https://www.europarl.europa.eu/cmsdata/196377/AI%20HLEG_Ethics%20Guidelines%20for%20Trustworthy%20AI.pdf

European Commission. (2020). *Ethics guidelines for trustworthy AI.* Retrieved from https://ec.europa.eu/futurium/en/ai-alliance-consultation

European Commission. (2020). *White Paper on Artificial Intelligence: A European approach to excellence and trust.* Publications Office of the European Union.

European Commission. (2021). *Proposal for a Regulation laying down harmonised rules on artificial intelligence (Artificial Intelligence Act).* https://digital-strategy.ec.europa.eu

European Commission. (2021). *Proposal for a regulation laying down harmonized rules on artificial intelligence (Artificial Intelligence Act).* https://eur-lex.europa.eu

European Commission. (2021). *Proposal for a regulation laying down harmonized rules on artificial intelligence (Artificial Intelligence Act).* Retrieved from https://digital-strategy.ec.europa.eu

European Commission. (2021a). *Proposal for a regulation laying down harmonized rules on artificial intelligence (Artificial Intelligence Act).* Retrieved from https://eur-lex.europa.eu

European Commission. (2021b). *Artificial Intelligence Act: Proposal for a regulation laying down harmonized rules on artificial intelligence (Artificial Intelligence Act) and amending certain Union legislative acts.* https://ec.europa.eu

European Digital SME Alliance. (2025, April 23). From regulation to action: AI Act compliance for digital SMEs. *The European Digital SME Alliance.* https://www.digitalsme.eu/events/from-regulation-to-action-ai-act-compliance-for-digital-smes

European Economic and Social Committee. (2021, June). Boosting the use of artificial intelligence in Europe's micro, small and medium-sized enterprises. *European Economic and Social Committee.* https://www.eesc.europa.eu/sites/default/files/files/qe-02-21-953-en-n.pdf

European Parliament & Council. (2016). *Regulation (EU) 2016/679 (General Data Protection Regulation, GDPR).* Retrieved from https://eur-lex.europa.eu

European Union. (2024, July 25). AI innovation package & GenAI4EU. *European Union.* https://digital-strategy.ec.europa.eu/en/factpages/ai-innovation-package

European Union. (2024, June 13). Artificial Intelligence Act (Regulation (EU) 2024/1689). *European Commission*. https://artificialintelligenceact.eu

European Union. (2025, February 19). Small businesses' guide to the AI Act. *European Union*. https://artificialintelligenceact.eu/small-businesses-guide-to-the-ai-act/

European Union. (2025, May 2). AI regulatory sandboxes for AI: Member State overview. *EU Artificial Intelligence Act*. https://artificialintelligenceact.eu/ai-regulatory-sandbox-approaches-eu-member-state-overview/

European Union. (n.d.). EU AI Act compliance checker. *European Union*. https://artificialintelligenceact.eu/assessment/eu-ai-act-compliance-checker/

FATF. (2021). *Opportunities and challenges of new technologies for AML/CFT*. Financial Action Task Force. https://www.fatf-gafi.org

Fedele, A., Punzi, C., & Tramacere, S. (2024). The ALTAI checklist as a tool to assess ethical and legal implications for a trustworthy AI development in education. *Computer Law & Security Review*, *53*, 105986. DOI: 10.1016/j.clsr.2024.105986

Federal Reserve Board. (2021). *AI in financial services: Risks, benefits, and regulatory considerations*. https://www.federalreserve.gov

Feldstein, S. (2019). *The global expansion of AI surveillance*. Carnegie Endowment for International Peace.

Feldstein, S. (2024). Evaluating Europe's push to enact AI regulations: How will this influence global norms? *Democratization*, *31*(5), 1049–1066. DOI: 10.1080/13510347.2023.2196068

Feng, Y., & Sexton, J. (2022). Ethical AI governance in China: Balancing state control and global engagement. *AI & Society*, *37*(4), 1012–1025. DOI: 10.1007/s00146-021-01290-3

Ferrara, E. (2020). Disinformation and social bot operations in the run-up to the 2020 US election. *Harvard Kennedy School Misinformation Review*, *1*(3), 1–12. DOI: 10.37016/mr-2020-022

Ferretti, A., & Vayena, E. (2022). In the Shadow of Privacy: Overlooked Ethical Concerns in COVID-19 Digital Epidemiology. *Epidemics*, *41*, 100652. DOI: 10.1016/j.epidem.2022.100652

Financial Stability Board. (2017). *Artificial intelligence and machine learning in financial services*. https://www.fsb.org

Financial Times. (2025). *Business school teaching case study: can benefit corporations solve AI governance challenges?* Retrieved from https://www.ft.com/content/3eeb7661-2b63-4339-b446-d1f6e45a2558

Floridi, L. (2018). *The ethics of artificial intelligence.* Oxford University Press.

Floridi, L. (2021). *Ethics, governance, and policies in AI.* Springer.

Floridi, L., Buttaboni, C., Hine, E., Novelli, C., Schroder, T., & Shanklin, G. (2025). *Open-Source AI made in the EU: Why it is a Good Idea.* DOI: 10.1007/s11023-025-09728-x

Floridi, L., & Chiriatti, M. (2020). GPT-3: Its nature, scope, limits, and consequences. *Minds and Machines, 30*(4), 681–694. DOI: 10.1007/s11023-020-09548-1

Floridi, L., & Cowls, J. (2019). A unified framework of five principles for AI in society. *Harvard Data Science Review, 1*(1). Advance online publication. DOI: 10.1162/99608f92.8cd550d1

Floridi, L., Cowls, J., Beltrametti, M., Chatila, R., Chazerand, P., Dignum, V., Luetge, C., Madelin, R., Pagallo, U., Rossi, F., Schafer, B., Valcke, P., & Vayena, E. (2018). AI4People—An ethical framework for a good AI society: Opportunities, risks, principles, and recommendations. *Minds and Machines, 28*(4), 689–707. DOI: 10.1007/s11023-018-9482-5 PMID: 30930541

Fontes, F.. (2022). *On surveillance misuse.* Unknown Journal.

Ford, M. (2015). *Rise of the robots: Technology and the threat of a jobless future.* Basic Books.

Fractal. (2021). *Transforming business with ethical AI practices.* Fractal. https://fractal.ai/casestudies/transforming-business-with-ethical-ai-practices/

Frank, M. R., Autor, D., Bessen, J. E., Brynjolfsson, E., Cebrian, M., Deming, D. J., Feldman, M., Groh, M., Lobo, J., Moro, E., Wang, D., Youn, H., & Rahwan, I. (2019). Toward understanding the impact of artificial intelligence on labor. *Proceedings of the National Academy of Sciences of the United States of America, 116*(14), 6531–6539. DOI: 10.1073/pnas.1900949116 PMID: 30910965

Frazier, A. E., Nelson, T., Kedron, P., Shook, E., Dodge, S., Murray, A., Goodchild, M., Battersby, S., Blanford, J. I., Claramunt, C., Holler, J., Koylu, C., Lee, A., Manson, S., Şalap-Ayça, S., Wilson, J. P., Zhao, B., Bennett, L., Cabrera, C., & Xu, W. (2025). Rethinking GIScience Education in an Age of Disruptions. *Transactions in GIS, 29*(2), e70048. Advance online publication. DOI: 10.1111/tgis.70048

Frey, C. B., & Osborne, M. A. (2017). The future of employment: How susceptible are jobs to computerization? *Technological Forecasting and Social Change, 114*, 254–280. DOI: 10.1016/j.techfore.2016.08.019

Friedman, B., Kahn, P. H., & Borning, A. (2006). Value sensitive design and information systems. In *Human-Computer Interaction in Management Information Systems* (pp. 348–372). M.E. Sharpe.

Fukizi, K. Y. (2023). Collaborative Decision-Making Assistant for Healthcare Professionals: A Human-Centered AI Prototype Powered by Azure Open AI. *COMPASS 2023 - Proceedings of the ACM SIGCAS/SIGCHI Conference on Computing and Sustainable Societies*. DOI: 10.1145/3588001.3609370

Fukuda-Parr, S., & Gibbons, E. (2021). Emerging consensus on 'ethical AI' : Human rights critique of stakeholder guidelines. *Global Policy, 12*(S6), 32–44. DOI: 10.1111/1758-5899.12965

Gabriel, I. (2020). Artificial intelligence, values, and alignment. *Minds and Machines, 30*(3), 411–437. DOI: 10.1007/s11023-020-09539-2

Gaftandzhieva, S., Talukder, A., Gohain, N., Hussain, S., Theodorou, P., Salal, Y. K., & Doneva, R. (2022). Exploring online activities to predict the final grade of student. *Mathematics, 10*(20), 3758. DOI: 10.3390/math10203758

Gao, S. (2021). Geospatial Artificial Intelligence (GeoAI). *Geography (Sheffield, England)*. Advance online publication. DOI: 10.1093/obo/9780199874002-0228

Gartner, Inc. (2023). More than 80% of enterprises will have used generative AI APIs or deployed generative AI-enabled applications by 2026. Retrieved from: https://www.gartner.com/en/newsroom/press-releases/2023-10-11-gartner-says-more-than-80-percent-of-enterprises-will-have-used-generative-ai-apis-or-deployed-generative-ai-enabled-applications-by-2026?utm_source=chatgpt.com

Gasser, U. (2020). *The role of AI education and public awareness in ethical AI development*. Springer.

Gasser, U., Almeida, V. A., & Palfrey, J. (2017). Accountability in algorithmic decision-making. *Communications of the ACM, 61*(6), 56–62.

Gasser, U., Humenik, M., Kahn, L., & Puschmann, C. (2020). The rise of AI and the need for global governance. *Journal of Global Governance, 26*(4), 509–523. DOI: 10.1057/s41310-020-00099-9

Gentelet, K., & Mizrahi, S. K. (2024). A Human-Centered Approach to AI Governance: Operationalizing Human Rights Through Citizen Participation. In *Human-Centered AI* (pp. 215–230). Chapman and Hall/CRC. DOI: 10.1201/9781003320791-24

Georgiadou, K. (2024). Solidarity, Gender, Dialogue in the AI era. In *Kritische Pädagogik und Bildungsforschung: Anschlüsse an Paulo Freire* (pp. 477–489). Springer Fachmedien Wiesbaden. DOI: 10.1007/978-3-658-41415-3_23

Ghamisi, P., Yu, W., Marinoni, A., Gevaert, C.M., Persello, C., Selvakumaran, S., Girotto, M., Horton, B.P., Rufin, P., Hostert, P., Pacifici, F., & Atkinson, P.M. (2024). Responsible AI for Earth Observation. ArXiv, abs/2405.20868.

Ghandour, A., Woodford, B. J., & Abusaimeh, H. (2024). Ethical Considerations in the use of ChatGPT: An exploration through the lens of five moral dimensions. *IEEE Access : Practical Innovations, Open Solutions*, *12*, 60682–60693. DOI: 10.1109/ACCESS.2024.3394243

Ghosh, I., & Sanyal, M. K. (2021). Introspecting predictability of market fear in Indian context during COVID-19 pandemic : An integrated approach of applied predictive modelling and explainable AI. *International Journal of Information Management Data Insights*, *1*(2), 100039. DOI: 10.1016/j.jjimei.2021.100039

Gillespie, T. (2017). Algorithmic culture. *Media Culture & Society*, *39*(3), 336–347. DOI: 10.1177/0163443717702978

Gilpin, L. H., Caruana, R., Gehrke, J., & Koch, P. (2018). Explaining explanations: An overview of interpretability of machine learning. *Proceedings of the 2018 ICML Workshop on Human Interpretability in Machine Learning*, 1–15. DOI: 10.1109/DSAA.2018.00018

Global Legal Insights. (2025). AI, machine learning & big data laws 2025. *Global Legal Insights*. https://www.globallegalinsights.com/practice-areas/ai-machine-learning-and-big-data-laws-and-regulations/singapore

Global Partnership on Artificial Intelligence. (2023). Responsible AI working group recommendations. *Global Partnership on Artificial Intelligence*. https://gpai.ai/projects/responsible-ai/

Goddard, K., Roudsari, A., & Wyatt, J. C. (2012). Automation bias: A systematic review of frequency, effect mediators, and mitigators. *Journal of the American Medical Informatics Association : JAMIA*, *19*(1), 121–127. DOI: 10.1136/amiajnl-2011-000089 PMID: 21685142

Goirand, M., Austin, E., & Clay-Williams, R. (2024). Bringing clarity and transparency to the consultative process underpinning the implementation of an ethics framework for AI-based healthcare applications: A qualitative study. *AI and Ethics*, •••, 1–21.

Golda, A.. (2024). *Privacy and security concerns in generative AI: A comprehensive survey*. IEEE., DOI: 10.1109/ACCESS.2024.3381611

Göllner, S., & Tropmann-Frick, M. (2023a). Bridging the Gap between Theory and Practice: Towards Responsible AI Evaluation. *CEUR Workshop Proceedings*, •••, 3580.

Göllner, S., & Tropmann-Frick, M. (2023b). VERIFAI - A Step Towards Evaluating the Responsibility of AI-Systems. *Lecture Notes in Informatics (LNI), Proceedings - Series of the Gesellschaft Fur Informatik (GI), P-331*. DOI: 10.18420/BTW2023-63

Gondwe, G. (2025). *Can AI outsmart fake news?* Emerging Media.

Gongane, V. U., Munot, M. V., & Anuse, A. D. (2024). Explainable AI techniques for detection of fake news and hate speech. *Journal of Computational Social Science*, *7*(1), 587–623. Advance online publication. DOI: 10.1007/s42001-024-00248-9

Gonik, J., Le, J., Viswanathan, A., & Zhang, Y. (2020). Cybergan: Generating high-fidelity cybersecurity data with generative adversarial networks. *Accelerating Space Commerce, Exploration, and New Discovery Conference, ASCEND 2020*.

Goodchild, M. F. (1992). Geographical information science. *International Journal of Geographical Information Systems*, *6*(1), 31–45. DOI: 10.1080/02693799208901893

Goodchild, M. F. (2007). Citizens as Sensors: The World of Volunteered Geography. *GeoJournal*, *69*(4), 211–221. DOI: 10.1007/s10708-007-9111-y

Goodchild, M. F. (2009). Geographic information systems and science: Today and tomorrow. *Annals of GIS*, *15*(1), 3–9. DOI: 10.1080/19475680903250715

Goodchild, M. F. (2020). GIScience, geography, and ethics. *Annals of GIS*, *26*(1), 1–9. DOI: 10.1080/19475683.2020.1734952

Goodfellow, I., Bengio, Y., & Courville, A. (2016). *Deep learning*. MIT Press.

Goodman, B., & Flaxman, S. (2017). European Union regulations on algorithmic decision-making and a "right to explanation.". *AI Magazine*, *38*(3), 50–57. DOI: 10.1609/aimag.v38i3.2741

Google, A. I. (2021). Explainable AI in financial applications: Enhancing transparency in credit scoring and risk assessment. *Google AI Blog*. https://ai.googleblog.com

Gordon, J.-S. • Nyholm, S. (2021). Ethics of Artificial Intelligence, Internet Encyclopedia of Philosophy, 2021

Government Digital Service. (2025, February 10). A playbook for the UK government. *UK Government.* https://www.gov.uk/government/publications/ai-playbook-for-the-uk-government

Government of Canada. (2019). *Directive on Automated Decision-Making.* Treasury Board Secretariat of Canada. https://www.canada.ca/en/treasury-board-secretariat/services/automated-decision-making-directive.html

Government of Canada. (2023, March 13). The Artificial Intelligence and Data Act (AIDA) – Companion document. *Innovation, Science and Economic Development Canada.* https://ised-isde.canada.ca/site/innovation-better-canada/en/artificial-intelligence-and-data-act-aida-companion-document

GPAI. (2021). Global Partnership on AI. Retrieved from https://gpai.ai

Grant, N., & Weise, K. (2023). In AI race, Microsoft and Google choose speed over caution. *The New York Times, 7.*

Green, B., & Chen, Y. (2019). The principles and limits of algorithm-in-the-loop decision making. *Proceedings of the ACM on human-computer interaction, 3*(CSCW), 1-24. DOI: 10.1145/3359152

Gridchina, N., Savvina, N., & Zavyalov, S. (2023, June). Topical issues of the use of artificial intelligence technologies in education and law. In 2023 3rd International Conference on Technology Enhanced Learning in Higher Education (TELE) (pp. 8-11). IEEE.

Grönroos, C. (2008). Adopting a service business logic in relational business-to-business marketing : Value creation, interaction and joint value co-creation. *Otago forum, 2*(9), 269-287.

Grönroos, C. (2011). A service perspective on business relationships : The value creation, interaction and marketing interface. *Industrial Marketing Management, 40*(2), 240–247. DOI: 10.1016/j.indmarman.2010.06.036

Guess, A., Nyhan, B., & Reifler, J. (2020). Exposure to untrustworthy websites in the 2016 US election. *Nature Human Behaviour, 4*(5), 472–480. DOI: 10.1038/s41562-020-0833-x

Guidotti, R., Monreale, A., Ruggieri, S., Turini, F., Giannotti, F., & Pedreschi, D. (2018). A survey of methods for explaining black box models. *ACM Computing Surveys, 51*(5), 1–42. DOI: 10.1145/3236009

Gümüş, A. E., & Gürkaynak, N. (2024). Setting the research agenda: Emotion recognition tools for AI-generated marketing and CRM. In Proceedings of the Medical Technologies Congress. Muğla, Türkiye. https://biyoklinikder.org/TIPTEKNO24-Bildiriler/61.pdf

Gunning, D., Stefik, M., Choi, J., Miller, T., Stumpf, S., & Yang, G. Z. (2019). XAI—Explainable artificial intelligence. *Science Robotics*, *4*(37), eaay7120. Advance online publication. DOI: 10.1126/scirobotics.aay7120

Guo, F., Li, T., & Cunningham, C. J. (2025, May). One year in the classroom with ChatGPT: Empirical insights and transformative impacts. []). Frontiers Media SA.]. *Frontiers in Education*, *10*, 1574477. DOI: 10.3389/feduc.2025.1574477

Guo, L., Flathmann, C., Anaraky, R., McNeese, N., & Knijnenburg, B. (2022, June). The effect of recommendation source and justification on professional development recommendations for high school teachers. In *Proceedings of the 33rd ACM conference on hypertext and social media* (pp. 175-185). DOI: 10.1145/3511095.3531280

Guo, Y., Shi, J., & Tu, Y. (2022). Ethical risks in AI-enabled financial services: A review and research agenda. *Journal of Business Ethics*, *180*(3), 813–834.

Gupta, J., & An, S. (2023). *Shashi Shekhar*. Towards Responsible Spatial Data Science and Geo-AI., DOI: 10.1145/3607947.3608060

Gursoy, S. (2020). The role of education in the ethical implications of AI technology. *International Journal of Artificial Intelligence*, *19*(2), 67–75.

Haefner, N., Wincent, J., Parida, V., & Gassmann, O. (2021). Artificial intelligence and innovation management : A review, framework, and research agenda . *Technological Forecasting and Social Change*, *162*, 120392. DOI: 10.1016/j.techfore.2020.120392

Hajli, N.. (2021). Social bots and the spread of disinformation. *British Journal of Management*. Advance online publication. DOI: 10.1111/1467-8551.12554

Hanemaayer, A. (2021). Don't touch my stuff: Historicising resistance to AI and algorithmic computer technologies in medicine. *Interdisciplinary Science Reviews*, *46*(1-2), 126–137. DOI: 10.1080/03080188.2020.1840222

Hanny, D., & Resch, B. (2025). Multimodal GeoAI: An integrated spatio-temporal topic-sentiment model for the analysis of geo-social media posts for disaster management. *International Journal of Applied Earth Observation and Geoinformation*, *139*, 104540–104540. DOI: 10.1016/j.jag.2025.104540

Hao, K. (2019). Apple Card is being investigated for discriminating against women. *MIT Technology Review*. Retrieved from https://www.technologyreview.com/2019/11/10/132086/apple-card-discrimination-goldman-sachs/

Harper, E., & Millard, J. (2024). Global Approaches to Digital Sovereignty: Legal Mechanisms for Managing Data and Digital Infrastructure. *Legal Studies in Digital Age*, *3*(4), 36–46.

Harris, S., & Jirotka, M. (2019). The ethical implications of AI and how to address them: A policy perspective. *Journal of AI & Ethics*, *1*(1), 37–53.

Hartzog, W. (2018). *Privacy's blueprint: The battle to control the design of new technologies*. Harvard University Press.

Haugeland, J., Craver, C. F., & Klein, C. (Eds.). (2023). *Mind Design III: Philosophy, Psychology, and Artificial Intelligence*. MIT press.

Hawkins, W., & Mittelstadt, B. (2023). *The ethical ambiguity of AI data enrichment : Measuring gaps in research ethics norms and practices*. 261-270.

Hayes, C. M. (2023). Generative artificial intelligence and copyright : Both sides of the black box. *Available at SSRN* 4517799. DOI: 10.2139/ssrn.4517799

Hazeleger, W., Aerts, J.P.M., Bauer, P. *et al.* (2024). Digital twins of the Earth with and for humans. *Commun Earth Environ 5*, 463 DOI: 10.1038/s43247-024-01626-x

Hendriks, A., Karhunmaa, K., & Delvenne, P. (2025). Shaping the future: A conceptual review of sociotechnical imaginaries. *Futures*, *103607*, 103607. Advance online publication. DOI: 10.1016/j.futures.2025.103607

Henry, J. V. (2024). Theorizing 'The Gap'twenty years later: Global development, design, and speculative ethics in edtech research. *Postdigital Science and Education*, *6*(1), 231–258. DOI: 10.1007/s42438-023-00429-1

Hermanto, Y. B., & Srimulyani, V. A. (2022). The influences of corporates justice on worker accomplishment using dimension of corporates citizenship behavior as mediation. *Sustainability*, *14*(20), 20322. DOI: 10.3390/su142013322

He, Y. (2024). Artificial intelligence and religious freedom: Divergent paths converging on economic expansion. *Humanities & Social Sciences Communications*, *11*(1), 1–12. DOI: 10.1057/s41599-024-02642-0

Hill, J. E., Harris, C., & Clegg, A. (2024). Methods for using Bing's AI-powered search engine for data extraction for a systematic review. *Research Synthesis Methods*, *15*(2), 347–353. DOI: 10.1002/jrsm.1689 PMID: 38066713

Hochmair, H. H., Juhász, L., & Kemp, T. (2024, November). Juhász, L., & Kemp, T. (2024). Correctness Comparison of ChatGPT-4, Gemini, Claude-3, and Copilot for Spatial Tasks. *Transactions in GIS*, *28*(7), 2219–2231. Advance online publication. DOI: 10.1111/tgis.13233

Hochmair, H. H., Juhász, L., & Li, H. (2025). Advancing AI-Driven Geospatial Analysis and Data Generation: Methods, Applications and Future Directions. *ISPRS International Journal of Geo-Information*, *14*(2), 56. DOI: 10.3390/ijgi14020056

Hollanek, T., Pi, Y., Fiorini, C., Vignali, V., Peters, D., & Drage, E. (2025, May 22). A toolkit for compliance, a toolkit for justice: Drawing on cross-sectoral expertise to develop a pro-justice EU AI Act toolkit [Preprint]. *arXiv*. https://doi.org//arXiv .2505.17165DOI: 10.1145/3715275.3732078

Holmes, W., Bialik, M., & Fadel, C. (2019). *Artificial Intelligence in Education: Promises and Implications for Teaching and Learning*. Center for Curriculum Redesign.

Holmes, W., Persson, J., Chounta, I.-A., Wasson, B., & Dimitrova, V. (2022). *Artificial intelligence and education : A critical view through the lens of human rights, democracy and the rule of law*. Council of Europe.

Holstein, K., Wortman Vaughan, J., Wallach, H., Daumé, H.III, & Karlen, D. (2019). Improving fairness in machine learning systems: What do industry practitioners need to know? In *Proceedings of the 2019 CHI Conference on Human Factors in Computing Systems* (pp. 1-13). ACM. DOI: 10.1145/3290605.3300830

Holten Møller, N., Shklovski, I., & Hildebrandt, T. T. (2020). *Shifting concepts of value : Designing algorithmic decision-support systems for public services*. 1-12.

Hossain, M. E., & Islam, M. A. (2024). AI and the Future of Education: Philosophical Questions about the Role of Artificial Intelligence in the Classroom. *International Journal of Research and Innovation in Social Science*, *8*(IIIS, 3s), 5541–5547. DOI: 10.47772/IJRISS.2024.803419S

Houfaf-Khoufaf, W., Touya, G., & Le Guilcher, A. (2021). Geographically masking addresses to study COVID-19 clusters. *Cartography and Geographic Information Science*, *51*(2), 242–256. DOI: 10.1080/15230406.2021.1977709

Howard, P. N. (2016). *Computational propaganda*. Oxford Internet Institute.

Hsu, H. Y., Yu, K. C., Wu, J. L., & Hsu, K. S. (2024, November). Sentiment Analysis of Social Support on Treads of Social Media Using Deep Learning Approach. In *2024 IEEE International Conference on E-health Networking, Application & Services (HealthCom)* (pp. 1-3). IEEE.

Huang, C., Mees, O., Zeng, A., & Burgard, W. (2023). *Visual Language Maps for Robot Navigation*. ArXiv.org. /arXiv.2210.05714DOI: 10.1109/ICRA48891.2023.10160969

Huang, X., Tu, Z., Ye, X., & Goodchild, M. (2025). *The Role of Open-Source LLMs in Shaping the Future of GeoAI*. ArXiv.org. https://arxiv.org/abs/2504.17833

Huang, C., Zhang, Z., Mao, B., & Yao, X. (2022). An Overview of Artificial intelligence ethics. *IEEE Transactions on Artificial Intelligence*, *4*(4), 799–819. DOI: 10.1109/TAI.2022.3194503

Huang, K., Joshi, A., Dun, S., & Hamilton, N. (2024). AI regulations. In *Generative AI security: theories and practices* (pp. 61–98). Springer Nature Switzerland. DOI: 10.1007/978-3-031-54252-7_3

Huang, N., Qiu, S., Yang, S., & Deng, R. (2021). Ethics of a leaders and corporates citizenship behavior: Mediation of trust and psychological well-being. *Psychology Research and Behavior Management*, *14*, 705–724. DOI: 10.2147/PRBM.S311856 PMID: 34104011

*Human Technology Institute* (2025, February). In their words: Perspectives and experiences of SMEs using generative AI. *Human Technology Institute, University of Technology Sydney*. https://www.saam.com.au/wp-content/uploads/2025/02/SAAM_In-their-words_perspectives-and-experiences-of-SMEs-using-AI-report-1.pdf

Hu, T., Liu, T., Divyacharan Jarugumalli, V. S., Cheng, S., & Deng, C. (2025). FAIR principles in workflows: A GIScience workflow management system for reproducible and replicable studies. *International Journal of Applied Earth Observation and Geoinformation*, *138*, 104477. DOI: 10.1016/j.jag.2025.104477

Hutson, J., & Plate, D. (2024). The Algorithm of Fear: Unpacking Prejudice Against AI and the Mistrust of Technology. Journal of Innovation and Technology, 2024.

Hu, Y., Gao, S., Lunga, D., Li, W., Newsam, S., & Bhaduri, B. (2019). GeoAI at ACM SIGSPATIAL. *SIGSPATIAL Special*, *11*(2), 5–15. DOI: 10.1145/3377000.3377002

Hu, Y., Goodchild, M., Zhu, A. X., Yuan, M., Aydin, O., Bhaduri, B., Gao, S., Li, W., Lunga, D., & Newsam, S. (2024). A five-year milestone: Reflections on advances and limitations in GeoAI research. *Annals of GIS*, *30*(1), 1–14. DOI: 10.1080/19475683.2024.2309866

IBM. (2020). *AI Fairness 360 Toolkit*. IBM. https://www.ibm.com/blogs/research/2020/09/ai-fairness-360/

IEEE. (2019). *Ethically aligned design: A vision for prioritizing human well-being with autonomous and intelligent systems*. IEEE Standards Association.

IEEE. (2019). *Ethically aligned design: A vision for prioritizing human well-being with autonomous and intelligent systems*. IEEE Standards Association. https://ethicsinaction.ieee.org

IEEE. (2020). *Ethically aligned design: A vision for prioritizing human well-being with autonomous and intelligent systems*. Retrieved from https://ethicsinaction.ieee.org

Ige, A. B., Chukwurah, N., Idemudia, C., & Adebayo, V. I. (2022). Ethical considerations in data governance: Balancing privacy, security, and transparency in data management. Journal of Ethical Data Practices.[Year unavailable].[Epub ahead of print].

IndiaAI Subcommittee on AI Governance. (2025, January 6). Report on AI governance and guidelines development. *IndiaAI*. https://indiaai.gov.in/article/report-on-ai-governance-guidelines-development

Infocomm Media Development Authority. (2020). *AI Governance and Ethics Framework: Ensuring responsible AI use in Singapore*. https://www.imda.gov.sg

InfoComm Media Development Authority. (2025, May 28). Starter kit for safety testing of LLM-based applications. *IMDA*. https://www.imda.gov.sg/-/media/imda/files/about/emerging-tech-and-research/artificial-intelligence/large-language-model-starter-kit.pdf

Information and Communication Technology Division, Government of the People's Republic of Bangladesh. (2020, March). National strategy for artificial intelligence 2020–2025. *ICT Division, Government of Bangladesh*. https://file-rangpur.portal.gov.bd/files/pbs2.dinajpur.gov.bd/files/1885c0a0_28a4_4fcc_8d4d_dcd7ce23fd8b/7b684f19a15dfcd0f542382764572486.pdf

Information and Communication Technology Division, Government of the People's Republic of Malaysia. (2024, September). The national guidelines on AI governance & ethics. *Ministry of Science, Technology & Innovation, Malaysia*. https://mastic.mosti.gov.my/publication/the-national-guidelines-on-ai-governance-ethics/

Information Systems Audit and Control Association. (2024, October 18). Understanding the EU AI Act: Requirements and next steps. *Information Systems Audit and Control Association*. https://www.isaca.org/resources/white-papers/2024/understanding-the-eu-ai-act

Iniesta, R. (2023). The human role to guarantee an ethical AI in healthcare: A five-facts approach. *AI and Ethics*. Advance online publication. DOI: 10.1007/s43681-023-00353-x

Inova, A. I. Solutions. (2025). Inova BLDR platform. *Inova AI Solutions*. https://www.inovasolutions.ai/solutions-by-industry/

ISO/IEC. (2023). *What Is ISO/IEC 42001 Artificial Intelligence Management Systems Accreditation? Derived from*: https://anab.ansi.org/accreditation/iso-iec-42001-artificial-intelligence-management-systems/?gad_source=1&gad_campaignid=22984523464&gbraid=0AAAADvqK-UdvgmTQ_ZwQ5ecti0sfslqO&gclid=CjwKCAjw6P3GBhBVEiwAJPjmLvjZ2kqWRyUYoyv2UVYNeCIhYR4SDOTzbr0CkRBXrcpajgFGOoMi7hoCbPoQAvD_BwE

Jackaria, P. M., Hajan, B. H., Mastul, A. R. H., & Sali, F. Z. (2024). Generation AI in a Reimagined Classroom: Challenges, Opportunities and Implications to Education. Exploring Youth Studies in the Age of AI, 174-185.

Jacob Kirksey, J., & Elefante, J. (2024). Familiar faces in high school: How having the same peers from year-to-year links to student absenteeism. [JESPAR]. *Journal of Education for Students Placed at Risk*, *29*(1), 23–42. DOI: 10.1080/10824669.2022.2102982

Jacobs, G., Van Houdt, F., & Coons, G. (2024). Studying Surveillance AI-cologies in Public Safety: How AI Is in the World and the World in AI. *Surveillance & Society*, *22*(2), 160–178. DOI: 10.24908/ss.v22i2.16104

Janowicz, K. (2023). Philosophical Foundations of GeoAI: Exploring Sustainability, Diversity, and Bias in GeoAI and Spatial Data Science. *ArXiv (Cornell University)*. /arxiv.2304.06508DOI: 10.1201/9781003308423-3

Janowicz, K., Gao, S., McKenzie, G., Hu, Y., & Bhaduri, B. (2019). GeoAI: Spatially explicit artificial intelligence techniques for geographic knowledge discovery and beyond. *International Journal of Geographical Information Science*, *34*(4), 625–636. DOI: 10.1080/13658816.2019.1684500

Jarrahi, M. H. (2018). Artificial intelligence and the future of work : Human-AI symbiosis in organizational decision making. *Business Horizons*, *61*(4), 577–586. DOI: 10.1016/j.bushor.2018.03.007

Jasanoff, S. (2004). *States of Knowledge: The Co-production of Science and the Social Order*. Routledge Taylor & Francis Group. DOI: 10.4324/9780203413845

Jasanoff, S., & Kim, S. (2015). *Dreamscapes of modernity: sociotechnical imaginaries and the fabrication of power*. The University of Chicago Press. DOI: 10.7208/chicago/9780226276663.001.0001

Jauhiainen, J. S., & Garagorry Guerra, A. (2025). Generative AI in education: ChatGPT-4 in evaluating students' written responses. *Innovations in Education and Teaching International, 62*(4), 1377–1394. DOI: 10.1080/14703297.2024.2422337

Jha, A., & Singh, S. R. (2025). Bridging the Divide: Capacity Building for AI Adoption in Developing Countries. In *AI Strategies for Social Entrepreneurship and Sustainable Economic Development* (pp. 133-150). IGI Global Scientific Publishing.

Jiang, H., Li, J., Zhao, P., Zeng, F., Xiao, Z., & Iyengar, A. (2021). Location Privacy-preserving Mechanisms in Location-based Services. *ACM Computing Surveys, 54*(1), 1–36. DOI: 10.1145/3423165

Jio, W., Lemi, K., & Kant, S. (2024). Does Corporates Commitment act as a Mediating Element among Ethics of a leaders on Corporates Citizenship Conducts in Ethiopian health sector Facilities. *Journal of Corporate and Administrative Studies, 16*(2), 18–33.

Jobin, A., Ienca, M., & Vayena, E. (2019). The global landscape of AI ethics guidelines. *Nature Machine Intelligence, 1*(9), 389–399. DOI: 10.1038/s42256-019-0088-2

Jobin, A., Ienca, M., & Vayena, E. (2023). Worldwide AI ethics: A review of 200 guidelines and recommendations. *Patterns (New York, N.Y.), 4*(10), 100705. https://www.sciencedirect.com/science/article/pii/S2666389923002416

Johnson, N. F., Zhao, G., Hunsader, E., Meng, J., Ravindar, A., Carran, S., & Tivnan, B. (2013). Abrupt rise of new machine ecology beyond human response time. *Scientific Reports, 3*(1), 1–7. DOI: 10.1038/srep02627

Joswig, T., & Kurz, W. (2025, February 23). Regulatory and compliance requirements for SMEs operating AI systems through data centers in the EU, with a focus on data protection challenges in Germany. *Journal of Next-Generation Research, 1*(2), 4–44. DOI: 10.70792/jngr5.0.v1i2.89

Kalaivani, S., Nasreen Banu, D., Parvin Begum, I., & Sivasamy, J. (2023). Cognitive Robotics: Integrating Artificial Intelligence and Embodied Intelligence for Advanced Problem Solving. *Journal of Artificial Intelligence and Capsule Networks, 5*(4). Advance online publication. DOI: 10.36548/jaicn.2023.4.002

Kamiran, F., & Calders, T. (2012). Data preprocessing techniques for classification without discrimination. *Knowledge and Information Systems, 33*(1), 1–33. DOI: 10.1007/s10115-011-0463-8

Kan, C. H. (2024). Artificial intelligence (ai) in the age of democracy and human rights: Normative challenges and regulatory perspectives. *International Journal of Eurasian Education and Culture, 9*(25), 145–166.

Kang, Y., Gao, S., & Roth, R. E. (2024). Artificial intelligence studies in cartography: A review and synthesis of methods, applications, and ethics. *Cartography and Geographic Information Science*, *51*(4), 599–630. DOI: 10.1080/15230406.2023.2295943

Kapantai, E.. (2021). A unified taxonomical framework on disinformation. *New Media & Society*. Advance online publication. DOI: 10.1177/1461444820959296

Kapoor, A., & Sindwani, R. (2025). Artificial Intelligence as a Catalyst for Advancing Contactless Payments: Unveiling Key Adoption Drivers. In *AI Marketing and Ethical Considerations in Consumer Engagement* (pp. 321–338). IGI Global Scientific Publishing. DOI: 10.4018/979-8-3373-3476-9.ch016

Kapoor, A., Sindwani, R., & Goel, M. (2023a). Evaluating mobile wallet acceptance factors using best worst method. *International Journal of Process Management and Benchmarking*, *13*(4), 449–469. DOI: 10.1504/IJPMB.2023.129822

Kapoor, A., Sindwani, R., & Goel, M. (2023b). Exploring quick commerce service experience: A moderated mediated investigation using SEM and fsQCA. *Total Quality Management & Business Excellence*, *34*(13-14), 1896–1919. DOI: 10.1080/14783363.2023.2213653

Kapoor, A., Sindwani, R., & Goel, M. (2024a). Mobile wallets: Theoretical and empirical analysis. *Global Business Review*, *25*(2, suppl), S211–S228. DOI: 10.1177/0972150920961254

Kapoor, A., Sindwani, R., & Goel, M. (2024b). Exploring the symmetric and asymmetric effects of mobile wallet service experience on brand loyalty in India: Evidence from SEM and fsQCA. *Journal of Financial Services Marketing*, *29*(4), 1524–1541. DOI: 10.1057/s41264-024-00280-w

Kapoor, A., Sindwani, R., & Goel, M. (2025). A novel framework for understanding the interplay between the mobile wallet service quality dimensions and loyalty intention. *The TQM Journal*, *37*(6), 1716–1741. DOI: 10.1108/TQM-12-2023-0422

Kapoor, A., Sindwani, R., Goel, M., & Shankar, A. (2022). Mobile wallet adoption intention amid COVID-19 pandemic outbreak: A novel conceptual framework. *Computers & Industrial Engineering*, *172*, 108646. DOI: 10.1016/j.cie.2022.108646 PMID: 36118531

Karnouskos, S. (2020). Artificial intelligence in digital media: The era of deepfakes. *IEEE Transactions on Technology and Society*, *1*(3), 138–147. DOI: 10.1109/TTS.2020.3001312

Karpova, E. A. (2023). AI in Health Ethical Review: A Value Design Methodology. *Chelovek*, *34*(3), 129. Advance online publication. DOI: 10.31857/S023620070026109-6

Kasneci, E., Seßler, K., Küchemann, S., Bannert, M., Dementieva, D., Fischer, F., Gasser, U., Groh, G., Günnemann, S., Hüllermeier, E., Krusche, S., Kutyniok, G., Michaeli, T., Nerdel, C., Pfeffer, J., Poquet, O., Sailer, M., Schmidt, A., Seidel, T., & Kasneci, G. (2023). ChatGPT for good? On opportunities and challenges of large language models for education. *Learning and Individual Differences*, *103*, 102274. DOI: 10.1016/j.lindif.2023.102274

Kathirvel, A., Gobalakrinshan, N., & Naren, A. (2025). " Cyber security in industrial automation using AI, in AI-Enhanced Cybersecurity for Industrial Automation. *IGI Global*. https://www.igi-global.com/gateway/chapter/379640

Kathirvel, A., Gobinath, V. M., Sathya, A. K., Naren, A., & Raj, A. (2025a). The Role of Robotics in Smart Manufacturing: Increasing Efficiency and Reducing Costs, in The Quantum AI Era of Neuromarketing, *IGI Global*. https://www.igi-global.com/gateway/chapter/365528

Kathirvel, A., Naren, K., & Vianny, M. (2025b). Strategies for Combating Criminal Use and Abuse of Artificial Intelligence" In Avoiding Ad Fraud and Supporting Brand Safety: Programmatic Advertising Solutions. *IGI Global*. https://www.igi-global.com/gateway/chapter/368386

Kathole, A. B., Vhatkar, K. N., Raut, R., Patil, S. D., & Jadhav, A. (2023). Security Issues in Generative Adversarial Networks. In *Generative Adversarial Networks and Deep Learning*. Theory and Applications., DOI: 10.1201/9781003203964-5

Kathuria, Y. Ruhani, Vandana, Tyagi, M., & Jain, V. (2024, October). Protecting data privacy in the age of AI: A comparative analysis of legal approaches across different jurisdictions. In *AIP Conference Proceedings* (Vol. 3220, No. 1, p. 040007). AIP Publishing LLC.

Kayyali, M. (2025a). Ethical Implications of Generative AI in Education: Privacy, Bias, and Integrity. In Transformative AI Practices for Personalized Learning Strategies (pp. 185-218). IGI Global Scientific Publishing.

Kayyali, M. (2025b). Ethical Considerations in Higher Education: Accountability Navigating Complexities. In Accountability in Higher Education: Navigating Current Issues and Trends (pp. 145-184). IGI Global Scientific Publishing.

Kayyali, M. (2022). Equity, Equality, Diversity, and Inclusion as Key Performance Indicators in Higher Education. *Online Submission*, *3*(5), 1–3.

Kayyali, M. (2024). Future directions towards sustainable educational equity: Policies and practices. In *Exploring Educational Equity at the Intersection of Policy and Practice* (pp. 81–100). IGI Global. DOI: 10.4018/979-8-3693-1614-6.ch005

Kempton, A. M., & Vassilakopoulou, P. (2021). Accountability, Transparency and Explainability in AI for Healthcare. In *8th International Conference on Infrastructures in Healthcare.*

Keßler, C., & McKenzie, G. (2017). A geoprivacy manifesto. *Transactions in GIS*, *22*(1), 3–19. DOI: 10.1111/tgis.12305

Keymakr. (2024). *Regional and International AI Regulations and Laws in 2024.* Retrieved from https://keymakr.com/blog/regional-and-international-ai-regulations-and-laws-in-2024/

Khalifa, M., & Albadawy, M. (2024). Artificial Intelligence for Clinical Prediction: Exploring Key Domains and Essential Functions. In *Computer Methods and Programs in Biomedicine Update* (Vol. 5). DOI: 10.1016/j.cmpbup.2024.100148

Khalifa, M., Albadawy, M., & Iqbal, U. (2024). Advancing clinical decision support: The role of artificial intelligence across six domains. In *Computer Methods and Programs in Biomedicine Update* (Vol. 5). DOI: 10.1016/j.cmpbup.2024.100142

Khalifa, M. (2022, June). The role of artificial intelligence in the education process of political science field. In *2022 ASU International Conference in Emerging Technologies for Sustainability and Intelligent Systems (ICETSIS)* (pp. 409-416). IEEE.

Khan, A. A., Hussain, T., Al Siyabi, M., Al Shibli, A., Hussain, S., & Pillai, J. R. (2024). Challenges and Prospects of Cutting-Edge Technology in Education. *Studies on Education, Science, and Technology*, *2024*, 104.

Khan, M. M., Shah, N., Shaikh, N., Thabet, A., & Belkhair, S. (2024). Towards secure and trusted AI in healthcare: A systematic review of emerging innovations and ethical challenges. *International Journal of Medical Informatics*, •••, 105780. PMID: 39753062

Khan, U. A., & Alamäki, A. (2023). Designing an Ethical and Secure Pain Estimation System Using AI Sandbox for Contactless Healthcare. *International Journal of Online and Biomedical Engineering*, *19*(15), 166–201. Advance online publication. DOI: 10.3991/ijoe.v19i15.43663

Khazanah Research Institute. (2025, January). AI governance in Malaysia: Risks, challenges and pathways forward. *Khazanah Research Institute.* https://www.krinstitute.org/assets/contentMS/img/template/editor/KRI%20AIIG%20Report%20Final_v4.7%20[upload].pdf

Kin, L. C., Chin, A. S., & Seah, C. (2025). AI, machine learning & big data laws and regulations 2025 – Singapore. In AI, machine learning & big data laws & regulations 2025. *Global Legal Insights*. https://www.globallegalinsights.com/practice-areas/ai-machine-learning-and-big-data-laws-and-regulations/singapore

Kiran Deep Singh, & Singh, P. (2024). Fog Cloud Computing and IoT Integration for AI enabled Autonomous Systems in Robotics. *EAI Endorsed Transactions on AI and Robotics, 3*. DOI: 10.4108/airo.3617

Kitchin, R. (2016). The ethics of smart cities and urban science. *Philosophical Transactions. Series A, Mathematical, Physical, and Engineering Sciences, 374*(2083), 20160115. DOI: 10.1098/rsta.2016.0115

Kleinberg, J., Mullainathan, S., & Raghavan, M. (2016). Inherent trade-offs in the fair determination of risk scores. *arXiv preprint arXiv:1609.05807.*

Koehl, S. (2025). Designing AI Software for Large Classroom Engagement (Doctoral dissertation).

Komljenovic, J. (2022). The future of value in digitalised higher education: Why data privacy should not be our biggest concern. *Higher Education, 83*(1), 119–135. DOI: 10.1007/s10734-020-00639-7 PMID: 33230347

Korinek, A., & Stiglitz, J. E. (2021). Steering technological progress: AI, automation, and economic growth. *NBER Working Paper No. 28825*. DOI: 10.3386/w28825

Kozodoi, N., Jacob, J., & Lessmann, S. (2022). Fairness in credit scoring: Assessment, implementation and profit implications. *European Journal of Operational Research, 297*(3), 1083–1094. DOI: 10.1016/j.ejor.2021.06.023

KPMG International, & the University of Queensland. (2023, April 2). Trust in artificial intelligence: A global study 2023. *KPMG International*. https://assets.kpmg.com/content/dam/kpmg/xx/pdf/2023/09/trust-in-ai-global-study-2023.pdf

Kroll, J. A., Huey, J. D., & Barocas, S. (2020). The risks of keyless entry: The ethical implications of AI-driven systems in access control. *Journal of AI Ethics, 1*(2), 123–134. DOI: 10.1007/s43681-020-00008-w

Kroll, J. A., Huey, J., Barocas, S., Felten, E. W., Reidenberg, J. R., Robinson, D. G., & Yu, H. (2017). Accountable algorithms. *University of Pennsylvania Law Review, 165*(3), 633–705.

Krotzek, J. L. (2019). Inside the voter's mind: The effect of psychometric microtargeting on feelings toward and propensity to vote for a candidate. *International Journal of Communication*, •••, 2019.

Kumar, V., & Reinartz, W. (2016). Creating enduring customer value. *Journal of Marketing*, *80*(6), 36–68. DOI: 10.1509/jm.15.0414

La Torre, M., Chiappini, H., Anasetti, M., & Siri, M. (2020). Artificial intelligence and business ethics: A review of the financial services industry. *Journal of Business Ethics*, *164*(3), 593 614.

LaForce, S., & Bright, D. (2022). Are we there yet? The developing state of mobile access equity. *Assistive Technology*, *34*(6), 628–636. DOI: 10.1080/10400435.2021.1926372 PMID: 34048330

Larsen, B. C. (2021, July). A framework for understanding AI-induced field change: How AI technologies are legitimized and institutionalized. In *Proceedings of the 2021 AAAI/ACM Conference on AI, Ethics, and Society* (pp. 683-694). DOI: 10.1145/3461702.3462591

Lartey, D., & Law, K. M. (2025). Artificial intelligence adoption in urban planning governance: A systematic review of advancements in decision-making, and policy making. *Landscape and Urban Planning*, *258*, 105337. DOI: 10.1016/j.landurbplan.2025.105337

Lata, P. (2024). Towards Equitable Learning: Exploring Artificial Intelligence in Inclusive Education.

Lazaro, C., & Rizzi, M. (2022). Predictive analytics and governance: A new sociotechnical imaginary for uncertain futures. *International Journal of Law in Context*, *19*(1), 1–21. DOI: 10.1017/S1744552322000477

LeBlanc, P. (2021). *Students first: Equity, access, and opportunity in higher education*. Harvard Education Press.

Lee, M. K., Kusbit, D., Kahng, A., Kim, J. T., Yuan, X., Chan, A., See, D., Noothigattu, R., Lee, S., & Psomas, A. (2019). WeBuildAI: Participatory framework for algorithmic governance. *Proceedings of the ACM on human-computer interaction, 3*(CSCW), 1-35.

Lens. (2025). *Lens*. Retrieved October. https://www.lens.org/lens/

Liang, F., & Lee, K. (2021). China's approach to AI governance: Balancing control and innovation. *AI & Society*, *36*(4), 847–860.

Liew, J. (2024). Translating Vision into Reality for Early Education & Development. *Early Education and Development*, *35*(1), 1–2. Advance online publication. DOI: 10.1080/10409289.2024.2287684

Lin, L. (2024, July 16). China puts power of state behind AI—and risks strangling it. *The Wall Street Journal*. https://www.wsj.com/tech/china-puts-power-of-state-behind-aiand-risks-strangling-it-f045e11d

Lin, T. A., & Chen, P. H. C. (2022). Artificial intelligence in a structurally unjust society. Feminist Philosophy Quarterly, 8(3/4).

Lipton, Z. C. (2018). The mythos of model interpretability. *ACM Queue; Tomorrow's Computing Today*, *16*(3), 31–57. DOI: 10.1145/3236386.3241340

Liu, I. F. C., & Edmondson, D. (2023, August). China: New interim measures to regulate generative AI. *Baker & McKenzie*. https://insightplus.bakermckenzie.com/bm/attachment_dw.action?attdocparam

Liu, Z., & Luo, L. (2024, March). Using Artificial Intelligence for Intelligent Ideological and Political Education Teaching. In *2024 International Conference on Interactive Intelligent Systems and Techniques (IIST)* (pp. 137-144). IEEE.

Li, W., & Arundel, S. T. (2022). GeoAI and the future of spatial analytics. In New thinking in GIScience, Singapore: Springer Nature Singapore, pp. 151-158. DOI: 10.31223/X5DT3J

Li, W., Batty, M., & Goodchild, M. F. (2019). Real-time GIS for smart cities. *International Journal of Geographical Information Science*, *34*(2), 311–324. DOI: 10.1080/13658816.2019.1673397

Li, W., & Hsu, C.-Y. (2018). GeoAI: Where machine learning and big data converge in geospatial science. *Big Data and Cognitive Computing*, *2*(4), 22. DOI: 10.3390/bdcc2040022

Li, W., & Hsu, C.-Y. (2022). GeoAI for Large-Scale Image Analysis and Machine Vision: Recent Progress of Artificial Intelligence in Geography. *ISPRS International Journal of Geo-Information*, *11*(7), 385. DOI: 10.3390/ijgi11070385

Li, W., Wang, C., & Croitoru, A. (2020). GeoAI: Enabling the understanding of geospatial big data. *Computers, Environment and Urban Systems*, *81*, 101472. DOI: 10.1016/j.compenvurbsys.2020.101472

Longo, L., Brcic, M., Cabitza, F., Choi, J., Confalonieri, R., Del Ser, J., Guidotti, R., Hayashi, Y., Herrera, F., & Holzinger, A. (2024). Explainable Artificial Intelligence (XAI) 2.0 : A manifesto of open challenges and interdisciplinary research directions. *Information Fusion*, *106*, 102301. DOI: 10.1016/j.inffus.2024.102301

Lorestani, A. M., Ranbaduge, T., & Rakotoarivelo, T. (2024). *Privacy risk in Geo-Data: A survey*. ArXiv.org. https://arxiv.org/abs/2402.03612

Luckin, R., Holmes, W., Griffiths, M., & Forcier, L. B. (2016). *Intelligence Unleashed: An argument for AI in Education*. Pearson. Retrieved from UCL Discovery UCL Discovery

Lumme, E. M. (2024). A Systems-Theoretic Approach for the Analysis of AI Ethics. *Communications in Computer and Information Science, 1958 CCIS*. DOI: 10.1007/978-3-031-49215-0_6

Lütz, F. (2024, March). The AI Act, gender equality and non-discrimination: what role for the AI office? In *ERA Forum* (Vol. 25, No. 1, pp. 79-95). Berlin/Heidelberg: Springer Berlin Heidelberg. DOI: 10.1007/s12027-024-00785-w

Mager, A., & Katzenbach, C. (2021). Future imaginaries in the making and governing of digital technology: Multiple, contested, commodified. *New Media & Society*, *23*(2), 223–236. DOI: 10.1177/1461444820929321

Maghsudi, S., Lan, A., Xu, J., & van Der Schaar, M. (2021). Personalized education in the artificial intelligence era: What to expect next. *IEEE Signal Processing Magazine*, *38*(3), 37–50. DOI: 10.1109/MSP.2021.3055032

Mai, G., Xie, Y., Jia, X., Lao, N., Rao, J., Zhu, Q., Liu, Z., Chiang, Y.-Y., & Jiao, J. (2025). Towards the next generation of Geospatial Artificial Intelligence. *International Journal of Applied Earth Observation and Geoinformation*, *136*, 104368. DOI: 10.1016/j.jag.2025.104368

Mak, K.-K., & Pichika, M. R. (2019). Artificial intelligence in drug development : Present status and future prospects. *Drug Discovery Today*, *24*(3), 773–780. DOI: 10.1016/j.drudis.2018.11.014

Malaysia National AI Office [NAIO]. (2024, September). The national guidelines on AI governance & ethics. *Malaysia National AI Office*. https://mastic.mosti.gov.my/publication/the-national-guidelines-on-ai-governance-ethics/

Mamud, N. A., Kamil, N. L. M., & Hanafi, H. (2023). Fostering nurses'organisational citizenship behaviour through contextual elements. *Malaysian Journal of Public Health Medicine*, *23*(2), 36–46. DOI: 10.37268/mjphm/vol.23/no.2/art.1734

Manhiça, R., Santos, A., & Cravino, J. (2022, June). The use of artificial intelligence in learning management systems in the context of higher education: Systematic literature review. In *2022 17th Iberian Conference on Information Systems and Technologies (CISTI)* (pp. 1–6). IEEE. DOI: 10.23919/CISTI54936.2022.9837481

Manyika, J. (2022). Getting AI right: Introductory notes on AI & society. *Daedalus*, *151*(2), 5–27. DOI: 10.1162/daed_e_01897

Mao, H., Hu, Y., Kar, B., Gao, S., McKenzie, G., & Geo, A. I. (2017). Workshop Report: The 1st ACM SIGSPATIAL International Workshop on GeoAI: @AI and Deep Learning for Geographic Knowledge Discovery: Redondo Beach, CA, USA-November 7, 2016. ACM Sigspatial Spec. 2017, 9, 25.

Marasinghe, R., Yigitcanlar, T., Mayere, S., Washington, T., & Limb, M. (2024). Towards Responsible Urban Geospatial AI: Insights From the White and Grey Literatures. *Journal of Geovisualization and Spatial Analysis*, *8*(2), 24. Advance online publication. DOI: 10.1007/s41651-024-00184-2

Markovsky, I., & Dörfler, F. (2021). Behavioral systems theory in data-driven analysis, signal processing, and control. *Annual Reviews in Control*, *48*, 42–64. DOI: 10.1016/j.arcontrol.2021.09.005

Marres, N., Castelle, M., Gobbo, B., Poletti, C., & Tripp, J. (2024). AI as super-controversy: Eliciting AI and society controversies with an extended expert community in the U.K. *Big Data & Society*, *11*(2), 20539517241255103. Advance online publication. DOI: 10.1177/20539517241255103

Martin, K. D., & Murphy, P. E. (2017). The role of data privacy in marketing. *Journal of the Academy of Marketing Science*, *45*(2), 135–155. DOI: 10.1007/s11747-016-0495-4

Maslej, N., Fattorini, L., Perrault, R., Parli, V., Reuel, A., Brynjolfsson, E., . . . Clark, J. (2024). Artificial intelligence index report 2024. *arXiv preprint arXiv:2405.19522.*

Matuszczyk, N., Barnes, C. R., Gupta, R., Ozel, B., & Mitra, A. (2025). *From Bias to Accountability: How the EU AI Act Confronts Challenges in European GeoAI Auditing*. ArXiv.org. https://arxiv.org/abs/2505.18236

Mbiazi, D., Bhange, M., Babaei, M., Sheth, I., & Kenfack, P. J. (2023). Survey on AI Ethics: A Socio-technical Perspective. *arXiv preprint arXiv:2311.17228.* https://arxiv.org/abs/2311.17228

McGraw, D. K. (2024). Ethical responsibility in the design of artificial intelligence (AI) systems. *International Journal of Responsibility*, *7*(1). Advance online publication. DOI: 10.62365/2576-0955.1114

McKenzie, G., & Zhang, H. (2023). Platial k-anonymity: Improving location anonymity through temporal popularity signatures. In *12th International Conference on Geographic Information Science (GIScience 2023),* pp. 9-1. Schloss Dagstuhl–Leibniz-Zentrum für Informatik.

McKenzie, G., Zhang, H., & Gambs, S. (2023). Privacy and Ethics in GeoAI. In *Gao, S., Hu, Y., & Li, W. [eds.] Handbook of Geospatial Artificial Intelligence. CRC Press Ebooks*, pp. 388–405. DOI: 10.1201/9781003308423-19

Mckinsey. (2025). The state of AI: How organizations are rewiring to capture value. *Quantum Black AI by Mckinsey.* https://www.mckinsey.com/~/media/mckinsey/business%20functions/quantumblack/our%20insights/the%20state%20of%20ai/2025/the-state-of-ai-how-organizations-are-rewiring-to-capture-value_final.pdf

Mehrabi, N., Morstatter, F., Saxena, N., Lerman, K., & Galstyan, A. (2021). A survey on bias and fairness in machine learning. *ACM Computing Surveys, 54*(6), 1–35. DOI: 10.1145/3457607

Meszaros, J., Minari, J., & Huys, I. (2022). The future regulation of artificial intelligence systems in healthcare services and medical research in the European Union. *Frontiers in Genetics, 13*, 927721. Advance online publication. DOI: 10.3389/fgene.2022.927721 PMID: 36267404

Michael, M. (2017). Enacting big futures, little futures: Toward an ecology of futures. *The Sociological Review, 65*(3), 509–524. DOI: 10.1111/1467-954X.12444

Microsoft & PwC. (2019). *Responsible AI in financial services: Avoiding pitfalls and ensuring fairness.* PwC. https://www.pwc.com

Ministry of Economy. Trade and Industry & Ministry of Internal Affairs and Communications. (2024, April 19). AI guidelines for business Ver 1.0. *METI & MIC.* https://www.meti.go.jp/english/press/2024/0419_002.html

Ministry of Education, Government of India. (2020). *National education policy 2020.* Government of India. Retrieved from https://www.education.gov.in/nep/

Ministry of Electronics and Information Technology, Government of India. (2023, August 11). *Digital Personal Data Protection Act, 2023* [Act No. 22 of 2023]. *MeitY.* https://www.meity.gov.in/static/uploads/2024/06/2bf1f0e9f04e6fb4f8fef35e82c42aa5.pdf

Ministry of IT & Telecom. Pakistan. (2023, May). Draft national artificial intelligence policy. *Ministry of IT & Telecom, Pakistan.* https://moitt.gov.pk/SiteImage/Misc/files/National%20AI%20Policy%20Consultation%20Draft%20V1.pdf

Ministry of Science and Technology of China. (2021). *AI ethical norms and governance guidelines.* Retrieved from http://www.most.gov.cn

Ministry of Science and Technology of the People's Republic of China. (2021). *Ethical norms for the new generation artificial intelligence*. Government of China. Retrieved from https://www.most.gov.cn

Ministry of State for Artificial Intelligence, Digital Economy and Remote Work Applications, United Arab Emirates. (2022, December). AI ethics principles and guidelines. *UAE Government AI Office*. https://ai.gov.ae/wp-content/uploads/2023/03/MOCAI-AI-Ethics-EN-1.pdf

Ministry of State for Artificial Intelligence, Digital Economy and Remote Work Applications, United Arab Emirates. (2024, June 10). The UAE charter for the development and use of artificial intelligence. *UAE Government AI Office*. https://ai.gov.ae/wp-content/uploads/2024/07/UAEAI-Methaq-EN2-3.pdf

Ministry of State for Artificial Intelligence. Digital Economy and Remote Work Applications (UAE Government AI Office). (2025). AI ethics self-assessment tool. *UAE Government AI Office*. https://ai.gov.ae/aiethics-self-assessment/

Miragoli, M. (2024). Conformism, Ignorance & Injustice: AI as a Tool of Epistemic Oppression. *Episteme (Edinburgh)*, •••, 1–19.

Miranda-Pascual, À., Guerra-Balboa, P., Parra-Arnau, J., Forné, J., & Strufe, T. (2023). SoK: Differentially private publication of trajectory data. *Proceedings on Privacy Enhancing Technologies*.

Mirzani, Y. (2024). *Artificial intelligence and innovation management : Transformative forces shaping organizational creativity. EPRA International Journal of Multidisciplinary Research*. IJMR.

Mittelstadt, B. (2019). Principles alone cannot guarantee ethical AI. *Nature Machine Intelligence*, *1*(11), 501–507. DOI: 10.1038/s42256-019-0114-4

Mittelstadt, B. D., Allo, P., Taddeo, M., Wachter, S., & Floridi, L. (2016). The ethics of algorithms: Mapping the debate. *Big Data & Society*, *3*(2), 1–21. DOI: 10.1177/2053951716679679

Mittelstadt, B. D., Russell, C., & Wachter, S. (2016). Explaining explanations in AI. *Proceedings of the 2016 AAAI Conference on Artificial Intelligence*, 100-107.

Mohammed Bin Rashid School of Government. (2025, April). *The artificial intelligence SMEs ecosystem in the UAE*. Mohammed Bin Rashid School of Government. https://mbrsg.ae/ar/the-artificial-intelligence-smes-ecosystem-in_the-uae

Mohseni, S.. (2020). Machine learning explanations to prevent overtrust. *arXiv*. https://arxiv.org/abs/2007.12358

Moltubakk Kempton, A., & Vassilakopoulou, P. (2021). *Accountability*. Transparency and Explainability in AI for Healthcare.

Monlezun, D. J. (2023). Chapter 8 - AI+health ethics: moral interoperability and pluralism. In *The Thinking Healthcare System.*

Mooghali, M., Stroud, A. M., Yoo, D. W., & Barry, B. A. & ... (2023). Barriers and Facilitators to Trustworthy and Ethical AI-enabled Medical Care From Patient and Healthcare Provider Perspectives: A Literature Review. *MedRxiv*.

Morandín-Ahuerma, F. (2023). *Montreal declaration for responsible AI: 10 principles and 59 recommendations.*

Morin, O. (2025, June 13). What's inside the EU AI Act and what it means for your privacy. *Investopedia*. https://www.investopedia.com/eu-ai-act-11737033

Morley, J., Floridi, L., Kinsey, L., & Elhalal, A. (2021). From what to how: An initial review of publicly available AI ethics tools, methods, and research to translate principles into practices. *Science and Engineering Ethics*, *27*(1), 1–26. DOI: 10.1007/s11948-020-00239-y PMID: 33492516

Morozov, E. (2013). *To save everything, click here: smart machines, dumb humans, and the myth of tecnological perfectionism*. Perseus Books.

Morpurgo, F., & Occhipinti, C. (2023). Let the AI assisted medicine remain human Developing a care centered model of ethical decision making in AI based healthcare. *CEUR Workshop Proceedings*, •••, 3486.

Moussawi, S., Deng, X., & Joshi, K. D. (2024). AI and Discrimination: Sources of Algorithmic Biases. *The Data Base for Advances in Information Systems*, *55*(4), 6–11. DOI: 10.1145/3701613.3701615

Mozur, P. (2018). Inside China's dystopian dreams: AI, shame and lots of cameras. *The New York Times*. https://www.nytimes.com/2018/07/08/business/china-surveillance-technology.html

Mügge, D. (2024). EU AI sovereignity: For whom, to what end and to whose benefit? *Journal of European Public Policy*, *31*(8), 2200–2225. DOI: 10.1080/13501763.2024.2318475

Mühlhoff, R. (2023). Predictive privacy: Collective data protection in the context of AI and big data. *Big Data & Society*. Advance online publication. DOI: 10.1177/20539517231166886

Mühlhoff, R. (2025). *The Ethics of AI*. Bristol University Press., DOI: 10.51952/9781529249262

Müller, V. C. (2020). Ethics of artificial intelligence and robotics, In Zalta, E.N. (2020). The Stanford Encyclopedia of Philosophy, 2020

Mumford, R. (2024). *The Incorporation of an Ethics Framework in the Development of Emerging Information Technology Systems*. Salve Regina University.

Munjal, K., & Bhatia, R. (2022). A systematic review of homomorphic encryption and its contributions in healthcare industry. *Complex & Intelligent Systems*, 1–28.

Murray, R.. (2024). *Chilling effects on discourse due to surveillance*. Unknown Journal.

Mutambara, A. G. (2025). *Artificial intelligence: A driver of inclusive development and shared prosperity for the Global South*. CRC Press. DOI: 10.1201/9781003511014

Nambisan, S., Lyytinen, K., & Yoo, Y. (2020). *Handbook of digital innovation*. Edward Elgar Publishing. DOI: 10.4337/9781788119986

Nambisan, S., Wright, M., & Feldman, M. (2019). The digital transformation of innovation and entrepreneurship : Progress, challenges and key themes. *Research Policy*, *48*(8), 103773. DOI: 10.1016/j.respol.2019.03.018

Narayanan, A., & Chen, N. (2018). Certification and auditing of AI systems: Lessons from finance. *Communications of the ACM*, *61*(12), 54–57.

Naren, K., Kathirvel, A., & Saravanan, N. (2025). Entrepreneurship and Business Culture in Rural Area in Digital Disruption in Hospitality, Sustainable Hospitality: A Global Perspective on People, Planet, and Profit. *Emerald Publishing Limited*. https://bookstore.emerald.com/digital-disruption-in-hospitality-sustainable-hospitality-hb-9781836623618.html

Nasarian, E., Alizadehsani, R., Acharya, U. R., & Tsui, K. L. (2024). Designing interpretable ML system to enhance trust in healthcare: A systematic review to proposed responsible clinician-AI-collaboration framework. *Information Fusion*, *108*, 102412. DOI: 10.1016/j.inffus.2024.102412

Nasir, S., Khan, R. A., & Bai, S. (2024). Ethical Framework for Harnessing the Power of AI in Healthcare and Beyond. *IEEE Access : Practical Innovations, Open Solutions*, *12*, 31014–31035. Advance online publication. DOI: 10.1109/ACCESS.2024.3369912

National Institute of Standards and Technology (NIST). (2023, January 26). AI risk management framework (AI RMF) 1.0. *NIST*. https://doi.org/DOI: 10.6028/NIST.AI.100-1

Natorski, M. (2024). Compromise in multilateral negotiations and the global regulation of artificial intelligence. *Democratization, 31*(5), 1091–1116. DOI: 10.1080/13510347.2024.2351583

Ng, D. T. K., Leung, J. K. L., Chu, S. K. W., & Qiao, M. S. (2021). Conceptualizing AI literacy: An exploratory review. *Computers and Education Artificial Intelligence, 2*, 100041. Advance online publication. DOI: 10.1016/j.caeai.2021.100041

Nguyen, D. Q.. (2021). DeepfakeNet: Efficient deepfake detection method. *International Journal of Advanced Computer Science and Applications.*

Nicolau, M., Perez-Liebana, D., O'Neill, M., & Brabazon, A. (2016). Evolutionary behavior tree approaches for navigating platform games. *IEEE Transactions on Computational Intelligence and AI in Games, 9*(3), 227–238. DOI: 10.1109/TCIAIG.2016.2543661

Nikolenko, S. I. (2021). Synthetic Data for Deep Learning. In *Springer optimization and its applications.* Springer International Publishing., DOI: 10.1007/978-3-030-75178-4

Nizza, U. (2024, July 5). Assessing the impact of the European AI Act on innovation dynamics. *Northwestern University Law School.* https://www.law.northwestern.edu/research-faculty/clbe/events/standardization/documents/nizza_assessing_impact_ai_act_innovation.pdf

Noble, S. U. (2021). *Algorithms of oppression: How search engines reinforce racism.* NYU Press.

Normann, R., & Ramirez, R. (1993). From value chain to value constellation : Designing interactive strategy. *Harvard Business Review, 71*(4), 65–77.

Norman, W. (2013). *Stakeholder theory.* International Encyclopedia of Ethics.

Norton Rose Fulbright. (2024, February 14). Singapore proposes governance framework for generative AI. Norton Rose Fulbright. https://www.nortonrosefulbright.com/en/inside-fintech/blog/2024/02/singapore-proposes-governance-framework-for-generative-ai

Nourhashim, H. (2024, November 22). Malaysia's advancing AI governance: Driving contextual AI innovation amid trade-offs and opportunities. *AI Singapore.* https://aisingapore.org/wp-content/uploads/2025/03/Malaysias-Advancing-AI-Governance-Driving-Contextual-AI-Innovation-Amid-Trade-offs-and-Opportunities.pdf

Ntoutsi, E., Fafalios, P., Gadiraju, U., Iosifidis, V., Nejdl, W., Vidal, M., Ruggieri, S., Turini, F., Papadopoulos, S., Krasanakis, E., Kompatsiaris, I., Kinder-Kurlanda, K., Wagner, C., Karimi, F., Fernandez, M., Alani, H., Berendt, B., Kruegel, T., Heinze, C., & Staab, S. (2020). Bias in Data-driven Artificial Intelligence systems—An Introductory Survey. *Wiley Interdisciplinary Reviews. Data Mining and Knowledge Discovery*, *10*(3), 1–14. DOI: 10.1002/widm.1356

O'Flaherty, M. (2020). *Artificial intelligence and human rights: A global framework*. Oxford University Press.

O'Neil, C. (2016). *Weapons of math destruction: How big data increases inequality and threatens democracy*. Crown Publishing Group.

Obermeyer, Z., Powers, B., Vogeli, C., & Mullainathan, S. (2019). Dissecting racial bias in an algorithm used to manage the health of populations. *Science*, *366*(6464), 447–453. DOI: 10.1126/science.aax2342

Ochoa, K. S. (2024). Can Artificial Intelligence Mark the Next Architectural Revolution? Design Exploration in the Realm of Generative Algorithms and Search Engines. Decoding Cultural Heritage: A Critical Dissection and Taxonomy of Human Creativity Through Digital Tools, 3.

Odeh, A. H., & Al Hattab, M. (2023, November). AI Methods Used for Spam Detection in Social Systems-An Overview. In *2023 Tenth International Conference on Social Networks Analysis, Management and Security (SNAMS)* (pp. 1-8). IEEE.

OECD AI Principles. (2019). AI Principles. Retrieved from: https://www.oecd.org/en/topics/sub-issues/ai-principles.html?utm_source=chatgpt.com

OECD. (2019a). *Preparing for the future of work: Policies to facilitate smooth transitions*. Organisation for Economic Co-operation and Development.

OECD. (2019a). *Principles on AI*. Retrieved from https://www.oecd.org/going-digital/ai/principles

OECD. (2019b). *OECD AI principles: Recommendations of the council on artificial intelligence*. Retrieved from https://www.oecd.org/going-digital/ai/principles/

OECD. (2019b). *OECD principles on AI*. Organisation for Economic Co-operation and Development. https://www.oecd.org/going-digital/ai/principles/

OECD. (2020). *OECD Principles on Artificial Intelligence*. https://www.oecd.org/going-digital/ai/principles/

OECD. (2024). *AI Principles Overview*. Organisation for Economic Co-operation and Development. Retrieved from https://oecd.ai/en/ai-principles

Office of Science and Technology Policy, The White House. (2022, October 4). Blueprint for an AI Bill of Rights: Making automated systems work for the American people. *OSTP*. https://bidenwhitehouse.archives.gov/ostp/ai-bill-of-rights/

Office of the Australian Information Commissioner (OAIC). (2024, January 26). Notifiable Data Breaches report: January to June 2024. *OAIC*. https://www.oaic.gov.au/privacy/notifiable-data-breaches/notifiable-data-breaches-publications/notifiable-data-breaches-report-january-to-june-2024/

Ofqual. (2020, August 13). *A-level results: Almost 40% of teacher assessments downgraded*. The Guardian. https://www.theguardian.com/education/2020/aug/13/almost-40-of-english-students-have-a-level-results-downgraded

Olorunsogo, T., Adenyi, A. O., Okolo, C. A., & Babawarun, O. (2024). Ethical considerations in AI-enhanced medical decision support systems: A review. *World Journal of Advanced Engineering Technology and Sciences*, *11*(1), 329–336. Advance online publication. DOI: 10.30574/wjaets.2024.11.1.0061

Oluoch, I. (2024). Crossing Boundaries: The Ethics of AI and Geographic Information Technologies. *ISPRS International Journal of Geo-Information*, *13*(3), 87. DOI: 10.3390/ijgi13030087

Openshaw, S., & Openshaw, C. (1997). *Artificial Intelligence in Geography*. John Wiley & Sons, Inc.

Organisation for Economic Co-operation and Development (OECD). (2019). *OECD principles on artificial intelligence*. Retrieved from https://www.oecd.org

Organisation for Economic Co-operation and Development (OECD). (2021). *AI principles: Recommendations on AI governance for the financial sector*. OECD AI Policy Observatory. https://oecd.ai

Organisation for Economic Co-operation and Development [OECD]. (2024, September 16). SME digitalisation to manage shocks and transitions: 2024 OECD D4SME survey. *OECD-Digital for SMEs*. https://www.oecd.org/content/dam/oecd/en/publications/reports/2024/09/sme-digitalisation-to-manage-shocks-and-transitions_735fc44d/eb4ec9ac-en.pdf

Organisation for Economic Co-operation and Development. (2019, May; updated 2024, May 3). *OECD principles on artificial intelligence*. *OECD*. https://www.oecd.org/going-digital/ai/principles/

Organisation for Economic Co-operation and Development. (2023). AI Policy Observatory: Bridging the SME governance gap. *OECD.AI*. https://oecd.ai/en/

Organisation for Economic Co-operation and Development. (2023). *The state of implementation of the OECD AI Principles four years on.* https://www.oecd.org/en/publications/the-state-of-implementation-of-the-oecd-ai-principles-four-years-on_835641c9-en.html

Organisation for Economic Co-operation and Development. (2024). *AI principles.* OECD. https://www.oecd.org/en/topics/sub-issues/ai-principles.html

Organisation for Economic Co-operation and Development. (2025, June). Scoping note for an AI policy toolkit to support economies in realising AI benefits. *OECD.* https://www.oecd.org/content/dam/oecd/en/events/2025/06/mcm/MCM-2025-Scoping-Note-for-an-AI-Policy-Toolkit-to-Support-Economies-in-Realising-AI-Benefits.pdf

*OriginLab - Origin and OriginPro - data analysis and graphing software.* Retrieved October, 2025, from https://www.originlab.com/

Ortiz-Vázquez, A., Llorens, S., & Salanova, M. (2025). Development and Validation of the Encouraging Culture of organization Scale (POC-S). *Psicothema*, *37*(1), 80.

Owen, R., Stilgoe, J., Macnaghten, P., Gorman, M., Fisher, E., & Guston, D. (2013). A framework for responsible innovation. *Responsible innovation: managing the responsible emergence of science and innovation in society*, 27-50.

Ozmen Garibay, O., Winslow, B., Andolina, S., Antona, M., Bodenschatz, A., Coursaris, C., Falco, G., Fiore, S. M., Garibay, I., Grieman, K., Havens, J. C., Jirotka, M., Kacorri, H., Karwowski, W., Kider, J., Konstan, J., Koon, S., Lopez-Gonzalez, M., Maifeld-Carucci, I., & Xu, W. (2023). Six human-centered artificial intelligence grand challenges. *International Journal of Human-Computer Interaction*, *39*(3), 391–437. DOI: 10.1080/10447318.2022.2153320

Pagano, T. P., Loureiro, R. B., Lisboa, F. V. N., Peixoto, R. M., Guimarães, G. A. S., Cruz, G. O. R., Araujo, M. M., Santos, L. L., Cruz, M. A. S., Oliveira, E. L. S., Winkler, I., & Nascimento, E. G. S. (2023). Bias and Unfairness in Machine Learning Models: A Systematic Review on Datasets, Tools, Fairness Metrics, and Identification and Mitigation Methods. *Big Data and Cognitive Computing*, *7*(1), 15. DOI: 10.3390/bdcc7010015

Paltieli, G. (2022). The political imaginary of National AI Strategies. *AI & Society*, *37*(4), 1613–1624. DOI: 10.1007/s00146-021-01258-1

Pandey, A. (2025, June 18). Digital divide: MSMEs struggle to tap into AI revolution as high costs block access. *Economic Times*. https://economictimes.indiatimes.com/small-biz/sme-sector/digital-divide-msmes-struggle-to-tap-into-ai-revolution-as-high-costs-block-access/articleshow/121924294.cms

Panindre, P., Mandal, A., Paradkar, M., & Kumar, S. (2023, May). Ai-based detection of signs of depression from physiological data obtained from health trackers. In 2023 2nd International Conference on Applied Artificial Intelligence and Computing (ICAAIC) (pp. 644-649). IEEE.

Pantanowitz, L., Hanna, M., Pantanowitz, J., Lennerz, J., Henricks, W. H., Shen, P., & Rashidi, H. H. (2024). Regulatory aspects of AI-ML. *Modern Pathology*, •••, 100609. DOI: 10.1016/j.modpat.2024.100609 PMID: 39260776

Papernot, N., McDaniel, P., Goodfellow, I., Jha, S., Celik, Z. B., & Swami, A. (2018). Practical black-box attacks against machine learning. *Proceedings of the 2017 ACM on Asia Conference on Computer and Communications Security*, 506-519. DOI: 10.1145/3052973.3053009

Parasuraman, R., & Riley, V. (1997). Humans and automation: Use, misuse, disuse, abuse. *Human Factors*, *39*(2), 230–253. DOI: 10.1518/001872097778543886

Parinandi, S., Crosson, J., Peterson, K., & Nadarevic, S. (2024). Investigating the politics and content of US State artificial intelligence legislation. *Business and Politics*, *26*(2), 240–262. DOI: 10.1017/bap.2023.40

Pariser, E. (2011). *The filter bubble: How the new personalized web is changing what we read and how we think*. Penguin Books.

Parson, E. T. A., Fyshe, A., & Lizotte, D. (2019). Artificial Intelligence's Societal Impacts, Governance, and Ethics. *UCLA School of Law, Public Law Research Paper*.

Partnership on AI. (2021). Ethical principles for AI development. Retrieved from https://www.partnershiponai.org

Pasquale, F. (2015). *The black box society: The secret algorithms that control money and information*. Harvard University Press. DOI: 10.4159/harvard.9780674736061

Pasquale, F. (2020). *New laws of robotics: Defending human expertise in the age of AI*. Harvard University Press.

Pastor-Escuredo, D., Treleaven, P., & Vinuesa, R. (2022). An ethical framework for artificial intelligence and sustainable cities. *AI*, *3*(4), 961–974. DOI: 10.3390/ai3040057

Patel, P., Moz-Ruiz, D., Garcia, R., Chatterjee, A., Morreale, P., & Burnett, M. (2024). From Workshops to Classrooms: Faculty Experiences with Implementing Inclusive Design Principles. *SIGCSE 2024 - Proceedings of the 55th ACM Technical Symposium on Computer Science Education, 1*. DOI: 10.1145/3626252.3630861

Payne, A. F., Storbacka, K., & Frow, P. (2008). Managing the co-creation of value. *Journal of the Academy of Marketing Science, 36*(1), 83–96. DOI: 10.1007/s11747-007-0070-0

Peña Acuña, B., Fernández-Aller, C., Zavrsnik, A., Sarf, P., Ramus Cvetkova, I., Binark, M., & Ramos, A. (2025). *Artificial Intelligence and Digital Rights*. Opportunities and Challenges.

PepsiCo. (2023). *Case Study: AI Integration and Ethical Considerations at PepsiCo*. AI Expert Network. https://aiexpert.network/case-study-ai-integration-and-ethical-considerations-at-pepsico/

Peralta, A. (2024). Best Practices for Managing Privacy and Security of Cloud-Based Student Data in Primary and Secondary Education.

Personal Data Protection Commission. Singapore & Infocomm Media Development Authority. (2019, January; 2020, January; 2024, May). Model AI Governance Framework (1st ed., 2019; 2nd ed., 2020; Generative AI extension, 2024). *PDPC & IMDA*. https://www.pdpc.gov.sg/help-and-resources/2020/01/second-edition-of-model-artificial-intelligence-governance-framework

Petersson, L., Vincent, K., Svedberg, P., Nygren, J. M., & Larsson, I. (2023). Ethical Perspectives on Implementing AI to Predict Mortality Risk in Emergency Department Patients: A Qualitative Study. *Studies in Health Technology and Informatics, 302*. Advance online publication. DOI: 10.3233/SHTI230234 PMID: 37203776

Peters, U. (2023). Extended implicit bias: When the metaphysics and ethics of implicit bias collide. *Erkenntnis, 88*(8), 3457–3478. DOI: 10.1007/s10670-021-00511-9 PMID: 35035000

Pistor, K. (2020). Rules of finance and the AI challenge. *Journal of International Economic Law, 23*(3), 579–596.

Plummer, D. (2024). Ethical AI and the Future of Healthcare: Combining Academic Theory and Industry Practice to Ensure Patient-Centered Care. *European Journal of Technology, 8*(1), 1–13. Advance online publication. DOI: 10.47672/ejt.1727

Polevikov, S. (2023). Advancing AI in healthcare: A comprehensive review of best practices. In *Clinica Chimica Acta* (Vol. 548). DOI: 10.1016/j.cca.2023.117519

Porter, M. E. (1985). Technology and competitive advantage. *The Journal of Business Strategy*, *5*(3), 60–78. DOI: 10.1108/eb039075

Pourzolfaghar, Z., Alfano, M., & Helfert, M. (2023). Application of ethical AI requirements to an AI solution use-case in healthcare domain. *American Journal of Business*, *38*(3), 112–128. Advance online publication. DOI: 10.1108/AJB-12-2022-0201

Pousttchi, K., Gleiss, A., Buzzi, B., & Kohlhagen, M. (2019). *Technology impact types for digital transformation. 1*, 487-494.

Prahalad, C. (2004). *The future of competition : Co-creating unique value with customers*. Harvard Business Rress.

Prahalad, C. K., & Ramaswamy, V. (2004). Co-creation experiences : The next practice in value creation. *Journal of Interactive Marketing*, *18*(3), 5–14. DOI: 10.1002/dir.20015

Prasad, S., Chanussot, J., & Li, J. (2024). *Advances in Machine Learning and Image Analysis for GeoAI*. Elsevier., DOI: 10.1016/C2022-0-00952-1

Praveenraj, J.. (2023). *XAI and decision-interpretation methods*. Unknown Journal.

Proietti, S., & Magnani, R. (2025, January 14). Assessing AI adoption and digitalization in SMEs: A framework for implementation [Preprint]. *arXiv*. https://doi.org//arXiv.2501.08184DOI: 10.32388/0N53P4

Prorok, M., & Takacs, I. (2023). Legal and Moral Issues Related to Artificial Intelligence Technology in Business. *SISY 2023 - IEEE 21st International Symposium on Intelligent Systems and Informatics, Proceedings*. DOI: 10.1109/SISY60376.2023.10417881

Prove, P. (2024). Freedom of expression, and the protection of journalists in conflict situations. *Media Development*, *70*(3).

Purbahapsari, A. F., & Batoarung, I. B. (2022). *Geospatial Artificial Intelligence for Early Detection of Forest and Land Fires*. KnE Social Sciences. DOI: 10.18502/kss.v7i9.10947

PwC Middle East. (2023). Gulf AI readiness report: Private sector governance gaps. *PwC Middle East*. https://www.pwc.com/m1/en/publications/ai-readiness-middle-east.html

Qiu, Q., Li, W., Liu, J., Tao, L., Hao, M., Wang, Y., & Xie, Z. (2024). AutoGeo-FAIR: A framework for evaluating the sharing degree of open earth surface system data based on extended FAIR principles. *Journal of Environmental Management, 372*, 123432. Advance online publication. DOI: 10.1016/j.jenvman.2024.123432

Raghavan, M., Barocas, S., Kleinberg, J., & Levy, K. (2020). Mitigating bias in algorithmic hiring: Evaluating claims and practices. *Proceedings of the 2020 Conference on Fairness, Accountability, and Transparency*, 469–481. DOI: 10.1145/3351095.3372828

Rahwan, I. (2018). Society-in-the-loop : Programming the algorithmic social contract. *Ethics and Information Technology, 20*(1), 5–14. DOI: 10.1007/s10676-017-9430-8

Rahwan, I., Cebrian, M., Obradovich, N., Bongard, J., Bonnefon, J. F., Breazeal, C., Crandall, J. W., Christakis, N. A., Couzin, I. D., Jackson, M. O., Jennings, N. R., Kamar, E., Kloumann, I. M., Larochelle, H., Lazer, D., McElreath, R., Mislove, A., Parkes, D. C., Pentland, A. S., & Wellman, M. (2019). Machine behaviour. *Nature, 568*(7753), 477–486. DOI: 10.1038/s41586-019-1138-y PMID: 31019318

Raj, A., & Arora, L. Girija, Sanjay Surendranath, Kapoor, S., Pradhan, D., & Shetgaonkar, A. (2025). *AI and Generative AI Transforming Disaster Management: A Survey of Damage Assessment and Response Techniques.* ArXiv.org. https://arxiv.org/abs/2505.08202v1

Raji, I. D., Smart, A., White, R. N., Mitchell, M., Gebru, T., Hutchinson, B., Smith-Loud, J., Theron, D., & Barnes, P. (2020). *Closing the AI accountability gap : Defining an end-to-end framework for internal algorithmic auditing*. 33-44.

Raji, I. D., Gebru, T., Mitchell, M., Buolamwini, J., Lee, J., & Denton, E. (2020). Saving face: Investigating the ethical concerns of facial recognition auditing. *Proceedings of the 2020 Conference on Fairness, Accountability, and Transparency*, 92-100. DOI: 10.1145/3375627.3375820

Raley, R., & Rhee, J. (2023). Critical AI: A Field in Formation. *American Literature, 95*(2), 185–204. Advance online publication. DOI: 10.1215/00029831-10575021

Ramanathan, A., Vijaya Rani, K. R., Bhupathyraaj, M., Chacko, L., & Alharbi, H. F. (2023). An Ethical Outlook on the Applications of Artificial Intelligence in Medicine and Pharmacy. In *Artificial Intelligence in Pharmaceutical Sciences*. DOI: 10.1201/9781003343981-10

Ramanayake, R., Wicke, P., & Nallur, V. (2023). Immune moral models? Pro-social rule breaking as a moral enhancement approach for ethical AI. *AI & Society, 38*(2), 801–813. Advance online publication. DOI: 10.1007/s00146-022-01478-z PMID: 35645466

Rana, J., Gutierrez, P. L., & Oldroyd, J. C. (2023). Quantitative methods. In *Global encyclopedia of public administration, public policy, and governance* (pp. 11202–11207). Springer.

Rane, N. (2023). ChatGPT and similar generative artificial intelligence (AI) for smart industry : Role, challenges and opportunities for industry 4.0, industry 5.0 and society 5.0. *Challenges and Opportunities for Industry, 4.*

Rao, J., Gao, S., Mai, G., & Janowicz, K. (2023). Building Privacy-Preserving and Secure Geospatial Artificial Intelligence Foundation Models (Vision Paper). In *Proceedings of the 31st ACM International Conference on Advances in Geographic Information Systems.* pp. 1-4. DOI: 10.1145/3589132.3625611

Rask, M., & Shimizu, K. (2024, February). Beyond the Average: Exploring the Potential and Challenges of Large Language Models in Social Science Research. In *2024 International Conference on Artificial Intelligence, Computer, Data Sciences and Applications (ACDSA)* (pp. 1-5). IEEE.

Rawls, J. (1971). *A theory of justice.* Belknap Press of Harvard University Press. DOI: 10.4159/9780674042605

Raza, K. (2020). Artificial intelligence against COVID-19: a meta-analysis of current research. In Hassanien, A., Dey, N., & Elghamrawy, S. (Eds.), *Big Data Analytics and Artificial Intelligence Against COVID-19: Innovation Vision and Approach* (pp. 165–176). Springer International Publishing. DOI: 10.1007/978-3-030-55258-9_10

Research, I. B. M. (2020). *AI ethics and bias: Trustworthy AI for financial decision-making.* IBM. https://research.ibm.com

Reserve Bank of India. (2022). *Guidelines on digital lending.* Government of India. Retrieved from https://www.rbi.org.in

Reuters. (2025, April 8). Europe wants to lighten AI compliance burden for startups. *Reuters.* https://www.reuters.com/world/europe/europe-wants-lighten-ai-compliance-burden-startups-2025-04-08

Ribeiro, M. T., Singh, S., & Guestrin, C. (2016). *"Why should I trust you?" Explaining the predictions of any classifier.* In *Proceedings of the 22nd ACM SIGKDD International Conference on Knowledge Discovery and Data Mining* (pp. 1135-1144). ACM. DOI: 10.1145/2939672.2939778

Ribeiro, M. T., Singh, S., & Guestrin, C. (2016). *"Why should I trust you?" Explaining the predictions of any classifier. Proceedings of the 22nd ACM SIGKDD International Conference on Knowledge Discovery and Data Mining*, 1135-1144.

Ribeiro, M. T., Singh, S., & Guestrin, C. (2016). Model-agnostic interpretability of machine learning. *arXiv preprint arXiv:1606.05386*.

Richardson, R., Schultz, J. M., & Crawford, K. (2019). Dirty data, bad predictions: How civil rights violations impact police data, predictive policing systems, and justice. *New York University Law Review Online*, *94*, 192–233.

Richter, V., Katzenbach, C., & Schäfer, M. S. (2023). Imaginaries of artificial intelligence. In *Handbook of critical studies of artificial intelligence* (pp. 209–223). Edward Elgar Publishing., DOI: 10.4337/9781803928562.00024

Richter, V., Katzenbach, C., & Zeng, J. (2025). Negotiating AI(s) futures: Competing imaginaries of AI by stakeholders in the US, China, and Germany. *JCOM, Journal of Science Communication*, *24*(2), A08. DOI: 10.22323/2.24020208

Ridzuan, N. N., Masri, M., Anshari, M., Fitriyani, N. L., & Syafrudin, M. (2024). AI in the financial sector: The line between innovation, regulation and ethical responsibility. *Information (Basel)*, *15*(8), 432. DOI: 10.3390/info15080432

Rissanen, M. (2020). Translational health technology and system schemes: Enhancing the dynamics of health informatics. *Health Information Science and Systems*, *8*(1), 39. DOI: 10.1007/s13755-020-00133-5 PMID: 33194173

Rivi -Tipei, I. (2023). Ethical Considerations in Artificial Intelligence Development and Deployment. *SCIENTIA MORALITAS International Journal of Multidisciplinary Research*, 2, 1-15. https://www.scientiamoralitas.com/index.php/sm/article/view/236

Roberts, H., Hine, E., Taddeo, M., & Floridi, L. (2024). Global AI governance: Barriers and pathways forward. *International Affairs*, *100*(3), 1275–1286. DOI: 10.1093/ia/iiae073

Roff, H. M. (2021). Lethal autonomous weapons and the future of warfare: Ethics, policy, and governance. *Ethics & International Affairs*, *35*(1), 45–55. DOI: 10.1017/S0892679420000970

Romero Moreno, F. (2024). Generative AI and deepfakes: A human rights approach to tackling harmful content. *International Review of Law Computers & Technology*, *38*(3), 297–326. DOI: 10.1080/13600869.2024.2324540

Roozenbeek, J., Schneider, C. R., Dryhurst, S., Kerr, J., Freeman, A. L. J., Recchia, G., van der Bles, A. M., & van der Linden, S. (2020). Susceptibility to misinformation about COVID-19. *Royal Society Open Science*, *7*(10), 201199. Advance online publication. DOI: 10.1098/rsos.201199 PMID: 33204475

Roumate, F. (2024). AI and technological sovereignty. In *Artificial Intelligence and the New World Order: New weapons, New Wars and a New Balance of Power* (pp. 59–65). Springer Nature Switzerland. DOI: 10.1007/978-3-031-50312-2_5

Royal College of Surgeons of England Bulletin. (2021), "Bias in medical artificial intelligence", derived from: https://publishing.rcseng.ac.uk/doi/full/10.1308/rcsbull.2021.111

Russell, S. (2021). *Human compatible: Artificial intelligence and the problem of control*. Penguin.

Russell, S., & Norvig, P. (2020). *Artificial intelligence: A modern approach* (4th ed.). Pearson.

Sadek, M., Calvo, R. A., & Mougenot, C. (2024). Designing value-sensitive AI: A critical review and recommendations for socio-technical design processes. *AI and Ethics*, *4*(4), 949–967. DOI: 10.1007/s43681-023-00373-7

Saeidnia, H. R., Hosseini, E., Lund, B., Tehrani, M. A., Zaker, S., & Molaei, S. (2025). Artificial intelligence in the battle against disinformation and misinformation: A systematic review of challenges and approaches. *Knowledge and Information Systems*, *67*(4), 1–20. DOI: 10.1007/s10115-024-02337-7

Sahithya, V., Sujithra, M., Senthilkumar, B., Shreenithi, M., & Priya, H. (2024). *A Guide to Ethical Geospatial Practices*. Advances in Geospatial Technologies Book Series., DOI: 10.4018/979-8-3693-6381-2.ch002

Said, F., & Nabilah, F. (2024, November). Future of Malaysia's AI governance. *Institute of Strategic & International Studies Malaysia*. https://www.isis.org.my/wp-content/uploads/2024/12/AI-Governance.pdf

Sakhare, N. N., Limkar, S., Mahadik, R. V., Phursule, R., Godbole, A., Shirkande, S. T., & Patange, A. (2023). Ethical considerations of AI applications in medicine: A policy framework for responsible deployment. In *Journal of Krishna Institute of Medical Sciences University* (Vol. 12, Issue 4).

Salajan, F. D., Barnes, T. L., & Becker, A. (2024). Comparative regionalism cases of artificial intelligence governance in education: The Caribbean Community and the European Union. *The Round Table*, *113*(1), 71–85. Advance online publication. DOI: 10.1080/00358533.2024.2307793

Salami, A. O. (2024). Artificial intelligence, digital colonialism, and the implications for Africa's future development. *Data & Policy*, *6*, e67. DOI: 10.1017/dap.2024.75

Saleem, S. A. M., & Naseem, S. M. B. (2023, December). A Comprehensive Review of the Artificial Intelligence (AI) shift in E-commerce. In 2023 6th International Conference on Advances in Science and Technology (ICAST) (pp. 51-56). IEEE.

Salman A.H. Selmy, Kuche, D. E., & Yang, Y. (2024). Geospatial Data: Acquisition, Applications and Challenges. *IntechOpen EBooks*. DOI: 10.5772/intechopen.1006635

Sancin, V., & Bobnar, L. (2024). The Right to Freedom of Expression in the Era of Artificial Intelligence Systems. *Pravni Letopis*, 109.

Sanghvi, H., Bhavsar, R., Hundlani, V., Gohil, L., Vyas, T., Nair, A., Desai, S., Jadav, N. K., Tanwar, S., Sharma, R., & Yamsani, N. (2024). MetaHate : AI-based hate speech detection for secured online gaming in metaverse using blockchain. *Security and Privacy*, *7*(2), e343. Advance online publication. DOI: 10.1002/spy2.343

Sanjeetha, M. B. F., & Abeygunawardhana, P. K. (2024). AI in Smart Devices and Services: A Systematic Review of Social Impacts and Ethical Implications. *TechRxiv*. DOI: DOI: 10.36227/techrxiv.172671793.34847626/v1

Sarabhai, S., Chakraborty, M., Batra, M., Kler, R., Banerjee, S., & Mishra, S. (2023, November). Using AI and Machine Learning to Predict Consumer Buying Behavior: Insights from Behavioral Economics in Case of Alcoholic Beverages. In 2023 3rd International Conference on Technological Advancements in Computational Sciences (ICTACS) (pp. 980-986). IEEE.

Saraswat, D., Bhattacharya, P., Verma, A., Prasad, V. K., Tanwar, S., Sharma, G., Bokoro, P. N., & Sharma, R. (2022). Explainable AI for healthcare 5.0: Opportunities and challenges. *IEEE Access : Practical Innovations, Open Solutions*, *10*, 84486–84517. DOI: 10.1109/ACCESS.2022.3197671

Sargiotis, D. (2024). Fostering Ethical and Inclusive AI: A Human-Centric Paradigm for Social Impact. Available at *SSRN* 4879372. DOI: 10.2139/ssrn.4879372

Saudi Data & AI Authority. (2024, January). Generative AI guidelines for government and public use. *SDAIA*. https://sdaia.gov.sa/en/SDAIA/about/Files/GenAIGuidelinesForGovernmentENCompressed.pdf (SDAIA Government Version, SDAIA Public Version)

Saudi Data& AI Authority. (2023, September). AI ethics principles. *SDAIA*. https://sdaia.gov.sa/en/SDAIA/about/Documents/ai-principles.pdf

Saura, J. R., Ribeiro-Soriano, D., & Palacios-Marqués, D. (2022). Assessing behavioral data science privacy issues in government artificial intelligence deployment. *Government Information Quarterly*, *39*(4), 101679. DOI: 10.1016/j.giq.2022.101679

Sayles, J. (2024). Capacity Building for AI Actors. In *Principles of AI Governance and Model Risk Management: Master the Techniques for Ethical and Transparent AI Systems* (pp. 321–331). Apress. DOI: 10.1007/979-8-8688-0983-5_16

Schoenfeld, A. H. (2022). Why are learning and teaching mathematics so difficult? In *Handbook of cognitive mathematics* (pp. 1–35). Springer International Publishing.

Scholz, T. (2017). *Uberworked and underpaid: How workers are disrupting the digital economy*. Polity Press.

Schultz, M. D., Clegg, M., Hofstetter, R., & Seele, P. (2024). Algorithms and dehumanization: A definition and avoidance model. *AI & Society*, •••, 1–21.

Securiti. (2025, January 16). An overview of Malaysia's National Guidelines on AI governance and ethics. *Securiti.AI*. https://securiti.ai/malaysia-national-guidelines-on-ai-governance-and-thics

Segarra-Blasco, A., Tomàs-Porres, J., & Teruel, M. (2025). AI, robots and innovation in European SMEs. *Small Business Economics*, *65*(1), 1–27. DOI: 10.1007/s11187-025-01017-2

Selwyn, N. (2019). What's the problem with learning analytics? *Journal of Learning Analytics*, *6*(3), 11–19. DOI: 10.18608/jla.2019.63.3

Setiawan, A. (2024). Intelligent Systems For Disabilities Technology For A More Inclusive Life. *Available at SSRN* 5062273.

Sevinç, T. (2024). Global solidarity. *Journal of Global Ethics*, *20*(3), 308–316. DOI: 10.1080/17449626.2024.2422364

Shafik, W. (2023). Introduction to ChatGPT. In *Advanced Applications of Generative AI and Natural Language Processing Models*. DOI: 10.4018/979-8-3693-0502-7.ch001

Shafik, W. (2024b). Artificial intelligence and Blockchain technology enabling cybersecurity in telehealth systems. In *Artificial Intelligence and Blockchain Technology in Modern Telehealth Systems*. DOI: 10.1049/PBHE061E_ch11

Shafik, W. (2024h). Generative AI for Social Good and Sustainable Development. In *Generative AI: Current Trends and Applications* (pp. 185–217). Springer. DOI: 10.1007/978-981-97-8460-8_10

Shafik, W. (2024j). The Role of Artificial Intelligence in the Emerging Digital Economy Era. *Artificial Intelligence Enabled Management: An Emerging Economy Perspective*, 33.

Shafik, W. (2024m). Towards Trustworthy and Explainable AI Educational Systems. In *Explainable AI for Education: Recent Trends and Challenges* (pp. 17–41). Springer. DOI: 10.1007/978-3-031-72410-7_2

Shafik, W. (2025a). Human-Computer Interaction (HCI) Technologies in Socially-Enabled Artificial Intelligence. In *Future of Digital Technology and AI in Social Sectors* (pp. 121–150). IGI Global.

Shafik, W. (2025b). Quantum Computing and Generative Adversarial Networks (GANs): Ethics, Privacy, and Security. In *Quantum AI and its Applications in Blockchain Technology* (pp. 111–156). IGI Global Scientific Publishing.

Shafik, W., Mojtaba Matinkhah, S., & Sanda, M. N. (2020). Network Resource Management Drives Machine Learning: A Survey and Future Research Direction. *Journal of Communications Technology, Electronics and Computer Science, 2020*(0). https://doi.org/http://dx.doi.org/10.22385/jctecs.v30i0.312

Shafik, W. (2024a). Agile Security Strategies. In *Smart and Agile Cybersecurity for IoT and IIoT Environments* (pp. 82–101). IGI Global. DOI: 10.4018/979-8-3693-3451-5.ch004

Shafik, W. (2024c). Artificial intelligence and internet of things roles in sustainable next-generation manufacturing: An overview of emerging trends in Industry 6.0. *Advances in Business Information Systems and Analytics*, *6*(0), 207–239. DOI: 10.4018/979-8-3693-3140-8.ch012

Shafik, W. (2024d). Artificial Intelligence and Machine Learning with Cyber Ethics for the Future World. In *Future Communication Systems Using Artificial Intelligence, Internet of Things and Data Science* (pp. 110–130). CRC Press., DOI: 10.1201/9781032648309-9

Shafik, W. (2024e). Deep Learning Impacts in the Field of Artificial Intelligence. In *Deep Learning Concepts in Operations Research* (pp. 9–26). Auerbach Publications., DOI: 10.1201/9781003433309-2

Shafik, W. (2024f). Ethical Use of Machine Learning Techniques in Smart Cities. In *Ethical Artificial Intelligence in Power Electronics* (pp. 21–47). CRC Press. DOI: 10.1201/9781032648323-3

Shafik, W. (2024i). Machine Learning for Advanced Wireless Communication. In *Microwave Devices and Circuits for Advanced Wireless Communication* (pp. 252–278). CRC Press., DOI: 10.1201/9781032656021-13

Shafik, W. (2024k). Toward a More Ethical Future of Artificial Intelligence and Data Science. In *The Ethical Frontier of AI and Data Analysis* (pp. 362–388). IGI Global., DOI: 10.4018/979-8-3693-2964-1.ch022

Shahriari, K., & Shahriari, M. (2017). *IEEE standard review—Ethically aligned design : A vision for prioritizing human wellbeing with artificial intelligence and autonomous systems*. 197-201.

Shahvaroughi Farahani, M., & Ghasemi, G. (2024). *Artificial Intelligence and Inequality: Challenges and Opportunities*. Qeios., DOI: 10.32388/7HWUZ2

Shao, C., Ciampaglia, G. L., Varol, O., Yang, K.-C., Flammini, A., & Menczer, F. (2018). Spread of low-credibility content by social bots. *Nature Communications*, *9*(1), 4787. DOI: 10.1038/s41467-018-06930-7 PMID: 30459415

Sharkey, N. (2018). The impact of gender and race bias in AI systems. *International Journal of Ethics in AI*, *1*(1), 20–35. DOI: 10.1108/IJEA-2018-0004

Sharma, H. K., & Tailor, R. M. (2023, September). Exploring the Potential of AI for Urban Crime Prediction in India: A Case Study of Indian Metropolitan Cities. In 2023 6th International Conference on Contemporary Computing and Informatics (IC3I) (Vol. 6, pp. 1047-1051). IEEE.

Sharma, A. K., & Sharma, R. (2024). Navigating the Ethical Landscape: Implementing Machine Learning in Smart Healthcare Informatics. *Indian Journal of Community Health*, *36*(1), 149–152. DOI: 10.47203/IJCH.2024.v36i01.024

Sharma, B., Sharma, L., Lal, C., & Roy, S. (2023). Anomaly based network intrusion detection for IoT attacks using deep learning technique. *Computers & Electrical Engineering*, *107*, 108626. Advance online publication. DOI: 10.1016/j.compeleceng.2023.108626

Sharma, S., & Gupta, H. (2021). Ethical implications of AI in emerging economies. *AI and Ethics*, *1*(3), 305–316. DOI: 10.1007/s43681-020-00004-w

Sharma, T., Yadav, D., & Chouhan, R. (2020). A survey on privacy and security concerns in AI applications. *Journal of Information Security and Applications*, *53*, 102562. DOI: 10.1016/j.jisa.2020.102562

Sharma, V., Mishra, N., Kukreja, V., Alkhayyat, A., & Elngar, A. A. (2023). Framework for Evaluating Ethics in AI. *International Conference on Innovative Data Communication Technologies and Application, ICIDCA 2023 - Proceedings*. DOI: 10.1109/ICIDCA56705.2023.10099747

Shen, J., & Anderson, M. (2024). Ethical Challenges and Solutions of Generative AI. *Informatics (MDPI), 11*(3), 58. https://www.mdpi.com/2227-9709/11/3/58. DOI: 10.3390/informatics11030058

Shi, M., Janowicz, K., Verstegen, J., Currier, K., Wiedemann, N., Mai, G., Majic, I., Liu, Z., & Zhu, R. (2025). Geography for AI sustainability and sustainability for GeoAI. *Cartography and Geographic Information Science, 52*(4), 331–349. DOI: 10.1080/15230406.2025.2479796

Shneiderman, B. (2020). Human-centered artificial intelligence: Reliable, safe & trustworthy. *ArXiv*. https://arxiv.org/abs/2002.04087

Shoaib, M. R.. (2023). Deepfakes, misinformation, and disinformation. *arXiv*. https://arxiv.org/abs/2311.17394

Shokoor, F., Shafik, W., & Matinkhah, S. M. (2022). Overview of 5G & Beyond Security. *EAI Endorsed Transactions on Internet of Things, 8*(30), e2. Advance online publication. DOI: 10.4108/eetiot.v8i30.1624

Shu, K., Bhattacharjee, A., Alatawi, F., Nazer, T. H., Ding, K., Karami, M., & Liu, H. (2020). Combating disinformation in a social media age. *Wiley Interdisciplinary Reviews. Data Mining and Knowledge Discovery, 10*(6), e1385. Advance online publication. DOI: 10.1002/widm.1385

Shu, K., Wang, S., Liu, H., & Tang, J. (2020). Fake news detection on social media: A data mining perspective. *ACM Computing Surveys, 53*(3), 1–38. DOI: 10.1145/2988545

Siau, K., & Wang, W. (2020). Artificial Intelligence (AI) Ethics: Ethics of AI and Ethical AI. *Journal of Database Management, 31*(2), 74–87. DOI: 10.4018/JDM.2020040105

Simonsen, J., & Robertson, T. (2013). *Routledge international handbook of participatory design* (Vol. 711). Routledge New York.

Singh, L. (2022). Automated Kantian Ethics: A Faithful Implementation. *Lecture Notes in Computer Science (Including Subseries Lecture Notes in Artificial Intelligence and Lecture Notes in Bioinformatics), 13404 LNAI*. DOI: 10.1007/978-3-031-15791-2_16

Singh, A., & Vashist, H. (2025). Ethics of a leaders in health-care organizations–a scoping review. *Leadership in Health Services*, *40*(1), 206–148.

Singhal, A., Neveditsin, N., Tanveer, H., & Mago, V. (2024). Toward Fairness, Accountability, Transparency, and Ethics in AI for Social Media and Health Care: Scoping Review. *JMIR Medical Informatics*, *12*(1), e50048. DOI: 10.2196/50048 PMID: 38568737

Siswadi, Y., Jufrizen, J., Saripuddin, J., Farisi, S., & Sari, M. (2023). Culture of organization and Corporates Citizenship Behavior: The Mediating Function of Learning Organizations and Corporates Commitment. *Jurnal Riset Bisnis Dan Manajemen*, *16*(1), 81–82. DOI: 10.23969/jrbm.v16i1.7184

Sjoding, M. W., Dickson, R. P., Iwashyna, T. J., Gay, S. E., & Valley, T. S. (2020). Racial bias in pulse oximetry measurement. *The New England Journal of Medicine*, *383*(25), 2477–2478. DOI: 10.1056/NEJMc2029240 PMID: 33326721

Skadden, Arps, Slate, Meagher & Flom LLP. (2024, April 5). Utah becomes first state to enact AI-centric consumer protection law. *Skadden*. https://www.skadden.com/insights/publications/2024/04/utah-becomes-first-state

Smith, S. R., & Hyman, D. (2020). Public engagement and AI: The role of education and awareness. *AI & Society*, *35*(1), 5–15. DOI: 10.1007/s00146-019-00938-9

Solakis, K., Katsoni, V., Mahmoud, A. B., & Grigoriou, N. (2024). Factors affecting value co-creation through artificial intelligence in tourism : A general literature review. *Journal ofTourism Futures*, *10*(1), 116–130. DOI: 10.1108/JTF-06-2021-0157

Soomro, R. B., Al-Rahmi, W. M., Dahri, N. A., Almuqren, L., Al-Mogren, A. S., & Aldaijy, A. (2025). A SEM–ANN analysis to examine impact of artificial intelligence technologies on sustainable performance of SMEs. *Scientific Reports*, *15*(1), 5438. DOI: 10.1038/s41598-025-86464-3 PMID: 39948417

Soudi, M. S., & Bauters, M. (2024). AI guidelines and ethical readiness inside SMEs: A review and recommendations. *Digital Society : Ethics, Socio-Legal and Governance of Digital Technology*, *3*(1), 3. DOI: 10.1007/s44206-024-00087-1

Specht, D. (2020). Mapping Crisis: Participation, Datafication and Humanitarianism in the Age of Digital Mapping. In *University of London Press eBooks*. University of Lodz Press., DOI: 10.14296/920.9781912250387

Stahl, B. C., & Wright, D. (2018). *Ethical deployment of AI surveillance*. Unknown Journal.

Stappers, J. (2025, February 11). Artificial Intelligence and compliance: Preparing for the future of AI governance, risk, and compliance. *Navex*. https://www.navex.com/en-us/blog/article/artificial-intelligence-and-compliance-preparing-for-the-future-of-ai-governance-risk-and-compliance/

Stilgoe, J., Owen, R., & Macnaghten, P. (2020). Developing a framework for responsible innovation. In *The ethics of nanotechnology, geoengineering, and clean energy* (pp. 347–359). Routledge. DOI: 10.4324/9781003075028-22

Stockinger, A., Schäfer, S., & Lecheler, S. (2025). Navigating the gray areas of content moderation: Professional moderators' perspectives on uncivil user comments and the role of (AI-based) technological tools. *new media & society, 27*(3), 1215-1234.

Strickland, E. (2019). IBM Watson, heal thyself: How AI fell short in the healthcare revolution. *IEEE Spectrum*. Retrieved from https://spectrum.ieee.org

Stylianides, C., Alexandropoulou, C. A., Sulaiman, W., Panagiotopoulos, I., Kleanthous, S., Dimitrakopoulos, G., Constantinou, I., Vassiliou, S., Garcia, F., Politi, E., Ntalaperas, D., Papageorgiou, X., Ioannides, N., Palazis, L., Pattichis, C. S., & Panayides, A. S. (2023). AI-based solutions for predicting sepsis in ICUs. *2023 IEEE EMBS Special Topic Conference on Data Science and Engineering in Healthcare, Medicine and Biology, IEEECONF 2023*. DOI: 10.1109/IEEECONF58974.2023.10404309

Subasi, A., Ozaltin, O., Mitra, A., Subasi, M. E., & Sarirete, A. (2023). Trustworthy artificial intelligence in healthcare. In *Accelerating Strategic Changes for Digital Transformation in the Healthcare Industry: Volume 2* (Vol. 2). DOI: 10.1016/B978-0-443-15299-3.00015-4

Suchman, L. (2023). The uncontroversial 'thingness' of AI. *Big Data & Society, 10*(2), 20539517231206794. Advance online publication. DOI: 10.1177/20539517231206794

Sun, G., Cai, S., Yu, H., Maharjan, S., Chang, V., Du, X., & Guizani, M. (2019). Location privacy preservation for mobile users in location-based services. *IEEE Access : Practical Innovations, Open Solutions, 7*, 87425–87438. DOI: 10.1109/ACCESS.2019.2925571

Suresh, H., & Guttag, J. (2021). *A framework for understanding sources of harm throughout the machine learning life cycle*. 1-9.

Suresh, H., & Guttag, J. V. (2019). A Framework for Understanding Unintended Consequences of Machine Learning. *Arxiv.org*. https://arxiv.org/abs/1901.10002

Suresh, H., & Guttag, J. (2021). *A Framework for Understanding Sources of Harm throughout the Machine Learning Life Cycle*. Equity and Access in Algorithms, Mechanisms, and Optimization., DOI: 10.1145/3465416.3483305

Suresh, H., & Guttag, J. V. (2019). A framework for understanding unintended consequences of machine learning. *Communications of the ACM*, *64*(11), 62–71.

Susskind, D. (2020). *A world without work: Technology, automation, and how we should respond*. Metropolitan Books.

Szymkowiak, A., Melović, B., Dabić, M., Jeganathan, K., & Kundi, G. S. (2021). Information technology and Gen Z: The role of teachers, the internet, and technology in the education of young people. *Technology in Society*, *65*, 101565. DOI: 10.1016/j.techsoc.2021.101565

Taddeo, M., & Floridi, L. (2018). How AI can be a force for good. *Science*, *361*(6404), 751–752. DOI: 10.1126/science.aat5991

Tang, L., Li, J., & Fantus, S. (2023). Medical artificial intelligence ethics: A systematic review of empirical studies. *Digital Health*, *9*, 20552076231186064. DOI: 10.1177/20552076231186064 PMID: 37434728

Taylor, C. (2004). *Modern Social Imaginaries*. Duke University Press. DOI: 10.1215/9780822385806

Tennison, J. (2024). Why We Need Inclusive Data Governance in the Age of AI. *The Role of Governance in Unleashing the Value of Data*, 74.

Teo, S. A. (2025). The Unbearable Likeness of Being: How Artificial Intelligence Challenges the Social Ontology of International Human Rights Law. *The Journal of Cross-Disciplinary Research in Computational Law*.

The UK's Technology Trade Association. (2023, March 29). UK government unveils its approach to regulating AI with white paper. *TechUK*. https://www.techuk.org/resource/uk-government-unveils-its-approach-to-regulating-ai.html

The UK's Technology Trade Association. (2025). AI adoption for SME, research insights and practical guidance. *TechUK*. https://www.techuk.org/what-we-deliver/events/ai-adoption-for-smes-research-insights-practical-guidance.html

Theodorou, A., & Dignum, V. (2020). Towards ethical and socio-legal governance in AI. *Nature Machine Intelligence*, *2*(1), 10–12. DOI: 10.1038/s42256-019-0136-y

Tidjon, L. N., & Khomh, F. (2023). The Different Faces of AI Ethics Across the World: A Principle-to-Practice Gap Analysis. *IEEE Transactions on Artificial Intelligence*, *4*(4), 820–839. Advance online publication. DOI: 10.1109/TAI.2022.3225132

Timmermans, J., Stahl, B. C., Ikonen, V., & Bozdag, E. (2010). The ethics of cloud computing: A conceptual review. *Proceedings of the IEEE Second International Conference Cloud Computing Technology and Science* (pp. 614-620). IEEE Press. DOI: 10.1109/CloudCom.2010.59

Tiwari, V. (2023). *Explainable AI techniques for transparency*. Unknown Journal.

Tobler, W. R. (1970). A computer movie simulating urban growth in the Detroit region. *Economic Geography*, *46*(supplement), 234–240. DOI: 10.2307/143141

Topol, E. (2019). *Deep medicine: How artificial intelligence can make healthcare human again*. Basic Books.

Topol, E. J. (2019). High-performance medicine : The convergence of human and artificial intelligence. *Nature Medicine*, *25*(1), 44–56. DOI: 10.1038/s41591-018-0300-7

Torabian, J. E. (2022). Education and Culture: The Instrumentalisation. In *Wealth, Values, Culture & Education: Reviving the essentials for equality & sustainability* (pp. 91–135). Springer International Publishing. DOI: 10.1007/978-3-030-92893-3_4

Treacy, S. (2023). Mechanisms and Constraints Underpinning Ethically Aligned Artificial Intelligence Systems: An Exploration of Key Performance Areas. *Human-Centric Intelligent Systems*, *3*(3), 189–196. Advance online publication. DOI: 10.1007/s44230-023-00036-0

Tuia, D., Schindler, K., Demir, B., & Zhu, X. X. Mrinalini Kochupillai, Sašo Džeroski, Jan, Hoos, H. H., Frate, F. D., Mihai Datcu, Markl, V., Bertrand Le Saux, Schneider, R., & Gustau Camps-Valls. (2024). Artificial Intelligence to Advance Earth Observation: A review of models, recent trends, and pathways forward. *IEEE Geoscience and Remote Sensing Magazine*, pp. 2–25._DOI: 10.1109/MGRS.2024.3425961

Tursunalieva, A., Alexander, D. L. J., Dunne, R., Li, J., Riera, L., & Zhao, Y. (2024). Making Sense of Machine Learning: A Review of Interpretation Techniques and Their Applications. *Applied Sciences, 14*(2), pag. 496. DOI: 10.3390/app14020496

U.S. Chamber of Commerce Technology Engagement Center. (2024, September 15). *Empowering small business: The impact of technology on U.S. small business*. U.S. Chamber of Commerce. https://www.uschamber.com/assets/documents/Impact-of-Technology-on-Small-Business-Report-2024.pdf

U.S. Chamber of Commerce. (2024, September 15). Empowering small business: The impact of technology on U.S. small business. *U.S. Chamber of Commerce*. https://www.uschamber.com/assets/documents/Impact-of-Technology-on-Small-Business-Report-2024.pdf

U.S. Department of Treasury. (2022). *Financial Stability Oversight Council report on digital asset financial stability risks and regulation*. U.S. Department of Treasury. https://home.treasury.gov

U.S. Federal Trade Commission. (2021). *Aiming for truth, fairness, and equity in your company's use of AI*. FTC Guidance. https://www.ftc.gov

Udegbe, F. C., Ebulue, O. R., Ebulue, C. C., & Ekesiobi, C. S. (2024). The role of artificial intelligence in healthcare: A systematic review of applications and challenges. *International Medical Science Research Journal*, *4*(4), 500–508. DOI: 10.51594/imsrj.v4i4.1052

Ughulu, J. (2024). Ethics of a leaders in modern organizations: Navigating complexity and promoting integrity. *International Journal of Economics. Corporate and Management Investigation*, *8*(5), 48–62.

UK Financial Conduct Authority. (2022). *The role of artificial intelligence in financial services: Ethical and regulatory challenges*. FCA Discussion Paper. https://www.fca.org.uk

UK Government Office for Artificial Intelligence. (2023, March 29). A pro-innovation approach to AI regulation. *UK Government*. https://www.gov.uk/government/publications/ai-regulation-a-pro-innovation-approach/white-paper

Ulnicane, I. (2025). Governance fix? Power and politics in controversies about governing generative AI. *Policy and Society*, *44*(1), 70–84. DOI: 10.1093/polsoc/puae022

UNESCO. (2021). *Recommendation on the ethics of artificial intelligence*. https://unesdoc.unesco.org/ark:/48223/pf0000377897

UNESCO. (2021). *Recommendation on the Ethics of Artificial Intelligence*. United Nations Educational, Scientific and Cultural Organization. Retrieved from https://unesdoc.unesco.org/ark:/48223/pf0000380455

UNESCO. (2023). *Global AI Ethics and Governance Observatory*. https://www.unesco.org/ethics-ai/en

UNESCO. (n.d.). *Towards an Ethics of Artificial Intelligence*. Retrieved from https://www.un.org/en/chronicle/article/towards-ethics-artificial-intelligence

UNESCO's Recommendation on the Ethics of AI. (2021). Recommendation on the Ethics of Artificial Intelligence. Retrieved from: https://www.unesco.org/en/legal-affairs/recommendation-ethics-artificial-intelligence?utm_source=chatgpt.com

UN-GGIM. *(2020). Future trends in geospatial information management: the five to ten year vision THIRD EDITION.* https://ggim.un.org/meetings/GGIM-committee/10thSession/documents/Future_Trends_Report_THIRD_EDITION_digital_accessible.pdf

United Nations Department of Economic and Social Affairs. (2024, April). United Nations system-wide white paper on AI governance. *UNDESA*. https://unsceb.org/sites/default/files/2024-04/United%20Nations%20System%20White%20Paper%20on%20AI%20Governance.pdf

United Nations Educational, Scientific and Cultural Organization (UNESCO). (2021). *Recommendation on the ethics of artificial intelligence*. Retrieved from https://www.unesco.org

United Nations Educational, Scientific and Cultural Organization. (2021). Recommendation on the ethics of artificial intelligence. *The United Nations Educational, Scientific and Cultural Organization*. https://unesdoc.unesco.org/ark:/48223/pf0000381137

United Nations Educational, Scientific and Cultural Organization. (2023). ethics of artificial intelligence. *UNESCO*. https://www.unesco.org/en/artificial-intelligence/recommendation-ethics

United Nations Educational, Scientific and Cultural Organization. (2023a). Ethical impact assessment: A tool of the recommendation on the ethics of artificial intelligence. *UNESCO*. https://unesdoc.unesco.org/ark:/48223/pf0000386276

United Nations Educational, Scientific and Cultural Organization. (2023b). Readiness assessment methodology (RAM): A tool of the Recommendation on the ethics of artificial intelligence. *UNESCO*. https://unesdoc.unesco.org/ark:/48223/pf0000385198

United Nations. (2021). *The role of AI in achieving the Sustainable Development Goals*. Retrieved from https://www.un.org

United Nations. (2024, September). Governing AI for humanity. *High-level Advisory Body on Artificial Intelligence, Office of the Secretary-General*. https://www.un.org/sites/un2.un.org/files/governing_ai_for_humanity_final_report_en.pdf

Unwin, P. T. H., & Unwin, T. (2017). *Reclaiming information and communication technologies for development*. Oxford University Press. DOI: 10.1093/oso/9780198795292.001.0001

Vakkuri, V. (2019). AI governance and civil society: A model for ethical collaboration. *AI & Society*, *34*(4), 481–497. DOI: 10.1007/s00146-019-00890-x

Vallabhaneni, A. S., Perla, A., Regalla, R. R., & Kumari, N. (2024). The Power of Personalization: AI-Driven Recommendations. In Minds Unveiled (pp. 111-127). Productivity Press.

van Dijck, J., van Es, K., Helmond, A., & van der Vlist, F. (2025). *Governing the Digital Society*. Amsterdam University Press., DOI: 10.5117/9789048562718

Van Eck, N. J., & Waltman, L. (2022). *VOSviewer* manual (Version 1.6.18). Leiden University, CWTS. Retrieved from https://www.vosviewer.com/documentation/Manual_VOSviewer_1.6.18.pdf

Vargo, S. L., & Lusch, R. F. (2004). Evolving to a new dominant logic for marketing. *Journal of Marketing*, *68*(1), 1–17. DOI: 10.1509/jmkg.68.1.1.24036

Vargo, S. L., & Lusch, R. F. (2008). Service-dominant logic : Continuing the evolution. *Journal of the Academy of Marketing Science*, *36*(1), 1–10. DOI: 10.1007/s11747-007-0069-6

Vasant, P., & Weber, G. W. (Eds.). (2021). *Artificial intelligence and big data analytics for smart finance: Methods, technologies and applications*. Springer.

Veale, M., & Binns, R. (2017). Fairer machine learning in the real world: Mitigating discrimination without collecting sensitive data. *Big Data & Society*, *4*(2), 1–17. DOI: 10.1177/2053951717743530

Veale, M., & Borgesius, F. Z. (2021). Demystifying the draft EU Artificial Intelligence Act. *Computer Law & Security Review*, *43*, 105632. DOI: 10.1016/j.clsr.2021.105632

Veluru, S. (2024). *Risks of large-scale data breaches in AI systems*. Unknown Journal.

Vesna, L., Sawale, P. S., Kaul, P., Pal, S., & Murthy, B. S. N. V. R. (2025). Digital divide in AI-powered education: Challenges and solutions for equitable learning. *Journal of Information Systems Engineering & Management*, *10*(21), 1–9.

Vianny, M., Ayyaswamy, K., Rithik, G., Kathirvel, N., & Raj, A. (2025). Private AI an Overview, In AI-Driven Breakthroughs in Antimicrobial Resistance. *IGI Global*. https://igi-global.com/gateway/chapter/374868

Villani, C., Bonnet, Y., Berthet, C., Levin, F., Schoenauer, M., Cornut, A. C., & Rondepierre, B. (2018). *Donner un sens à l'intelligence artificielle : Pour une stratégie nationale et européenne*. Conseil national du numérique.

Vincent, J. (2025). *AI Accountability: How Doing The Right Thing Builds A Better Business*. Forbes. https://www.forbes.com/sites/bruceweinstein/2025/01/29/ai-accountability-how-doing-the-right-thing-builds-a-better-business/

Voigt, P., & Von dem Bussche, A. (2017). The eu general data protection regulation (gdpr). *A practical guide, 1st ed., Cham: Springer International Publishing, 10*(3152676), 10-5555.

Voigt, P., & Von dem Bussche, A. (2017). *The EU general data protection regulation (GDPR): A practical guide*. Springer. DOI: 10.1007/978-3-319-57959-7

Vold, T., & Eikset, A. S. (2024, November). Use of AI in Education-A Case Study from Norway. In 2024 21st International Conference on Information Technology Based Higher Education and Training (ITHET) (pp. 1-5). IEEE.

Volinz, L. (2025). Authoritarian Surveillance: An Introduction. *Surveillance & Society, 23*(1), 112–116. DOI: 10.24908/ss.v23i1.18975

Von Eschenbach, M. (2021). *Black-box AI and public trust.* Unknown Journal.

von Schomberg, R. (2013). A vision of responsible research and innovation. In Owen, R., Heintz, M., & Bessant, J. (Eds.), *Responsible innovation: Managing the responsible emergence of science and innovation in society* (pp. 51–74). John Wiley & Sons. DOI: 10.1002/9781118551424.ch3

VoPham, T., Hart, J. E., Laden, F., & Chiang, Y.-Y. VoPham. (2018). Emerging trends in geospatial artificial intelligence (geoAI): Potential applications for environmental epidemiology. *Environmental Health : A Global Access Science Source, 17*(1), 40. Advance online publication. DOI: 10.1186/s12940-018-0386-x

Vosoughi, S., Roy, D., & Aral, S. (2018). The spread of true and false news online. *Science, 359*(6380), 1146–1151. DOI: 10.1126/science.aap9559

Vozna, A., & Costantini, S. (2025). Ethical, legal, and societal dimensions of AI-Driven social robots in elderly healthcare. *Intelligenza Artificiale, 19*(1), 30–40. DOI: 10.1177/17248035241310192

Vulpe, S.-N., Rughini , R., urcanu, D., & Rosner, D. (2024). AI and cybersecurity: A risk society perspective. *Frontiers of Computer Science, 6*, 1462250. Advance online publication. DOI: 10.3389/fcomp.2024.1462250

Wachter, S., Mittelstadt, B., & Floridi, L. (2017). Why a right to explanation of automated decision-making does not exist in the General Data Protection Regulation. *International Data Privacy Law, 7*(2), 76–99. DOI: 10.1093/idpl/ipx005

Wachter, S., Mittelstadt, B., & Russell, C. (2017). Counterfactual explanations without opening the black box: Automated decisions and the GDPR. *SSRN, 31*, 841. DOI: 10.2139/ssrn.3063289

Wahltinez, O., Cheung, A., Alcantara, R., Cheung, D., Daswani, M., Erlinger, A., Lee, M., Yawalkar, P., Lê, P., Navarro, O. P., Brenner, M. P., & Murphy, K. (2022). COVID-19 Open-Data a global-scale spatially granular meta-dataset for coronavirus disease. *Scientific Data*, *9*(1), 162. Advance online publication. DOI: 10.1038/s41597-022-01263-z

Walke, F., Klopfers, L., & Winkler, T. (2025). *Leveraging generative AI and ChatGPT in SMEs: A grounded model*. University of Hagen.

Wang, J., & Fan, W. (2025). The effect of ChatGPT on students' learning performance, learning perception, and higher-order thinking: Insights from a meta-analysis. *Humanities & Social Sciences Communications*, *12*(1), 1–21. DOI: 10.1057/s41599-025-04787-y

Wang, J., Mao, W., & Wenjie, W. (2023). The Ethics of Artificial Intelligence: Sociopolitical and Legal Dimensions. *Interdisciplinary Studies in Society, Law, and Politics*, *2*(2), 27–32. DOI: 10.61838/kman.isslp.2.2.6

Wang, Q., Gao, J., Lin, W., & Yuan, Y. (2019). Learning from synthetic data for crowd counting in the wild. In *Proceedings of the IEEE/CVF conference on computer vision and pattern recognition*, pp. 8198–8207. DOI: 10.1109/CVPR.2019.00839

Wang, S., Anselin, L., Bhaduri, B., Crosby, C., Goodchild, M. F., Liu, Y., & Nyerges, T. L. (2013). CyberGIS software: A synthetic review and integration roadmap. *International Journal of Geographical Information Science*, *27*(11), 2122–2145. DOI: 10.1080/13658816.2013.776049

Wang, S., Hu, T., Xiao, H., Li, Y., Zhang, C., Ning, H., Zhu, R., Li, Z., & Ye, X. (2024). GPT, large language models (LLMs) and generative artificial intelligence (GAI) models in geospatial science: A systematic review. *International Journal of Digital Earth*, *17*(1), 2353122. Advance online publication. DOI: 10.1080/17538947.2024.2353122

Weber, L. C. (2024, April 25). Ethical deployment of AI in Malaysia: Navigating societal dynamics and technological innovation. SSRN *Electronic Journal*. https://doi.org/DOI: 10.2139/ssrn.4807537

White & Case. (2025, June 13). AI Watch: Global regulatory tracker – Japan. *White & Case*. https://www.whitecase.com/insight-our-thinking/ai-watch-global-regulatory-tracker-japan

White House Office of Science and Technology Policy (OSTP). (2022). *Blueprint for an AI Bill of Rights: Making automated systems work for the American people*. Retrieved from https://www.whitehouse.gov

Whittaker, M., Crawford, K., Dobbe, R., Fried, G., Kaziunas, E., Mathur, V., . . . Schwartz, O. (2018). AI now report 2018. *AI Now Institute*. Retrieved from https://ainowinstitute.org

Whittlestone, J., & Clarke, S. (2022). *AI Challenges for Society and Ethics*. The Oxford Handbook of AI Governance., DOI: 10.1093/oxfordhb/9780197579329.013.3

Wilkinson, M. D., Dumontier, M., & Aalbersberg, Ij. J., Appleton, G., Axton, M., Baak, A., Blomberg, N., Boiten, J.-W., da Silva Santos, L. B., Bourne, P. E., Bouwman, J., Brookes, A. J., Clark, T., Crosas, M., Dillo, I., Dumon, O., Edmunds, S., Evelo, C. T., Finkers, R., & Gonzalez-Beltran, A. (2016). The FAIR Guiding Principles for Scientific Data Management and Stewardship. *Scientific Data, 3*(1). https://www.nature.com/articles/sdata201618

Williamson, B., & Eynon, R. (2020). Historical threads, missing links, and future directions in AI in education. *Learning, Media and Technology, 45*(3), 223–235. DOI: 10.1080/17439884.2020.1798995

Willingham, D. T. (2021). *Why don't students like school?: A cognitive scientist answers questions about how the mind works and what it means for the classroom*. John Wiley & Sons.

Wilson, H. J., & Daugherty, P. R. (2018). Collaborative intelligence : Humans and AI are joining forces. *Harvard Business Review, 96*(4), 114–123.

Winfield, A. F. T., & Jirotka, M. (2018). Ethical governance is essential to building trust in robotics and artificial intelligence systems. *Philosophical Transactions. Series A, Mathematical, Physical, and Engineering Sciences, 376*(2133), 20180085. DOI: 10.1098/rsta.2018.0085

Wixom, B. H., Beath, C. M., & Owens, L. (2023). *Data is everybody's business: The fundamentals of data monetization*. Mit Press. DOI: 10.7551/mitpress/14867.001.0001

Wong, K. K. (2024, June). Blended Learning and AI: Enhancing Teaching and Learning in Higher Education. In International Conference on Blended Learning (pp. 39-61). Singapore: Springer Nature Singapore. DOI: 10.1007/978-981-97-4442-8_4

Wong, J., Morgan, D., Straub, V. J., Hashem, Y., & Bright, J. (2025). Key challenges for the participatory governance of AI in public administration. In *Handbook on Governance and Data Science* (pp. 179–197). Edward Elgar Publishing. DOI: 10.4337/9781035301348.00017

World Bank Group. (2024, December 5). Global trends in AI governance: Evolving country approaches. *World Bank Group*. https://documents1.worldbank.org/curated/en/099120224205026271/pdf/P1786161ad76ca0ae1ba3b1558ca4ff88ba.pdf

World Bank. (2019). Small and Medium Enterprises (SMEs) Finance. *World Bank*. https://www.worldbank.org/en/topic/smefinance

World Economic Forum. (2020). *AI governance: A holistic approach to implementing ethical AI*. World Economic Forum. https://www.weforum.org

World Economic Forum. (2020a). *The future of jobs report 2020*. Retrieved from https://www.weforum.org/reports/the-future-of-jobs-report-2020

World Economic Forum. (2020b). *A model for ethical AI*. Retrieved from https://www.weforum.org/stories/2020/01/tech-companies-ethics-responsible-ai-microsoft/

World Economic Forum. (2021). *9 ethical AI principles for organizations to follow*. Retrieved from https://www.weforum.org/stories/2021/06/ethical-principles-for-ai/

World Economic Forum. (2023). AI governance alliance. *World Economic Forum*. https://initiatives.weforum.org/ai-governance-alliance/home

World Economic Forum. (2024). AI governance alliance briefing paper series. *World Economic Forum*. https://www3.weforum.org/docs/WEF_AI_Governance_Alliance_Briefing_Paper_Series_2024.pdf

World Health Organization. (2021). *Ethics and governance of artificial intelligence for health*. https://www.who.int/publications/i/item/9789240029200

World Health Organization. (2023). *WHO outlines considerations for regulation of artificial intelligence for health*. https://www.who.int/news/item/19-10-2023-who-outlines-considerations-for-regulation-of-artificial-intelligence-for-health

World Health Organization. (2024). WHO releases AI ethics and governance guidance for large multi-modal models. *World Health Organization (WHO)*. https://www.who.int/news/item/18-01-2024-who-releases-ai-ethics-and-governance-guidance-for-large-multi-modal-models

Wright, E., Zhou, J., Lindsay, D., Przhedetsky, L., Chen, F., & Davison, A. (2022). SMEs and explainable AI: Australian case studies. *Computer & law*. https://www7.austlii.edu.au/au/journals/ANZCompuLawJl/2022/4.pdf

Wu, Y. (2023, June 27). How to interpret China's first effort to regulate generative AI measures. *China Briefing*. https://www.china-briefing.com/news/how-to-interpret-chinas-first-effort-to-regulate-generative-ai-measures

Xing, J., & Sieber, R. (2023). The challenges of integrating explainable artificial intelligence into GeoAI. *Transactions in GIS*, *27*(3), 626–645. Advance online publication. DOI: 10.1111/tgis.13045

Yanamala, A. K. Y., & Suryadevara, S. (2024). Navigating data protection challenges in the era of artificial intelligence: A comprehensive review. *Revista de Inteligencia Artificial en Medicina, 15*(1), 113–146.

Yang, A., & Li, B. (2025, May 29). AI Watch: Global regulatory tracker – China. *White & Case*. https://www.whitecase.com/insight-our-thinking/ai-watch-global-regulatory-tracker-china

Yeung, K. (2020). Recommendation of the council on artificial intelligence (OECD). *International legal materials, 59*(1), 27-34.

Ye, X., Du, J., Li, X., Shaw, S.-L., Fu, Y., Dong, X., Zhang, Z., & Wu, L. (2025). Human-centered GeoAI foundation models: Where GeoAI meets human dynamics. *Urban Informatics, 4*(1), 2. Advance online publication. DOI: 10.1007/s44212-025-00067-x

Yusuf, S., Durodola, R., Ocran, G., Abubakar, J., Amarichi, Z., Echere, A., & Paul-Adeleye, A. (2024, September). Challenges and opportunities in AI and digital transformation for SMEs: A cross-continental perspective. *World Journal of Advanced Research and Review*. https://www.researchgate.net/publication/383873269_Challenges_and_opportunities_in_AI_and_digital_transformation_for_SMEs_A_cross-_continental_perspective

Zellers, R., Holtzman, A., Rashkin, H., Bisk, Y., Farhadi, A., Roesner, F., & Choi, Y. (2019). Defending against neural fake news. *Advances in Neural Information Processing Systems, 32*, 9051–9062.

Zemel, R., Wu, Y., Swersky, K., Pitassi, T., & Dwork, C. (2013). Learning fair representations. *Proceedings of the 30th International Conference on Machine Learning*, 325–333.

Zeng, Y., Lu, E., & Huangfu, C. (2019). *Linking artificial intelligence principles. arXiv preprint* arXiv:1812.04814.

Zeng, Y., Lu, E., & Huangfu, C. (2019). Linking artificial intelligence principles. *Science and Engineering Ethics, 25*(2), 419–434. DOI: 10.1007/s11948-018-0031-0

Zeng, Z. (2020). *Artificial intelligence, privacy, and ethics: A global perspective.* Springer.

Zhang, B., & Dafoe, A. (2020). Artificial intelligence: American attitudes and trends. *Center for the Governance of AI, Future of Humanity Institute, University of Oxford.*

Zhang, G., Zhang, S., Man, Z., Cui, C., & Hu, W. (2024). Location Privacy Protection in Edge Computing: Co-Design of Differential Privacy and Offloading Mode. *Electronics (Basel), 13*(13), 2668–2668. DOI: 10.3390/electronics13132668

Zhang, J., & Goodchild, M. F. (2002). Uncertainty in Geographical Information. In *CRC Press eBooks*. Informa., DOI: 10.1201/b12624

Zhao, A., Jiang, N., Wang, C., Liu, S., & Qiu, K. (2023). Synchronization Optimization of Chaotic Laser Based on Generative Adversarial Network. *Guangxue Xuebao. Acta Optica Sinica, 43*(1). Advance online publication. DOI: 10.3788/AOS220994

Zheng, Z., & Sieber, R. (2021). Putting humans back in the loop of machine learning in Canadian smart cities. *Transactions in GIS, 26*(1), 8–24. DOI: 10.1111/tgis.12869

Zhou, B., Zou, L., Mostafavi, A., Lin, B., Yang, M., Gharaibeh, N., Cai, H., Abedin, J., & Mandal, D. (2022). VictimFinder: Harvesting rescue requests in disaster response from social media with BERT. *Computers, Environment and Urban Systems, 95*, 101824. DOI: 10.1016/j.compenvurbsys.2022.101824

Zhou, X., & Zafarani, R. (2020). A survey of fake news detection: Methods, evaluation, and challenges. *ACM Computing Surveys, 53*(5), 1–40. DOI: 10.1145/3395046

Zhu, A.-X., & Turner, M. (2022). How is the Third Law of Geography different? *Annals of GIS, 28*(1), 57–67. DOI: 10.1080/19475683.2022.2026467

Zhu, J., Ni, P., & Wang, G. (2020). Activity minimization of misinformation influence in social networks. *IEEE Transactions on Computational Social Systems, 7*(4), 897–906. DOI: 10.1109/TCSS.2020.2997188

Zhuo, T. Y.. (2023). Exploring AI ethics of ChatGPT. *arXiv*. https://arxiv.org/abs/2301.12867

Zhu, X. X., Tuia, D., Mou, L., Xia, G.-S., Zhang, L., Xu, F., & Fraundorfer, F. (2017). Deep learning in remote sensing: A comprehensive review and list of resources. *IEEE Geoscience and Remote Sensing Magazine, 5*(4), 8–36. DOI: 10.1109/MGRS.2017.2762307

Zuboff, S. (2019). *Surveillance capitalism and the challenge of collective action. 28*(1), 10-29.

Zuboff, S. (2019). *The age of surveillance capitalism: The fight for a human future at the new frontier of power*. PublicAffairs.

# About the Contributors

**Hakikur Rahman** is an academic with over 30 years of experience, having served leading educational institutions in Bangladesh and abroad. He has established various ICT4D projects funded by organizations like the ADB, EU, UNDP, and the World Bank. Currently, he is a Professor and Chair, CSE, International Standard University (ISU), Bangladesh and Director at the Ansted University Sustainability Research Institute, Malaysia. Prior to this he was a Professor and the Dean of School of Engineering at the Presidency University. Previously, he was a Professor at the Asian University of Bangladesh, Chair of the CSE Department at BRAC University, and a Post-Doctoral Researcher at the University of Minho, Portugal. Dr. Rahman has authored and edited over 25 books, contributed more than 50 book chapters, and published over 200 articles on topics such as ICTs, knowledge management, open innovation, and data mining. He holds two PhDs—one in Computer Engineering from Ansted University and another in ICT from Empresarial University of Costa Rica—and completed his postdoctoral research at the University of Minho in 2013. He is the Chief Editor of the International Journal of Information Communication Technologies and Human Development (IJICTHD) and the International Journal of Knowledge Management (IJKM).

***

**Kashvi Chaturvedi** is presently enrolled at Ajeenkya DY Patil University to pursue a B.Tech in Artificial Intelligence and Data Science. Her technological skills are demonstrated by her knowledge on languages like Python, C++, C. Her Semiconductors Research Paper examined developments in microchip technology, while her work on the Comparative Analysis of BJT Configurations offered insights into electronic circuits. Kashvi is now researching on Hyperloop's potential to transform high-speed transportation. She recently took part in the IIT Madras-hosted Global Hyperloop Competition 2025, where she offered creative suggestions for this futuristic form of transportation. In addition, she is investigating the possibilities of Small Modular Reactors (SMRs) and Ethical dimensions of AI in sustainable energy solutions.

**Susanta Das** is a seasoned educator with a proven record of teaching and mentoring diverse students, working at multiple institutions, and providing service to the various initiatives of organizations. He received his Ph.D. and M.A. degrees from Western Michigan University (WMU), USA, and M.Sc. degree from Banaras Hindu University (BHU), India all in Physics. He continued his research as a Marie-Curie post-doctoral fellow at Stockholm University, Stockholm, Sweden on beam diagnostics for the DESIREE (Double ElectroStatic Ion Ring ExpEriment) in the project DITANET (Diagnostic Techniques for particle Accelerators – a Marie-Curie initial training NETwork), at the Indian Institute of Science Education and Research-Kolkata, India on high-pressure physics as a Project Scientist-B, and at the University of Electro-Communications, Tokyo, Japan on ion-surface interactions as a Post-doctoral fellow. Throughout his career, Dr. Das worked and collaborated with many researchers, Ph.D., master, and visiting students from different countries. He further visited many countries to discuss research and an international conference participant (UK, Italy, Germany, Greece, Belgium, Bulgaria, Romania, Brazil etc.). He co-authored several research articles in WoS/Scopus indexed international journals, conference proceedings, and scholarly book chapters published by renowned international publishing houses. Beyond his research endeavors, Dr. Das has a storied history of service to academic and administrative committees, exhibiting his commitment to the growth and development of educational institutions. His experience includes tenure at Central University South Bihar, Sri Sri University, and P.K. University before assuming his current role at Ajeenkya DY Patil University. He received the Marie-Curie post-doctoral fellowship in Sweden, Science Academies' Summer Research Fellowship in India, Gwen Frostic Doctoral Fellowship, Department Graduate Research and Creative Scholar Award by WMU, and Leo R. Parpart Doctoral Fellowship by Dept. of Physics, WMU, among many others, throughout his academic journey. At present, Dr. Das continues to delve into cutting-edge fields, with a keen interest in nanotechnology, UN SDGs, and data science. His multifaceted contributions, spanning teaching, research, and administrative leadership, showcase a dedicated professional who is instrumental in advancing the vision and mission of the institutions he serves.

**Rashid Hashmi** is a senior TV journalist, academician, and researcher with over 22 years of professional experience in broadcast journalism and simultaneous academic engagement. He currently serves as Professor of Practice at the Sharda School of Media, Film & Entertainment, Sharda University, where he leads TV production, manages the in-house news platforms Sharda TV and Sharda Samachar, and mentors students in anchoring, reporting, digital storytelling, and podcasting. Academic & Research Contributions Prof. Hashmi is a Gold Medalist in Master's in Mass Communication and is pursuing his Ph.D. from IFTM University, with a research focus on "Emergence of Digital Medium and AI: Navigating Challenges in a Changing Media Landscape." He has an impressive academic and research portfolio that includes: 6 Book Chapters in IGI-GLOBAL 5 National Book Chapters 4 International Conference Papers presented in the last 10 months 2 Upcoming Research Papers in Scopus-indexed journals 2 Upcoming Books: Anchoring, Reporting: Paradigm Shift from Traditional to Digital Media Indian Democracy in 25 Years of New Century Prof. Hashmi actively participates in national and international forums, and his articles are published in leading newspapers and magazines, reflecting his continuous engagement with media trends and public issues. Patents Prof. Hashmi is also an innovator with two significant IP contributions: Patent Title: System & Method for AI-Driven Advances in Autonomous Vehicle Technology Application No: 202441074339 Filed On: 01/10/2024 Published: 11/10/2024 Focus: Real-time decision-making in autonomous vehicles using AI and sensor fusion. Design Patent Title: IoT-Based Solar Panel Energy Generation Monitoring Device Design Registration No: 436930-001 Date of Issue: 06/01/2025 Focus: Real-time monitoring of solar panel energy output with smart grid integration. Professional Career Prof. Hashmi has worked with major news channels such as: India News – Executive Editor & Prime-Time Anchor FM News – Editor-in-Chief & Prime-Time Anchor Total News – Anchor and Reporter ETV – Copy Editor and Reporter He has conducted in-depth coverage of major events like the 26/11 Mumbai Attacks, Delhi serial blasts, elections, and border conflicts. His exclusive interviews span political leaders (Gen. V.K. Singh, Amar Singh), Bollywood stars (Kangana Ranaut, Alia Bhatt), and sports icons (Yuvraj Singh). Awards & Recognition ENBA Award 2022 – Best Anchor (Pakistan Political Crisis) ENBA Award 2023 – Best Anchor (Sri Lanka Crisis) Over 20 prominent awards for excellence in TV journalism Institutional Leadership at Sharda University Initiated podcasting culture at SSMFE Spearheaded Media Fusion Summit & Media Fest featuring prominent anchors and influencers Signed MoU with the Newspaper Association of India Facilitated industry visits, newsroom simulations, and live anchoring workshops Teaches subjects including Anchoring, TV Journalism, Digital Integration, BARC Research, and Podcast Production Digital Media Presence Prof. Hashmi is also a successful digital content creator and runs the YouTube channel 'Rashtravadi Rashid Official' with 44 lakh+ views and 20k+ subscribers, significantly contributing to youth engagement and institutional outreach.

**S. Ida Evangeline**, presently working as an Assistant Professor in the Department of Electrical and Electronics Engineering at the Government College of Engineering, Tirunelveli, Tamilnadu, India. She obtained her Bachelor's in Electrical and Electronics Engineering from Manonmaniam Sundaranar University in 2003, and Master's in Power Systems Engineering from Anna University, Chennai, Tamilnadu, India in 2005. She completed her PhD degree in 2023 at Anna University, Chennai, Tamilnadu, India. She has published six research papers in international journals. Her current research interests include optimal power flow, renewable energy, soft computing, machine learning, and deep learning.

**Ghinwa Jalloul** is the founder and CEO of Sydney Institute of Technology & innovation (SITI) and an Adjunct Fellow at the University of Technology Sydney. With a PhD in Information Technology and foundational training in mathematics and computer science, she has led transformative work in academia, government and international development since the early 1990s. As a former Member of the Lebanese Parliament, she founded and chaired its first information Technology Committee and spearheaded the landmark E-transaction and Privacy Law laying the groundwork for national policies on cybersecurity, fintech and digital privacy of personal data. Today, Dr. Jalloul focuses on closing the digital and ethical gaps in AI adoption. Through SITI, she leads responsible AI training programs, develops AI ethics toolkits for SMEs and champions the Ethics4AI platform to foster inclusive, values aligned innovation. She has directed large scale digital literacy and entrepreneurship programs across the Middle East and North Africa in collaboration with UNESCO, UNITAR, and regional ministries. Her scholarly contributions span peer reviewed publications, book chapters, books and global reports. Recognized with awards such as the UNESCO creativity Award, the Mohammed Bin Rashid Arabic Language Prize and Named a Global Women Peace Ambassador by the Women's Federation of World Peace Organization, Dr. Jalloul continues to advocate for inclusive ethically grounded AI transformation globally.

**Yousef Jalloul** is an expert in AI implementation with extensive experience across diverse industries. As the Founder and CEO of Inova AI Solutions, he brings a wealth of practical knowledge and hands-on expertise to help organizations navigate the complexities of AI integration. His work focuses on developing scalable, no-code AI ethical solutions that empower businesses to adopt intelligent technologies with confidence, efficiency, and measurable impact.

**Jayashree K**, currently working as an Assistant Professor at Department of Management Studies Dayananda Sagar College of Engineering, Bengaluru with a

Domain Expertise in Marketing. Has 8 years of teaching experience, and is qualified with MBA, M. Com, and State Eligibility Test and pursuing PhD.

**Rishikaysh Kaakandikar** is a distinguished academician, author, and researcher in the field of finance and management. Currently serving as an Associate Professor at SaiBalaji International Institute of Management Sciences (SBIIMS), Pune, he has over 13 years of experience in academia and more than a year in the corporate sector. His academic journey is marked by a deep passion for knowledge dissemination and research-driven education. Dr. Kaakandikar holds a Ph.D. in Management and an MBA in Finance, and his scholarly contributions have significantly shaped the landscape of finance education. A prolific researcher, he has authored more than 50 research papers, many of which have been published in Scopus, ABDC, and UGC Care-listed journals. His research interests include financial management, fintech, blockchain technology, mutual funds, AI in HR, and the impact of digital disruption on business ecosystems. In addition to his academic research, Dr. Kaakandikar is a renowned author with multiple books to his credit, such as: AI-Powered HR Finance: Transforming Workforce Management and Financial Strategies in the Digital Age TechHumanize: The Fintech Evolution in HR – Innovations, Challenges, and Future Perspectives Business Research Methods These publications reflect his commitment to exploring the intersection of technology and finance, offering practical insights for students, educators, and industry professionals alike. Dr. Kaakandikar's contributions have earned him several accolades, including the Best Teacher Award by the Institute of Scholars (InSc). He is frequently invited to deliver guest lectures, keynote speeches, and workshops at national and international platforms, where he shares his expertise on emerging trends in finance, AI, and digital transformation. An active member of the global academic community, he maintains a strong online presence on platforms such as Google Scholar, ResearchGate, and ORCID, where his work continues to inspire students and researchers worldwide. Through his dedication, scholarly rigor, and student-centric approach, Dr. Rishikaysh Kaakandikar stands as a guiding light in the field of modern management education.

**Ashwarya Kapoor** received her masters degree from University of Delhi. Currently, she is pursuing PhD from J.C. Bose University of Science & Technology, YMCA, Faridabad, India. Her research interest is management of financial services. She has presented many papers in international conferences.

**Wasswa Shafik** (Member, IEEE) received a Bachelor of Science degree in information technology from Ndejje University, Uganda, a Master of Engineering degree in Information Technology (Computer and Communication Networks) from Yazd University, Iran, and a PhD in Computer Science with the School of Digital

Science, Universiti Brunei Darussalam, Brunei Darussalam. He is also the Founder and Research Director of the Dig Connectivity Research Laboratory (DCRLab) after serving as a Research Associate at the Network Interconnectivity Research Laboratory at Yazd University. Prior to this, he worked as a Community Data Analyst at Population Services International (PSI-Uganda), a Community Data Officer at the Programme for Accessible Health Communication (PACE-Uganda), a Research Assistant at the Socio-Economic Data Centre (SEDC-Uganda), Uganda, an Assistant Data Officer at TechnoServe, Kampala, Uganda, and Asmaah Charity Organization as Cheif Executive Officer. He has hundreds of publications with renowned publishers. His research interests include Artificial Intelligence, Neural Networks, Computer Vision, Smart Agriculture, Digital Health, and Green Computing.

**Kamaljeet Kaur Siddhu** has done Ph.D in Marketing. She is a Fellow Member of the Institute of Company Secretaries of India. She is Alumni of IIM Kashipur. Basically, a Master's in commerce, she is also a Law graduate and has done several PG diploma and certification courses including PG. in Psychology, New Labor Codes, certified CSR & ESG Professional, ISO 9001:2015 Lead Auditing, etc. She has more than 20 years' experience in the field of Corporate Affairs, Legal & HR. She is currently working as Executive Vice President of Ajeenkya D Y Patil Group and earlier worked as Director – Technical Campus at D.Y. Patil Knowledge City, Lohegaon. Few of her previous assignments were as Director- Corporate Affairs with a Private University, Vice- President – Corporate Affairs, Legal & HR with City Corporation Limited (Makers of Amanora Park Town); Legal Head & Company Secretary with Manufacturing Construction, and IT Companies. She is also a Member of Arbitration, Mediation & Conciliation Network and has assisted Dr. K. R. Chandratre, past president of ICSI in writing a book on "Sections- 397 & 398 of the Companies Act- Oppression and Mismanagement and "Cases and Material on Company Law. She is a Division Director in D-125 of Toastmasters, two times Triple Crown Award Winner and Division level International Speech Contest winner in District 98 of Toastmasters. She is also engaged as Visiting Faculty for teaching Management subjects to MBA Students; and has done paper presentations in various National Seminars and has acted as External Examiner for various Management Institutes. She has served as President of Rotary Club of Pune Hadapsar (2018-19) and works actively for Rotary District 3131 doing social work in the areas of Literacy, Environment and Women Empowerment. She is currently Co-Director for Peace Building & Internal Conflict in Rotary District 3131. She is also a freelance career and psychological counsellor. Her other skills include Counselling, Behavior Therapy, Competency Mapping, Transactional analysis & Neuro Linguistic Programming, Institutional and Corporate Trainings, Legal Advisory, POCSO awareness trainings, Management Consulting, POSH awareness trainings, Drafting of Company Manuals, Risk Management, Corporate Governance, CSR Policy, Labor Law Compliances; Profitability Management, QMS, EMS, HSE awareness training, train the trainer and other tailormade training programs for individuals as well as organizations. She has conducted several workshops and coahing programs such as Teachers training Program as per new Education Policy, "Women Empowerment", Invincible Leader Program, Leaderships skills, Negotiation skills, Transactional Analysis, Motivation, Diversity & Inclusion, Customer Handling Techniques, Conflict Management, Etiquettes, Empathy, Conflict Resolution, Goal setting, Law of attraction, Neuro Linguistic Programming, Cognitive behavior therapy, Career Guidance, Time Management, 7 habits of highly effective people, Selling Skills, POSH, POCSO, Stress Management, Brand Management, Communication Skills, Anger Management, Motivation Essential Life skills and tailor-made trainings for Individuals, Institution as well as Corporates.

**Rajiv Sindwani** holds PhD degree in the area of service quality management. Presently he is working in the Department of Management Studies at J C Bose University of Science and Technology, YMCA, Faridabad, India. He is having more than 15 years of experience in various sectors including consumer durables, banking and education. He has published many papers in reputed international journals and also presented papers in various international conferences.

**Shradha Sinkar** is a passionate MBA student specializing in Finance at SaiBalaji International Institute of Management Sciences (SBIIMS), Pune. With a strong foundation in financial management, investment analysis, and corporate finance, she is developing the analytical and strategic skills necessary for the dynamic world of finance. Shradha actively engages in academic competitions, finance-related workshops, and institutional activities, reflecting her commitment to learning and professional growth. She aspires to build a successful career in financial consulting or investment banking, contributing to data-driven and ethical financial decision-making in organizations.

# Index

## A

## S



## T



## U



## V